D0824744

Handbook of
MARINE MINERAL DEPOSITS

 *Marine Science Series*

The CRC Marine Science Series is dedicated to providing state-of-the-art coverage of important topics in marine biology, marine chemistry, marine geology, and physical oceanography. The Series includes volumes that focus on the synthesis of recent advances in marine science.

CRC MARINE SCIENCE SERIES

SERIES EDITORS

Michael J. Kennish, Ph.D.
Peter L. Lutz, Ph.D.

PUBLISHED TITLES

Handbook of

MARINE MINERAL DEPOSITS

Edited by

David S. Cronan, Ph.D., D.Sc., FIMM

Professor of Marine Geochemistry
T.H. Huxley School of Environment,
Earth Sciences and Engineering
Royal School of Mines
Imperial College, University of London
London, England

CRC Press
Boca Raton London New York Washington, D.C.

Library of Congress Cataloging-in-Publication Data

Handbook of marine mineral deposits / editor, David S. Cronan.
      p. cm. — (Marine science series)
    Includes bibliographical references and index.
    ISBN 0-8493-8429-X (alk. paper)
    1. Marine mineral resources. I. Cronan, D. S. (David Spencer)
  II. Series
  TN264.C73   1999
  553′.09162—dc21   99-37077
                                                                                CIP

About the Editor

Professor David Cronan was educated in England at Durham (Castle), Oxford (Keble College), and London (Imperial College) Universities, receiving his doctorate in marine geochemistry at the last of these in 1967, with a thesis on marine manganese nodules. Following this, he worked as a marine geologist with the British Geological Survey for two years, and then as an Assistant and, subsequently, Associate Professor of Sedimentary Geochemistry at the University of Ottawa, Canada, for a further four years. He returned to Imperial College in 1973 to take over the direction of applied marine geochemistry in the then Department of Geology, supervising more than 25 Ph.D. students since that time.

Throughout his career, Professor Cronan has been concerned mainly with research on deep sea minerals, principally manganese nodules and hydrothermal deposits. He has participated in or led more than 20 research cruises, has presented papers and/or chaired sessions at more than 50 major international scientific meetings, has made several television and radio broadcasts, and has more than 135 publications to his name, including two books. He was awarded a D.Sc. by Durham University in 1985 for his work on marine minerals.

In recent years, Professor Cronan's main scientific interest has been in the Exclusive Economic Zones set up under the Law of the Sea Convention, the minerals in them, and the conditions under which they might be extracted. He has acted as an adviser to both the United Nations Development Programme and the International Seabed Authority and has served on or chaired several U.K. government or learned society committees dealing with his area of expertise. As well as lecturing on marine minerals at Imperial College, he has also taught his subject in the Sea Use course at the London School of Economics.

Previous editorial work undertaken by Professor Cronan includes membership on the editorial boards of *Geochimica et Cosmochimica Acta*, *Marine Geology*, *Marine Mining*, and *Marine Georesources and Geotechnology*. He also edited the book *Sedimentation and Mineral Deposits in the South-western Pacific*, published by Academic Press in 1986.

Contributors

Michael Bau, Ph.D.
GeoForschungsZentrum Potsdam
Potsdam, Germany

Michael Cruickshank, Ph.D.
Marine Minerals Technology Center
University of Hawaii
Honolulu, Hawaii

Richard H. T. Garnett, Ph.D., M.B.A.
President, Valrik Enterprises, Inc.
Oakville, Ontario
Canada

Prof. Dr. Peter Halbach
Free University Berlin
Department of Economic and
  Environmental Geology
Berlin, Germany

Mark D. Hannington, Ph.D.
Geological Survey of Canada
Ottawa, Ontario
Canada

James R. Hein, Ph.D.
U.S. Geological Survey
Menlo Park, California

Prof. Dr. Peter M. Herzig
Lehrstuhl für Lagerstättenlehre
Institut für Mineralogie
T. U. Bergakademie Freiberg
Freiberg, Germany

Pratima Jauhari, Ph.D.
Geological Oceanography Division
National Institute of Oceanography
Dona Paula, Goa, India

Jung-Keuk Kang, Ph.D.
Korea Ocean Research and
  Development Institute
Ansan, Seoul, Korea

Andrea Koschinsky, Ph.D.
Free University Berlin
Department of Economic and
  Environmental Geology
Berlin, Germany

Dr. Hermann Rudolf Kudrass
Bundesanstalt für Geowissenschaften
  und Rohstoffe
Hannover, Germany

Frank T. Manheim, Ph.D.
U.S. Geological Survey
Woods Hole, Massachusetts

Dr. Vesna Marchig
Bundesanstalt für Geowissenschaften
  und Rohstoffe
Hannover, Germany

Dr. M. Moammar
Faculty of Marine Science
King-Abdulaziz University
Jeddah, Saudi Arabia

Charles L. Morgan, Ph.D.
Morgan & Associates
Mililani, Hawaii

Ute Münch, M.Sc.
Free University Berlin
Department of Economic and
  Environmental Geology
Berlin, Germany

J.N. Pattan, Ph.D.
Geological Oceanography Division
National Institute of Oceanography
Dona Paula, Goa, India

G. Victor Rajamanickam, Ph.D.
Department of Earth Sciences
Tamil University
Thanjavur, Tamil Nadu, India

Leanne Roberts, B.Sc.
U.S. Geological Survey
Menlo Park, California

Dr. J.C. Scholten
Institut für Geowissenschaften
Universität Kiel
Kiel, Germany

Dr. D. Garbe-Schönberg
Institut für Geowissenschaften
Universität Kiel
Kiel, Germany

Dr. Ulrich von Stackelberg
Bundesanstalt für Geowissenschaften und
  Rohstoffe
Hannover, Germany

Prof. Dr. P. Stoffers
Institut für Geowissenschaften
Universität Kiel
Kiel, Germany

John C. Wiltshire, Ph.D.
Hawaii Undersea Research Laboratory
University of Hawaii
Honolulu, Hawaii

Wyss W.-S. Yim, D.Sc. (London)
Department of Earth Sciences
The University of Hong Kong
Hong Kong, SAR, China

Contents

PART I: Placer Minerals

PART II: Ferromanganese Oxide Minerals

PART III: Hydrothermal Minerals

Preface

Marine minerals have been in the public consciousness to a greater or lesser extent for some 40 years now. Of course, individual marine mineral deposits, such as placers, were mined before then, but these were small-scale operations which attracted little international attention. It was in the 1950s that the work of John Mero on manganese nodules, particularly Pacific manganese nodules, focused public attention on the potential of mineral resources from the sea. Since that time, interest in marine minerals has waxed and waned, dependent upon a number of factors, and at the end of the 20th century no deep-sea mining operations are yet under way. However, most authorities accept that during the 21st century deep-sea mineral resources are likely to be exploited.

The history of mankind's interest in deep-sea manganese nodules provides a good example of how public perception of marine minerals has varied over the past half century. Mero[1] considered that the elements Mn, Ni, Cu, and Co would be of prime economic value in manganese nodules, with a variety of other elements being possibly recoverable as accessory resources. He considered that even if less than 10% of the nodule deposits on the Pacific Ocean floor were minable, there would be sufficient supplies of many metals to last for thousands of years. However, Mero's beliefs regarding the value of manganese nodules fell at the optimistic end of a spectrum of opinion on this matter, which ranges from optimistic to pessimistic. Interest in manganese nodules by mining companies commenced around the mid-1960s and generally built up during the 1970s, coincident with the deliberations of the Third United Nations Law of the Sea Conference. However, the outcome of that Conference in 1982, which was widely regarded as unfavorable for the mining industry, coupled with a general downturn in metal prices, led to a lessening of mining company interest in the nodules. In the meantime, however, several government-backed national consortia, for example those of Japan and India, had become interested in manganese nodule mining, and the work of these entities expanded as exploration and evaluation of the deposits by mining companies declined. Unhappiness about some of the deep-sea mining provisions in the 1982 Law of the Sea Convention was partly responsible for some of the major industrialized nations not signing it until Part 11, that part dealing with deep-sea mining, was substantially amended in the agreement of 28th July 1994. This ameliorated some of what were considered to be the more onerous provisions as far as deep-sea mining was concerned. The Convention entered into force in November 1994. Consequent to the Convention taking effect, the International Seabed Authority was established. This is the organ responsible for administering the mineral resources of the deep seabed, i.e., the seafloor outside national jurisdiction. This body is in the process of drafting regulations concerning future deep-sea mining, including provisions to protect the marine environment. Verlaan[2] has provided a useful summary of the development of marine resources in relation to the Law of the Sea.

During the 1980s, interest in marine minerals in Exclusive Economic Zones started to increase. An important consequence of the Third Law of the Sea Conference, even before the Convention came into force, was the general acceptance of a 200-nautical-mile Exclusive Economic Zone in which the adjacent coastal state could claim jurisdiction over, *inter alia,* seafloor mineral deposits. Exclusive Economic Zones had been unilaterally claimed previously by some states, but these had not been recognized by many others. The 1982 Law of the Sea Convention more or less adopted the regime for the continental shelf established under the 1958 Geneva Convention, and the Exclusive Economic Zone concept is generally recognized now, even by states that have not signed the Law of the Sea Convention. The minerals found in Exclusive Economic Zones include all those that occur on the deep seabed, albeit in different proportions, but also include a number of minerals *not* found in the deep sea, such as placers and aggregates. Marine mining is already taking place in Exclusive Economic Zones, but not yet in the deep sea, outside national jurisdiction.

At the end of the 20th century, the outlook for marine minerals remains rather unclear. Well-established Exclusive Economic Zone mining activities, such as aggregate and placer extraction, will continue. Significant marine phosphorite mining would seem to be unlikely in the foreseeable future in view of the extremely large reserves of phosphate minerals on land, and, for this reason, they are not dealt with in this book. Manganese nodule mining is likely to commence some time in the 21st century, although it is not possible to give a precise estimate as to when. It will depend upon many factors: economic, technological, and political. What is clear, however, is that so much momentum has built up in the move toward manganese nodule mining that the concept is unlikely to be just abandoned. The outlook for mining of submarine hydrothermal deposits is even more uncertain than that of manganese nodules. Interest in these first emerged in relation to copper and zinc sulfide deposits in the Red Sea in the 1970s, and since that time many more hydrothermal deposits have been discovered in submarine volcanic settings elsewhere. Some of these deposits are extremely high grade and, if they occurred on land, would probably be considered to be economic mineral deposits. However, their occurrence not on land but on the seabed introduces a large element of uncertainty into their future exploitability. The highly localized nature of the deposits, coupled with a high degree of variability in their composition, means that each deposit will have to be evaluated on its own merits, and few, if any, generalizations will be possible.

The present volume has been structured to reflect current and perceived future interest in marine mineral deposits. It is divided into three parts. Part I deals with the placer mineral deposits currently being mined on beaches and in Exclusive Economic Zones. Part II deals with marine ferromanganese oxide deposits, both manganese nodules and cobalt-rich manganese crusts, largely in the international mine sites of the deep sea, but also in the case of crusts in Exclusive Economic Zones. Part III deals with submarine hydrothermal mineral deposits, first on mid-ocean ridges, then in subduction-related environments, and finally in the Red Sea, where the deposits were first discovered and where their economic potential still remains greatest.

Many of the contributors were friends and colleagues of Robby Moore, to whom this book is dedicated, through long association with his annual Underwater Mining Institute. Some are scientists from developing countries, whose cause Robby always championed both through the UMI and elsewhere.

David S. Cronan
Imperial College
October 1999

REFERENCES

1. Mero, J.L., *The Mineral Resources of the Sea*, Elsevier, Amsterdam, 1965.
2. Verlaan, P., Law of the Sea and development of marine resources: a challenge for scientists, *Ocean Challenge*, 8, 2, 47, 1998.

Dedication

PROFESSOR J. ROBERT MOORE

Robby Moore was a man with many dear friends and admirers. His professional life as a marine economic geologist spanned 44 years, of which 28 were enjoyed as an academician and teacher, first at the University of Wisconsin, Madison, then the University of Alaska, Fairbanks, and subsequently at the University of Texas, Austin. He was a pioneer in the field of marine minerals; he founded and directed the Marine Research Laboratory at Madison and the now flourishing Underwater Mining Institute which, in December 1994 at Monterey, California, held its 25th Annual Meeting. He, unfortunately, was too ill to attend and succumbed to cancer in the following March.

Robby was the driving force behind the International Marine Minerals Society, an organization of professional aficionados in marine minerals with members from more than 20 countries throughout the world. The Society presented Robby its "Outstanding Career Achievement Award" in 1989 in recognition of his leadership and many contributions to marine mining. At its 29th meeting in Toronto, Canada, the Directors of the Society presented the first annual J. Robert Moore Award for Initiatives in Marine Mining to six of his senior colleagues, a gesture that says much for the esteem in which he is still held.

In 1980, he became founding editor of the international peer reviewed journal *Marine Mining*, published by Taylor and Francis, carrying it with optimism and vigor through some mighty lean times in the late 1980s. More recently, the journal was combined with *Marine Geotechnology* to form the present *Marine Georesources and Geotechnology* of which Robby, struggling with his illness, actively remained Co-Editor-in-Chief through Volume 12.

Robby's early years in the field with the oil industry made him many friends and gave him a good understanding of the realities of minerals economics, which he constantly imparted to his students as his interests became more focused on hard minerals. As an educator, he was universally held in the highest regard by both undergraduates and graduate students for his logical and articulate lectures, and for his insight into marine research, based on his exceptional experience. His field expeditions to Alaska and Hawaii will long be remembered by those students lucky enough to accompany him on these exciting and innovative research activities. He was a prolific writer and communicator, keeping in close touch with his colleagues and students on all aspects of his work and interests. In his later days, he involved himself with giving back to nature, and he and his wife, Dorothy Taylor Moore, who survives him, planted thousands of young trees on their property in Missouri.

A passage from Dunbar's classic text *Historical Geology* was read, at Robby's request, at his burial service by Dr. Patrick L. Parker, a longtime friend and colleague.

"It is always necessary to close a lecture on Geology in humility. On the ship Earth which bears us into immensity toward an end which God alone knows, we are steerage passengers. We are emigrants who know only their own misfortune. The least ignorant among us, the most daring, the most reckless, ask ourselves questions; we demand when the voyage of humanity began, how long it will last, how the ship goes, why do its deck and hull vibrate, why do sounds sometimes come up from the hold and go out by the hatchway; we ask what secrets do the depths of the strange vessel conceal and we suffer from never knowing the secrets …"

Robby was such a seeker after truth, and a humble one at that, epitomized best perhaps in the words of his young staff assistant at Austin, Natalie Potts, who said, "It's just amazing that he was such a really important and internationally known man and he was so down to earth and easy to talk to."

It is to the memory of Professor J. Robert Moore, friend, colleague, and mentor, that this volume is dedicated.

M. Cruickshank

Part I

Placer Minerals

1 Marine Placer Deposits and Sea-Level Changes

Hermann Rudolf Kudrass

ABSTRACT

The origin of placer deposits is closely related to Pleistocene glacio-eustatic changes of sea level. Fluviatile placer deposits of gold and cassiterite (tin oxide) on the inner shelf originated during glacial periods of falling sea level, when rejuvenated fluviatile erosion concentrated these heavy minerals in lag sediments.

Other economically important placer minerals such as rutile, ilmenite (titanium), magnetite (iron), and monazite (rare earth elements) are predominantly concentrated by the panning system of the surf zone. Beach placer deposits containing these minerals were formed on the middle shelf during glacial periods of low sea level. Most of these deposits were destroyed by the transgressing sea. Parts of them were moved onshore, and the wealth of placer deposits along many present shorelines is predominantly a result of transgressive beach-barrier migration. Only a few shelf deposits were large enough to survive the transgression as disseminated shelf deposits.

Eluvial placer deposits of gold, cassiterite, and phosphorite may also be concentrated by submarine erosion.

1.1 PLACER MINERALS — THEIR USES AND REGIONAL OCCURRENCE

Marine placer deposits are exploited in many places for a great variety of commodities (Table 1.1).[1] In these deposits economically valuable minerals have been mechanically concentrated in rivers and along beaches because of their higher density (>3.2 gcm^{-3}) compared to the bulk of detrital minerals, which consist mostly of quartz and feldspar with a density of 2.7 gcm^{-3}. These placer minerals, sometimes called heavy minerals, are derived by weathering from continental rocks of mostly volcanic, plutonic, or metamorphic origin and have a broad compositional range.

The placer minerals rutile and ilmenite are the main sources of titanium and are or have been mined from beach sand in southeast and southwest Australia, in east South Africa, south India, Mozambique, Senegal, Brazil, and Florida. Titanium is used as an alloy or, more commonly in its oxide form, as a pigment for brilliant white-colored paint.

The minerals zircon, garnet, sillimanite, and monazite are frequently recovered as by-products of mining ilmenite-bearing sands. They are used as refractory or foundry sand (zircon, sillimanite), as abrasives (garnet, zircon), or as raw material (monazite) for rare earth elements (cerium, lanthanum, neodymium) and thorium, which are recovered for various purposes (catalysts in refining crude oil, monitor coloring, lamps, radioactive uses, etc.).

The iron–titanium-rich placer mineral magnetite has been mined in large quantities from the northwestern coast of New Zealand (North Island), Indonesia (Java), the Philippines (Luzon), and Japan (Hokkaido). In Japan this type of magnetite was added to other iron ores and used to prolong the durability of the expensive refractory lining of furnaces.

TABLE 1.1
Heavy Minerals

Mineral	Density	Composition	Value (1991) kg U.S. Dollar	Main Use	Occurrence of Marine Deposits
Gold	15.0–19.3	Au	11,700	ornament	Alaska (Nome) New Zealand (South Island)
Diamond	3.5	C	~ 40,000	jewels, cutting	South Africa, Namibia
Cassiterite	6.8–7.1	SnO_2	2.7	metal coating	Indonesia (Sunda Shelf), Malaysia (Lumut), Thailand (Phuket),
Rutile	4.2	TiO_2	0.7	pigment, metal	Australia (Queensland, New South Wales, Eneabba)
Ilmenite	4.5–5.0	$FeTiO_3$	0.08	pigment	South Africa (Richards Bay), India (Tamil Nadu, Kerala), Australia (Eneabba, Bunbury), Sri Lanka (Pulmoddai), Senegal, Florida, Madagascar
Magnetite	5.2–6.5	Fe_3O_4	0.03	steel	New Zealand (North Island), Indonesia (Java) Philippines (Luzon), Japan (Hokkaido)
Zircon	4.2–4.9	$ZrSiO_4$	0.4	refractory, foundry sand	Australia (Queensland, Eneabba, Bunbury) India (Tamil Nadu), Florida, Brazil
Garnet	3.5–4.2	$Fe_3Al_2(SiO_4)_3$	0.2	abrasive	Australia (Eneabba, Bunbury), India (Tamil Nadu)
Monazite	4.9–5.3	$(Ce,Y)PO_4$	0.6	catalyst for oil refining	Australia (Eneabba, Bunbury), India (Tamil Nadu), Brazil (Sao Paulo, Bahia)
Sillimanite	3.2	Al_2SiO_5	0.3	refractory	India (Tamil Nadu)
Apatite	3.2	$Ca_5(F,OH)(PO_4)_3$	0.04	fertilizer	Peru, Chile

The fluviatile placer mineral cassiterite, a tin oxide, is recovered from nearshore and offshore sediments in Indonesia, Malaysia, and Thailand, where about one third of the world's production derives. This metal is used as corrosion-resistant plating for steel and in alloys.

Gold is mostly also recovered from fluviatile placer deposits. However, it sometimes occurs in beach placer deposits (New Zealand, Alaska), and besides being used for ornaments and bullion its consumption for industrial purposes is increasing.

Diamonds are mined in beach and shelf sediments along the west coast of South Africa and Namibia and are used as jewels and for industrial cutting processes.

Phosphorite, consisting of varieties of the heavy mineral apatite, is not a detrital mineral like other heavy minerals, but is authigenically formed in sediments of upwelling areas. Phosphorite is used mainly in the phosphate fertilizer industry.

1.2 PLEISTOCENE SEA-LEVEL CHANGES

About 100,000-years-long cycles of accumulation and partial disintegration of continental ice caps have caused large fluctuations of the Pleistocene sea level during the last 1.6 million years. The last cycle especially determined the distribution and preservation of marine placer deposits. During this cycle the maximum extent of ice 18,000 years ago caused a lowering of sea level by 120 m (Figure 1.1).[2] Starting 15,000 years ago, the ice sheet began to melt[3] and sea level rapidly rose at an average rate of 1.3 m/100 years, reaching or slightly surpassing its present position about 6000 years ago (Figure 1.1). On a 50 km-broad shelf with a constant slope and a slope break at 150 m, the transgressing sea would have moved the strandline by an average of 4.3 m per year. Changes in shelf slope and rates of sea level rise (see below) would have varied this transgressive speed.

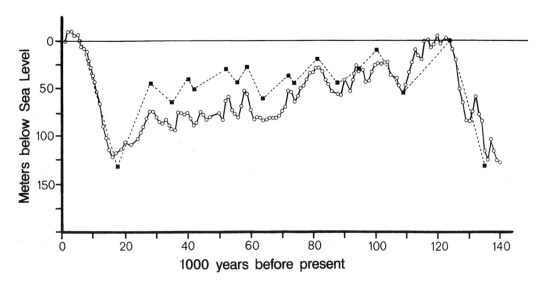

FIGURE 1.1 Sea level estimated from terraces in New Guinea (squares) and using stable isotope data of foraminifera (circles). (From Shackleton, N. J., Oxygen isotopes, ice volume and sea level, *Quat. Sci. Res.*, 6, 183, 1987.)

Analysis of drowned corals from the insular shelf of Barbados shows that sea level did not rise constantly, but that two periods of accelerated rise, centered at 12,000 and 9000 years, are separated by a period with small changes when sea level reached a position of about 50 m below its present position.[3] A pronounced terrace near the 50 m isobath reported from several stable shelf areas[4,5] may have been formed during this period, but the terrace may also be related to the average intermediate sea level position during the 70,000-year-long period preceding the last glacial maximum (Figure 1.1).

Effects of the high sea level of the last interglacial period, lasting from 125,000 to 100,000 years ago, are better known. Beach placers from this or earlier periods have been mined along the southeast Australian coast.[6] Along active continental margins, placer deposits which originated during periods of high sea level may be uplifted and thus have been saved from erosion by transgressions.

1.3 ORIGIN OF MARINE PLACER DEPOSITS

Three generic types of placer deposits are known from shelf areas: disseminated beach placers, drowned fluviatile placers, and eluvial placers. Gold and cassiterite, which are heavy minerals with a high density, are concentrated in fluviatile and eluvial placers; beach placers usually contain the light heavy minerals,[7] but may also incorporate small amounts of fine-grained gold and cassiterite.

1.3.1 FLUVIATILE PLACERS

Most gold and cassiterite placer deposits of the shelf originated by fluviatile processes during several cycles of sea level changes.[8-10] Because the two minerals have a high density (>6 gcm^{-3}), they are only transported during periods of high fluviatile run off, when gravel is also mobilized and moves as bed load along river beds. During this transport, effects of kinetic sieving[11] and hydraulic sorting[12] concentrate these minerals in the coarse sand and gravel overlying the bottom of river channels. Consequently, gold and cassiterite placer deposits are confined to the bottom fill of river beds and usually occur only within some kilometers of their respective source rocks.

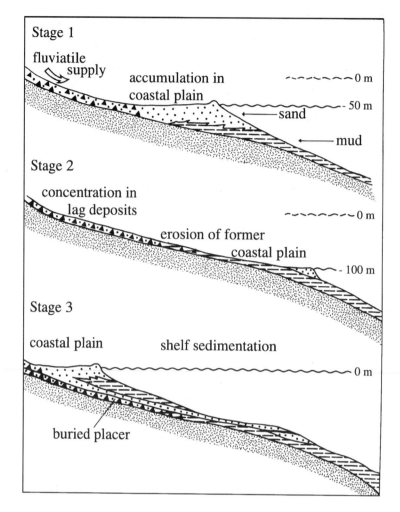

FIGURE 1.2 Conceptual model for the origin of fluviatile placer deposits containing gold and cassiterite (indicated by triangles) on the shelf during medium, low, and high sea level.

During periods of lowered sea level, these fluviatile placer deposits may have extended to the present shelf (Figure 1.2, Stage 1). A further lowering of the sea level induced their erosion and that of the fluviatile deposits located further upstream. This erosion upgraded the placer deposits in the resulting lag sediments and transported them slightly seaward (Figure 1.2, Stage 2). A subsequent rise of sea level produced a cover of transgressive sand and shelf mud, burying the placer deposits (Figure 1.2, Stage 3). Repeated changes of sea level in combination with fluviatile deposition and erosion can lead to a considerable accumulation of heavy minerals, as for instance in the "tin-valleys" of the Indonesian Sunda shelf extending from the islands of Bangka, Belitung, and Kundur (Figure 1.3).[8,13]

1.3.2 BEACH PLACERS

Heavy minerals with a specific density of less than 6 gcm^{-3}, also called light heavy minerals, cannot be concentrated by fluviatile transportation processes, because the density difference compared to the bulk sediment consisting of quartz and feldspar, with a specific density of 2.7 gcm^{-3}, is not great enough to cause different transport modes in the mostly turbulent fluviatile hydraulic regime. These heavy minerals particularly rutile, ilmenite, magnetite, monazite, zircon, sillimanite, and

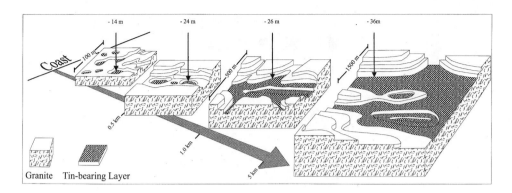

FIGURE 1.3 Cassiterite-rich sediments at the bottom of a submarine valley cut into granitic basement offshore from the island of Kundur/Indonesia. For clarification, the overburden, made up mainly of marine muds, is removed. The schematic diagram is based on results of seismic profiling and drilling; note the different horizontal scales. (Modified from Sujitno, S., Some notes of offshore exploration for tin in Indonesia, 1966–1976, *Committee Co-ord. Joint Prosp. Min. Res. in Asian Offshore Areas, Techn. Bull.*, 11, 169, 1977.)

garnet can be concentrated, however, by the huge panning system of sandy beaches.[14,15] In the surf zone especially, film-sizing (Figure 1.4),[9,16] and to lesser extent sorting by rollability[17] and "shear sorting,"[18] tend to separate minerals according to their density, size, and shape. By "shear sorting," larger, especially light minerals are pushed upward in the stirred-up sand flow of the surfzone. Once at the surface, the large grains are more easily mobilized by rolling than the small grains, which are easily trapped in the interstices of the sandy surface. These processes can be observed and studied on almost every sandy beach. Heavy minerals, and especially the economically valuable heavy minerals, are concentrated by the combination of these processes in the upper part of the beach, where wind may erode them and form heavy-mineral-rich coastal dune deposits.

The concentration of heavy minerals in the beach ridge sand is usually much higher than in the adjacent dunes. However, the total amount of heavy minerals occurring in the dunes is much higher than that in the beach sand. During extended glacial periods, sea level fluctuated around the middle shelf where, in case of sufficient supply of heavy-mineral-bearing fluviatile sand, large placer deposits could develop in beach ridges and dunes (Figure 1.5, Stage 1).

When sea level was falling during a regression (Figure 1.5, Stage 2), the placer deposits along the rivers were eroded. However, because these deposits had usually formed at some distance from the river mouth, most of them were preserved. During a transgression, fluviatile sand was trapped in the drowned river channels and could not reach the shoreline. With the rising sea level, the sediment-starved beach barriers started to migrate across the shelf.[19] The amount of sand stored on the shelf, the intensity of longshore currents, the wind direction and wave climate, the morphology and width of the shelf are all parameters determining the dispersal of sand during a transgression.[9] In the case of large sand volumes, usually found near the mouths of major rivers in combination with a wide outer to middle shelf, the transgressing sea could not move the total amount of sand stored in the coastal system. Under these conditions, the transgression resulted in an extensive mixture of the available mobile sand, and former highly concentrated beach and dune deposits became widely spread and disseminated. The ilmenite–rutile–zircon placer deposit on the middle shelf of northern Mozambique, which is related to the Zambezi River, is one example of this type of disseminated placer deposit (Figures 1.6 and 1.7).[5]

In the case of a moderate to low volume of sand stored in the coastal system and a narrow steep shelf, the transgressing sea moved the beach barriers by storm washover processes like a bulldozer across the shelf. Intense storms, directed toward the coast, cause a temporary rise of sea level along the affected shoreline. The elevated sea may break through the beach barrier, eroding and transporting vast amounts of beach sand, which is spread as a washover-fan on the landward

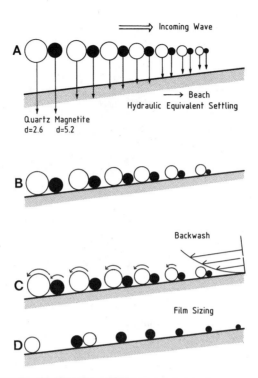

FIGURE 1.4 Sorting of minerals with different densities in the surf zone. (Modified from Seibold, E. and Berger, W. H., *The Seafloor: An Introduction to Marine Geology,* Springer Verlag, Berlin, 1982, 288.) The incoming wave transports various grain-sizes of light and heavy minerals (A), which settle corresponding to their hydraulic equivalent diameter (B). In the thin water-layer of the backwash, current velocities fall off steeply near the water–sand interface, and the larger light-mineral grains are more easily removed than the smaller heavy-mineral grains (C). This film sizing produces a lag deposit rich in heavy minerals (D). (From Kudrass, H. R., Sedimentary models to estimate the heavy-mineral potential of shelf sediments, in *Marine Minerals*, Teleki, P.G. et al. (Eds.), Reidel Publishing Company, Dordrecht, 1987, 39. With permission.)

coastal plain. Due to the vertical differentiation of heavy-mineral assemblages within the beach barrier, valuable heavy minerals are preferentially contained in the landward migrating sand, while the other heavy minerals are predominantly left behind in the trailing sand sheet covering the transgressed shelf (Figure 1.5). The sand finally arriving at the modern coast line (Figure 1.5, Stage 3) was an ideal substratum for further differentiation in the surf zone, which led to the formation of probably the most important beach and dune placer deposits, such as those on the southeast coast of Australia.[9]

1.3.3 ELUVIAL PLACERS

Concentration of detrital placer minerals by systems other than these in the surfzone is confined to minerals with a high density, i.e., gold and cassiterite. They may, for example, be concentrated by submarine erosion of submerged fluviatile sediments. The tin placer deposits at the bottom of tidal scour basins in the Strait of Malacca are one example of eluvial placers.[20] However, this placer type is rare and usually contains only small reserves.

In contrast, most of the phosphorite deposits containing varieties of apatite are concentrated by submarine currents. This mineral forms by authigenic precipitation in fine-grained sediments deposited below high-productivity zones of upwelling areas.[21] Changes of sea level in combination with alterations of submarine currents and the erosional wave base may result in winnowing of the fine-grained mud and an enrichment of the sand-size apatite grains.

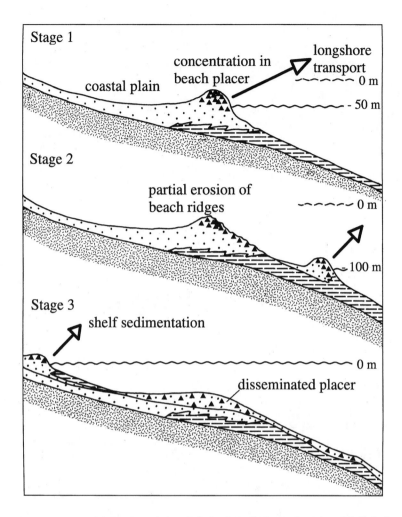

FIGURE 1.5 Conceptual model for the origin of disseminated placer deposits with light heavy minerals (indicated by triangles) on the shelf during medium, low, and high sea level.

1.4 CONCLUSIONS

Sea level changes are the prime factor influencing placer mineral concentration processes and the regional distribution and preservation of placer deposits on the modern shelf. The three genetic types of placer deposits are related differently to the change of sea level.

1. Fluviatile placer minerals such as gold and cassiterite are transported and concentrated on the shelf during periods of falling sea level occurring during interglacial-to-glacial transitions.
2. Beach placers, on the contrary, originate during periods of stable or slightly fluctuating sea level, typical during extended intermediate glacial periods. Most of these beach placer deposits are subject to dispersion or shoreward removal during periods of rising sea level during interglacial periods. The wealth of placer deposits along present shorelines is largely a result of transgressive beach-barrier migration, by which much of the shelf sand with its preconcentrated heavy-mineral assemblages was moved to its present coastal position. This lateral and vertical migration was effective, especially during the last two transitions from glacial to interglacial periods.

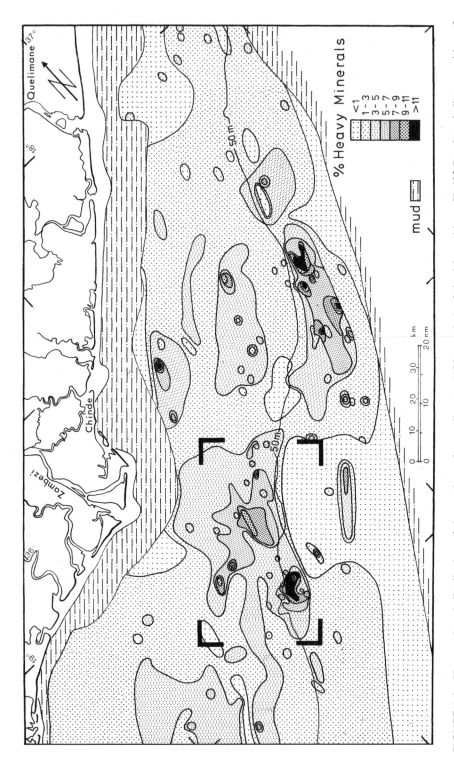

FIGURE 1.6 Heavy-mineral distribution of the sandy surface sediments (456 samples) of the Mozambique Shelf.[5] Brackets indicate position of Figure 1.7. (From Kudrass, H. R., Sedimentary models to estimate the heavy-mineral potential of shelf sediments, in *Marine Minerals*, Teleki, P.G. et al. (Eds.), Reidel Publishing Company, Dordrecht, 1987, 39. With permission.)

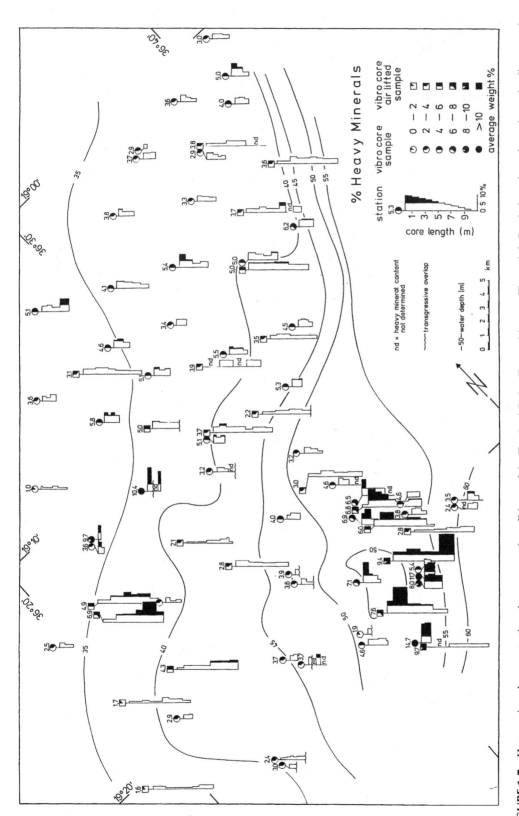

FIGURE 1.7 Heavy-mineral concentrations in core samples offshore of the Zambezi Delta. For position see Figure 1.6. Symbols and numbers above the diagrams show the average heavy-mineral concentration for each core. The 50 m terrace contains a large disseminated heavy-mineral deposit with approximately 50 mio t ilmenite, 4 mio t zircon, and 1 mio t rutile.[5] (From Kudrass, H. R., Sedimentary models to estimate the heavy-mineral potential of shelf sediments, in *Marine Minerals*, Teleki, P.G. et al. (Eds.), Reidel Publishing Company, Dordrecht, 1987, 39. With permission.)

3. Eluvial placer deposits are caused by sea-level-related shifts from depositional to erosional submarine environments resulting in local lag sediments enriched in gold, cassiterite, and — in upwelling areas — phosphorite.

Tectonic uplift can preserve beach placer deposits from interglacial periods by moving them out of the marine regime.

REFERENCES

1. Harben, P. W. and Bates, R. L., *Industrial Minerals — Geology and World Deposits*, Industrial Minerals Div., Metal Bull., London, 1990.
2. Shackleton, N. J., Oxygen isotopes, ice volume and sea level, *Quat. Sci. Res.,* 6, 183, 1987.
3. Fairbanks, R. G., A 17,000-year glacio-eustatic sea level record: influence of glacial melting rates on the Younger Dryas event and deep-ocean circulation, *Nature,* 342, 637, 1989.
4. McMaster, R. L., Lachance, T. P., and Ashraf, A., Continental shelf geomorphic features off Portuguese Guinea, Guinea, and Sierra Leone (West-Africa), *Amer. Assoc. Petrol. Geol. Bull.,* 54, 158, 1970.
5. Beiersdorf, H., Kudrass, H. R., and Stackelberg, U. von., Placer deposits of ilmenite and zircon on the Zambezi shelf, *Geol. Jahrb.,* D-36, 5, 1980.
6. Baxter, J. L., Heavy mineral sand deposits of western Australia. Australasian Institute of Mining and Metallurgy, Monograph 14, *Geology of the Mineral Deposits of Australia and Papua New Guinea,* 2, 1587, 1990.
7. Emery, K. O. and Noakes, L. C., Economic placer deposits of the continental shelf. *Committee for Co-ord. Joint. Prosp. for Min. Res. in Asian Offshore Areas, Techn. Bull.,* 1, 95, 1968.
8. Sujitno, S., Some notes of offshore exploration for tin in Indonesia, 1966–1976, *Committee Co-ord. Joint Prosp. Min. Res. in Asian Offshore Areas, Techn. Bull.,* 11, 169, 1977.
9. Kudrass, H. R., Sedimentary models to estimate the heavy-mineral potential of shelf sediments, in *Marine Minerals,* Teleki, P.G. et al. (Eds.), Reidel Publishing Company, Dordrecht, 1987, 39.
10. Jury, A. P. and Hancock, P. M., Alluvial gold deposits and mining opportunities on the west coast, South Island, New Zealand, in *Mineral Deposits of New Zealand,* Kear, D. (Ed.), Austral. Inst. Mining Metallurgy, Victoria, 1989, 147.
11. Middleton, G. V., Experimental studies related to problems of flysch sedimentation, in *Flysch Sedimentology in North America,* Lajoie, J. (Ed.), Geol. Assoc. Canada, Spec. Publ., 7, 253, 1970.
12. Slingerland, R., Role of hydraulic sorting in the origin of fluvial placers, *Jour. Sed. Petrol.,* 54, 1, 137, 1984.
13. Aleva, G. J. J., Aspects of the historical and physical geology of the Sunda Shelf essential to the exploration of submarine tin placers, *Geol. Mijnbouw,* 52, 2, 79, 1973.
14. Engelhardt, W. von., Über die Schwermineralsande der Ostseeküste zwischen Warnemünde und Darßer Ort und ihre Bildung durch die Brandung, *Zeitschrift angew. Min.,* 1, 1, 30, 1937.
15. Komar, P. D. and Wang, C., Processes of selective grain transport and the formation of beach placers, *Jour. Geol.,* 92, 637, 1984.
16. Seibold, E. and Berger, W. H., *The Seafloor: An Introduction to Marine Geology,* Springer Verlag, Berlin, 1982, 288.
17. Veenstra, H. J. and Winkelmolen, A. M., Size, shape and density sorting around two barrier islands along the north coast of Holland, *Geol. Mijnbouw,* 55, 87, 1976.
18. Sallenger, A. H., Inverse grading and hydraulic equivalence in grain-flow deposits, *Jour. Sed. Petrol.,* 49, 2, 553, 1979.
19. Swift, D. J. P., Barrier-island genesis: evidence from the central Atlantic shelf, eastern U.S.A., *Sediment. Geol.,* 14, 1, 1, 1975.
20. Kudrass, H. R. and Schlüter, H. U., Development of cassiterite-bearing sediments and their relation to Late Pleistocene sea-level changes in the Straits of Malacca, *Mar. Geol.,* 120, 175, 1994.
21. Burnett, W. C. and Riggs, S. R. (Eds.) *Phosphate Deposits of the World, Vol. 3, Neogene to Modern Phosphorites,* Cambridge University Press, Cambridge, 1990, 1.

2 Light Heavy Minerals on the Indian Continental Shelf, Including Beaches

G. Victor Rajamanickam

ABSTRACT

The Indian subcontinent is well endowed with placer minerals on and off both its east and west coasts. This is the result of a combination of factors, including abundant accessory minerals in the rocks of the hinterland, subtropical erosion and transportation processes, stable coastlines, and offshore wind systems which allow beach and offshore concentration to take place. Rising sea level has also played a part in the landward transport of offshore placers. Ilmenite, rutile, magnetite, zircon, garnet, and monazite are among the most important minerals present, and these have been selectively mined at a number of locations. Large-scale offshore mining, however, must await the establishment of a secure legal framework within which the exploitation can take place, including laws designed to protect the environment.

2.1 INTRODUCTION

Beach placers are the products of the ebb and flow of tides, waves, and inshore currents. Once the minerals reach the shore through the transporting medium of streams, wind, rain, and littoral currents, they undergo a series of transformations according to the nature of the shore. The shore can be classified, on the basis of the profiles of equilibrium, as abraded, accumulative, or stable. *Abraded coastlines* are characterized by high scarps or cliffs and are not favorable sites for the deposition of sediments. *Accumulative coastlines* allow the deposition of very fine sediments, predominantly of quartz with low specific gravity. In order to be enriched by heavy minerals such as beach placers, a coastline's conditions should be neither erosive nor accumulative. Such conditions prevail on stable shores (Figure 2.1), such as those off many parts of India.

2.2 ENVIRONMENTS OF PLACER DEPOSITION

2.2.1 SOURCE

The availability of placer minerals is very much linked to the source rock. The supply of placer minerals can be from primary mineral deposits, from the release of accessory minerals embedded in the rocks, or from older placer-rich materials. Native metals, such as gold, platinum, and other valuable minerals, like diamond, cassiterite, and wolframite, are generally liberated through the weathering of primary deposits, whereas minerals such as monazite, zircon, rutile, and ilmenite are made available in the course of the weathering of silicate rocks. Old placers can supply any type of minerals according to the nature of accumulation. Generally enriched placers, in minable quantities, are found close to their primary source. Some examples are given in Table 2.1.

0-8493-8429-X/00/$0.00+$.50
© 2000 by CRC Press LLC

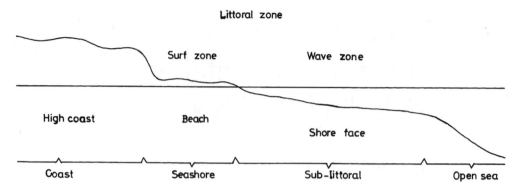

FIGURE 2.1 Diagram of the structure of a stable coastline.

TABLE 2.1

Typical Associations of Placer Minerals Arising from the Destruction of Certain Groups of Rocks and Their Associated Deposits

Rock	Valuable Minerals	Dominant Accessory Minerals
Granitoid formation	Ilmenite, rutile, zircon, beryl, monazite	Garnets, magnetite, sphene, apatite, pyroxene, amphiboles
The same with pegmatites	Cassiterite, wolframite, tantalite, spodumene, columbite, thorite, topaz, beryl	Fluorite, tourmaline
The same with skarns	Magnetite, scheelite, cassiterite	Garnets (zonal), diopside, wollastonite, vesuvianite, hedenbergite, actinolite, tremolite, scapolite
The same with hydrothermal deposits	Gold, cinnabar, wolframite, cassiterite	Barite, siderite
Plagiogranite–syenite formation	Rutile, ilmenite, zircon, corundum	Magnetite, apatite, orthite, garnets, monazite, columbite, eudialyte, loparite, sphene, perovskite, aegirine, fluorite, pyroxenes, amphiboles
Gabbro-diabase formation	Ilmenite, leucoxene titanomagnetite	Diopside-augite, hypersthene, amphiboles, apatite, spinel
Pyroxenite formation (serpentinites, dunites, peridotites, pyroxenites)	Platinum, iridosmium, ilmenite, titanomagnetite	Olivine, bronzite, diopside-augite, chrome-diopside, magnetite, orthorhombic pyroxene, titanaugite
Peridotite formation	Diamond, rutile, ilmenite	Spinel, chromopicotite, pyrope, chrome-diopside, magnetite, orthorhombic pyroxene, titanaugite, phlogopite
Ultrabasic–alkaline formation with carbonates	Pyrochlore, apatite, ilmenite, titanomagnetite, tantaloniobates	Magnetite, orthite, forsterite, sphene, titanaugite, phlogopite perovskite, amphibole, anatase, spinel

Because the individual minerals are deposited according to definite habit and size, the nature of placers of a particular mineral, from different primary sources, will be different. The ilmenite supplied from a basaltic source will be fine grained with a needle shape and lack of an exsolution phenomenon, whereas the ilmenite supplied from acidic rocks will be granular and intergrown with magnetite. Other placer minerals are found to have characteristic zoning, overgrowths, outgrowths, etc., inherited from their original formation.

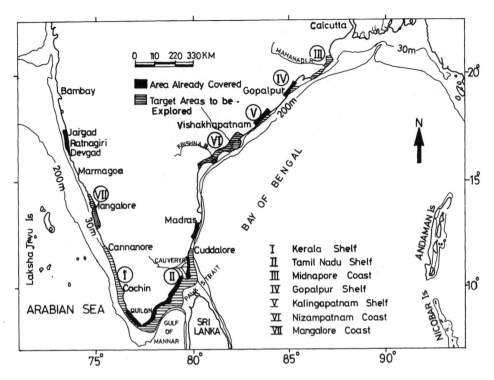

FIGURE 2.2 Target areas for exploration of heavy mineral placers in India.

2.2.2 TRANSPORT

Under humid conditions liberation of mineral grains is rapid, and their removal by rain wash is fast. Once grains reach the regular transporting medium of rivers or the sea, they undergo sorting and segregation.

Placers can be formed where the stream load changes due to a sudden change in its flow velocity. The placers are sometimes trapped in channels and shoals. In such deposits, one may see a mixture of coarse and fine grains, on some occasions in cross-bedded form. During floods, the heavier grains are sometimes deposited in disseminated form and covered by thick fine-grained sediments. Economically viable placer deposits may not form in such flood plain environments.

In many cases, for example, the occurrence of monazite, garnet, zircon, and ilmenite deposits along the Tamil Nadu and Kerala coasts of India, the main transport is from offshore, where the heavy minerals must have been deposited during a previous low sea level stand. During rising sea level, these deposits have been transported shoreward for further concentration. Rajamanickam[1] has attributed the rich ilmenite deposits of the Konkan coast of India to landward migration of offshore minerals. Keeping in mind the probable source rocks distributed around the country and the suitable drainage conditions with suitable transporting agents, target areas for placer mineral exploration have been selected in Figure 2.2.

2.3 PLACER MINERAL DEPOSITS OFF INDIA

In the last few decades exploration for and exploitation of placer minerals has been active off both the east and west coasts of India (Figure 2.2). India is endowed with a coastline of over 6500 km, hosting some of the world's largest and richest beach placer deposits. These include ilmenite (278 million tonnes), garnet (86 million tonnes), zircon (18 million tonnes), rutile (13 million tonnes),

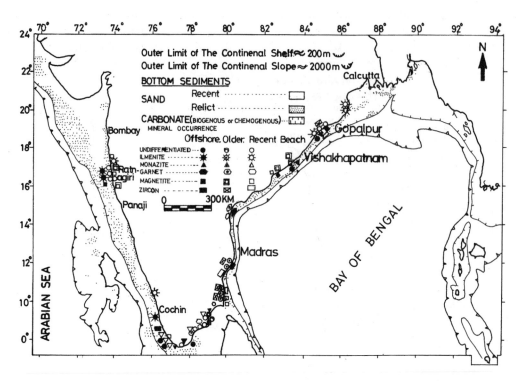

FIGURE 2.3   Map showing the terrigenous mineral occurrences on the continental margins of India.

monazite (7 million tonnes), and sillimanite (84 million tonnes). The major deposits are found along the coast of Chavara (Kerala), in a few places off the Ratnagiri district (Maharashtra), along the beaches of Manavalakurichi (Tamil Nadu), Kakinada (Andhra Pradesh), and Puri (Orissa).[2] The type of placers so far recognized and evaluated as economically viable deposits are shown in Figure 2.3. Details on the reserves and concentration of heavy minerals are outlined below.

2.3.1   ILMENITE AND RUTILE

Ilmenite and rutile, components of beach sand deposits, are found all around India from the Ratnagiri coast in the west, to the Orissa coast in the east. These minerals are concentrated in three well-defined zones: (1) over a stretch of 22 km between Neendakara and Kayankulam, Quilon district, Kerala; (2) over a stretch of 6 km from the mouth of the Valliyar River to Kolachal in Manavalakurichi in the Kanyakumari district of Tamil Nadu; and (3) on the Chatrapur coast, over an area of 26 sq km in the Ganjam district of Orissa.

The Quilon deposit is characterized by the highest concentration of heavy minerals. The heavy mineral contents of the Quilon and Manavalakurichi deposits vary between 65 to 85% and 50 to 70%, respectively, whereas the Chatrapur coast deposit has only 20 to 25% heavy minerals. These deposits contain an average of 9.4% ilmenite, 0.40% rutile, 3.30% sillimanite, 0.33% zircon, and 0.29% monazite.

The reserves of ilmenite and rutile in the Quilon district are estimated to be 17 and 1.25 million tonnes (MT), respectively. In Manavalakurichi, 4.00 million tonnes of ilmenite and 0.8 million tonnes of rutile are estimated to be present. The Orissa coast has reserves of 22 million tonnes of ilmenite and 0.8 million tonnes of rutile.

A new deposit of Ti-rich beach sand has been identified in the Tanjore district, Tamil Nadu, extending for more than 12 km along the Cauvery River Delta between Sirkali and Kaveripattinam.

When compared to other deposits being mined by the Indian Rare Earths Company at Manavalakurichi, these deposits contain a higher content of zircon, along with monazite and ilmenite.

A recent survey of Chilka Lake between Malud and Puri, Orissa, has shown the presence of heavy mineral concentrations in the range of 5 to 15%. The heavy mineral assemblage, in order of abundance, includes ilmenite, sillimanite, garnet, rutile, monazite, and the polymineralic rock pyribole. Ilmenite dominates, comprising 40 to 50% of the total heavies.[3]

The distribution of total heavies in the offshore to 40-m depth off Chavara and Manavalakurichi ranges from 4 to 12%. Opaques, monazite, and zircon are found closer to the shore, whereas minerals such as sillimanite, hypersthene, and hornblende increase offshore. It is inferred that these deposits reflect the concentration of placers along the Chavara and Manavalakurichi coast by means of wave action combined with currents and geomorphic factors.[4]

2.3.2 GARNET

In India total recoverable reserves of garnets are placed at 39.5 million tonnes, of which proved, probable, and possible categories are 17.4, 12.8, and 9.3 million tonnes, respectively. Garnet placer deposits occur in Kerala, Orissa, and Tamil Nadu States in the amounts of 0.74, 0.14, and 16 million tonnes, respectively.

About 32% of the recoverable reserves of garnet sands are located in the Ganjam district of Orissa. Garnet is also recovered as a co-product, along with other heavy minerals, during the processing of beach sands in Tamil Nadu, which is the single largest producer.[5]

In the Chavara and Quilon regions of the Kerala coast, garnet is seen in association with monazite, ilmenite, zircon, kyanite, and sillimanite. It is produced as a by-product after the separation of monazite and ilmenite. It is present as 5 to 8% of the total heavies.

Along the Chatrapur coast, near Gopalpur, Orissa state, garnet is being mined from the foreshore area of Aryapalli by the Orissa Sand Complex of Indian Rare Earths Ltd. It is associated with an assemblage of ilmenite, monazite, zircon, rutile, and sillimanite. Recently, the Atomic Minerals Division of the Government of India has estimated a reserve of 230 million tonnes of sands with 20 to 25 wt% of heavies along this coastal tract. Garnet is found to comprise about 10 to 12% of the total heavy minerals.

In Tamil Nadu, garnet occurs along with ilmenite, monazite, and kyanite along the coast of Manavalakurichi to Kanyakumari. However, in Tamil Nadu it is also found to be associated with ilmenite, zircon, and kyanite from Tuticorin to the Valinokkam coast and from Velanganni to Thirumullaivasal. Coarse garnets of more than 1 mm in size are being mined in the river beds in the Musiri and Thuraiyur regions in the Tiruchirapalli district. The percentage of garnet mined is 20 to 30% along the river bed, while it is only 10 to 15% along the coast.

2.3.3 ZIRCON

Zircon occurs along with other heavy minerals in beach sands along the coasts of Kerala, Tamil Nadu, and Orissa.

Zircon occurs in Kerala between the Neendakara and Kayankulam areas along an approximately 22 km stretch of coast. The percentage of zircon mined is from 5 to 8% of the total sands. Zircon is found to be in association with ilmenite in 60 to 70% of sands, with rutile in 4 to 7%, with monazite in 0.5 to 1%, with sillimanite in 4 to 8%, and with leucoxene in 1 to 1.5% of sands.

Zircon occurs for about 6 km extending from the mouth of the Valliyar River to Kolachal in Manavalakurichi and adjoining coastal tracts of the Kanyakumari district in Tamil Nadu. The percentage of zircon is about 4 to 6% of the total sands. In these sands, 45 to 55% is associated with ilmenite, 2 to 3% is associated with rutile, 7 to 14% is associated with garnet, 3 to 4% is associated with monazite, 2 to 3% is associated with sillimanite, and 0.5 to 1% is associated with leucoxene.

On the Chatrapur coast, on a 20-km-long stretch, deposits of zircon comprising 0.33% of the total sands are reported. The beach sands of this coast also contain 9.4% ilmenite, 0.40% rutile, 3.30% sillimanite, 6.8% garnet, and 0.29% monazite.[5]

The overall reserves of zircon in Tamil Nadu, Kerala, and Orissa are inferred to be 0.598 million tonnes, 0.134 million tonnes, and 0.720 million tonnes, respectively.

2.4 REGIONAL DISTRIBUTION OF PLACER DEPOSITS OFF INDIA

2.4.1 WEST COAST

Studies reveal that the concentrations of heavy minerals are higher along the Kerala coast than along other parts of the Indian coast.[6,7] Rao and Rao[8] have carried out a detailed study of ilmenite concentrates from Chavara and Manavalakurichi. Similar types of deposits with varying mineral composition have also been identified in Ratnagiri, Maharashtra coast.[9-11] Siddiquie et al.[12] have pointed out that the distribution of sediments and the concentration of heavy minerals in the placers off Ratnagiri are apparently controlled by three factors — the source rocks, streams and their entry into the arcuate bays, and waves and currents operating in the bays.

By both geological and geophysical methods (shallow seismic), Siddiquie et al.[13] have undertaken exploratory surveys for offshore ilmenite placers off the Konkan coast. Magnetic anomalies of high frequency, low amplitude, and short wave length are attributed to the distribution of placers. The surveys indicate the presence of ilmenite sand on the sea bed over an area of 96 sq km, and the sands contain 11 to 57% ilmenite, titanomagnetite, and hematite. The thickness of the ilmenite-bearing sands ranges from 2 to 10 m, and these extend to water depths of about 20 m. Ilmenite placers were found to extend approximately 2 to 5 km offshore. The reserves of ilmenites are inferred to be 12.5 million tonnes. Thus, the probable reserves in the area will be many times more than the onshore reserves of 4 million tonnes.[11]

Rajamanickam[1] undertook detailed placer mineral exploration in the Jaigad, Ambwah, and Varvada bays of the Konkan coast. The salient findings are sizable placer deposits of magnetite and ilmenite covering an area of about 24.3 sq km, 8 sq km, 4.02 sq km, in Jaigad, Ambwah, and Varvada bays, respectively. The probable reserve of ilmenite to a subsurface depth of 1 m in these areas is 21.08 million tonnes.

Ali et al.[14] have suggested that beach placers in the Bhatya and Purnagad sectors of the Ratnagiri district contain 40 to 60 wt% magnetite and 25 to 35 wt% ilmenite/hematite.

After a reconnaissance survey, Dhanunjaya Rao et al.,[15] showed the region between Mangalore and Cochin as the potential site for placer mineral exploration. The heavy minerals' content of its beach sands varies from 3.45 to 36%, with ilmenite (1.2 to 16.2%) and garnet (0.03 to 11.3%). The distribution of these minerals is thought to be controlled possibly by the activity of long shore currents.

The Geological Survey of India has assessed a reserve of heavy mineral deposits in the Varkala and Chavara sectors of Kerala, in 0 to 4 m of water, to be 0.953 and 0.461 million tonnes of ilmenite, 0.313 and 0.207 million tonnes of sillimanite, and 0.058 and 0.060 million tonnes of zircon.[16] Babu and Thirivikramji[17] consider a number of factors, such as promontories/sea cliffs, nature of shore processes, and long shore currents, as causes of the formation of modern placers along Kerala beaches.

By the application of Q mode factor analysis, Rajamanickam[18] has studied the provenance of ilmenite-rich placers off Maharashtra and attributed their origin to an offshore source in addition to Deccan basalts. From that study it was further inferred that economically viable placer deposits can only be located in bays situated away from present-day rivers. Additionally, Rajamanickam et al.[19] have reported that the ilmenites off the Konkan coast contain 55% TiO_2 and that associated magnetites have a maximum of 55% Fe_2O_3 and 25% TiO_2. It was concluded that the ilmenite-rich sands are paleo-placers, because the mineral concentrate and the nature of the coastal rocks do not match well.

Gujar[20] has made an exhaustive study of nearshore placer sands off the south Konkan coast, Maharashtra, and reported the distribution of offshore placer sands to a water depth of 17 to 19 m by echosounding, shallow seismic profiling, and gravity core sampling. He has inferred reserves of 2.26 million tonnes of magnetite and 0.99 million tonnes of ilmenite in a total sand-covered area of 13.15 sq km with a concentration of 1.78 to 6.33% ilmenite and 4 to 21% magnetite to a subsurface depth of 1 m. In addition, he has reported the comparability of the chemical composition of these ilmenites with the ilmenites mined from other parts of India (Table 2.2).

2.4.2 EAST COAST

The occurrences of considerable quantities of placer mineral reserves in the Visakhapatnam, Yarda, Waltair, and Bhimunipatnam coastal areas of Andhra Pradesh[21,22] (Figure 2.2, V) and the coastal areas of Orissa[23] (Figure 2.2, IV) have been reported. Rao[24] has concluded that beach erosion and beach regression play a significant role in affecting the concentration of well sorted heavy mineral sands along this coast.

The coastal stretch between Kakinada and the Tandava River in East Godavari district, Andhra Pradesh (Figure 2.2, V), contains heavy mineral enrichment, primarily ilmenite, rutile, zircon, garnet, monazite, magnetite, and appreciable amounts of pyriboles to a subseafloor depth of 12 m. The grade of ilmenite present in the sand and its chemistry (with TiO_2 46 to 48% and a negligible amount of impurities such as Cr and V) have made it suitable for the titanium slag industry.[25]

The Kalingapatnam coast near Visakhapatnam has a promising placer mineral sand deposit of Recent age. The deposit, situated in the Srikakulam district, Andhra Pradesh, occurs between the Vamsadhara River to the south and Meghavaram village to the north. The length and average width of the deposit are 25 km and 700 m, respectively, to a maximum explored depth of 14 m. Evaluation of the deposit has established a reserve of 18 million tonnes of heavy minerals with a working grade of 13.10%. Ilmenite, sillimanite, and garnet constitute 90% of the total mineral assemblage, with reserves of 5.85 million tonnes, 6.30 million tonnes, and 4.80 million tonnes, respectively.[26]

An area of 1900 sq km of the inner continental shelf off Kalingapatnam–Sonapurapeta near Visakhapatnam was surveyed. A preliminary appraisal based on a 2.5 km × 2 km grid sampling made with a vibrocorer has indicated that in the Baruva–Kapasukuddi offshore sector between the 8 to 18 m isobaths, the top 2 m of sediment is estimated to have a potential of about 6 million tonnes of heavy minerals. By contrast, in the Bavanapadu–Rattikonda offshore sector, between the 8 to 14 m isobaths, the top 1.5 m of sediment is estimated to contain about 4 million tonnes of heavy minerals. In the Kidsingi–Rattikonda offshore sector, between the 20- to 30-m isobaths, the top 2 m of sediment is estimated to contain about 3.6 million tonnes. Finally, in the Maruvada–Kambalaraidupeta area, between the 20 to 32 m isobath, the top 1.5 m of sediment is estimated to contain about 3.7 million tonnes of heavies.[27]

Ramamohana Rao et al.[28] have reported the occurrence of thin layers of black sand in the inland stream channels in the Visakhapatnam–Bhimunipattinam coastal areas of Andhra Pradesh, and the appearance of many unstable heavy minerals in the deposits. The source is the nearby highland area. Anon.[29] has estimated the reserves of beach placers on the Visakhapatnam–Bhimunipattinam coast to be 0.027 million tonnes of magnetite, 0.005 million tonnes of garnet, and 5.799 million tonnes of ilmenite.

Venkateswarlu et al.[30] have delineated a palaeo-depression, from palaeo-bathymetry using shallow seismic studies, off Bhimunipattinam. The disposition of this feature in relation to the seabed sediment distribution suggests it is significant as a prospective zone for the accumulation of placer sands at depth. Further, Dhanunjaya Rao et al.[31] have noted the presence of higher percentages of heavy minerals such as ilmenite, rutile, monazite, and magnetite in the Krishna–Godavari delta sediments than in sediments from the Pennar delta of Andhra Pradesh.

TABLE 2.2
Percentages of Various Constituents in Minable Ilmenites from Placer Sands off India

	Ratnagiri Dt. India			Central TN Coast			Southern TN Coast			Orissa	SW TN Coast		SW Kerala Coast		S. Konkan Coast, Maharashtra				Northern TN Coast
	Jaigad	Ambwah	Varvada	South Sector	Central Sector	North Sector	Kanya-Kumari	Kallar-Vaippar	Lean Zone	Gopalpur	Manavalakurchi		Quilon		Wada Vetye	Ambolgarh	Rajapur	Vijaydurg	
	1	2	3	4	5	6	7	8	9	10	11	12	13	14	15	16	17	18	19
TiO_2	47.96	51.38	52.30	43.56	43.43	40.95	47.63	42.61	43.20	50.10	54.30	54.10	60.30	59.80	40.39	43.95	45.80	46.60	38.61
FeO	27.55	27.82	30.81	39.31	38.21	35.03	20.50	25.50	22.85	34.10	26.0	25.60	9.70	10.90	35.92	30.30	40.55	45.22	27.73
Fe_2O_3	15.66	16.54	11.58	12.57	13.56	18.83	26.51	27.60	29.15	12.76	15.5	15.30	24.80	23.70	16.60	17.50	16.96	16.40	13.07
P_2O_5	0.89	0.41	0.64	0.46	0.44	0.26	0.46	0.48	0.52	0.03	0.26	—	0.17	—	0.22	0.15	0.20	0.26	0.24
Cr_2O_3	0.06	0.09	0.10	0.05	0.06	0.03	0.03	0.03	0.02	0.05	0.07	0.09	0.14	0.15	0.02	0.01	0.01	0.01	—
V_2O_5	—	—	—	0.17	—	—	0.38	0.32	0.50	0.24	0.20	0.23	0.26	0.23	0.02	0.02	0.02	0.02	—
Al_2O_3	1.10	1.18	1.09	1.74	1.60	2.50	0.10	0.09	0.07	0.60	1.10	—	1.00	—	1.80	1.20	1.01	0.85	10.62
MnO	0.35	0.32	0.39	0.31	0.31	0.30	0.31	0.13	0.39	0.55	0.40	—	0.40	—	0.40	0.38	0.40	0.49	0.20
CaO	0.61	0.16	0.03	0.61	0.33	0.20	1.35	1.32	1.24	—	0.08	—	0.15	—	0.61	0.27	0.19	0.64	2.38
MgO	2.52	0.87	1.06	0.31	0.13	0.11	0.38	0.49	0.31	0.75	0.85	—	0.65	—	1.90	1.40	0.66	1.40	0.78
SiO_2	2.93	1.88	1.83	0.13	0.61	0.37	0.92	1.10	0.45	0.80	1.40	—	1.40	—	3.41	2.60	2.87	3.25	1.84
Nb_2O_5	—	—	—	—	—	—	0.03	0.05	0.05	—	—	—	—	—	—	—	—	—	—
SnO_2	0.02	—	—	0.00	0.00	0.00	—	—	—	—	0.06	—	0.01	—	—	—	—	—	—
CoO	0.02	0.02	0.02	0.03	0.03	0.02	—	—	—	—	0.08	—	—	—	0.01	0.01	0.01	0.01	—
WO_3	—	—	—	—	—	—	—	—	—	—	0.005	—	—	—	—	—	—	—	—
Au	—	—	—	—	—	—	—	—	—	—	0.006	—	0.004	—	—	—	—	—	—
Pt	—	—	—	—	—	—	—	—	—	—	—	—	0.02	—	—	—	—	—	—
Cd	—	—	—	—	—	—	0.01	0.02	0.01	—	—	—	—	—	—	—	—	—	—
CuO	0.06	0.04	0.06	0.26	0.09	0.10	—	—	—	—	0.01	—	—	—	0.03	0.01	0.004	0.02	—
ZnO	0.03	0.04	0.04	0.68	0.01	0.01	—	—	—	—	—	—	—	—	0.01	0.01	0.009	0.01	—
Li_2O	0.01	0.00	0.00	0.02	0.01	0.01	0.01	—	—	—	—	—	—	—	—	—	—	—	—
NiO	0.06	0.04	0.05	0.04	0.04	0.02	—	—	—	—	0.04	—	—	—	0.01	0.01	0.009	0.01	—
PbO	0.00	0.02	0.01	0.01	0.00	0.00	0.03	—	—	—	—	—	—	—	—	—	—	—	—
Na_2O	—	—	—	—	—	—	—	—	—	—	—	—	—	—	—	—	—	—	3.37
K_2O	—	—	—	—	—	—	—	—	—	—	—	—	—	—	—	—	—	—	0.67

1 to 3 — Rajamanickam[1]; 4 to 6 — Chandrasekar[42]; 7 to 9 — Angusamy[44]; 11 to 14 — Sinha[56]; 10, 15 to 18 — Gujar[20]; 19 — Mohan[52]

The coastal tract from Machchali–Sunnapalli in the south, to Bendi Creek in the north through Bavanapadu, Srikakulam district, Andhra Pradesh, contains sizable reserves of heavy minerals. Important heavy minerals identified in order of abundance are ilmenite, garnet, sillimanite, rutile, leucoxene, zircon, and monazite. Estimated reserves are 29.78 million tonnes of total heavy minerals at a working grade of 14.38%, contained in 207 million tonnes of sand. Ilmenite, the predominant mineral at 34.89 wt%, has been estimated at 10.40 million tonnes. Reserves of other important economic minerals are garnet (10.04 million tonnes), sillimanite (7.37 million tonnes), rutile (0.39 million tonnes), leucoxene (0.35 million tonnes), and zircon (0.21 million tonnes). Monazite constitutes about 1.00% of heavy minerals, while other minerals, such as kyanite, pyriboles, and magnetite, have grades of 2.50% and <0.10%. The ilmenite fraction contains 51% of TiO_2 (Table 2.2), with low CaO, MgO, P_2O_5, V_2O_5, and Cr percentages, and hence is very well suited for both titanium slag and the synthetic industry.[32]

The coastal area between Koyyam and the Nagavali River, Srikakulam district, Andhra Pradesh, spans over 20 km in length with an average width of 800 m. Here, the heavy mineral suite consists of ilmenite, sillimanite, garnet, rutile, leucoxene, zircon, and monazite, in that order of abundance. Preliminary studies reveal reserves of 6.20 million tonnes of total heavy minerals at a working grade of 10%. The sand has been estimated to be of the order of 62 million tonnes. Ilmenite comprises 2 million tonnes at 32% of the total heavies. The reserves of other economically important minerals that can be worked as by- and co-products are garnet 1.50 million tonnes, sillimanite 1.90 million tonnes, rutile 0.15 million tonnes, leucoxene 0.15 million tonnes, and zircon 0.08 million tonnes. Monazite is less than 1.00% of the total heavy minerals. The rutile content, when compared to the other heavy mineral deposits along the East Coast, has been found to be high. The chemical analysis of ilmenite reveals that it contains 50% TiO_2 (Table 2.2), with significantly low levels of CaO, MgO, V_2O_5, and Cr.[33]

Beach placer deposits are also reported to occur in the Tinneveli, Ramnad, and Tanjore districts of Tamil Nadu.[24,34-38]

Setty and Rajamanickam[39] have studied the heavy mineral suite in the shelf sediments off the Madras coast and have divided it into two heavy mineral zones — a northern augite-rich zone and a southern augite-poor zone. Further, they also recorded that there is little or no topaz, tourmaline, sillimanite, staurolite, or monazite in these sediments.

Heavy mineral analysis of the surficial, as well as the vibrocore, subsamples from the inner shelf off the coast between Sonapurapeta, Andhra Pradesh, and Malud, Orissa, over a stretch of 84 km, reveals that the sand body has more than 5% heavy mineral content and occupies about 632 sq km on the surface, reducing to about 232 sq km at 1 m below the surface. The Geological Survey of India has estimated that along the Orissa coast, (Figure 2.2, IV) at depths of 0 to 1 m, a reserve of 17.28 million tonnes of ilmenite, 6.8 million tonnes of sillimanite, 4.86 million tonnes of garnet, and 1.62 million tonnes of monazite, zircon, and rutile occurs.[40] Ilmenite, hematite, garnet, monazite, zircon, rutile, magnetite, sillimanite, pyroxene, and amphibole from the beach sands of Ekakula, the Gahiramatha coast, and Orissa are reported here for the first time. Their total concentrations vary from 26.4 to 100%. This heavy mineral placer deposit has a good economic potential for monazite, zircon, and radioactive rare-earth elements besides titanium.[41]

Indian Mineral Year Book[42] has reported that the Chatrapur coast, stretching for 18 km over an area of 26 sq km in the Ganjam district of Orissa, contains very rich placer deposits with 9.4% ilmenite, 0.4% rutile, 0.33% zircon, 6.8% garnet, and 0.29% monazite. Further, Chandrasekar[43] has undertaken detailed exploratory work on beach placers in the Central Tamil Nadu coast and assessed the reserves of zircon, garnet, and ilmenite as 1.74, 2.45, and 5.97 million tonnes in enriched, and 0.48, 1.22, and 0.45 million tonnes in lean zones, respectively. Additionally, Angusamy et al.[44] and Angusamy[45] have estimated the reserves of zircon in the amount of 0.49 million tonnes, garnet 6.13 million tonnes, ilmenite 6.01 million tonnes, magnetite 0.10 million tonnes, and monazite 0.84 million tonnes in the Kanyakumari area of Tamil Nadu.

The heavy mineral suite at Manavalakurichi near Kanyakumari/Tamil Nadu (Figure 2.2, II) consists of ilmenite (15.5%), sillimanite (2.6%), garnet (2.4%), zircon (1.2%), rutile (0.9%), monazite (0.7%), leucoxene (0.7%), and kyanite (0.1%), and the assemblage is strikingly uniform over an investigated area of 1.42 sq km, indicating a common provenance throughout. A 7-km-long recessed headland-bounded bay between Muttam and Kolachal is essentially a closed littoral cell containing high concentrations of heavy minerals (15.6 to 39.2% with a thickness of 7.5 m), in contrast to the Pillaithoppu–Rajakkamangalam and Rajakkamangalam–Pallam blocks, containing lower heavy mineral concentrations (5.7 to 6.3%), located on the straight coastal segment in the southeast beyond the Muttam promontory. Major variations in the heavy mineral concentrations occurring over short distances within the bay are ascribable to local sorting processes rather than a regional variation in the sand-source mineralogy. Heavy mineral concentration decreases from 48 to 25% at the surface to 19.7 to 6.8% at a depth of 7.5 m.

The ilmenite from Manavalakurichi in Tamil Nadu is in the size range of 177 to 125 microns, and the garnet and sillimanite 422 to 120 microns. Zircon, monazite, and rutile are finer, falling in the range of 177 to 105 microns. The partially altered ilmenites contain 54 to 57% TiO_2 with higher FeO (20.58 to 34.50%) than Fe_2O_3 (7.69 to 19.63%). The heavy mineral resources are placed at 8.30 million tonnes, and more than 40 to 50% of the deposit may still be available for exploitation.[46]

Ilmenite-dominant placer heavy mineral concentrations of up to 31% occur in the unconsolidated, fossiliferous, fine-to-medium beach and dune sands along a 5000-m-long stretch of coast at Pillaithoppu–Rajakkamangalam, Kanyakumari district, Tamil Nadu. Ilmenite constitutes 50% of the heavy minerals, followed by sillimanite (28.3%), garnet (10.8%), zircon (4.7%), rutile (3.6%), monazite (1.74%), and pyribole (0.32%). Predominance of sillimanite over garnet in the heavy mineral suite contrasts with the adjoining Manavalakurichi beach placer deposit where garnet is the most abundant mineral after ilmenite. Mineralization is observed from the surface to a depth of 9 m over a width of 100 to 350 m.

More placer sands in the vicinity of Pillaithoppu–Rajakkamangalam, Kanyakumari district, Tamil Nadu, have been identified as potential heavy mineral reserves by virtue of their low slime content (9.58%) and moderate carbonate content (20.35%). Here, the heavy mineral reserves are estimated at 1.21 million tonnes in the "indicated" category at an average grade of 6.32% and a thickness of 6.35 m and could serve as an additional source for the existing heavy mineral separation plant at Manavalakurichi.[47]

Loveson[48] has investigated sea level variation and the heavy mineral accumulation along the southern Tamil Nadu beaches through satellite imagery and aerial photos. On the basis of this study, the enrichment of heavy minerals is found to be in two physiographic lowlands: (1) the Kallar to Vaippar area, and (2) Manappad to Kanyakumari.

The distribution and concentration of different detrital minerals has been studied in the sediments of the Gadilam River,[49] Vaippar River,[50] Palar River,[51] and their basins in order to evaluate the provenance of beach placer concentrations on the East Coast of India. These studies revealed that the contribution of terrigenous sediments to the placers is meager, confirming a source external to the beaches off these rivers by shoreward migration of sediments during rising sea level.

Chandrasekar and Rajamanickam[52] have evaluated the total reserves of zircon on the beaches of central Tamil Nadu to be 1.74 million tonnes. This could lead to a profitable exploitation along with other minerals like ilmenite, magnetite, and kyanite as by-products.

Mohan[53] has identified buried placer mineral concentration below 3 to 4 m on beach ridges from south of Pondicherry to north of Chennai on the Tamil Nadu coast. The estimated probable reserves of magnetite are 5.10 million tonnes and ilmenite 5.71 million tonnes. Further, on the basis of detailed mineralogical and textural studies, Mohan[54] has proposed that the increment of heavy minerals in the Parangipettai beach and nearshore environments of Tamil Nadu is due to the addition of minerals from sources such as palaeo-sediments and the contribution from the present-day Vellar River.

Angusamy and Rajamanickam[55] have classified the coastal stretch of the Mandapam to Kany-akumari region, Tamil Nadu, into five mineralogical provinces on the basis of the abundance of heavy minerals. The distribution of these minerals is controlled by geomorphology, tectonic regime, and littoral processes. Further, Udayaganesan and Rajamanickam[56] studied the depositional environment of Vaippar River sediments, Tamil Nadu, and showed that the difference between river and beach sediments may be due to their derivation from different sources, confirming a decoupling between the riverine and beach environments off eastern India.

2.5 SUMMARY AND CONCLUSIONS

India is endowed with rich placer deposits of monazite, magnetite, ilmenite, zircon, garnet, sillimanite, and kyanite. However, only ilmenite and magnetite have been investigated in detail in the offshore deposits. All other deposits off the country are kept in the status of inferred reserves. The exploitation of such placer minerals may not be possible until their offshore continuity has been thoroughly investigated. In order to strengthen the establishment of mineral industries on the basis of coastal placer deposits, it is necessary for the Indian government to take up a detailed investigation in the nearshore area, especially along the Kerala, Tamil Nadu, Andhra Pradesh, and Orissa coasts. However, until the framing of offshore mining laws is completed in India, it may not be possible to go ahead with offshore mining for placer minerals. Furthermore, before the mining of offshore placers proceeds, the environmental implications of the mining must be evaluated.

REFERENCES

1. Rajamanickam, G. V., *Geological Investigations of Offshore Heavy Mineral Placers of Konkan Coast, Maharashtra, India,* Ph.D. thesis, Indian School of Mines, Dhanbad, India unpublished, 1983, 258.
2. Mir Azam Ali, Viswanathan, G., and Krishnan, S., Beach and inland sand placer deposits of India — influence of geology, geomorphology and late Quaternary sea level oscillations in their formation, in *National Symp. Late Quaternary Geology and Sea Level Changes and Annual Convention of Geological Society of India,* Cochin, Abst. Vol., 1998, 53.
3. Roy, A. K., Ravi, G. S., and Mir Azam Ali, Textural characteristics and heavy mineral content of shelf sediments off Chilka lake, Orissa, in *National Symp. Late Quaternary Geology and Sea Level Changes and Annual Convention of Geological Society of India,* Cochin, Abst. Vol., 1998, 55.
4. Prakash, T. N., Kurien, N. P., and Felix, J., Transport and concentration of placers — a wave dominated process, in *National Symp. Late Quaternary Geology and Sea Level Changes and Annual Convention of Geological Society of India,* Cochin, Abst. Vol., 1998, 56.
5. Gujar, A. R., Nagendranath, B., and Banerjee, R., Marine minerals: the Indian perspective, *Marine Mining,* 7, 317, 1988.
6. Tipper, G. H., The monazite sands of Travancore, *Records of Geological Survey of India,* 44, 195, 1914.
7. Brown, J. C. and Dey, A. K., *India's Mineral Wealth,* Oxford University Press, New Delhi, 1955, 761.
8. Rao, N. K. and Rao, G. V. U., Intergrowths in ilmenite of the beach sands of Kerala, *Mining Magazine,* 35, 118, 1965.
9. Krishnan, M. S. and Roy, B. C., Titanium, *Records of Geological Survey of India,* 78, 1, 1945.
10. Roy, B. C., Ilmenite sand along Ratnagiri coast, Bombay, *Records of Geological Survey of India,* 87, 438, 1958.
11. Mane, R. B. and Gawade, M. K., Reports of the prospecting of ilmenite beach sands of Ratnagiri district, unpublished report, Directorate of Geology and Mines, Govt. Maharashtra, Nagpur, 1984, 15.
12. Siddiquie, H. N., Rajamanickam, G. V., and Almeida, F., Offshore ilmenite placers of Ratnagiri, Konkan coast, Maharashtra, India, *Marine Mining,* 2, 91, 1979.
13. Siddiquie, H. N., Rajamanickam, G. V., Gujar, A. R., and Ramana, M. V., Geological and geophysical exploration for offshore ilmenite placers off the Konkan coast, Maharashtra, India, in *Proc. 14th Offshore Technology Conference,* Houston, Texas, 1982, 749.

14. Ali, M. A., Naidu, P. S., and Manjunath, Y. S., Studies on the beach placers of Ratnagiri district, Maharashtra, India, in *Exploration and Research for Atomic Minerals,* Atomic Minerals Division, Hyderabad, 2, 167, 1989.

15. Dhanunjaya Rao G., Nature, distribution and evaluation of heavy minerals in the beach sand deposits between Mangalore and Cochin, West Coast of India, in *Exploration and Research for Atomic Minerals,* Atomic Minerals Division, Hyderabad, 3, 157, 1990.

16. Senthiappan, M., Michael, G. P., Charlu, T. K., Jayakumar, R., Abdulla, N. M., Kumaran, K., Unnikrishnan, E., and Bhatt, K. K., Multi-mineral placer deposits off Kerala coast, *Geological Survey of India News Letter,* 3, 2, 1992.

17. Babu, D. S. and Thrivikramaji, K. P., Paleogeographic interpretation of Kerala beach placers, south west coast of India, *Indian Journal of Marine Science,* 22, 203, 1993.

18. Rajamanickam, G. V., Provenance of the sediments inferred from transparent heavy minerals on the inner shelf of Central Maharashtra, in *Proc. Second South Asia Geological Congress,* Wijayanaanda, N. P., Cooray, P. G., and Mosley, P. (Eds.), GEOSAS II, Colombo, Sri Lanka, 1998, 309.

19. Rajamanickam, G. V., Varma, O. P., and Gujar, A. R., Ilmenite placer deposits in the bays of Jaigad, Ambwah, Varvada, Maharashtra, India, in *Proc. Second South Asia Geological Congress,* Wijayanaanda, N. P., Cooray, P. G., and Mosley, P. (Eds.), GEOSAS II, Colombo, Sri Lanka, 1995, 325.

20. Gujar, A. R., *Heavy Mineral Placers in the Nearshore Areas of South Konkan Maharashtra: Their Nature, Distribution, Origin and Economic Evaluation,* Ph.D. thesis submitted to Tamil University, Thanjavur, Unpublished, 1996, 234.

21. Mahadevan, C. and Rao, P. R., Evolution of Visakapatnam beach, Andhra Univ., *Memoir Oceanography,* 2, 33, 1958.

22. Borreswara Rao, C. and Lafond, E. C., Study of the deposition of heavy mineral sands at the confluences of some rivers along the east coast of India, Andhra Univ., *Memoir Oceanography,* 2, 48, 1958.

23. Officers of the Geological Survey of India, The economic geology and mineral resources of Orissa, Orissa Govt. Press, Cuttack, 1949, 131.

24. Rao, T. K., *Records of Geological Survey of India,* V.92, Part I, General Report, 39, 1957.

25. Dikshitulu, G. R., Desapathi, T., Nageswara Rao, M., Krishnan, S., and Mir Azam Ali, Heavy mineral placer deposit of Kakinada–Tandava river confluence, east Godavari district, Andhra Pradesh, in *Proc. National Symp. Late Quaternary Geology and Sea Level Changes and Annual Convention Geological Society of India,* Cochin, Abst. Vol., 1998, 54.

26. Panda, N. K., Murthy, P. V. V. S. S., Sahoo, P., Ravi, G. S., and Mir Azam Ali, Evolution of Kalingapatnam coast and its influence on placer mineral concentration, east coast of India, in *Proc. National Symp. Late Quaternary Geology and Sea Level Changes and Annual Convention Geological Society of India,* Cochin, Abst. Vol., 1998, 57.

27. Rao, B. R., Mohapatra, G. P., Vaz, G. G., Reddy, D. R. S., Hari Prasad, M., Misra, U. S., Raju, D. C. L., and Shankar, J., Inner shelf placer sands off north Andhra Coast, *Marine Wing News Letter, Geological Survey of India,* VIII, 11, 1992.

28. Ramamohana Rao, T., Shanmukha Rao, C. H., and Sanyasi Rao, K., Textural analysis and mineralogy of the black sand deposits of Visakhapatnam–Bhimunipatnam coast, Andhra Pradesh, India, *Journal of Geological Society of India,* 23, 284, 1982.

29. Anon., Denver portable trommel jig placer unit, *Mining Journal,* 104, 344, 1983.

30. Venkateswarlu, P. D., Chatterjee, P., Sengupta, B. J., Brahman, C. V., and Kar, Y. R., Shallow seismic investigations for placer minerals off Gopalpur (Orissa), east coast of India, *Indian Journal of Marine Science,* 18, 134, 1989.

31. Dhanunjaya Rao, G., Krishnaian Setty, and Raminaidu, C. H., Heavy mineral content and textural characteristics of coastal sands in the Krishna–Godavari, Gosthani–Champavati, and Pennar River deltas of Andhra Pradesh, India: a comparative study, in *Exploration and Research for Atomic Minerals,* Atomic Minerals Division, Hyderabad, 2, 147, 1989.

32. Rao, A. Y., Rao, A. P., Ravi, G. S., and Mir Azam Ali, Placer ilmenite deposits of Bhavanapadu coast, Srikakulam district, Andhra Pradesh, in *Proc. National Symp. Late Quaternary Geology and Sea Level Changes and Annual Convention Geological Society of India,* Cochin, Abst. Vol., 1998, 62.

33. Murthy, P. V. V. S. S., Anil Kumar, V., Ravi, G. S., and Mir Azam Ali, Studies on heavy mineral resources of Koyyam coast, Srikakulam district, Andhra Pradesh, in *Proc. National Symp. Late Quaternary Geology and Sea Level Changes and Annual Convention Geological Society of India,* Cochin, Abst. Vol., 1998, 63.

34. Jacob, K., Ilmenite and garnet sands of Chowghat (west coast), Tinnevelly, Ramnad and Tanjore coasts (east coast) *Records of Geological Survey of India,* 82, 527, 1956.

35. Arogyaswamy, R. N. P. and Panicker, M. R. K., Ilmenite bearing placers of Karaikkal beach, *Records of Geological Survey of India,* 96, 82, 1971.

36. Ahmed, E., *Coastal Geomorphology of India,* Orient Longman, Delhi, 1972, 260.

37. Jacob, K., Raman, P., and Damodaran, S., Report on the investigation for heavy mineral sands in the coastal parts of Tirunelveli district, Tamil Nadu state, Directorate of Geology, Tamil Nadu — A field report, 1977.

38. Krishnan, M. S., *Geology of India and Burma,* CBS Publishers, New Delhi, 268, 1982.

39. Setty, M. G. A. P. and Rajamanickam, G. V., Heavy mineral suite in the shelf sediments of Madras coast, *Journal of Indian Geophysical Union,* 10, 93, 1972.

40. Sengupta, R., Khalil, S. M., Rakshit, S. Deb, Roy, D. K., Sinha, J. K., Mitra, S. K., Majumdar, S., Raghav, S., and Bhattacharyya, S., Multimineral placer deposits in the inner shelf off Orissa coast, *Geological Survey of India, Special Publication,* 135, 1992.

41. Acharya, B. C., Panigrathy, P. K., Nayak, B. B., and Sahoo, R. K., Heavy mineral placer deposits of Ekakula beach, Gahiramatha coast, Orissa, India, *Resource Geology,* 48, 2, 125, 1998.

42. *Indian Minerals Year Book,* Published by Controller of Publications, Govt. of India, New Delhi, 369, 1995.

43. Chandrasekar, N., *Beach Placer Mineral Exploration along the Central Tamil Nadu Coast,* Ph.D. thesis, Madurai Kamaraj University, Madurai, unpublished 293, 1992.

44. Angusamy, N., Geetha, S., and Victor Rajamanickam, G., Beach placer minerals exploration along the coast between Mandapam and Kanyakumari, DOD Project Report, (Unpublished) 50, 1992.

45. Angusamy, N., *A Study of Beach Placers between Mandapam and Kanyakumari, Tamil Nadu, India,* Ph.D. thesis, submitted to Bharathidasan University, Tiruchirapally, unpublished 1995, 155.

46. Chandrasekaran, S., Viswanathan, G., Murugan, C., Thippeswamy, S., Ratna Reddy, N., and Krishnan, S., Ilmenite-rich beach placer deposits at Manavalakurichi, Tamil Nadu, in *Proc. National Symp. Late Quaternary Geology and Sea Level Changes and Annual Convention Geological Society of India,* Cochin, Abst. Vol., 60, 1998.

47. Viswanathan, G., Chandrasekaran, S., Thippeswamy, S., Murugan, C., Ratna Reddy, N., Krishnan, S., and Mir Azam Ali, Heavy mineral assemblage in the beach placer sand of Pillaithoppu–Rajakka-mangalam area of Kanyakumari district, Tamil Nadu, in *Proc. National Symp. Late Quaternary Geology and Sea Level Change and Annual Convention Geological Society of India,* Cochin, Abst. Vol., 61, 1998.

48. Loveson, V. J., *Geological and Geomorphological Investigations Related to Sea Level Variation and Heavy Mineral Accumulation along the Southern Tamil Nadu Beaches, India,* Ph.D. thesis, Madurai Kamaraj University, Madurai, unpublished, 1994, 223.

49. Muthukrishnan, N., *A Study of Detrital Minerals from the Sediments of Gadilam River Basin,* Tamil Nadu, M.Phil. thesis, Tamil University, Thanjavur, unpublished, 1993, 193.

50. Udayaganesan, P., *A Study of Detrital Minerals from the Sediments of Vaippar Basin, Tamil Nadu,* M.Phil. thesis, Tamil University, Thanjavur, unpublished, 118, 1993.

51. Rajasekar, T., *A Study on Detrital Minerals from the Sediments of Palar Basin, Tamil Nadu,* M.Phil. thesis, submitted to Tamil University, Thanjavur, unpublished, 108, 1994.

52. Chandrasekar, N. and Rajamanickam, G. V., Zircon placer deposits along the beaches of central Tamil Nadu, India, in *Second South Asia Geological Congress,* Wijayananda, N. P., Cooray, P. G., and Mosley, P. (Eds.), GEOSAS II, Colombo, Sri Lanka, 1997, 337.

53. Mohan, P. M., Identification of coastal placer deposits along the coast between Maduranthagam and Madras, report submitted to Department of Science and Technology, New Delhi, 1998.

54. Mohan, P. M., Distribution of heavy minerals in Parangipettai (Porto Novo) beach, Tamil Nadu, *Journal of Geological Society of India,* 46, 401, 1995.

55. Angusamy, N. and Rajamanickam, G. V., The distribution and nature of transparent heavy minerals along the beaches of southern Tamil Nadu, in *Proc. VIII and IX Conventions, Indian Geological Congress,* Rajamanickam, G. V. and Varma, O. P. (Eds.), University of Roorkee, Roorkee, 1996, 38.

56. Udayaganesan, P. and Rajamanickam, G. V., Grain size distribution and depositional environment of Vaippar River sediments, Tamil Nadu, in *Proc. VIII and IX Conventions, Indian Geological Congress,* Rajamanickam, G. V. and Varma, O. P. (Eds.), University of Roorkee, Roorkee, 1996, 44.

57. Sinha, R. K., *A Treatise on Industrial Minerals of India,* Allied Publishers, Bombay, 1967, 513.

3 Tin Placer Deposits on Continental Shelves

Wyss W.-S. Yim

ABSTRACT

A review of the geology of tin placer deposits on continental shelves is attempted. Under the heading of general characteristics, primary tin mineralization, distance of transport of cassiterite, dating and age of formation, concentration, preservation, and classification are examined. In the formation of giant deposits on continental shelves, the drowning of old river systems are recognized to be the most important by far. The tin placer deposits are divisible into eluvial deposits, colluvial deposits, alluvial deposits, beach deposits, and reworked mine tailings. Methods used for their exploration, including geophysics, drilling, chemical, and mineralogical, are presented. This is followed by a description of case studies of tin placer deposits on continental shelves including southeast Asia, southern China, Cornwall, and Australia. Finally, conclusions are drawn on the evolution of tin placer deposits through time, research and development, and future prospects. The lack of offshore tin exploitation outside southeast Asia is attributed to inadequate understanding of the distance of transport required to form economic tin placers, the localized nature of bedrock mineralization and its resultant placers, and limitations on the methods of prospecting used. Because "long" geological time periods are now known to be involved in the development of giant onshore tin placers, the same may well apply to their offshore counterparts.

3.1 INTRODUCTION

A *placer* is defined as a surficial mineral deposit formed by the mechanical concentration of mineral particles from weathered debris.[1] Cassiterite or tin dioxide, a stable heavy mineral,[2] is commonly found in such deposits. The bulk of the economic tin placer deposits occurring on continental shelves were formed under subaerial conditions prior to the submergence below sea level. Consequently, in terms of understanding the geology of tin placer deposits on continental shelves, it is essential to learn from their comparatively well-studied onshore counterparts.

Tin and gold are probably among the earliest placer deposits to have been utilized. However, placer deposits have been much neglected as a subject of study. In the textbook *Geology of Tin Deposits*,[3] placer deposits were recognized to form the major source of tin supply yet only 30 pages out of the total of 543 pages were devoted to aspects of them.

Locations of tin placer deposits on continental shelves are shown in Figure 3.1. When compared to their onshore counterparts, very little information is available on them. This is partly explained by the fact that much of the information is concealed in the reports of mining companies.

Both the terms *leads* and *deep leads* were introduced in alluvial mining[4] to refer to alluvial deposits containing the placer ore, which may be deeply buried by a barren overburden. However, no information on the separation depth between the two types can been found in the literature. This view is contradicted by some authors, e.g., Nye,[5] who stated that economic ore minerals need not be present within leads.

0-8493-8429-X/00/$0.00+$.50
© 2000 by CRC Press LLC

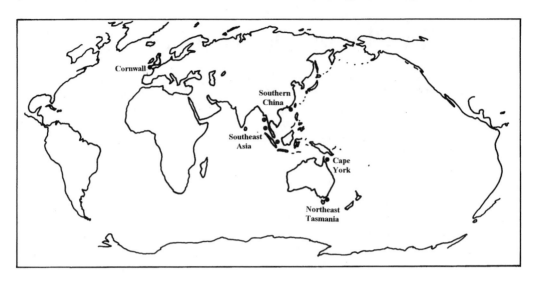

FIGURE 3.1   Locations of tin placer deposits on continental shelves.

Four aspects have been recognized in the genesis of tin placers in Indonesia[6] including:

1. The nature of the source rock containing the primary or secondary cassiterite.
2. The liberation of the cassiterite from the source rock without excessive comminution of the cassiterite grains.
3. The mechanical concentration of the now detrital cassiterite grains.
4. The protection of at least parts of the unconsolidated cassiterite concentration against attack by mechanical erosion.
5. An important additional factor in placer tin genesis requiring careful consideration is the stratigraphical control of all the above aspects, or in other words, the duration of geological time involved. For example, the timing and duration of exposure of the mineralized bedrock and the paleoenvironmental changes that have taken place are of importance. "Long" periods of geological time with favorable climatic conditions for deep weathering are thus helpful to tin placer formation.[7]

There have been only a few studies on the stratigraphical control of tin placer formation. Tin placers within the Older Sedimentary Cover of western Malaysia and Indonesia were suggested to be Late Pliocene to Early Pleistocene.[8] In a study of the phases of destruction of primary tin deposits and the location and order of development of tin placers in Cornwall,[9] placer deposits ranging from pre-Miocene to Holocene in age were indicated. The history of Indonesian placers was believed to extend back to Miocene and Pliocene times, but the upper limit is very young and may be within the limit of radiocarbon dating.[6] More recently, in a study of the genesis of Tasmanian tin placers,[7] a long geological history of events dating back to the unroofing of the mineralized granite in the late Paleozoic was demonstrated. This is contrary to the view that the global distribution of placer deposits is largely a product of variation, both at present and in the recent geological past, in geomorphological processes acting at the earth's surface.[10] The main reason why pre-Quaternary placers are not widely recognized is explained by the lack of suitable investigation methods in the past. This has, however, been changed by findings of the Ocean Drilling Program (formerly Deep Sea Drilling Project) over the last 30 years. Based on such developments, a better stratigraphical control on placer deposits[11] is now possible.

In this chapter, an attempt is made to review the geological aspects of tin placer deposits on continental shelves.

3.2 HISTORICAL BACKGROUND

Tin placers occurring on beaches on the coast of Cornwall in the United Kingdom were probably the first type of marine placers ever to be mined. "If one is willing to take the word of the antiquarians, it would appear that Phoenicians carried their tin from Cornwall to Cadiz, which from, say, 1000 to 200 BC remained the entrepot for most if not all of the Cornish product, receiving the tin from the Phoenician galleys. ..."[12] In the early days of tin mining, alluvial or beach mining prevailed to the exclusion of lode mining. However, beach mining is likely to be limited in scale for three reasons. First, there are only a few locations where tin lodes outcrop on the coast; second, because tin lodes are near vertical; and third, the tin lodes are resistant to wave attack. For these reasons the tin lodes are unlikely to be a major source of tin supply to form beach placers

E.T. Miles was the pioneer in offshore tin prospecting and large-scale submarine tin placer mining. In 1907, a wooden Banka-drill platform was set up on two small boats for prospecting off southern Thailand.[13] Exploring around Phuket Island, he discovered a large offshore tin placer from Thongka Harbour seaward for a distance of four kilometers adjacent to a coastal tin mine located on the shore, which was worked by a Chinese miner.[14] The first dredge began operations in 1908, increasing to five by 1911. All were of the bucket type. The grab dredge and the suction dredge were developed later for considerably greater depths in excess of 60 m.

Off Indonesia, the exploration for submarine tin placers on submerged valley floors 50 m below mean sea level on the Sunda Shelf began in the 1930s.[15] In 1938, the first seagoing bucket dredge started operations off Billiton (Belitung), Indonesia.[16] By the 1970s, a total of about 10 dredgers were in operation around the tin islands of Indonesia.

3.3 PRODUCTION

The tonnages and value of marine tin placer production in the past are difficult to estimate because the International Tin Council did not keep separate records for this category. For the period 1930 to 1965,[17] of the estimated world tin production of 5.4 million long tons (5.48 million metric tons), 74% or 4 million long tons (4.06 million metric tons) were from placer production. However, the proportion from marine tin placer production was not available. Based on published data,[18] the value of annual tin production from marine tin placers in 1968 was estimated to be US$24.2 million. This is second only to sand and gravel and ahead of diamonds in terms of all dredged materials. In terms of world tin production, it is probable that marine tin placers may not have been responsible for much more than 10% of the total to date.

Cassiterite was probably the first mineral to be exploited on the seabed for its contained metal, and tin remains the only nonferrous metal produced by the marine mining industry.[19] Over the period from 1967 to 1971 tin concentrate produced from marine mining showed increases annually from 1714 metric tons to 3426 metric tons. Estimates based on the total number of dredgers operating offshore and their capacities as a proportion of the total number of dredgers at work in Indonesia suggest that marine production may have increased from about 7250 metric tons in 1967 to perhaps 10,000 metric tons of tin concentrate in 1971. In 1971, there were four dredgers, including a suction dredger, in operation off Thailand with capacities ranging up to 260,000 m^3 a month and 10 operations off Indonesia with capacities ranging from 115,000 to 360,000 m^3 a month.[20] The majority of Indonesian dredgers have a capacity of about 230,000 m^3 a month.

It has been suggested that an increasing proportion of alluvial reserves in the southeast Asian tin field lie offshore.[21,22] However, the discovery of major tin fields in Brazil in the last two decades[23] and the collapse of tin metal prices in 1985 have led to a decline in interest in the search for offshore tin placers. By 1991, a total of only 24 tin dredgers operated in southeast Asia.[24] At the present day, offshore production is found only in Thailand and Indonesia.

3.4   GENERAL CHARACTERISTICS

In this section, topics relevant to the formation of the different types of offshore tin placers are examined. Because placers formed by the drowning of old river systems are by far the most important types, they are identical to onshore counterparts with the exception of the involvement of relative sea level changes and burial by transgressive marine deposits which are tin barren.

3.4.1   PRIMARY TIN MINERALIZATION

Tin is enriched in the continental crust through polycyclic events involving metamorphism, anatexis, and related tectonic–magmatic processes.[25] The world's major types of primary tin deposit have been classified into 11 groups,[26] including:

1. Disseminations other than those in placers and that are not included in the other major groups.
2. Pegmatites/aplites.
3. Skarns.
4. Hydrothermal breccias.
5. Deposits associated with greisenized and/or albitized country rock.
6. Stanniferous veins other than those of Group 5.
7. Lodes of the Cornish type.
8. Replacement (metasomatic) deposits that cannot be satisfactorily placed in any of the other groups.
9. Telescoped, mineralogically complex deposits (largely xenothermal or subvolcanic).
10. Deposits of the Mexican type (epithermal or fumarolic).
11. Stanniferous massive sulfide and massive iron oxide deposits.

A number of background generalizations can be made on the world's major types of primary tin deposits,[26] including:

1. Primary tin deposits have developed from the Precambrian to the Tertiary.
2. As the geological time-scale is ascended, tin deposits become more plentiful and more economically important.
3. The style of primary tin mineralization tends to vary with its age.
4. The world's tin deposits have a spotty distribution and are restricted to elongate zones (tin belts).[27]
5. Most primary tin deposits are spatially and probably genetically related to granitoids or their volcanic equivalents.
6. The overwhelming majority of tin deposits are associated with S-type granites[28] formed by the partial melting of sedimentary source materials or types closely akin to them.
7. The majority of the major types of primary tin deposits are associated with granitoid cusps and ridges.
8. Within a tinfield the centers of mineralization may occur at the intersections of two sets of preferred mineralization.
9. Faults have played major roles in determining the distribution, nature, and size of primary tin deposits and so have impounding bodies.
10. Cassiterite is the only tin mineral of major economic importance.

Primary tin deposits associated with greisens and disseminations have been noted to be of very limited economic importance but are the parents of important placers.[26] In the largest and most productive tinfield in the world, the main tin belt of the Malaysian Peninsula, greisen-bordered vein swarms are thought to be the major contributors of cassiterite to the placers.[22]

TABLE 3.1
Summary of Mohs' Scale of Hardness, Cleavage, Transportation Resistance, and Distribution Characteristics of Selected Heavy Minerals in Northeastern Tasmania

Mineral Name	Hardness	Cleavage	Transportation Resistance	Distribution Characteristics
Monazite	5	moderate	105–300	commonly egg-shaped, fairly widespread
Olivine	6½–7	none	250	usually well-rounded, restricted to near the source rock
Zircon	7½	imperfect	265	resistant to rounding, basaltic zircons may however be rounded by magmatic processes (J.D. Hollis, pers. comm.)
Chrysoberyl	8½	poor	300	found rounded in spite of hardness
Ilmenite	5–6	none	325	widespread moderately resistant mineral
Cassiterite	6–7	poor	360	brittle very dense mineral, found close to the source rock
Almandine	6½–7½	none	375	widespread resistant mineral
Magnetite	5½–6½	none	380	uncommon
Topaz	8	perfect	390	widespread mineral, decreases rapidly in grain size away from the source
Rutile	6–6½	perfect	455	uncommon
Pleonaste	7½–8	none	550	widespread resistant mineral
Chrome-spinel	7½–8	none	680	widespread resistant mineral
Corundum	9	none	750	widespread extremely resistant mineral

The transportation resistance is obtained by comparing the heavy minerals with compact hematite (100).[29]

Source: Yim, W.W.-S., Tin placer genesis in northeastern Tasmania, in *The Cainozoic in Australia: A Re-appraisal of the Evidence*, Williams, M.A.J., De Deckker, P., and Kershaw, A.P. (Eds.), Special Pub. 18, Geological Society of Australia, Sydney, 235, 1991. Reprinted with permission from the Geological Society of Australia Incorporated.

3.4.2 DISTANCE OF TRANSPORT OF CASSITERITE

Generally the greater the stability of a mineral, the greater the distance it may be transported from its source rock without material change of state.[4] The transportation resistance was proposed by Friese.[29] It is defined as the resistance of a mineral to mechanical action during fluvial transport and was determined by experimental comparison under laboratory conditions with compact hematite arbitrarily taken as 100. This property, which is primarily controlled by mineral hardness, cleavage, and tenacity, is an important factor, in addition to mineral density, in determining the distance of transport.

The distribution characteristics, hardness, cleavage, and transportation resistance of cassiterite and selected heavy minerals found in a study of heavy mineral provenance in the alluvial tinfield of northeastern Tasmania[7] is shown in Table 3.1. It can be seen that cassiterite, a brittle but very dense mineral with poor cleavage and moderate hardness, has a low transportation resistance of 360 in spite of its high chemical stability. Because of these characteristics, cassiterite is expected to occur close to the source rock. In contrast to cassiterite, corundum, an extremely hard (Mohs' scale = 9), chemically resistant, and moderately dense mineral, has the highest transportation resistance of 750. Consequently, corundum can be found at greater distances from its source than cassiterite.

Because of the high density of cassiterite, it is carried only a short distance from its source.[30] It is typically found to be concentrated in bed load deposits of rivers and streams. In an examination of economic placer deposits of the continental shelf by Emery and Noakes,[17] the median distance of transport from bedrock source for economic tin placer deposits is only 8 km. However, other undiscovered bedrock mineralization sources possibly contributed en route, and the dispersion could be even shorter.[11]

Major differences in the transportability between coarse and fine cassiterite in streams has been noted.[31] Fine cassiterite was considered to be too easily moved to be found in alluvial placers, while coarse cassiterite forms autochthonous placers because of low transportability. Such peculiarities of fine and coarse cassiterite transport may be used to explain the absence of alluvial tin placers that are distant from primary sources.

Three main factors were considered by Yim[32] to be important in determining the distance of cassiterite transport:

1. The size of cassiterite grains in the source rock. In fluvial regimes, coarse gravel-size cassiterite would not travel as far as fine cassiterite.
2. The hydrologic regime of the stream, particularly its gradient and discharge. A steep-gradient stream with high discharge would be expected to transport cassiterite farther from the source than a low-gradient stream with low discharge.
3. The affect of geological history on landscape evolution, termed *evolutionary geomorphology*.[33] Ancient placers would be expected to occur further from the source rock than young placers because of the greater likelihood of reworking. The tectonic, climatic, and vegetation changes which controlled placer development may be episodic.

Without careful consideration of the above factors, distance of transport figures are considered to be a distraction in the study of tin placers.[11] Nevertheless, it has been demonstrated relatively recently that, on the basis of particle size and trace element distribution of cassiterite in tin placers of northeastern Tasmania,[32] the bulk of economic placers were locally derived and do not normally exceed 1 km distance of transport downstream from the source rock. Possible exceptions are mass-flow deposits and where stream courses were able to maintain steep gradients for long distances. The 1 km distance is considerably less than the 7 to 10 km suggested for recoverable cassiterite of over 200 micron diameter on low river gradient.[34] It is also much less than the minimum distance of 10 km identified for tin-mine tailings transported by the Red River in Cornwall.[35]

3.4.3 DATING AND AGE OF FORMATION

One of the earliest publications devoted to the dating of tin placers was by Osberger.[36] For residual deposits in Indonesia, age information was obtained by:

1. Dating wood is still performed mainly by the radiocarbon method. Three infinite radio-carbon dates exceeding 46,000 to 55,000 years[37] and one age exceeding 250,000 years of a tree trunk[38] were given. It is, however, unclear how the age that exceeds 250,000 years was obtained because this is well in excess of the maximum limit of the radiocarbon method.
2. The geological position of the deposits.
3. The occurrence of fossils and artifacts.
4. Estimation of erosion rates.

Based on this evidence, the placers were believed to have begun forming probably in the Middle or Early Pleistocene.[37]

In Malaysia, tin placer formation was thought to have taken place from Late Pliocene to Middle Pleistocene times.[38] However, the evidence for age was based mainly on paleomagnetism and the stratigraphical context of the placers rather than on absolute dating. Because tin mineralization took place mainly in Mesozoic times in southeast Asia, the maximum age cannot exceed the age of mineralization. The use of radiocarbon dating for such deposits is also problematic not only because of the age limit of the method but also because pre-Holocene radiocarbon dates are known to show a young age bias.[39] Furthermore, the true age of tin placer deposition may exceed the age of plant remains. Also, paleomagnetic dating in alluvial sequences which are full of unconformities is unlikely to be reliable. The age of tin placer formation in Malaysia is therefore still unresolved.

In the past 20 years, there have been major advances in the understanding of the age of formation of tin placers particularly through the availability of a number of innovative dating methods and the continental erosion record obtainable from the results of the Ocean Drilling Program. U–Pb and Pb isotopic measurements of cassiterites[40] demonstrated that cassiterite can be used to directly date the age of tin mineralization, providing a maximum age of formation of the placer deposits. At the same time, minerals associated with the mineralization, such as zircon, may be dated by the U–Pb method,[41] the SHRIMP ion-probe method,[42] and the fission-track method.[43] More recently, the electron spin resonance method has also showed promise.[7,44] Based on these results, a more reliable age of formation of the tin placer deposits may be obtainable.

In the alluvial tinfield of northeastern Tasmania, fission-track dating of two physically distinctive types of alluvial zircons occurring together with cassiterites has yielded Devonian and Middle Eocene ages.[45] While the Devonian zircons are in agreement with the age of intrusion and mineralization of the granitoids, the Middle Eocene zircons revealed a basaltic source. The results show that the tin placers were formed by episodic recycling,[7] the last time causing mixing between the two types of zircons no earlier than the Middle Eocene. It is, however, clear that the original deposition of the cassiterite must have taken place prior to the Middle Eocene, perhaps as far back as the Permian.[7] The sea-surface temperature history for the Cainozoic period obtained from the oxygen-isotope study of formanifers from deep sea cores off the coast of Tasmania,[46] radiometric dating of basalts,[47,48] and paleobotanical studies of the alluvial deposits[49-52] have together provided a chronological framework showing that the deep leads or old river channels were buried during the Late Oligocene to the Early Miocene (Figure 3.2).[7]

In spite of the clear-cut age evidence for the age of tin placer deposits in northeastern Tasmania, it is not always possible in other tin placer areas in the world to obtain the same quality of stratigraphical control due to the lack of datable materials. In the case of southeast Asia, since the age of tin mineralization is Late Triassic to Jurassic, the maximum age of placer formation cannot be older than this. On the basis of the occurrence of tektites either within or on top of the placers, at least some may be younger than about 600,000 years, while others would pre-date this.[53]

In summary, the bulk of the tin placer deposits of the world are poorly dated. The lesson from the tinfield of northeastern Tasmania is that economic tin placer deposits can form at any time after the unroofing of mineralized granitoids. For "giant" deposits to develop, a much longer time scale than the Quaternary would be needed.

3.4.4 Concentration

The concept of hydraulic equivalence based on Stokes' Law was first introduced for heavy minerals transported and deposited underwater.[54] Subsequently, the concept was applied to develop a method of computing "hydraulic ratio," which measures directly the relative availability of heavy and light minerals of equivalent hydraulic value.[55] The idea was later taken up in geochemical studies[56-58] for the examination of the distribution of cassiterite in marine sediments in Mount's Bay, Cornwall.

Stokes' Law states that the settling velocity of a spherical particle varies with the square of the particle diameter with the effective density of the particle in the fluid, and inversely with the viscosity of the fluid. In other words, mineral particles of different specific gravity will be taken into suspension by a given current.

Stokes' Law

$$V = g/18 \cdot \rho_q - \rho_w/n \cdot d_q^2$$

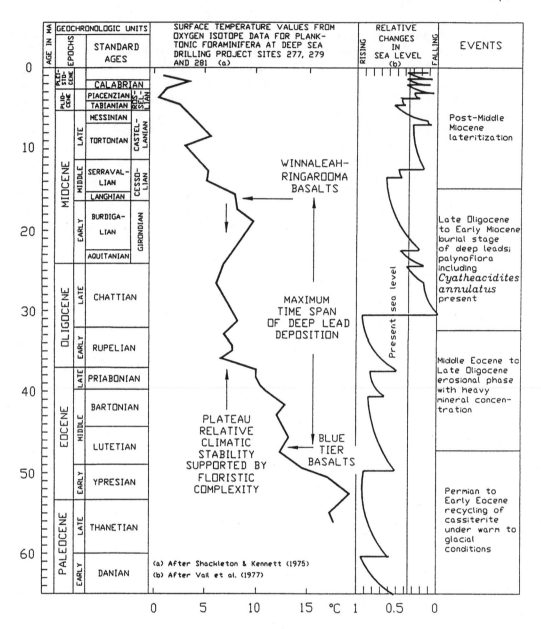

FIGURE 3.2 Possible relationship between Cainozoic sea-surface paleotemperatures, reconstructed from oxygen-isotope data for planktonic foraminifera at Deep Sea Drilling Project sites 277, 279, and 281; Cainozoic global changes in sea level and terrestrial events in northeastern Tasmania. (From Yim, W.W.-S., Tin placer genesis in northeastern Tasmania, in *The Cainozoic in Australia: A Re-appraisal of the Evidence*, Williams, M.A.J., De Deckker, P., and Kershaw, A.P. (Eds.), Special Pub. 18, Geological Society of Australia, Sydney, 235, 1991. With permission from the Geological Society of Australia Incorporated.)

where

V   = settling velocity of both grains in cm/sec

ρ_q = specific gravity of quartz

ρ_w = specific gravity of water

d_q = diameter of quartz grain in cm

n   = coefficient of viscosity of water

g   = acceleration due to gravity

and

$$V = g/18 \cdot \rho_c - \rho_w/n \cdot d_c^2$$

where
ρ_c = specific gravity of cassiterite
d_c = diameter of cassiterite grain

Since the settling velocity is constant

$$g/18 \cdot \rho_q - \rho_w/n \cdot d_c^2 = g/18 \cdot \rho_c - \rho_w/n \cdot d_c^2$$

$$(2.66 - 1.00) \, d_q^2 = (7.00 - 1.00) \, d_c^2$$

$$d_q = 1.901 \, d_c$$

Therefore, the equivalent diameter of quartz is equal to the known diameter of cassiterite multiplied by a factor of 1.901.

However, Stokes' Law is adhered to only under certain conditions, including:

1. The grain particles are spherical.
2. The mineral grains are pure or mono-mineralic.
3. Applies only to certain grain-size particles settling under free settling conditions. The minimum size of quartz spheres that will obey Stokes' Law at 18°C is below 50 microns when the Reynold's number Re = 0.1.[59]
4. The proportion of solids does not exceed 0.5% by volume.
5. The temperature remains constant during the period of settling.

Although the above conditions are difficult to achieve in nature, an approximation to hydraulic equivalence exists.[60] Thus sand-size cassiterite occurs in gravel-dominated deposits formed under a high-energy environment, such as in a steep-gradient river channel. Furthermore, because of the low mobility of cassiterite, it is always enriched in the basal sediments above the bedrock.

In a study of heavy mineral provenance and the genesis of stanniferous placers in northeastern Tasmania,[7] the main causes of cassiterite concentration were found to include:

1. The presence of an extensive area of bedrock mineralization.
2. The long geological history since the unroofing of the tin-bearing Devonian granites probably from the Carboniferous Period onward.
3. The existence of periglacial and warm climatic conditions at different times to facilitate cassiterite liberation from the source rock. Such conditions are favorable for both the physical and chemical disintegration of granite.
4. The involvement of fluvial processes in the recycling of cassiterite.
5. The existence of piedmont zones at the margin of the granite massifs to facilitate cassiterite deposition and concentration.

Each tinfield of the world, because of differences in the age and other characteristics of bedrock mineralization, age of subaerial exposure of the mineralized bedrock, and subsequent geological history, would be expected to show variability in the nature of the resultant tin placer deposits. The similarity is in the fluvial processes of concentration through selective or hydraulic sorting.[61]

3.4.5 Preservation

This subject was poorly understood until relatively recently due mainly to the problem of reliably dating tin placer deposits. Recent advances in understanding preservation of tin placers unfortunately occurred just after there was a major crash in international tin metal prices followed by a loss of interest by mining companies in exploring for the commodity. Prior to this, the general view was that tin placers were both formed and preserved in the Late Tertiary to Quaternary.

In the tinfield of northeastern Tasmania, unique circumstances exist to provide information on the age of preservation of the deep leads. The reworking of cassiterite has been demonstrated by the mixing between basaltic heavy mineral assemblages and cassiterite to be no earlier than the Middle Eocene.[45] The drastic fall in global sea level resulting from the formation of the Antarctica ice cap in the Oligocene (Figure 3.2) was responsible for causing accelerated erosion in the continental highlands and accelerated deposition in the continental basins to preserve the deep leads.[7]

3.4.6 Classification

In a summary of the world's major types of tin deposits,[62] tin placers were recognized as by far the most important source of tin. The tin placer deposits were categorized into ancient placers and modern placers. For the ancient category, the only examples given were the stanniferous bodies in metamorphosed sediments in Poland and Madagascar. For the modern category, in addition to whether they are submarine or subaerial, groupings were made into residual, eluvial, colluvial, and alluvial types, while subdivisions were made on the sites of accumulation into fluviatile, lacustrine, estuarine, and marine. The general characteristics of the main types of tin placer deposits are presented below.

3.4.6.1 Eluvial Deposits

This type of ore deposit, also known as residual or "kulit" deposits, has formed essentially *in situ* without major lateral transport or sorting of the minerals. On the basis of cassiterite grain-size distribution characteristics in southeast Asia,[63] both coarse (exceeding 1.4 mm) and fine (below 0.075 mm) samples of cassiterite are present but without any clear peak size.

With the exception of southeast Asia, references to eluvial tin deposits are lacking in the literature. In northeastern Tasmania, the explanation given[7] to account for this included:

1. The low economic potential of eluvial deposits, including inferior ore grade and reserves, in comparison to alluvial deposits.
2. Eluvial deposits have already been exhausted because of their thin overburden.
3. The unfavorable geomorphological history, such as a dominance of erosional activity on slopes with steep gradient and unfavorable paleoenvironmental conditions for deep weathering and elutriation of the residual soils.
4. The gradational change from eluvial to alluvial placers and their close association with each other make it difficult to draw a clear-cut separation between them.
5. The difficulty in distinguishing between disturbed samples of eluvial and alluvial deposits.

An additional factor may be the higher erosion rates in Tasmania than in southeast Asia.

3.4.6.2 Colluvial Deposits

This type of deposit is usually found at the foot of a slope or cliff and is gradational between eluvial deposits and alluvial deposits. It is also argued that if a deposit is truly colluvial in origin, both its ore grade and reserve will be limited and thus be of minor economic interest. The reasons for their lack of recognition may be similar to those given earlier for the eluvial deposits.

3.4.6.3 Alluvial Deposits

This type of deposit, derived from leads or deep leads of fluvial origin, is responsible for the bulk of tin production both onshore and offshore.

Based on the cassiterite grain-size distribution characteristics in southeast Asia,[63] alluvial deposits may be subdivided into two major types:

1. Washed-out or elutriated "kaksa" deposits where the coarser and heavier minerals remain close to the source, while the finer and lighter minerals are removed. There is a marked depletion in the finer sizes and a sharply peaked size distribution.
2. Transported or "mintjan" deposits, where the heavy minerals, after varying degrees of transport, have been trapped and retained in a suitable sedimentary environment. Cassiterite exceeding 0.425 mm and below 0.075 mm is lacking.

An examination of the theories involved in the genesis of the kaksa was made by Krol.[53] He concluded that residual concentration was the main agent responsible.

3.4.6.4 Beach Deposits

This type of deposit is even more localized in occurrence in comparison to the eluvial and colluvial deposits. Two examples of such deposits are at Cligga, Cornwall,[64] and Cable Bay in the Cape York tinfield (Figure 3.1).[3] Their limited extent, other than due to the nature of the primary deposits, may be linked to the brittle nature of cassiterite, leading to their destruction by mechanical means under the high-energy beach environment. Nevertheless, Cornwall, Bangka, and Billiton were given by Hosking[62] as examples of raised beach placers.

3.4.6.5 Reworked Mine Tailings

This type of deposit occurs in areas with a long history of mining activity where large-scale underground mining and inadequate mineral recovery techniques in the past have led to large quantities of tin-rich mine tailings entering the sea via streams.[35] For example, in Cornwall, tailings-derived beach placer and submarine placers have been recognized (Figure 3.3).[62] An additional category of tailings-derived estuarine placers[65,66] may also be added.

3.5 EXPLORATION METHODOLOGY

3.5.1 OFFSHORE FIELD METHODS

3.5.1.1 Geophysical

Of all the geophysical methods for offshore prospecting for tin placers, the seismic profiling method is by far the most widely used. The use of this method dates back to 1956 in Indonesia[67] and to 1962 in Thailand.[68] The original purpose was to provide an indication of the thickness and type of unconsolidated sediments on the seafloor and the depth of bedrock so that promising sites could be targeted for follow-up drilling.

The seismic method is based on measurements of seismic reflection and refraction to distinguish between sedimentary layers differing in their physical properties and overlying the acoustic basement. An explosion or a series of explosions are used to generate shock waves at different velocities propagating through the seabed. Hydrophones are used to record waves refracted through denser layers or reflected from the interfaces between layers. Waves are refracted back to the sensor only from layers that have increasing velocity signatures with depth.[69] The alternative sparker method

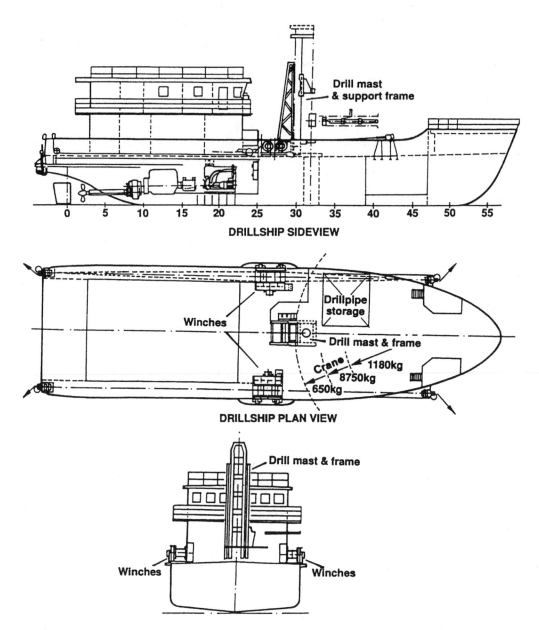

FIGURE 3.3 Schematic diagram of the drillship designed for tin exploration in Indonesia by Conrad–Stanen BV. Anonymous, Drillship for tin exploration in Indonesia, *Geodrilling Int.*, October 1997, 8, 1997. Reprinted with permission of Mining Journal Limited.)

is based on utilizing an electrical discharge. This allows for deeper penetration of the sea bottom but with less vertical resolution.

The seismic method has great advantages in providing continuous detailed, though indirect, stratigraphical profiling over large areas at relatively low cost.[22] Augmented by drilling, offshore seismic analysis can provide an understanding of stratigraphical relationship including the determination of:

1. Bedrock lithologies.
2. Large-scale sedimentary structures.
3. Sediment lithologies.
4. Submerged valley morphologies.
5. Gaseous organic layers.
6. Weathering surfaces.

By comparing seismic profiles with borehole information,[70,71] in [22] empirical relationships between different lithologies and their characteristic seismic patterns have been established:

1. Granite — strongly reflecting "peaks" (summits of hard masses of core boulders) surrounded by a homogeneous mass of weathered granite.
2. Sedimentary bedrock — a strongly reflecting irregular, impenetrable surface without "peaks" or a distinct weathering mantle.
3. Superficial marine mud — finely textured horizontal layering becoming more coarsely textured where sands predominate.
4. Clayey sediments — series of strong, continuous subhorizontal reflections of lower frequency than the soft mud layers.
5. Sandy or gravelly sediments — coarse "speckled" or "smudgy" texture with vague outlines and no definite continuous internal reflectors, due to small-scale diffraction patterns from the surface and within layers.
6. Sandy clay and clayey sand — similar either to sand or clay, depending on clay content, sand grain size, degree of compaction, and other factors likely to produce significant acoustic discontinuities.

Of the other geophysical methods used to assist offshore tin placer exploration, the magnetic induced polarization method has had limited success. This method is based on the assumption that cassiterite concentrations in placers are associated with magnetic minerals such as ilmenite.

3.5.1.2 Drilling

Numerous accounts of the drilling equipment useful for tin exploration are available; for example, Sujitno.[72] In practice, there are still major difficulties encountered using the methods available because tin is usually associated with gravel deposits which are too large for the drill casing. If the gravel is broken up, as in the case of the mechanized Bangka drill, there is the additional problem of estimating the result to determine the ore grade found in the hole because of the increase in volume. The vibrocoring method used in the past suffered from a lack of depth penetration because a casing was not inserted into the seabed. For example, the Hydrowerkstatten vibrocorer used off north Cornwall[73] has a maximum coring depth of only 4 m, while the Geodoff Mark II, with a counterflush core barrel used in southeast Asia, is capable of depths of 10 m or more. However, even encased vibrocoring, the method is unsuited to sampling compacted and cemented materials and/or materials dominated by sand and gravel. Drills using the counterflush or water-jet system appear to be the best because they are capable of taking a large volume of materials, including sand, tough clay, gravel, wood fragments, and partially cemented rock, relatively quickly. The recovery of cassiterite from these samples should therefore provide a better estimate of the ore grade.

A state-of-the-art drillship designed for offshore tin exploration by Conrad–Stanen BV has been reported to be nearing completion in Indonesia.[74] This drillship, shown in Figure 3.3, will be used for carrying out investigations of the seabed using counter flush drilling techniques to locate tin placer deposits of economic potential. Major features include an electrically governed hydraulic circuit for the drilling rig and ancillary plant with load sensing control of all functions, safeguarding the equipment and ensuring smooth and safe operation.

3.5.2 LABORATORY METHODS

3.5.2.1 Chemical Methods

Geochemical methods for tin determination developed mainly for tin exploration included the emission spectrographic method,[75] the X-ray fluorescence method,[76-79] the colorimetric method,[80] and the atomic absorption method.[81,82] All these were reviewed by Yim[73] in a study of the geochemistry of tin sediments off Cornwall, but only the latter two methods have been reported in the literature to have been used for offshore tin placer exploration.[56,82-84, etc.] All geochemical methods face the common problem of sample size representivity and different degrees of interferences confirmed by a comparison of tin analysis of ground and unground samples.[73] The erratic results for unground samples are explained by cassiterite segregation, making it difficult to use small sample size. Furthermore, the colorimetric method of Stanton and McDonald[80] is prone to iron interferences, while the ammonium iodide sublimation method used for the extraction of tin is affected by the presence of shells in marine sediments, resulting in lower assay values than the true ones.[85,86]

In order to reduce errors in the analysis, the strategy recommended includes:

1. Collection of statistically representative sample sizes using the Gy sampling formula[87] and a modification based on the concentration of the mineral.[88]
2. Use of a sample splitter such as the Jones' riffle to ensure that all the subsamples are representative.
3. Use of concentration methods to upgrade the tin content present in the samples. In sediment samples containing shells, the shells may be removed by glacial acetic acid.[73] Other concentration methods include panning and heavy liquid separations.
4. Use of grinding to minus 200 mesh in a Tema mill to ensure that the subsample used for the tin assay is representative.
5. Choose an interference-free method for assaying, such as the atomic absorption method, the X-ray fluorescence method, or the volumetric method.
6. Use of mineralogical methods as a means of confirming the results of the tin assay.

3.5.2.2 Mineralogical Methods

Mineralogical methods of tin analysis have a major advantage over chemical methods in that grains of cassiterite of different grain size may be examined microscopically to confirm their presence. However, such methods have a disadvantage in that they are relatively time consuming in comparison to the chemical methods.

For cassiterite concentration in the field, panning is a well-proven method. However, during offshore tin placer exploration the choice of mineralogical methods would depend on the type of drill used. In the case of the jet drill where a large quantity of sample is available, initial concentration by panning would be appropriate for two reasons: first, to reduce the amount of sample to be handled, and second, to reduce sampling error related to heavy mineral segregation. In other types of drilling where a much smaller volume of sample is available, it may be inappropriate to use panning because potentially useful sedimentological information may be lost.

Heavy liquid separations using tetra-bromo-ethane (SG = 2.96), methylene iodide (SG = 3.32), and Clerici's solution (SG = 4.3) have been an effective way of examining the liberation characteristics of offshore tin deposits derived from mine tailings in Cornwall[89] and for the examination

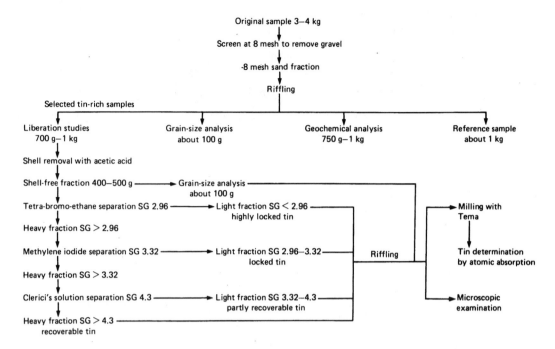

FIGURE 3.4 Flow chart of sample treatment used for the assessment of liberation characteristics of tin-bearing sands off north Cornwall. (From Yim, W.W.-S., Liberation studies on tin-bearing sands off north Cornwall, United Kingdom, *Mar. Mining*, 5, 87, 1984. Reprinted with permission of Taylor and Francis.)

of other heavy minerals associated with cassiterite in northeastern Tasmania.[7] Figures 3.4 and 3.5 show the flow chart of sample treatment used in these studies. In the case of the latter, magnetic separations of the heavy mineral fractions obtained using a Boxmag hand magnet have greatly assisted in the identification of the heavy minerals present. However, tetra-bromo-ethane produces carcinogenic fumes, while Clerici's solution is highly toxic, and it is essential to follow safe handling procedures in the laboratory.[90] In the last decade, the availability of sodium polytungstate, which is soluble in water, and the magnetic fluid separator[91] has meant that alternatives to the use of toxic heavy liquids are now available.

As a rapid confirmatory test of the presence of sand-sized cassiterite grains, density measurements[92] and microchemical tests[93,94] are available. The torsion microbalance developed by Berman[92] is an effective way of determining the specific gravities of mono-mineralic and composite cassiterite grains ranging in weight up to 75 mg. For confirming the presence of cassiterite, the tinning test is a widely used microchemical test. In this test, nascent hydrogen is produced by the reaction with dilute hydrochloric acid, which reduces the surface of the cassiterite grain to a grey matte coating.[94] For coated cassiterite grains, boiling in a mixture of nitric acid and hydrochloric acid is recommended as a preliminary step to remove the coating prior to carrying out the tinning test.

Mineralogical assessment methods are available to improve the assessment of cassiterite concentrates from ores.[95-97] These methods usually involve panning and/or heavy liquid separation followed by grain-size analysis and the counting of cassiterite grains. Care is needed to ensure that a statistically representative number of cassiterite grains is counted in each of the size fractions.

Hydraulic ratios between cassiterite and its associated light and heavy minerals may be determined using a semiautomatic particle-size analyzer.[98] The anomalous hydraulic ratios obtained may provide an indication of the distance of transport of cassiterite from the mineralized bedrock.

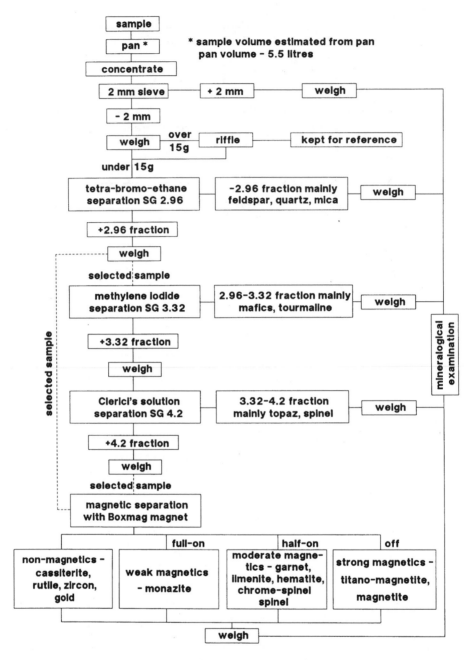

FIGURE 3.5 Flow chart of sample treatment procedures used for the heavy mineral provenance study in the tinfield of northeastern Tasmania after Yim[7] (Yim, W.W.-S., Tin placer genesis in northeastern Tasmania, in *The Cainozoic in Australia: A Re-appraisal of the Evidence*, Williams, M.A.J., De Deckker, P., and Kershaw, A.P. (Eds.), Special Pub. 18, Geological Society of Australia, Sydney, 235, 1991. Reprinted with permission from the Geological Society of Australia Incorporated).

3.6 CASE STUDIES

In this section, geological aspects of tin placers on continental shelves are examined according to localities.

3.6.1 SOUTHEAST ASIAN PLACERS

Based on offshore drilling and shallow seismic data,[16,99] five regional units on the Sunda Shelf have been recognized, including:

1. Younger Sedimentary Cover — This is the Holocene sediment cover with blanket-like deposits of neritic muds and beach sands up to 30 m in thickness.
2. Alluvial Complex (Young Alluvium) — This unit occurs disconformably beneath the Younger Sedimentary Cover. It consists of meander channel and floodplain deposits with last interglacial marine deposits disconformably beneath it. Peats commonly occur and the valleys are incised down to a 120-m maximum depth. The age is Pleistocene to Late Tertiary, and only minor amounts of tin are present.
3. Older Sedimentary Cover (Old Alluvium) — This unit lies disconformably below the Alluvial Complex and differs markedly from the overlying sediments. It comprises a proximal facies of subhorizontal, gravelly, poorly sorted granite wash close to the bedrock in a piedmont-fan environment. A Pliocene to Early Pleistocene age[38] and a Tertiary age[16] have been suggested. This stratigraphical unit is responsible for the bulk of the offshore tin placer production.
4. Transitional Unit — This is a poorly differentiated unit of variable lithology including peat and braided stream alluvium. A Middle Pleistocene age is suggested for this unit. Some tin is present.
5. Sundaland Regolith — This unit lies disconformably beneath the Older Sedimentary Cover. It is deeply weathered and is associated with laterites and bauxites. It is suggested to have been formed by pedogenesis under a seasonal savanna climate when sea levels were much lower during the Late Miocene to Early Pliocene.[38] Aleva[16] referred to this as the Sunda Peneplain and suggested a probable Late Cretaceous age.

The distribution of existing offshore tin deposits in southeast Asia and potentially new areas were reviewed by Hosking.[14] Figure 3.6 shows the existing and potential new offshore tin areas of southeast Asia and the general distribution of granites and tin belts.

3.6.1.1 Indonesia

The Indonesian tin islands, Bangka, Billiton, and Singkep, together with a number of other islands (Figure 3.6) are thought to be remnants of the submerged continuation of the Malaysian tin belt.[100] Since tin mining began in Indonesia in 1709, 75% of total production has come from Bangka.[22]

The methods of offshore tin exploration in Indonesia included geophysical prospecting and offshore drilling.[101,102] An exploration program was carried out in the extensive sea areas between Singkep and Bangka and around the Karimata Islands in the late 1960s.[16] Along the coast of Bangka, several features were recognized:

1. The present-day slope of the island has receded in comparison to the Sunda Peneplain.
2. Abrasion notches have been cut into the Sunda Peneplain by the present-day sea level.
3. Drowned alluvial fan deposits exist below the present-day sea level on a younger abrasion surface with a Riss–Wurm interglacial age.
4. Rejuvenated erosion and incision of valleys have formed the Alluvial Complex.

The geological framework was summarized by Aleva[16] in order of increasing age as:

1. Holocene — Deposition of the Younger Sedimentary Cover.
2. Wurm Glaciation — Soil formation.

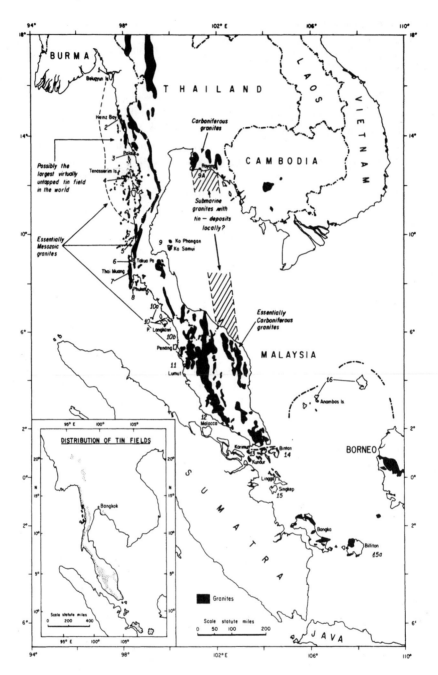

FIGURE 3.6 The offshore tin areas of southeast Asia, the general distribution of granites and tin belts. *1* Beluyun Island; *2* Heinze Basin; *3* Spider Island — beach placers; *4* Tenasserim Delta, Lampa and neighboring islands; *5* Ranong; *6* Takuopa — worked by suction dredge; *7* Thai Muang — beach placers; *8* Phuket — dredges on east and west coast, also illicit mining; *9* Ko Phangan and Ko Samui; *9A* Rayong — beach and submarine placers; *10* Langkawi Islands — beach placers; *10A* Ko Ra Wi and Ko La Ding; *11* Lumut-Dindings — beach and submarine placers; *12* Malacca — beach and submarine placers; *13* Karimun and Kundur; *14* Bintan; *15* The Tin islands Singkep, Bangka and Billiton; *15A* Between Billiton and Borneo; *16* Anambas and Natuna Islands. (From Hosking, K.F.G., X. The offshore tin deposits of southeast Asia, *Comm. Co-ord. Joint Prospecting Tech. Bull.*, 5, 112, 1971. Reprinted with permission from the Coordinating Committee for Coastal and Offshore Geoscience Programmes in East and Southeast Asia).

3. Riss–Wurm Interglacial — Younger planation surface referred to as an abrasion surface.
4. Pleistocene to Late Tertiary — Older Sedimentary Cover.
5. Upper Cretaceous (?) — Older planation surface referred to as the Sunda Peneplain.
6. Permian (and older?) — Folded sedimentary basement.

Although the ages were based largely on the general stratigraphical relationship, this provided a useful basis for understanding the distribution of tin placers on the continental shelf. Possible offshore placers shown in Figure 3.7 include:

1. Residual and elutriational deposits on the Sunda Peneplain.
2. Elutriational deposits in the Alluvial Complex.
3. Elutriational and residual deposits on the abrasion surface.
4. Disseminated cassiterite in the rejuvenated basement valleys.
5. Elutriational deposits in rejuvenated basement valleys.

3.6.1.2 Malaysia

The mining of offshore tin placers has taken place in Malaysia only on the west coast. In spite of much exploratory work carried out during the 1960s and 1970s, the scale of mining is small compared to that in Indonesia and Thailand. In a study of the genesis of tin placer deposits off the west coast of Malaysia by Batchelor,[22] two main areas were examined:

1. The Malacca State offshore area. This is one of the few sites in Malaysia with an offshore mining history. Two placer types were identified:
   a. Littoral placer deposits — Rich beach sands exceeding 0.4 kg/m^3 were once worked along a 20-km strip of coastline up to 1 km offshore.[103] They have been encountered at a depth of between 4 and 34 m below present sea level and most likely represent reworked deposits. However, they are limited in both size and grade and did not warrant mining interest.
   b. Fluviatile kaksa channel deposits — These are recognized as the most important placers in the Malacca area. They occur directly overlying weathered granite in buried fluvial paleochannels.
2. The Lumut–Dindings coastal and offshore areas. Stanniferous sand was found under mangrove muds and has been worked since 1910.[104] Tin-bearing beach sands and tin-bearing sands on the seabed were also reported by Scrivenor.[105] Five placer types were identified:
   a. Eluvial placers — Only one locality showing this type of deposit was identified. The deposits show *in situ* kaolinized feldspar megacrysts in a granitic matrix and grade upward into colluvium.
   b. Colluvial placers — These occur directly above bedrock as a moderately thick gravelly deposit along the slope and base of slope of the nearshore bedrock incline below the piedmont fans. They are traceable from near the scarp head down the incline over 600 m in horizontal distance. Over 50% of the cassiterite is greater than 40-mesh in grain size, and lateritic debris is present.
   c. Piedmont fan placers — These were formed by debris flow through ephemeral streams and sheet flood with the development of braided streams across the fan. Figure 3.8 shows the relationship between the stratigraphy of the sedimentary sequence and the granitic bedrock in the Lumut–Dindings nearshore area of Perak, Malaysia.[99]
   d. Channel placers — These were thought to represent linear channel deposits. They generally have lower gradients than the braided streams with a much higher proportion of muddy alluvium.

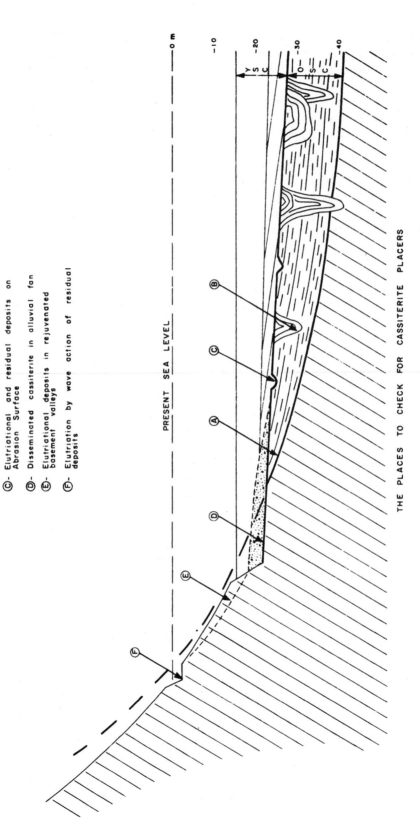

Ⓐ- Residual and elutriational deposits on Sunda Peneplain

Ⓑ- Elutriational deposits in Alluvial Complex

Ⓒ- Elutriational and residual deposits on Abrasion Surface

Ⓓ- Disseminated cassiterite in alluvial fan

Ⓔ- Elutriational deposits in rejuvenated basement valleys

Ⓕ- Elutriation by wave action of residual deposits

PRESENT SEA LEVEL

THE PLACES TO CHECK FOR CASSITERITE PLACERS

FIGURE 3.7 Six types of tin placer deposits on the Indonesian part of the Sunda Shelf. (From Aleva, G.J.J., Aspects of the historical and physical geology of the Sunda Shelf essential to the exploration of submarine tin placers, *Geol. Mijn.*, 52, 79, 1973. Reprinted with kind permission from Kluwer Academic Publishers.)

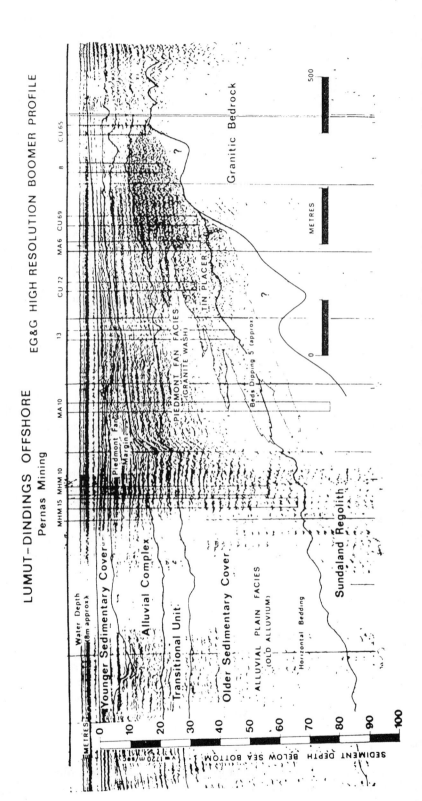

FIGURE 3.8 The relationship between the stratigraphy of the sedimentary sequence and the granitic bedrock in the Lumut–Dindings nearshore area of Perak, Malaysia. (From Batchelor, B.C., Geological characteristics of certain coastal and offshore placers as essential guides for tin exploration in Sundaland, southeast Asia, *Geol. Soc. Malaysia Bull.*, 11, 283, 1979. Reprinted with permission of the Geological Society of Malaysia.)

    e. Littoral placers — Only one such deposit is known between Mengkudu Bay and Cape Akuan. It is an elongated zone almost 900 m in length along the coast and extending 100 to 200 m from the shore.

Based on the offshore tin placers in the Malacca Strait, it can be seen that placers formed under subaerial environments are much more important than those formed by submarine processes.

3.6.1.3 Thailand

Four types of tin placer deposits have been identified in the offshore areas of southwest Thailand[106]:

1. Valley bottom placers. Unlike those offshore of the Indonesian tin islands, this type of placer occurs only infrequently. Humid tropical deep weathering was thought to have disintegrated the cassiterite-bearing granitic rocks followed by concentration through the removal of the lighter minerals by sheetwash and fluvial action. The flat morphology was thought to have enhanced the residual behavior of cassiterite, with economic deposits seldom more than 500 m from the primary source rock. They are buried by Holocene marine muds and clearly represent fossil placers.
2. Residual placers. This type of deposit is represented by that at Thai Muang, which is considered to be the collapsed outcrop of a high-grade cassiterite vein or contact zone, reworked by ocean waves and longshore currents. The deposit is 10 km long and 2 km wide, almost parallel to the shore, and geologically related to roof pendants of argillite enclosed in granite. The central zone contains very rich patches, locally with cassiterite of more than 10-mm diameter. Perpendicular to the strike of the deposit, the cassiterite diminishes in content and grain size quickly on both sides. The cassiterite is contained in a relatively thin clayey coarse-grained sand layer covering the weathered bedrock with a total sediment thickness ranging up to 6 m. These characteristics have made the deposit a prime target for illegal mining using divers.
3. Colluvial placers. This type of deposit is found both onshore and offshore on relatively steep slopes and appears to be related to colluvial/alluvial fans. The cassiterite, if present, is associated with successive layers of coarse sandy material and distributed throughout the entire profile rather than only in the bottom gravel.
4. Sheet flood or avalanche placers. This type of deposit is represented in the Ranong area. It is associated with a 10-km-wide coastal plain formed by planated country rock separated from granitic highlands by a fault scarp. The characteristics of the deposits are similar to type 3 except for the decrease in tin content and grain size away from the fault scarp. It appears to be an alluvial fan where episodic reworking may have taken place, and it is in the inshore areas where drowned tin placers occur.

Typical sections through the four types of deposits are shown in Figure 3.9. Since tin is distributed in deposits overlain by barren Holocene muds, the deposits are all clearly pre-Holocene in age. In the case of the type 1 deposits, a third last glacial or older age is suggested for the cassiterite enrichment at the bottom. The difference between type 3 and 4 deposits appears to be in the degree of reworking of the colluvial deposits. Type 3 deposits appear to have been less subjected to reworking by fluvial process than type 4.

3.6.1.4 Burma

Although very little is known about the deposits in Burma, the west coast from Moulmein in the north to Victoria Point in the extreme south has been deemed worthy of intensive exploration for offshore tin placers.[14] The potential of these Burmese waters for tin was indicated by tin recovery

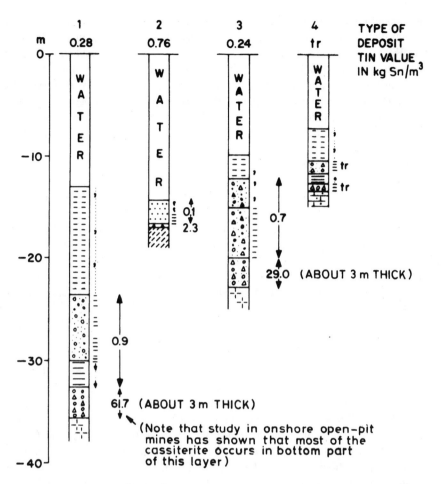

FIGURE 3.9 Typical sections through the four types of tin placer deposits in southern Thailand. *1* Valley bottom type; *2* Residual type; *3* Colluvial type; *4* Sheet flood or avalanche type. (From Aleva, G.J.J., Exploration for placer tin deposits offshore Thailand, *Proc. 11th Commonwealth Min. and Metall. Cong.*, Jones, M.J. (Ed.), Hong Kong 1978, Institution of Mining & Metallurgy, 59, 1979. Reprinted with permission of the Institution of Mining and Metallurgy.)

from the tideway of the Heiunze Basin,[107] from Belugyun Island near Moulmein, from the estuarine reaches of the Tenasserim River, and from the beach sands at Spider Island at the mouth of the Palauk River.[108] In addition, the belt of tin and tungsten mineralization traverses Lampi and a number of neighboring islands in the Mergui archipelago.[14]

3.6.2 SOUTHERN CHINA

Both beach[109] and nearshore placer deposits[110] are known in the Guangdong and Fukien provinces of southern China. Based on the small amount of information available, however, they are likely to be only of marginal commercial interest.

3.6.3 CORNWALL

Because of the generally steep stream gradient in Cornwall, tin-mine tailings discharged into streams were able to find their way into the sea. This was demonstrated by studies on the Red River and other streams covered by a geochemical reconnaissance survey in west Cornwall.[35,83] In the past, losses from the poor efficiency of mineral recovery techniques have been considerable.[111] This was demonstrated in the treatment of Cornish tin ores during the late 19th century and the early 20th century.[112] In 1890, the 15 mines operating in the Camborne–Redruth district sold 7131 tons of tin concentrate, while streamers recovering cassiterite from the tailings of these mines in the Red River sold 1730 tons of tin concentrate during the same period. The average loss in dressing tin during the early twentieth century must have exceeded 33%, and in two cases investigated,[112] it approached 60%. Reflecting this, the foreshore of the Gwithian Beach in St. Ives Bay has been worked for tin spasmodically in the past on a small scale.[113] The tin present in the beach sand appears to have been derived from mine tailings originating from the Red River.

From the early 1960s to 1985, submarine placers off Cornwall were prospected for by two mining companies and two academic institutions. Between 1962 and 1968, Coastal Prospecting Limited, a subsidiary of the Union Corporation, was involved in prospecting, beneficiation testing, and mining of tin sands in St. Ives Bay. In 1965, the Marine Mining Corporation came into being through its main constituents, Alpine Geophysical Associates Incorporated and Bessemer Securities Corporation, and prospected for tin in various offshore areas until 1985. At around the same time, research into the development of geological and geochemical techniques to assist in offshore tin prospecting was carried out by a group of staff and students at Imperial College London,[56,57,73,84,114,121] while limited offshore geophysical exploration was carried out by the Camborne School of Mines.[115]

Coastal Prospecting Limited investigated the tin sands in St. Ives Bay using a sparker survey,[114] foreshore and offshore drilling,[116] beneficiation testing,[116,117] and dredging.[113,116] The sparker survey was carried out to provide an indication of the sediment thickness. The Bay is covered by a continuous sheet of sand, which extends offshore as far as the line adjoining Godrevy Head and St. Ives Head (Figure 3.10). Immediately northwest of the mouth of the Red River and off the Hayle Estuary, the sediment is patchy due to rock outcrop, but there are depressions filled with sediment, which may be alluvial channels. In the latter area, the sediment over the bedrock depressions attains a thickness of greater than about 15 m. Near the low water mark over much of the coast, the sediment varies from about 9 m to over 12 m in thickness, and then thins out seaward to nothing at a depth of about 15 m. On the foreshore, a 10-cm Bangka drill was used, while for offshore drilling a 15-cm-diameter Bangka casing was used in conjunction with pumping equipment to retrieve the sands. Tin was found to be concentrated in the upper 1.5 m of the surficial seafloor sediments with only minor amounts at depth. The average grade of tin found was about 2% tin metal. However, beneficiation testing has confirmed that a high proportion of this tin is combined with other minerals and cannot be recovered.[117] Over a typical 28-day period, 8500 metric tons of sand at 0.23% metallic tin was processed yielding 14 metric tons of concentrate at a grade of 27.6% metallic tin, equivalent to a recovery of 36%.[113] However, there were problems in the dredging of the sand because the bay is affected by big Atlantic swells. Additionally, there was rapid wear and tear of the mining equipment, tidal limitations of Hayle Harbour, and days lost due to bad weather. During a good period the operation made a small profit, but overall it lost money and was closed down.[119]

When Union Corporation declined its interest in St. Ives Bay, the Marine Mining Corporation took over the lease. In 1966, an extensive sparker survey was carried out in 10 areas, followed by the selection of areas for vibrocoring. No appreciable quantity of tin was found. In the St. Agnes area, 71 vibrocores were collected in the Fal and Helford Estuaries, of which 26 bottomed on bedrock, while the remainder bottomed in gravels and clay. Tin was found in the upper 1.5 m of the surficial seafloor sediment with only minor quantities at depth. The vibrocoring identified two areas showing promise: first, the St. Agnes area, 1.2 km in width and 4 km in length running roughly parallel to the coast, and second, the Cligga Head area, which is separated from the St. Agnes area by barren ground comprising 2.5 km by 0.9 km of stanniferous sands. Although renewed mining in 1986 with an anticipated annual production of 750 metric tons of metallic tin was reported,[120] the crash of tin metal prices in 1985 saw the end of that venture.

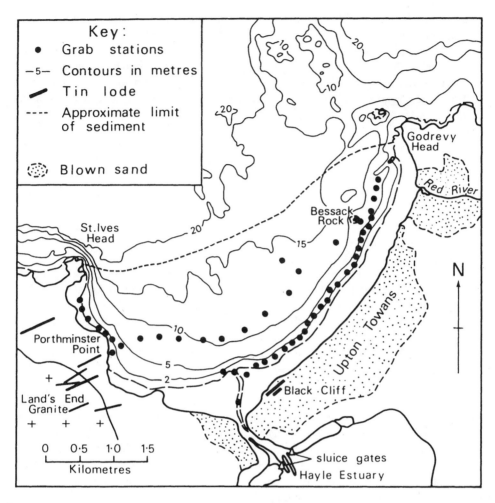

FIGURE 3.10 Bathymetry, the limit of sediment, location of grab sampling stations, and geology of St. Ives Bay, Cornwall. (From Yim, W.W.-S., Geochemical exploration for tin placers in St. Ives Bay, Cornwall, *Mar. Mining*, 2, 59, 1979. Reprinted with permission from Taylor and Francis.)

In Mount's Bay, a group of researchers at Imperial College carried out a number of offshore geochemical and geophysical studies in the 1960s and 1970s.[56,57,84,121] A main effort of this group was to determine whether geochemical anomalies occur in surficial sediments overlying submarine tin lodes. Elutriation experiments were carried out to study the hydraulic behavior of cassiterite particles in an attempt to explain and understand its natural distribution in order to assist offshore tin exploration. Problems that were revealed in the earlier work include:

1. The samples used for tin analysis were too small and were unrepresentative.[87]
2. The colorimetric method of tin determination[80] that had been used was inaccurate due to various problems.[85,86] These included the interference of shells and iron oxides in the ammonium iodide sublimation used to extract the tin.
3. The importance of tin mine tailings in the contamination of the surficial seafloor sediments was not fully realized. Taylor[113] considered that only the surficial 0.6 m of the seafloor sediments were minable in parts of St. Ives Bay. This surficial enrichment confirms a mine-tailings origin. What was considered to be an extremely good fit with a cubic trend surface for tin in Mounts Bay[84] is confirmative of the tin in the surficial

sediments being derived from mine tailings instead of erosion from tin lodes on the seabed or along the coast.

4. There were limitations in the method used in applying the hydraulic equivalence theory to natural sediments. For example, the tin content obtained through geochemical determination was assumed to be entirely in the form of mono-mineralic cassiterite grains. However, liberation studies indicated that around 70% of the tin is either in a highly locked or locked state.[89]

In 1972, a sampling program for the geochemical exploration of offshore tin deposits was initiated on the north coast of Cornwall by the group at Imperial College using a Shipek grab and a vibrocorer with a maximum penetration of 3 m.[73] The two main areas investigated were the coast off St. Agnes/Portreath and St. Ives Bay, with 238 and 48 grab samples collected, respectively. Only two vibrocores, the longest at 1.8 m were collected from the former area. Intertidal beach sediment sampling was also carried out in both areas. Tin determination on the samples was made on a shell-free basis to reduce sampling errors and errors resulting from the method of tin determination adopted.

In the St. Agnes/Portreath area, based on sparker profiles and vibrocoring results of the Marine Mining Corporation supplemented by information on areas of no sediment delineated by grabbing, a sediment thickness and distribution map was obtained (Figure 3.11).[122] However, because of the wide spacing of the sparker lines and the vibrocore stations, only general conclusions regarding sediment thickness could be made. The sand sheet can be seen to be fairly continuous except off St. Agnes Head and the area to the south. The sediments reach a maximum thickness of over 6 m off Porthowan and Chapel Porth and also appear to thicken at the present-day low water mark. Further offshore, the sediments attain a thickness in excess of 6 m only in the central part of the sediment sheet and about 0.5 km west-southwest of St. Agnes Head. The minus-8-mesh tin distribution is shown in Figure 3.12. The highest concentration found is in a linear belt about 15 to 18 m below the present low water mark, with one sample showing a metallic tin content of 7300 ppm. The tin size distribution data show clearly that the bulk of the tin is present in the minus-124-μm fraction. The presence of such a linear belt with tin enrichment is in agreement with the energy fence concept of Allen[123] in that tin deposited seaward of the energy fence under storm conditions cannot be re-entrained and is able to build up in concentration seaward of the fence. Based on the tin distribution found, a mine-tailings source of the tin is not in doubt. This is confirmed by the geochemical results on the two cores (Figure 3.13) in that both cores show enrichment in tin in the surficial 0.75 m.[122] The increase in tin content in the minus-90-μm fraction toward the base of core 046, however, requires explanation. One possibility is that the tin may have been derived from a non-tin mining source. It is not possible to verify this without further coring to greater depth. In the 15 beach sediment samples studied for tin distribution, the highest tin content was found in the innermost Chapel Porth sample (Figure 3.12) confirming that the high water mark was favorable for tin concentration.

The distribution of shell content and tin content in the minus-2057-μm of the St. Ives Bay grab samples is shown in Figure 3.14. It can be seen that the shell content decreases from west to east, while the tin content decreases from east to west. Both are in agreement with the sediment discharged into the bay via the Red River, which contains mine tailings from the Camborne–Redruth area. This study supported the findings of Coastal Prospecting Limited in that the richest tin area occurs off the mouth of the Red River. In addition to this area, the mouth region of the Hayle Estuary and the middle portion of the bay also show promise, but the grade found was lower than in the area off the mouth of the Red River. The overall grade in the surficial sediment is about 0.1% cassiterite.[118] The beach sediment sampling traverses, carried out under conditions of aggradation and degradation related to wind directional changes, were found to be important in determining the tin content in beach sediments.

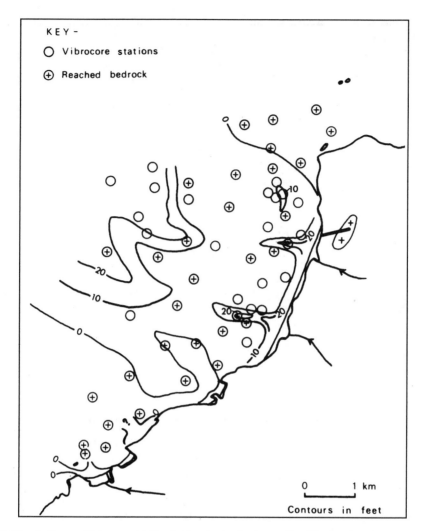

FIGURE 3.11 Sediment thickness and distribution in the St. Agnes Head/Portreath area of Cornwall based on cores, grabs, and sparker data. (Adapted from Yim, W.W.-S., *Some Aspects of the Geochemistry of Tin and Other Elements in Sediments off North Cornwall*, unpublished M.Phil. thesis, University of London, 1974. With permission.)

The main conclusion drawn from the St. Agnes/Portreath and St. Ives Bay studies[73] is that geochemical exploration techniques are useful in the assessment of offshore tin placers formed by mine tailings discharged into the sea.

3.6.4 AUSTRALIA

3.6.4.1 Tasmania

Ocean Mining A.G., an exploration affiliate of Ocean Science and Engineering Incorporated and the Anglo American Group, began investigation of the offshore tin potential of Tasmania in 1964.[124] A total of five offshore areas was under the companies' exploration license at the time, including Ringarooma Bay in northeastern Tasmania. Preliminary investigation up to the end of 1965 included running some 500 line miles of bathymetric traverses. Possible targets in the license area for exploration were:

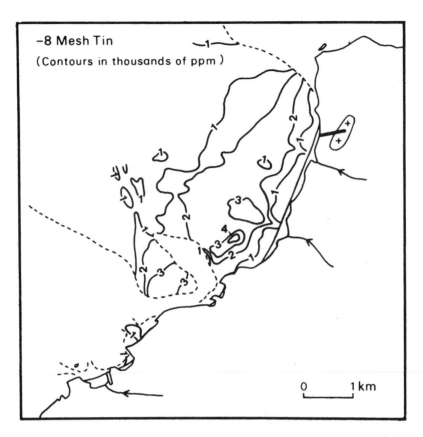

FIGURE 3.12 Minus-8-mesh (<2057 micron) tin distribution on a shell-free basis in surficial seafloor sediments of the St. Agnes/Portreath area, Cornwall. (Adapted from Yim, W.W.-S., *Some Aspects of the Geochemistry of Tin and Other Elements in Sediments off North Cornwall*, unpublished M.Phil. thesis, University of London, 1974. With permission.)

1. Submerged extension of present river channels.
2. Alluvial fans in which the heavy minerals may have been reworked by the more recent transgressions.
3. Submerged eluvial or kaksa deposits.
4. Extensions of onshore deep leads of Tertiary age.

In 1966, Ocean Mining AG began more detailed exploration divided into three phases:

Phase I — This was aimed at delineating alluvial target areas and locating those with mineral potential. Reconnaissance surveys were undertaken to determine the seafloor morphology, to determine the sediment thickness and the bedrock topography using seismic profiling, and preliminary sampling to show the sediment type.

Phase II — This was a follow-up survey of the promising areas identified in Phase I to permit the delimitation and preliminary evaluation of ore bodies which may occur. This phase involved the use of a new type of coring machine known as the Horton sampler with better capabilities for precise sampling of unconsolidated seafloor sediments than any other in existence at the time. The sampler was assembled aboard the research vessel *R.V. Wando River* in California by Ocean Science and Engineering Incorporated.

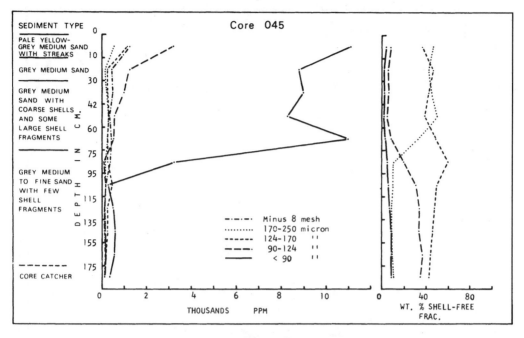

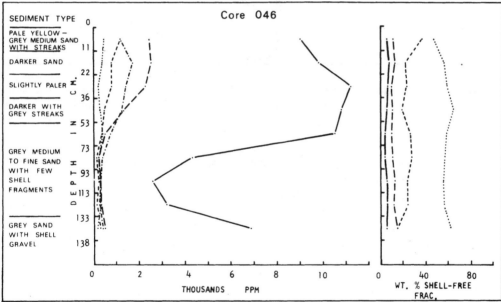

FIGURE 3.13 Sediment type and tin distribution in cores 045 and 046 from the St. Agnes/Portreath area of Cornwall. (Adapted from Yim, W.W.-S., *Some Aspects of the Geochemistry of Tin and Other Elements in Sediments off North Cornwall*, unpublished M.Phil. thesis, University of London, 1974. With permission.)

The sampler is a hydraulically driven vibratory corer designed to collect cores about 33 m long in maximum water depths of about 60 m. The 15-cm internal diameter drill string was made up of steel pipes, about 3 m long, which were coupled together internally. A total of 171 holes were drilled at intervals of about 1 km with an average core length of about 7 m and a maximum core length of about 26 m. The cores were

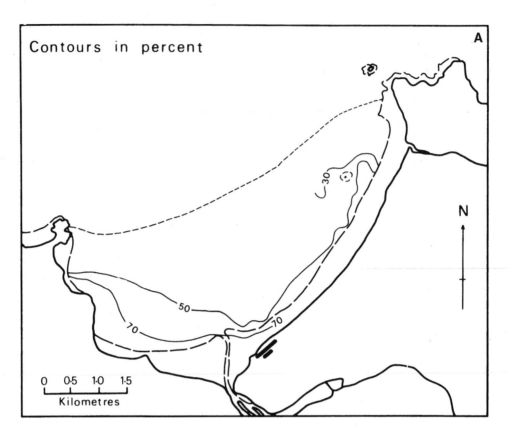

FIGURE 3.14 Shell content and total minus-8-mesh (<2057 micron) tin distribution on a shell-free basis in surficial seafloor sediments of St. Ives Bay, Cornwall. (Adapted from Yim, W.W.-S., *Some Aspects of the Geochemistry of Tin and Other Elements in Sediments off North Cornwall*, unpublished M.Phil. thesis, University of London, 1974. With permission.)

    hydraulically extruded and used for mineral dressing tests and to check assays for tin, titanium, and zirconium.

Phase III — This was the final evaluation of the Phase II results and studies on mining feasibility. The Ringarooma Bay was singled out as the area showing the most potential with a tin-ore reserve estimated at between 3300 and 4500 tons of tin metal.[125] Because both rutile and zircon are common heavy minerals associated with the cassiterite,[126] they helped to raise the value of the ore.

    The offshore data of Ocean Mining AG in Ringarooma Bay was reassessed periodically until 1983.[127] The most recent reassessments were by Hellyer Mining and Exploration Proprietory Limited.[128,129] The main area of interest occurs at a water depth of less than 35 m below mean sea level (Figure 3.15) where the maximum thickness of sediment is not more than about 20 m (Figure 3.16). Although the depth below seabed is less than those of deep leads occurring onshore, deep leads located in the Great Northern Plains were suggested by their bedrock depth to extend into Ringarooma Bay (Figure 3.17). Based on the distribution of metallic tin assay grades in borehole sections, the bulk of the tin concentration occurs at a water depth of between 33 and 42 m below mean sea level.[11] A number of northwesterly and northerly trending sediment-filled channels were identified.

    Based on the recent advance in understanding the genesis of onshore tin placers in northeastern Tasmania,[7] the offshore exploration work can be re-examined in a new light. For example, the method of boring used is unlikely to provide suitable sediment samples from adjacent to the bedrock.

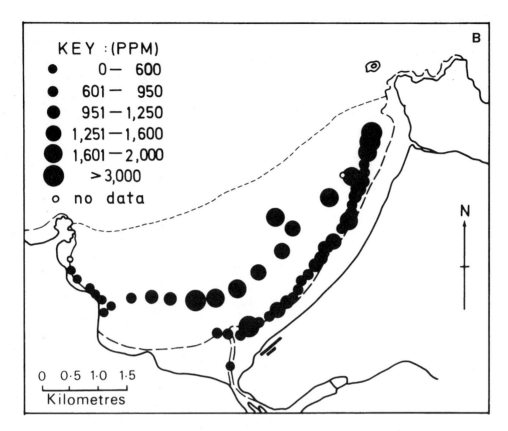

FIGURE 3.14   (continued)

The concept of deep leads with economic grades of cassiterite as being continuous for distances exceeding 1 km is invalid. Multiple local tin sources are possible. Furthermore, the 1-km borehole interval used by Ocean Mining AG[125] is grossly inadequate based on the known distance of transport of cassiterite required to form economic placers. In the Bass Strait, therefore, there is considerable optimism for finding tin shed from mineralized source rocks during periods of low sea level, providing that there is a recovery of tin metal prices to a level adequate to provide sufficient incentive to undertake the work.

In order to obtain age information on the Ringarooma Bay deposits, sediment samples from the three boreholes shown in Figure 3.15 were subjected to heavy mineral analysis.[7] The results (Table 3.1) confirmed the presence of three zircospilic suite minerals, spinel, corundum, and ilmenite, which were thought to have been derived from the Blue Tier basalts.[7] Based on this, the age of the offshore deep leads is likely to be similar to those of their onshore counterparts. They formed probably after the Middle Eocene but prior to the Middle Miocene.

3.6.4.2   Queensland

In Cable Bay of the Cape York tinfield, tin-rich sands are known to occur on the beach.[3]

3.7   CONCLUSIONS

From this review of tin placer deposits on continental shelves, it can be seen that, with the exception of Indonesia and Thailand where the coastal waters are comparatively calm, the exploitation of deposits in other parts of the world has not been economically viable. In addition to difficulties

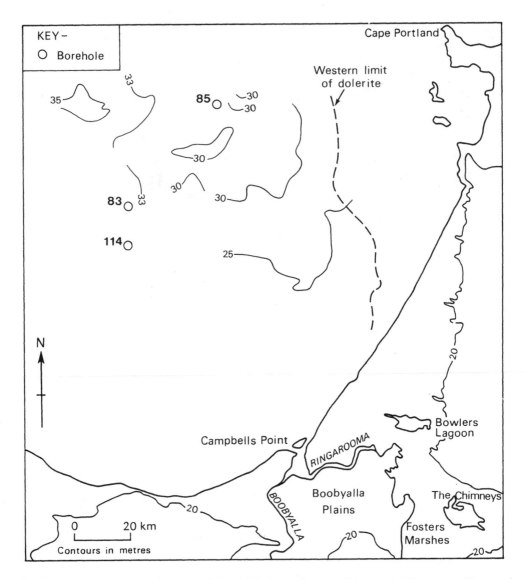

FIGURE 3.15 Bathymetry and western limit of dolerite in the area of interest in Ringarooma Bay, north-eastern Tasmania. Borehole locations referred to in the text are also shown. (From Yim, W.W.-S., Heavy mineral provenance and the genesis of stanniferous placers in northeastern Tasmania, unpublished Ph.D. thesis, University of Tasmania, 1990.)

caused by adverse physical environmental conditions, major contributing factors to the lack of offshore tin exploitation outside southeast Asia are the lack of understanding of the distance of transport required to form economic tin placers, the localized nature of bedrock tin mineralization and its resultant placers, and inadequacies in the methods of prospecting. The main conclusions drawn are summarized under three main headings.

3.7.1 EVOLUTION THROUGH TIME

The Mesozoic granites of southeast Asia have been responsible for the most productive tin placer deposits on continental shelves. Upper Paleozoic granites in Cornwall and Ringarooma Bay, Tasmania, on the other hand, have not been so productive.

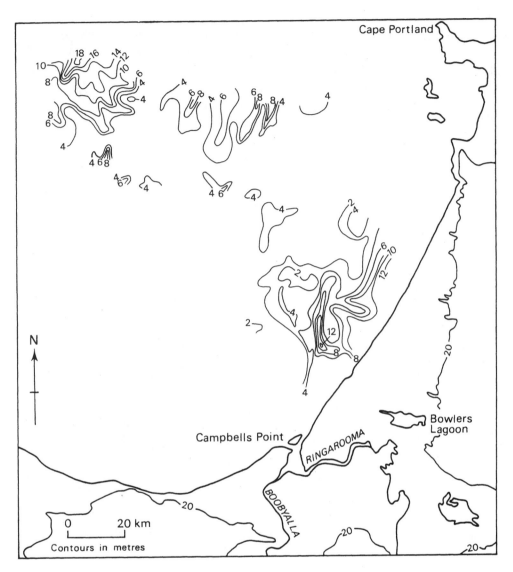

FIGURE 3.16 Sediment thickness in the area of interest in Ringarooma Bay, northeastern Tasmania. (From Yim, W.W.-S., Heavy mineral provenance and the genesis of stanniferous placers in northeastern Tasmania, unpublished Ph.D. thesis, University of Tasmania, 1990.)

At the present time, the onshore area where the genesis of tin placer deposits has been best worked out is in northeastern Tasmania.[7] Based on the stratigraphical control of the tin placers, it is possible to extend placer formation events back, at least in part, to the Permo-Carboniferous when fluvio-glacial conditions existed there. Very little is known, however, between this time and the reworking of the deep lead tin deposits shown by the mixing between cassiterite and the zircospilic assemblage of heavy minerals datable at Middle Eocene to Late Oligocene. While placers may be formed continuously over time, the fact remains that the Middle Eocene to Oligocene period appeared to be a critical period for the formation of giant placers in northeastern Tasmania. The long time scale involved in the formation of tin placers in northeastern Tasmania was pointed out[7] to be greater than that of any other unconsolidated tin placers in the world. In southeast Asia, tin placers within the Older Sedimentary Cover of Malaysia and Indonesia are Late Pliocene to

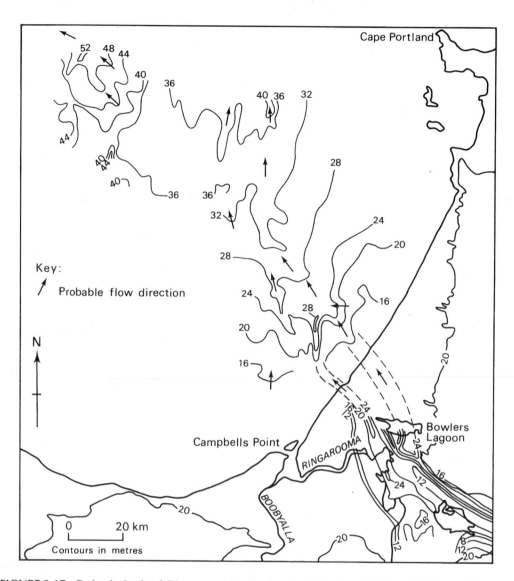

FIGURE 3.17 Bedrock depth of Ringarooma Bay and part of the Great Northern Plains, northeastern Tasmania. Interpreted depths to bedrock in Ringarooma Bay are in meters below chart datum. Depths to bedrock on the Great Northern Plains are in meters below mean sea level. (From Yim, W.W.-S., Heavy mineral provenance and the genesis of stanniferous placers in northeastern Tasmania, unpublished Ph.D. thesis, University of Tasmania, 1990.)

Early Pleistocene in age,[8] while the history of the Indonesian fluvial tin placers may extend back to Miocene and Pliocene times even though the upper limit of formation may be in the Quaternary.[6] However, the study of tin placers in southeast Asia has lacked the stratigraphical control available in northeastern Tasmania where the antiquity of landforms[130] through a long history of deep weathering and slow erosion rates is an essential element. Thus, the ages suggested for the formation of the southeast Asian placers are likely to be minimum ages. Because "long" geological time periods are now known to be involved in the development of giant onshore tin placers, the same may well apply to their offshore counterparts.

3.7.2 RECOMMENDATIONS FOR RESEARCH AND DEVELOPMENT

The following recommendations are made for future research and development:

1. Improvement of the method of drilling for offshore exploration. The drill should be capable of collecting partially cemented tin gravels and of reaching bedrock.
2. Use of a borehole spacing of 50 m or less.
3. Bulk sampling operations during a late exploration stage or pilot plant operation stage to supplement the information provided by boreholes.
4. Refinement of the stratigraphical control of the onshore and offshore tin placers in southeast Asia should be attempted. Thailand also possesses Cainozoic basalts containing the same zircospilic assemblage of heavy minerals found in Tasmania and eastern Australia. Therefore, studies on the Cainozoic stratigraphy in the onshore and offshore sequences would help to build a clearer picture of the genesis of the offshore tin placers of southeast Asia. Information on Cainozoic stratigraphy of shelf sequences from the oil and gas industry would also be valuable. Additionally, an Ocean Drilling Program site close to this tinfield is needed to provide sea-surface temperature history over the Cainozoic period.
5. Heavy mineral assemblages should be studied to provide information on recycling events. Minerals which are datable, such as zircon, sphene, apatite, glauconite, and cassiterite, should provide age information on such events.

3.7.3 FUTURE PROSPECTS

Since the collapse of international tin metal prices in 1985, there is little incentive to explore for and exploit tin placers on continental shelves in view of the relatively high costs compared to onshore mining. Based on the current state of knowledge, however, the proven reserves of tin metal from offshore tin placers can only increase, particularly because technological advancement in marine mining will continue. Innovative mining methods, such as the walking platform, will go some way to overcoming the rough sea conditions off the coast of Cornwall.[131] Alternatively, ore could be dredged during good weather and stockpiled for processing ashore, while scaling down the operations may be a means of ensuring that the mining is profitable. In Indonesia and Thailand, the traditional approach of dredgers will continue to be viable in the protected inshore areas. However, they will need to be modified to operate efficiently and economically at greater water depths as the shallow deposits are exhausted.

ACKNOWLEDGMENTS

I am grateful to Mr. D. Jennings, Drs. G.J.J. Aleva, M.R. Banks, D.A.F. Batchelor, the late R. Ford, the late S.S.F. Hui, R.H.T. Garnett, D. Taylor, J.S. Tooms, J.C. Van Moort, and Profs. D.S. Cronan and C.D. Ollier for their assistance during the different stages of this work. I would also like to thank the Times Higher Education Supplement for awarding a fellowship to enable me to study the tin mining industry in Malaysia during the 1970s, Mineral Resources Tasmania, and Amdex Mining Limited for assisting my studies in northeastern Tasmania. This chapter is a contribution to the International Geological Correlation Programme Project no. 396 "Continental shelves in the Quaternary."

REFERENCES

1. Bates, R.L. and Jackson, J.A. (Eds.), *Dictionary of Geological Terms*, 3rd ed., American Geological Institute, Anchor Press, New York, 1984.

2. Shilo, N.A., Placer-forming minerals and placer deposits, *Pacific Geol.*, 2, 29, 1970.

3. Taylor, R.G., *Geology of Tin Deposits*, Developments in Economic Geology, 11, Elsevier Publishing Company, Amsterdam, 1979.

4. Raeburn, C. and Milner, H.B., *Alluvial Prospecting*, Thomas Murby, London, 1925.

5. Nye, P.B., The sub-basaltic tin deposits of the Ringarooma valley, *Geol. Surv. Bull.*, Tasmania Department of Mines, 44, 1938.

6. Aleva, G.J.J., Indonesian fluvial cassiterite placers and their genetic environment, *J. Geol. Soc. London*, 142, 815, 1985.

7. Yim, W.W.-S., Tin placer genesis in northeastern Tasmania, in *The Cainozoic in Australia: A Reappraisal of the Evidence*, Williams, M.A.J., De Deckker, P., and Kershaw, A.P. (Eds.), Special Pub. 18, Geological Society of Australia, Sydney, 235, 1991.

8. Batchelor, B.C., Discontinuously rising late Cainozoic eustatic sea-levels, with special reference to Sundaland, southeast Asia, *Geol. Mijn.*, 58, 1, 1979.

9. Camm, G.S. and Hosking, K.F.G., Stanniferous placer development on an evolving landsurface with special reference to placers near St. Austell, Cornwall, *J. Geol. Soc. London*, 142, 803, 1985.

10. Sutherland, D.G., The transport and sorting of diamonds by fluvial and marine processes, *Econ. Geol.*, 77, 1613, 1982.

11. Yim, W.W.-S., Heavy mineral provenance and the genesis of stanniferous placers in northeastern Tasmania, unpublished Ph.D. thesis, University of Tasmania, 1990.

12. Lewis, G.R., *The Stannaries*, Harvard University Press, Cambridge, 1908.

13. Aranyakanon, P., 16. Offshore exploration for tin in Thailand, *Rep. 19th Session Comm. Co-ord. Joint Prospecting*, 10, 164, 1973.

14. Hosking, K.F.G., X. The offshore tin deposits of southeast Asia, *Comm. Co-ord. Joint Prospecting Tech. Bull.*, 5, 112, 1971.

15. Hails, J.R., Placer deposits, in *Handbook of Strata-bound and Stratiform Ore Deposits Vol. 3 Supergene and Surficial Ore Deposits, Textures and Fabrics*, Wolf, K.H. (Ed.), Elsevier Scientific, Amsterdam, 1976, 213.

16. Aleva, G.J.J., Aspects of the historical and physical geology of the Sunda Shelf essential to the exploration of submarine tin placers, *Geol. Mijn.*, 52, 79, 1973.

17. Emery, K.O. and Noakes, L.C., Economic placer deposits of the continental shelf, *Comm. Co-ord. Joint Prospecting Tech. Bull.*, 1, 95, 1968.

18. Cruickshank, M.J., Romanowitz, C.M., and Overall, M.P., Offshore mining — present and future, *Eng. Min. J.*, 169, 84, 1968.

19. Archer, A.A., Progress and prospects of marine mining, *Min. Mag.*, 150, 1974.

20. Archer, A.A., Economics of off-shore exploration and production of solid minerals on the continental shelf, *Ocean Manag.*, 1, 5, 1973.

21. Robertson, W., *Report on the World Tin Position with Projections for 1965 and 1970*, International Tin Council, London, 1965.

22. Batchelor, B.C., *Sundaland Tin Placer Genesis and Late Cainozoic Coastal and Offshore Stratigraphy in Western Malaysia and Indonesia*, unpublished Ph.D. thesis, University of Malaya, 1983.

23. Anonymous, *Tin Industry Empresas Brumadinho Basic Information*, unpublished report, Mineracao Brumadinho S/A, Sao Paolo, 1982.

24. Sutherland, D.G., Foreword, in *Alluvial Mining*, Institution of Mining & Metallurgy, Elsevier Applied Science, Barking, 1991.

25. Hutchinson, C.S. and Chakraborty, K.R., Tin: a mantle or crustal source?, *Geol. Soc. Malaysia Bull.*, 11, 71, 1979.

26. Hosking, K.F.G., The world's major types of tin deposit, in *Geology of Tin Deposits*, Hutchinson, C.S. (Ed.), United Nations, Economic and Social Commission for Asia and the Pacific, Springer Verlag, Berlin, 1988, 3.

27. Schuiling, R.D., Tin belts around the Atlantic Ocean: some aspects of the geochemistry of tin, *A Tech. Conf. on Tin*, London 1967, 531, 1967.

28. Chappell, B.W. and White, A.J.R. Two contrasting granite types, *Pacific Geol.*, 8, 173, 1974.

29. Friese, F.W., Untersuchung von mineralen auf abnutzbarkeit bei verfrachtung im wasser, *Miner. Petrog. Mitt.*, 41, 1, 1931.

30. Mackie, W., The principles that regulate the distribution of particles of heavy minerals in sedimentary rocks, as illustrated by the sandstone of the north-east of Scotland, *Trans. Edinburgh Geol. Soc.*, 11, 138, 1923.

31. Saks, S.E. and Gavshina, A.N., Stream deposition of cassiterite, *Lithologiya I Poleznye Iskopaemye*, 2, 129, 1976.

32. Yim, W.W.-S., Particle size and trace element distribution characteristics of cassiterite as an aid to provenance study of stanniferous placers in northeastern Tasmania, Australia, *J. Southeast Asian Earth Sci.*, 10, 131, 1994.

33. Ollier, C.D., Evolutionary geomorphology of Australia and Papua, New Guinea, *Trans. Inst. Br. Geog. New Series*, 4, 516, 1979.

34. Toh, E., Comparison of exploration for alluvial tin and gold, *Proc. 11th Commonwealth Min. and Metall. Cong.*, Jones, M.J. (Ed.), Hong Kong 1978, Institution of Mining & Metallurgy, 269, 1979.

35. Yim, W.W.-S., Geochemical investigations on fluvial sediments contaminated by tin-mine tailings, Cornwall, England, *Environ. Geol.*, 3, 245, 1981.

36. Osberger, R., Dating Indonesian cassiterite placers, *Min. Mag.*, 117, 260, 1967.

37. Cissarz, A. and Baum, F., Vorkommen und mineralinhalt der zinnerzlagerstatten von Bangka (Indonesien), *Geol. Jb.*, 77, 541, 1960.

38. Batchelor, D.A.F., Dating of Malaysian fluvial tin placers, *J. Southeast Asian Earth Sci.*, 2, 3, 1988.

39. Yim, W.W.-S., Ivanovich, M., and Yu, K.-F., Young age bias of radiocarbon dates in pre-Holocene deposits of Hong Kong and their implications for Pleistocene stratigraphy, *Geo-Marine Letters*, 10, 165, 1990.

40. Gulson, B.L. and Jones, M.T., Cassiterite: potential for direct dating of mineral deposits and a precise age for the Bushveld Complex granites, *Geol.*, 20, 355, 1992.

41. Krogh, T.E., A low contamination method for hydrothermal decomposition of zircon and extraction of U and Pb for isotopic age determinations, *Geochim. Cosmochim. Acta*, 37, 485, 1973.

42. Compston, W., Williams, I.S., and Meyer, C., U-Pb geochronology of zircons from lunar breccia 73217 using a sensitive high mass-resolution ion microprobe, *J. Geophys. Res.*, 89, supplement B525, 1984.

43. Gleadow, A.J.W., Hurford, A.J., and Quaife, R.D., Fission track dating of zircon: improved etching techniques, *Earth and Planet. Sci. Lett.*, 33, 273, 1976.

44. Yim, W.W.-S., ESR measurement of alluvial zircons as an aid to provenance determination in geology, in *Modern Applications of EPR/ESR from Biophysics to Materials Science*, Rudowicz, C.Z. (Ed.), Springer Verlag, Singapore, 1998, 128.

45. Yim, W.W.-S., Gleadow, A.J.W., and van Moort, J.C., Fission track dating of alluvial zircons and heavy mineral provenance in northeast Tasmania, *J. Geol. Soc. London*, 142, 351, 1985.

46. Shackleton, N.J. and Kennett, J.P., Palaeotemperature history of the Cenozoic and the initiation of Antarctica glaciation: oxygen and carbon isotope analyses in DSDP sites, in *Initial Reports of the Deep Sea Drilling Project*, Kennett, J.P., Houtz, R.E. et al. (Eds.), 29, U.S. Government Printing Office, Washington, D.C., 1975, 743.

47. Brown, A.V., Preliminary report on age determination of basalt samples from Ringarooma 1:50 000 sheet, unpublished report, Tasmania Department of Mines, 25, 1977.

48. Sutherland, F.L. and Wellman, P., 1986, Potassium–argon ages of Tertiary volcanic rocks, Tasmania, *Pap. Proc. Roy. Soc. Tasmania*, 120, 77, 1986.

49. Hill, R.S., *Nothofagus* macrofossils from the Tertiary of Tasmania, *Alcheringa*, 7, 169, 1983.

50. Bigwood, A.J. and Hill, R.S., Tertiary Araucarian macrofossils from Tasmania, *Aus. J. Bot.*, 33, 645-656.

51. Harris, W.K., *Palynological Examination of Samples from the Tin Leads of North East Tasmania for Utah Development Company*, unpublished report, South Australia Department of Mines, 60/15, 1965.

52. Hill, R.S. and Macphail, M.K., Reconstruction of the Oligocene vegetation at Pioneer, northeast Tasmania, *Alcheringa*, 7, 281, 1983.

53. Krol, G.L., Theories on the genesis of kaksa, *Geol. Mijn.*, 39, 437, 1960.

54. Rubey, W.W., The size distribution of heavy minerals within a water-laid sandstone, *J. Sed. Pet.*, 3, 3-29.

55. Rittenhouse, G., The transportation and deposition of heavy minerals, *Geol. Soc. Am. Bull.*, 54, 1725, 1943.

56. Ong, P.M., *Geochemical Investigation in Mount's Bay, Cornwall*, unpublished Ph.D. thesis, University of London, 1962.

57. Hazelhoff Roelfzema, B.H. and Tooms, J.S., Dispersion of cassiterite in the marine sediments of western Mount's Bay, Cornwall, in *A 2nd Tech. Conf. on Tin*, Fox, W. (Ed.), Bangkok 1969, 2, 489, 1970.

58. Tooms, J.S., Some aspects of exploration for marine mineral deposits, Proc. *9th Commonwealth Min. and Metall. Cong.*, Jones, M.J. (Ed.), London 1969, Institution of Mining & Metallurgy, 1970, 1.

59. Pryor, E.J., Blyth, N., and Eldridge, A., Purpose in fine sizing and comparison of methods, in *Symposium on Recent Development in Mineral Dressing*, Inst. Mining & Metallurgy, London, 1953.

60. Tourtelot, H.A., Hydraulic equivalence of grains of quartz and heavier minerals and implications for the study of placers, *United States Geol. Surv. Prof. Pap.*, 594F, 1968.

61. Slingerland, R., Role of hydraulic sorting in the origin of fluvial placers, *J. Sed. Pet.*, 54, 137, 1984.

62. Hosking, K.F.G., Primary mineral deposits, in *Geology of the Malay Peninsula*, Gobbet, D.H. and Hutchinson, C.S. (Eds.), Wiley-Interscience, New York, 1973, 335.

63. Taylor, D., Some thoughts on the development of the alluvial tinfields of Malay–Thai Peninsula, *Bull. Geol. Soc. Malaysia*, 19, 375, 1986.

64. Hosking, K.F.G., Some aspects of the stability of sulphides, and other normally unstable minerals of economic importance, in the lodes, boulders and pebbles of the Cornish beaches, *Camborne Sch. Mines Mag.*, 60, 11, 1960.

65. Yim, W.W.-S., *A Geochemical Investigation of the Distribution of Certain Elements in the Near-shore Marine Sediments of South-west Cornwall*, unpublished Diploma in Mineral Technology thesis, Camborne School of Mines, 1972.

66. Yim, W.W.-S., Heavy metal accumulation in estuarine sediments in a historical mining area of Cornwall, *Mar. Poll. Bull.*, 7, 147, 1976.

67. Van Overeem, A.J.A., Geological control of dredging operations on placer deposits in Billiton, *Geol. Mijn.*, 39, 458, 1960.

68. Beckman, W.C., Roberts, A.C., and Thompson, K.C., How underwater seismics aided Thailand tin exploration, *Engr. and Min. J.*, 163, 244, 1962.

69. MacDonald, E.H., *Alluvial Mining*, Chapman and Hall, London, 1983.

70. Ringis, J., *Interpretation of Reflection Seismic Profiler Surveys in the Lumut Area, West Peninsular Malaysia*, unpublished report, UNDP Tech. Support for Regional Prospecting in East Asia, 1976.

71. Bon, E.H., Exploration techniques employed in the Pulau Tujuh tin discovery, *Trans. Inst. Min. and Metall.*, 88, A13, 1979.

72. Sujitno, S., Offshore drilling for tin exploration in Indonesia, in *Drilling and Sampling Techniques in Tin Prospecting*, bin Hj Hassan, A.H. and Van Wees, H. (Eds.), Seatrad Centre, Ipoh, 1, 71, 1981.

73. Yim, W.W.-S., *Some Aspects of the Geochemistry of Tin and Other Elements in Sediments off North Cornwall*, unpublished M.Phil. thesis, University of London, 1974.

74. Anonymous, Drillship for tin exploration in Indonesia, *Geodrilling Int.*, October 1997, 8, 1997.

75. Nichol, I. and Henderson-Hamilton, J., A rapid quantitative spectrographic method for the analysis of rocks, soils and stream sediments, *Trans. Inst. Min. and Metall.*, 74, 955, 1965.

76. Bowie, S.H.U., Darnley, A.G., and Rhodes, J.R., Portable radioisotope X-ray fluorescence analyser, *Trans. Inst. Min. and Metall.*, 74, 361, 1964-5.

77. Darnley, A.G. and Leamy, C.C., The analysis of tin and copper ores using a portable radioisotope X-ray fluorescence analyser, in *Radioisotope Instruments in Indust. and Geophys.*, Int. Atom. Ener. Agency, Vienna, 1, 191, 1966.

78. Sweatman, T.R., Wong, Y.C., and Toong, K.S., Application of X-ray fluorescence analysis to the determination of tin in ores and concentrates, *Trans. Inst. Min. and Metall.*, 76, B149, 1967.

79. Garson, M.S. and Bateson, J.H., Possible use of the P.I.F. analyser in geochemical prospecting for tin, *Trans. Inst. Min. and Metall.*, 76, 165, 1967.

80. Stanton, R.E. and McDonald, A.J., Field determination of tin in soil and stream sediment surveys, *Trans, Inst. Min. and Metall.*, 71, 27, 1961-2.

81. Bowman, J.A., The determination of tin in ores and concentrates by atomic absorption spectrometry in the nitrous oxide-acetylene flame, *Anal. Chim. Acta.*, 42, 285, 1968.

82. Guru, S., *Geochemical Studies in South China Sea*, unpublished Diploma Imperial College thesis, University of London, 1972.

83. Hosking, K.F.G. and Ong, P.M., The distribution of tin and certain other heavy metals in the superficial portions of the Gwithian/Hayle beach of west Cornwall, *Trans. Roy. Soc. Cornwall*, 19, 351, 1963-4.

84. Hazelhoff Roelfzema, B.H., Geochemical dispersion of tin in marine sediments, Mount's Bay, Cornwall, *unpublished Ph.D. thesis*, University of London, 1968.

85. Yim, W.W.-S., Geochemical determination of tin in sediments off north Cornwall, *Trans. Inst. Min. and Metall.*, 84, B64, 1975.

86. Yim, W.W.-S., Rapid methods of tin determination for geochemical prospecting, *Geol. Soc. Malaysia Bull.*, 11, 375, 1979.

87. Gy, P., *Sampling of Particulate Materials: Theory and Practice*, Elsevier, Amsterdam, 1979.

88. Royle, A.G., Alluvial sampling formula and recent advances in alluvial deposit evaluation, *Trans. Inst. Min. and Metall.*, 95, B179, 1986.

89. Yim, W.W.-S., Liberation studies on tin-bearing sands off north Cornwall, United Kingdom, *Mar. Mining*, 5, 87, 1984.

90. Huaff, P.L. and Airey, J., The handling, hazards, and maintenance of heavy liquids in the geologic laboratory, *Geol. Surv. Circ.*, 827, United States Department of Interior, Reston, VA, 1980.

91. Walker, M.S. and Devernoe, A.L., Mineral separations using rotating magnetic fluids, *Int. J. Min. Process.*, 31, 195, 1991.

92. Berman, H., A torsion microbalance for the determination of specific gravities of minerals, *Am. Min.*, 24, 434, 1953.

93. Lever, R.R., Detection of small amounts of cassiterite in pan concentrates, *Min. and Chem. Eng. Rev.*, 57, 19, 1965.

94. Hosking, K.F.G., Practical aspects of the identification of cassiterite (SnO_2) by the "tinning test," *Geol. Soc. Malaysia Bull.*, 7, 17, 1974.

95. Ng, W.K. and Yong, S.K., Rapid semi-quantitative mineral analysis to improve efficiency in processing tin ores from West Malaysia, in *A 2nd Technical Conf. on Tin*, Bangkok 1969, 3, 1099, 1970.

96. Henley, K.J., Mineralogical assessment of a churn-drill core from an alluvial tin deposit, *Proc. Aust. Inst. Min. and Metall.*, 236, 43, 1970.

97. Henley, K.J., The quantitative mineralogical evaluation of alluvial ores, *Amdel Bull.*, 14, 20, 1972.

98. Yim, W.W.-S., Application of the Zeiss TGA 10 particle-size analyzer in the exploration of stanniferous placers, *Geol. Soc. Malaysia Bull.*, 20, 619, 1986.

99. Batchelor, B.C., Geological characteristics of certain coastal and offshore placers as essential guides for tin exploration in Sundaland, southeast Asia, *Geol. Soc. Malaysia Bull.*, 11, 283, 1979.

100. Van Bemmelen, R.W., *The Geology of Indonesia*, 2 volumes, Government Printing Office, The Hague, 1949.

101. Fick, L.J., Offshore prospecting, *Tech. Conf. on Tin*, International Tin Council, London, 1967.

102. Van Overeem, A.J.A., Offshore tin exploration in Indonesia, *Trans. Inst. Mining and Metall.*, 79, A81, 1970.

103. Warnford-Lock, C.G., *Mining in Malaya for Gold and Tin*, Crowther & Goodman, London, 1907.

104. Scrivenor, J.B. and Jones, W.R., *The Geology of South Perak, North Selangor and the Dindings*, Government Press, Kuala Lumpur,1919.

105. Scrivenor, J.B., *The Geology of Malayan Ore-deposits*, MacMillan and Company, London, 1928.

106. Aleva, G.J.J., Exploration for placer tin deposits offshore Thailand, *Proc. 11th Commonwealth Min. and Metall. Cong.*, Jones, M.J. (Ed.), Hong Kong 1978, Institution of Mining & Metallurgy, 59, 1979.

107. ECAFE, Tin ore resources of Asia and Australia, *Min. Res. Dev. Ser. United Nations ECAFE*, 23, 1964.

108. Jones, W.R., *Tinfields of the World*, Mining Publications, London, 1925.

109. Yim, W.W.-S. and Nau, P.S., Mineralogical assessment of a beach sediment concentrate from Pak Nai, Castle Peak, *Annals Geog. Geol. and Archaeo. Soc.*, The University of Hong Kong, 9, 5, 1981.

110. Ringis, J., personal communication, 1983.

111. Hosking, K.F.G., Problems associated with the application of geochemical methods of exploration in Cornwall, England, in *Geochemical Exploration*, Boyle, R.W. and McGerrigle, J.I. (Eds.), Canadian Institution of Mining & Metallurgy, 11, 176, 1971.

112. Thomas, W., Losses in the treatment of Cornish tin ores, *Trans. Corn. Inst. Min. Mech. and Metall. Engrs.*, 1, 56, 1913.

113. Taylor, J.T.M., Tin dredging off the coast of Cornwall, *Proc. 9th Commonwealth Min. Metall. Cong.*, Jones, M.J. (Ed.), London 1969, Institution of Mining & Metallurgy, 1, 1970.

114. Taylor Smith, D., *A Sparker Survey of St. Ives Bay, Cornwall*, unpublished Rep. Coastal Prospecting Ltd., 1962.

115. Atkinson, K., personal communication, 1972.
116. Lee, G.S., Prospecting for tin in the sands of St. Ives Bay, Cornwall, *Trans. Inst. Min. and Metall.*, 77, 49, 1968.
117. Penhale, J. and Hollick, C.T., Beneficiation testing of the St. Ives Bay, Cornwall, tin sands, *Trans. Inst. Min. and Metall.*, 77, A65, 1968.
118. Yim, W.W.-S., Geochemical exploration for tin placers in St. Ives Bay, Cornwall, *Mar. Mining*, 2, 59, 1979.
119. Horsfield, B. and Bennet Stone, P., *The Great Ocean Business*, Hodder & Stoughton, London, 1972.
120. Edwards, E. and Atkinson, K., *Ore Deposit Geology and Its Influence on Mineral Exploration*, Chapman and Hall, London, 1986.
121. Tooms, J.S., Taylor Smith, D., Nichol, I., Ong, P., and Wheildon, J., Geochemical and geophysical mineral exploration experiments in Mount's Bay, Cornwall, in *Submarine Geology and Geophysics*, Whittard, W.F. and Bradshaw, R. (Eds.), Colston Papers, 363, 1965.
122. Yim, W.W.-S., Geochemical exploration for offshore tin deposits in Cornwall, *Proc. 11th Commonwealth Min. and Metall. Cong.*, Jones, M.J. (Ed.), Hong Kong 1978, Institution of Mining & Metallurgy, 67, 1979.
123. Allen, J.R.L., *Physical Processes of Sedimentation*, George Allen and Unwin, London, 1970.
124. Lampietti, F.J., Davies, W., and Young, D.J., Prospecting for tin off Tasmania, *Min. Mag.*, 119, 160, 1968.
125. Ocean Mining AG, *TOE–JV Summary Ringarooma Bay, Tasmania*, unpublished report, 1969.
126. Everard, G., Examination of samples from Ocean Mining AG, *Tech. Rep. Tasmania Department of Mines*, 11, 127, 1966.
127. Jones, H.A. and Davies, P.J., Preliminary studies of offshore placer deposits, *Mar. Geol.*, 30, 243, 1979.
128. Hellyer Mining and Exploration pty. Ltd., *EL 42/80 Ringarooma Bay Annual Report 1981*, unpublished report, Tasmania Department of Mines, 1982.
129. Hellyer Mining and Exploration pty. Ltd., *EL 42/80 Ringarooma Bay Quarterly Exploration Progress Report for Period Ending 20 March, 1983*, unpublished report, Tasmania Department of Mines, 1983.
130. Ollier, C.D., *Ancient Landforms*, Belhaven Press, London, 1991.
131. McGuinness, W.T., personal communication, 1972.

4 Marine Placer Gold, with Particular Reference to Nome, Alaska

Richard H. T. Garnett

ABSTRACT

The gold first mined in 1900 on the present-day beach of Nome, Alaska, owed its existence to a combination of factors. Nearby, primary gold deposits had been eroded by glaciation, and the products were redeposited as terminal and side moraines. The glacial debris and contained, particulate gold were subjected to repeated faulting along a climatically exposed shoreline, with changing sea levels. Gold collected on submerged, buried paleobeaches and abrasion platforms cut into, and around, the moraines above bedrock. It was also concentrated in more widespread, thin, lag gravels. Glacial lithologies, mostly till, with locally high gold grades, now extend on the seabed for nearly 5 km offshore in water depths of 20 m and less.

Since its discovery, several imaginative, but short-lived attempts have succeeded in recovering a little gold from the seabed deposit. In 1986 a U.S. company, WestGold, imported a 33-ft³ bucket-ladder dredge, the world's largest, to mine reserves estimated from earlier drilling by other companies. The Bima had previously been operated as a cassiterite producer in very different, Indonesian waters for which it had been designed.

In Alaska, the sub-Arctic climate restricted the mining season to six months, and the operating conditions were extremely difficult for a conventional floating dredge. Very hard digging, combined with a long swell, restricted the excavation rate to less than half that anticipated. Environmental constraints and permitting procedures affected the project. High unit costs and the inability to extract ore selectively added to the problems. A decreasing gold price required a higher cutoff grade, which seriously fragmented the reserves. Time and financial constraints prevented additional drillhole sampling at the necessary density, and serious mining dilution was unavoidable. Recovery averaged 106% of that expected from reserve estimation, but varied widely. Maintenance of a complete metal balance from seabed to final production was essential for performance monitoring and correction, reserve reconciliation, and production planning.

In 1989 full-scale, on-site trials demonstrated the economic applicability of a track-mounted, highly selective mining system deployed on the seabed from an anchored barge, the beach, or sea-ice. With no dilution, very high recovered grades were attainable. Only the addition of a bucket-wheel was required to achieve the necessary digging rate.

The operation was terminated in 1990 and the leases surrendered. From 1987 to 1990 inclusive, the Bima's production totaled 118,078 fine oz. (3,673 kg), with an average recovered grade of 824 mg/m³. There has been no further offshore mining activity at Nome. Exploration for marine, placer gold off the coasts of other countries has not succeeded in identifying any economic deposits.

4.1 MARINE GOLD DEPOSITS

The simplest form of offshore gold concentration is in a buried fluvial channel which is usually covered by recent, transgressive, marine sediments. The highest gold grades invariably are at the base of the alluvial sequence against bedrock. Some upgrading may be evident at the paleoriver mouth, depending on the relative marine and riverine energy levels. Examples exist in the Philippines and other areas where tropical and subtropical weathering has eroded coastal, primary gold deposits.

In regions of arctic climate, more widespread gold-bearing source rocks were subjected in the Quaternary to massive weathering and grain reduction by glaciation. The resulting detrital sediments containing liberated gold were transported by high-energy glacial and fluvial processes to be distributed over relatively large regions. On any erosive coastline, the particulate gold was concentrated by marine and other agencies. At times of prolonged sea level stability, stillstand features were superimposed: beaches, erosional platforms, lag gravels, sand spits, mud-filled lagoons, and surficial, dendritic, gravel-filled channels. Any earlier, fluvial concentrations were distorted, destroyed, or buried. Secondary marine gold deposits of complex origin were formed, typified by sites in Alaska (the best documented), southern Chile and Argentina, New Zealand, and elsewhere.

Gold-bearing moraines and assemblages of glacial sediments subjected to coastal erosion and marine transport may be mobilized and reconstituted into a wide variety of mineralized geological features. Economic concentrations of gold, however, are rarely situated more than a few kilometers from the site of the intermediate, morainal accumulation. Because of its high density, depending on its fineness of 15 to 19.3 g/cm^3, gold is less mobile than other heavy minerals of equal particle size. Only very fine particles travel far under marine action, collecting in sand spits and sand banks, which are usually economically insignificant on any large scale. Beaches and abrasion platforms may comprise well-sorted sand and auriferous gravel, with gold grade and particle size increasing with depth toward a base. During any regressive phase, localized fluvial channels of limited width and depth may develop on the surface of marine features and exposed glacial deposits seaward of a previously created beach.

Any sediment-hosted gold exposed on the seabed can be upgraded by the winnowing action of agitated sea water, a marine deflation effect, to create a lag deposit. The thickness of the lag and the degree of gold concentration depend in part on the water depth, past and present. The seabed should be, or have been, for a long period within the range of influence of storms, the underwater effects of which remove the finer sediment particles. Auriferous lag concentrations develop over a range of sediments but especially over diamict and glacial till, creating a surficial armor of cobbles and gravels.

4.2 NOME MARINE GOLD DEPOSIT

In the northwest of the State of Alaska the city of Nome is situated at latitude 64°30'N and longitude 165°30'W. Nome is 100 km south of the Arctic Circle on the southern coast of the Seward peninsula where Norton Sound becomes part of the Bering Sea. The present day, WNW–ESE trending coastline on which Nome is sited forms the southern boundary of a 6-km-wide coastal plain at the foot of glaciated hills. A unique marine gold deposit is situated immediately offshore. It displays all the important features of offshore gold deposition and concentration and was worked commercially in the 1980s by a U.S. company, Western Gold Exploration and Mining Company, Limited Partnership (WestGold).

4.2.1 GEOLOGY

Various aspects of the Nome placer deposits, on and offshore, have been described by Nelson and Hopkins' and by Kaufman and Hopkins.[1,2] The detailed gold distribution within the marine sediments was also studied intensively by the staff of WestGold, benefiting from all existing exploration records.[3] All writers attribute the spatial distribution of gold in the sediments covering the coastal plain and offshore to glacial events combined with marine transgression and regression. During Pleistocene glaciation of the region, changing sea levels caused the Bering Sea to become intermittently a coastal plain joining the North American and Asian continents. Primary gold was released from hard rock sources by extreme frost action to be deposited in downslope colluvials and in fluvial channels which were incised into the coastal plain.[4] These preglacial deposits were later swept up, together with eroded bedrock, by advancing glaciers.

Major glacial effects included the erosion of what now remain as 350- to 400-m-high hills of calcareous metasediments of Cambrian to Ordovician age. The remnant heights of Anvil Mountain and of Newton Peak are situated within 7 to 12 km north of Nome, as shown in Figure 4.1. They represent the original source of most of the local placer gold. Additional gold from further afield was glacially transported with ice and detritus down the lower Snake River valley, as indicated by the arrows, onto and across the coastal plain. Smaller lobes of ice flowed from the southern faces of the nearby peaks. All coalesced into a piedmont ice field, the seaward flow of which was influenced by regional horst and graben structures, which have been identified by seismic surveys. Gold-bearing debris was bulldozed to at least 2 or 3 km beyond and to the south of the present shoreline to remain as glacial moraines and outwash deposits.[1] Further deformation by bedrock faulting and overthrusting caused by advancing ice in places created confused accumulations of debris.

Later, as the glaciers receded and the sea level rose, the morainal topography was moulded, leveled, and partially eroded by the coastal waters. The high storm energy, especially in the surf zone and shallow waters, reworked the component sediments. Marine invasion alternating with retreat of the sea acted on the exposed glacial till and drift, leaving relict, thin lag layers on the seafloor.[1] Strandline deposits and related abrasion platforms of varying maturity were created, and isostatic rebound resulted in a succession of raised beachlines up to 5 km inland. Some formed at lower elevations and are now offshore, having survived marine encroachment and rising sea level to different extents. In places they are now represented by discontinuous features separated by overlying, younger, fluvial channels.

The location and directional trend of the straight coastline at Nome was influenced by a similarly trending network of reactivated, arcuate faults of variable attitude. Some are shown in Figure 4.1. Metasedimentary basement rocks are overlain by a thick sequence of Pliocene to early Pleistocene marine facies. The glacial debris, where not reposing on the basement, lie on, and intrude into, the sequence. The result is a complex assemblage of glacial, fluvial, and marine sediments over an arcuate zone convex toward the sea. Figure 4.2 illustrates a typical north–south section through the zone, which extends along the coastline for about 15 km between the present mouths of the Penny and Snake Rivers. Its limits are defined by faults which bring basement rocks to the surface where they are discontinuously concealed by a veneer of sand and gravel.[3] The zone's center, referred to by WestGold as the Central Core, represents the middle front of the flattened, at one time topographically high, terminal moraine. A similar, but subsidiary, feature is the West Flank. Both are surrounded by the Marine Apron, which is itself enclosed by muds of the Marine Basin.

The different lithologies within the zone generally vary according to their distance from a related ice front and to their degree of elevational coincidence with a stillstand horizon.[5] Within the Central Core, the sediments comprise mostly till or diamict together with overlying and interbedded sand, gravel, and mud. Irregularly distributed lag gravels cover and surround areas of slightly anomalous seabed relief. Subtle bathymetric variations reflect underlying accumulations of glacial debris and superimposed stillstands. The entire sequence in places is repeated at depth. An example is provided in Figure 4.3.

Earlier marine transgressions are separated by the depositional products of glacial activity and stillstands. Seaward of the Central Core, the Marine Apron is a complex sequence of reworked material overlying marine muds. High energy sand and gravel facies form an apron of clastic sediments overlapping the edges of the Central Core and West Flank.[3] Narrow, dendritic, shallow

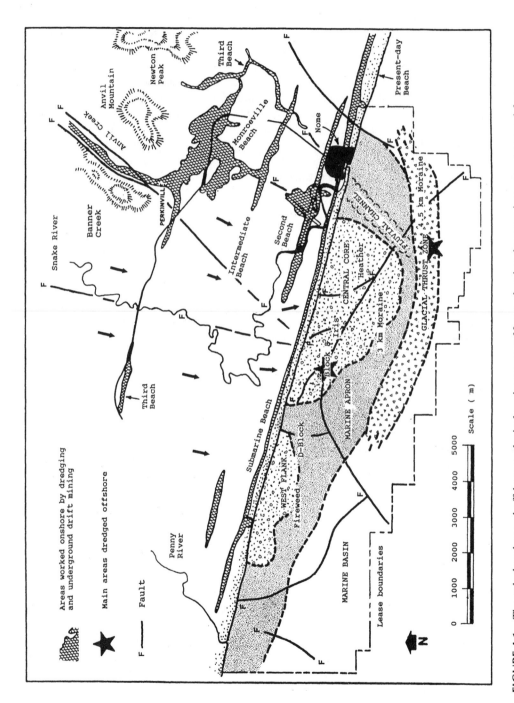

FIGURE 4.1   The paleobeaches and offshore geological environment at Nome, showing the more important structural features.

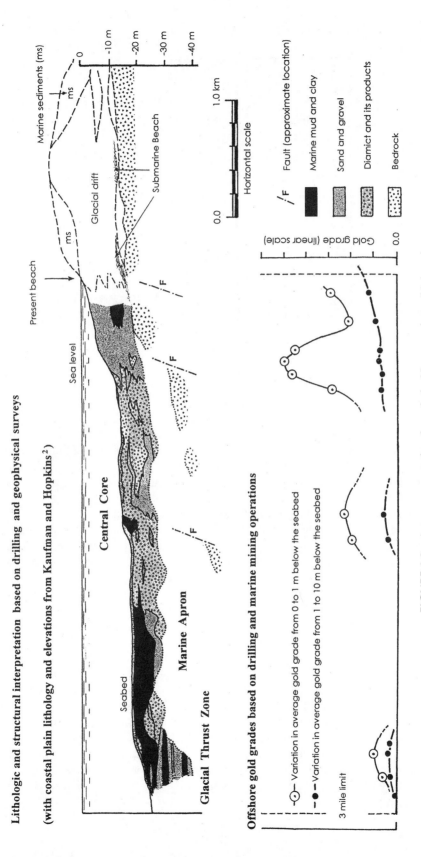

FIGURE 4.2 A transverse section through the Nome marine gold deposit.

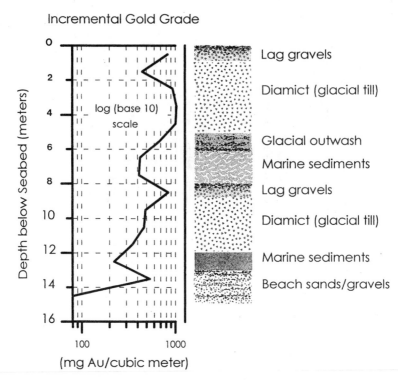

FIGURE 4.3 A typical grade and lithologic profile through the Central Core, offshore Nome.

channels remain from downcutting proglacial streams, originally flowing on exposed glacial sediments. The Glacial Thrust Zone is a mound of marine sediments subparallel to the Central Core, which resulted from muds being pushed and sheared ahead of the advancing ice mass. It is draped with sediments deposited either by a floating ice sheet with marine erratics or as proglacial outwash eroded and transported from the Central Core. A ubiquitous, thin covering of rippled sand is shown by side scan sonar surveys to be mobile under the influence of storms.

In places, relict alluvial channels and strandlines have survived the subsequent transgressive cycles. Some submerged beach lines have been identified up to 16 km offshore, but none of the drowned beaches is as well defined as the present day one. Onshore, paleobeaches, shown in Figure 4.1, exist between msl and +60 m. They have been major sources of mined gold and include the Third, Monroeville, Intermediate, and Second Beaches.[1,6] At an elevation of −10 to −12 m, the buried Submarine Beach has been mined onshore at up to 500 m inland of the existing coastline. It incorporates a broad, erratic abrasion platform extending below the present-day beach. This dual paleo feature extends offshore at the same elevation, approximately −15 m, as the preserved, buried, paleomouth of the Snake River, which lies on bedrock a little seaward of the present coastline.[4] Submarine Beach's main pay streak is up to 1 m thick with silt, sand, clay, and gravel containing gold, pyrite, garnet, magnetite, and ilmenite. Part of the Central Core includes Block 8 where fault-defined mounds of mixed sediments, hundreds of meters in lateral extent were repeatedly downthrown on the south and southwest and exposed to marine action. Figure 4.4 illustrates the importance of faulting in localizing the resulting paleobeach concentrations of gold. Such activity occurred during periods of varying sea level but predominantly at or below the same elevation as Submarine Beach. A discontinuous abrasion platform at the same horizon exists elsewhere in the Central Core and Marine Basin, with evidence of an irregular paleocoastline in plan.

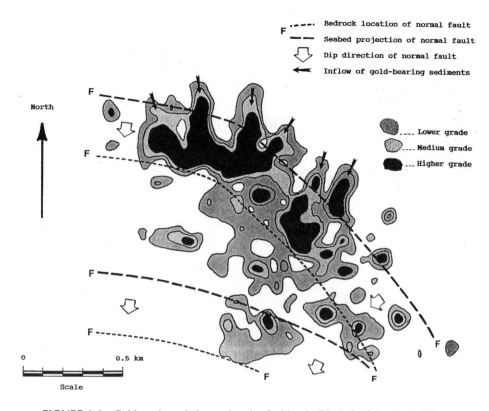

FIGURE 4.4   Gold grade variation and major faulting in Block 8 of the Central Core.

The offshore, unconsolidated sediments are very changeable over a few or tens of meters laterally and over less than a meter vertically. They are classifiable by generic terms such as glacial till, lag gravels, outwash, lagoonal mud, and beach sand, etc. Without genetic implications, the till may be termed diamict, with a qualifying description of the extent to which it has been subjected to washing and disintegration by exposure, usually on the seabed. Alternatively, their description may be based on quantified mixtures of their component silt, clay, sand, gravels, and cobbles, or in terms of their complete, disaggregated particle size frequency distributions.[7]

A lag gravel sequence within the Central Core usually comprises an upper surface armor of relict pebbles and semirounded cobbles, from the interstices of which small, lighter particles have been winnowed. Below, a layer of loose silt, sand, and some mixed gravel possesses an increasing silt content with greater depth and may rest on a clay horizon. This false bedrock is frequently the upper, slightly softened, surface of an underlying diamict or glacial till in which cobbles are embedded as clasts. Elsewhere, the lag sequence comprises other lithological components of a glacial moraine and its surrounding host sediments. The whole vertical section may be enclosed well within a 1-m interval and possesses a highly variable lithological content depending on the underlying source rock: silt and clay, 5 to 50%; sand, 30 to 75%; gravel, 5 to 30%; and cobbles and boulders, 0 to 30%.

To a depth of about 3 m below the seabed, 5 to 10% and 1 to 2% of the sediments exceed 100 and 300 mm in size, respectively. The clasts are semiangular, often slab-shaped, and siliceous. In places, boulders are common, up to 10%, in distinct fields. Erratic, rounded, granitic boulders averaging 1 m in diameter are encountered. Channels in outwash gravel carry less silt, approximately 10%, and sand with fairly well-sorted gravel, typically 40% exceeding about 20 mm in size and none exceeding about 150 mm. Beach sand and sand spit deposits consist of 90% extremely well-sorted silty sand with some 65% of the sediment being 0.106 to 0.212 mm. Silt represents about

10 to 15%. Some beaches display larger cobbles and boulders where they are incised into the Central Core and West Flank. Clay-filled glacial kettles and lagoonal mud accumulations have been preserved. In shallow waters all the surficial seafloor sediments exhibit depletion of particles of less than 0.125 mm in size.

4.2.2   GOLD DISTRIBUTION

The offshore area of anomalous gold mineralization is oriented parallel to the coast over an east–west distance of about 25 km. It extends from the foreshore to, in places, slightly beyond the 3-mile (4.8-km) State territorial limit into federal waters. The distribution reflects the footprint of glacial deposition with the surrounding mantle of more widely dispersed gold which has been redistributed off the moraines. Gold exists throughout at a low background grade of tens of mg/m³, with localized concentrations in almost every feature and environment except the mobile, surficial sand. The highest gold grades and coarsest gold particles are in sediments which lie between and on lobes of glacial material nearshore. Because of the complex geological history, the grade is considerably more erratic than those of most alluvial gold placers worldwide.

The highest and most consistent grades of several g/m³ occur within paleobeach areas in the Central Core, especially at an elevation similar to that of Submarine Beach. Displaying the coarsest gold particles of up to about 15 mm in size, the metal exists in buried back beach environments, typified by Block 8, and areas with coarse sediments up to boulder size. Channels in paleosurfaces of till supplied the beach, which also captured gold eroded and transported from a preexisting shoreline at a higher elevation inland. Figure 4.4 illustrates the importance of point sources of gold feeding beaches rejuvenated by repeated faulting.

Parts of the Central Core appear to have been specific sites of gold concentration maintained through a prolonged period of rising sea level. The result is a mineralized vertical section starting generally some distance above bedrock at about 30 m below msl. With gold grades increasing upward, the succession terminates at the seabed lag gravel. In Block 8, for example, with an average water depth of about 9 m, the following, typical peak grades and elevations may each represent a stillstand: 820 mg/m³ at 0 to 1 m below the seabed; 1120 mg/m³ at 3 to 4 m; 980 mg/m³ at 8 to 9 m; and 670 mg/m³ at 13 to 14 m. A similar profile is illustrated by Figure 4.3. In water depths of less than 7 m close to the present beach, other discrete areas have been estimated to contain 66,100 oz. of gold in the first 1 m below seabed, an additional 25,100 oz. down to 4 m, and a further 19,400 oz. down to 17 m.

Large areas of thin, auriferous, lag gravels overlying till on the Central Core in places grade laterally and vertically into concentrations on an old abrasion platform elongated subparallel to the present beach. Locally, the grade of the lag gravels may be very high, up to a few g/m³, but it decreases exponentially with depth and is variable laterally.[8] The highest grades generally occur in the upper 1 m of the sediments, especially in the top 300 to 500 mm, and at the basal interface with the false bedrock. They are most prevalent near the line of the paleobeach through Block 8 and the Heather and Iris Blocks shown in Figure 4.1. The West Flank includes the Fireweed area, and the 3-km Moraine lag gravel deposit lies further offshore.[3] The gold grade of the lag gravels declines with increasing distance offshore, as shown in Figure 4.2.

Within both the Central Core and the Marine Apron, gold preferentially occurs in a network of surficial, shallow, fluvial channels, 10 to 30 m in width and formed within proglacial outwash fans. These dendritic features are in places repeated at depth, reflecting the recurrent, ice-borne delivery and exposure of glacial debris. Their reworking into overlapping beach concentrations and superimposed lag gravels results in gold grades oscillating between zero and 5 g/m³ over distances of a few meters.[8] In some places, higher grade concentrations cross underlying lithological bound-aries.

Beneath the later glacial sediments and their derivatives, the paleocourse of the Snake River is identifiable offshore as the Fluvial Channel in Figure 4.1. Typical of an alluvial setting, the contained gold is concentrated with the coarsest gravels, at the base of the channel with the coarsest gravels. Beyond the Marine Apron the bathymetrically anomalous Glacial Thrust Zone comprises marine mud, sand, and some gravels capped by a lag gravel which is very thin but locally high grade. The zone includes three features, one of which is the 4.5 km Moraine.[3] There the grade typically declines from 2 to 0.5 g/m^3 at a depth of 0.5 m to a background of 20 to 30 mg/m^3 at 1.0 m below the seabed.

The surrounding Marine Basin consists of extensive marine muds with a very thin, fine sediment veneer winnowed by storm action off the Central Core and West Flank. Sea floor features, such as beaches, sand spits, and lagoons with their component mud, sand, and small pebbles, are identifiable by geophysical and bathymetric surveys in the Basin. In an aureole confined to within a few kilometers of the Central Core and West Flank, all such features contain anomalous, but subeconomic, amounts of finely sized gold.[3] Profitably recoverable gold is confined to areas which overlie the zone of glacial invasion.

The gold grade population, determined from drillhole samples, is approximately log-normal. It is complicated by the frequent superimposition of two or more populations, reflecting the multiplicity of effects which have concentrated gold in the area. The overall coefficient of variation is high, typically approaching 2.5, but varying from 1.5 to 3.5 for individual populations. The average gold particle size varies sympathetically with the grade. Its modal size is 425 to 300 microns, with the majority in a range of about 10 mm down to 75 microns. The average fineness is 93% for gold particles averaging 1.70 mm in size and 90% for those smaller than 150 microns. The sphericity varies with the host lithology, in places decreasing sympathetically with sediment particle size. It partially influences the gold's physical mobility. Garnet and pyrite are abundant in places as accessory minerals, especially on nearshore abrasion platforms where their concentration is indicative of the distance from a paleo backbeach. Magnetite and ilmenite constantly accompany the gold.

4.2.3 MINING HISTORY

Gold was discovered in 1900 on the present-day Nome beach, leading eventually to a major onshore mining industry based on bucket-ladder dredging.[6] The Alaska Gold Company dredged for many decades with half a dozen such units. Two dredges, with 9 ft^3 (255 l)-capacity buckets, were used until the early 1990s, one finally working just inshore on the Submarine Beach and the other a few kilometers inland on an alluvial channel. The total recorded onshore placer gold production from the immediate vicinity of Nome is over 5 million oz. In 1960 enaction of the U.S. Offshore Minerals Leasing Act allowed securement of title to the offshore for gold exploration and production. Almost all the Nome marine deposit is in Alaskan state waters, with only a small part of the Glacial Thrust Zone extending into federal U.S. waters beyond the 3-mile (4.83 km) limit. No federal leases have been awarded.

Six state leases, totaling 8722 ha, were acquired from Power Resources Corporation in 1985 by Inspiration Gold Inc. They were later transferred to WestGold, a company 50% owned by an associate of Minorco, with connections to the Anglo American Corporation Group of companies. From 1986 to 1990 inclusive, WestGold produced over 120,000 oz. of gold from the offshore, using a large dredge which was subsequently scrapped. The venture represents a unique and significant example of marine gold exploration and mining, which previously had succeeded only in the Philippines and has not been attempted anywhere else since.

4.2.4 OPERATING CONDITIONS

Nome's climate is typical of that prevailing in the Bering Sea. It is dominated by the wind circulation around the low pressure area lying over the Gulf of Alaska and a high pressure area over the Arctic Ocean. Winds blow from the east or northeast throughout the winter and from the west during most of the summer season. Air temperatures range from –41°C (excluding the chill factor) to 30°C.

The precipitation averages about 400 mm/year, half as snowfall, the rest as light rain. Sea water temperatures fall from around 17°C in early August to –1°C by the first week of November, when the average daily air temperature has dropped to below –7°C.

In June and July the almost constant wind averages 15 to 32 km/h, increasing to 15 to 40 km/h through the summer. It exceeds 16 km/h for 90% of the time. Wind initiates an average wave height of 2.0 m or more 15% of the time, lasting at least 12 h.[9] Storms centered well outside Norton Sound create a long period swell averaging 6 to 8 seconds and reaching 3.6 m in height during the latter half of the summer. Simultaneously, normal wind activity adds another 1.2 m to the wave height. Swells become exaggerated as they enter shallower water, and a large vessel may experience waves riding across its 3-m-high deck. In the average summer, Nome usually suffers the effects of three storms, defined as winds of 55 km/h or more blowing for longer than 6 hours, combined with wave heights exceeding 4.3 m. The 50- to 100-year storm, which may occur any year, produces 5.5 m waves combined with a prolonged 108-km/h wind.

Sea water depths increase steadily from the low water mark to become 17 to 20 m at the 3-mile (4.8 km) limit, and more slowly thereafter to a maximum of 27 m. The diurnal tidal range is about 0.5 m. All marine activities are determined by the presence of sea-ice, the arrival of which is indicated by the rate of fall of the water temperature. Ice develops on the open sea in early November and becomes shorefast in mid-to-late January. Overnight, a 100 mm or more thick cover of ice may suddenly form and render a vessel immobile. By March it is more than 1 m thick over 80 to 90% of the sea, accumulating locally due to ice rafting and pressure ridges. Thawing ice floes disappear in late May or early June.

4.2.5   EXPLORATION

Prior to WestGold's arrival off Nome in 1986, some offshore geophysical surveys and three drilling programs had already been completed.[10,11] During the winters of 1964 and 1969, sampling had been undertaken through the sea-ice using a Becker drill: 568 and 500 holes by Shell Oil Company and American Smelting and Refining Company (Asarco), respectively. These programs had been assisted by two marine seismic surveys by Shell in 1962 and 1966, with lines both parallel and perpendicular to the coastline and up to 6.5 km offshore.

In 1965 Ocean Mining AG collected 41 samples on the north shore of Norton Sound with a clamshell and airlift drill deployed from a 37-m-long supply vessel.[12] The maximum seabed penetration attained was 4.9 m, with an average of 2.0 m. The mean gold grade recorded was 45 mg/m³, with a maximum of 442 mg/m³. Only two samples revealed more than 260 mg/m³, each from an average penetration of 0.3 m. The program confirmed the existence of an offshore pale-ochannel at Bluff, 65 km east of Nome. It also successfully identified anomalous gold immediately offshore Nome and demonstrated the applicability of the clamshell technique to the practical reconnaissance of lag gravel deposits. Another sampling program, less successful in terms of the 56 completed drillholes, was undertaken by the U.S. Bureau of Mines in 1967, using Becker and Sonico drills. An additional 700 samples were collected by bottom grab samplers.

WestGold commenced dredging operations in 1986. The changing management was soon made aware of the need to provide a far better definition of the offshore resources than hitherto available. Promotion of the resources to reserve category was necessary for production.[6,7,13] The known, higher grade ground was either too close inshore to be safely accessible to a dredge or had not been sufficiently defined by Shell's and Asarco's work to allow the formulation of a mining plan. With considerable urgency, a drilling program was commenced off the ice but, for safety reasons, no further than about 1.6 km from the coastline.[12] The work was concentrated successfully on Block 8 in the Central Core. A Becker drill was used because it represented the most suitable system for sampling off the ice. The choice of machine would, it was hoped, also render the results compatible with those of the earlier Shell and Asarco programs.

During the summers of 1986 and 1987, 3400 line km of high-resolution geophysical data were acquired offshore Nome.[14] Seismic data were interpreted to provide facies interfaces and thicknesses, allowing faulting to be identified and profiles to be drawn. Simultaneous side scan sonar surveys, with a 3-mm penetration, revealed the seafloor's surficial character. All interpretations were aided by detailed bathymetric records to recognize the different offshore zones and their component geological features, which possessed a slight topographic signature. Subsequent West-Gold exploration programs were guided by the results, and drilling was used almost exclusively for reconnaissance and evaluation sampling.

WestGold's drillholes provided samples which, though physically disaggregated, revealed the character of the clastic sediments penetrated. Lithological records were later compared with similar observations as the ground was mined. All fully disintegrated samples were described in terms of standard classifications based on particle size. Drillholes yielded data at 1-m vertical intervals and at a horizontal frequency which depended on the drillhole spacing. The particle size distribution of gold and its fineness were recorded and sedimentological studies were undertaken. During subsequent production, systematic sampling of the dredge feed and of plant flowsheet products yielded metallurgical recoveries and metal balances. The final gold output was reconciled against that expected from exploration, and some of the experience gained led to changes in subsequent exploration procedures.[15]

4.2.6 SAMPLING SYSTEMS

The Sonico drill used off Nome by the U.S. Bureau of Mines in 1967 was a mechanically oscillated drill pipe with an induced, longitudinal vibration of 50 to 100 cycles/s. It possessed a double-walled casing of 52 and 102 mm inner and outer diameters, respectively, for reverse circulation sample retrieval. The average sediment penetration was 31.5 m/hole at a drilling rate of 0.67 m/min. The diameter (and consequent small sample size) was a disadvantage for evaluation purposes, and the drill was easily stopped by cobbles in consolidated ground.[16]

The machine used by WestGold and some of its predecessors, the Becker AP-1000 hammer drill, is a heavy-duty, percussive, reverse circulation unit designed for drilling in alluvial and glacial terrains.[12] The double-walled drill pipe is driven into the seafloor by the impact of a 2.5 t, Link Belt Model 180, double-acting, diesel, pile-driving hammer. It delivers 11 kJ of energy at each impact with a frequency of approximately 90 blows/min. As the pipe enters the seabed, compressed air and water are forced down the annulus. While the pipe is driven ahead the penetrated sediments are lifted up as a slurry which is continuously discharged into a cyclone to vent the air and reduce the velocity. The pipe, of 83-mm inside and 140-mm outside diameter, is coupled in 3-m lengths. A standard 8-tooth, crowd-out drill bit yielded on average about 70% of the theoretical volume from a drive interval of 1 m.[12]

All the Shell and Asarco drilling was conducted off the shorefast winter sea-ice, and rarely more than 1.5 km from the shore or in water deeper than about 9 m. Most was in less than 6 m of water. Shell's early 1964 program was very successful.[11,17] A total of 7436 m of sampled drill penetration was obtained at averages of 5.3 holes/d, 68.2 m per 12 h shift, and 13.1 m/hole. The total labor force numbered 23, and lost time amounted to 24%. The sample volume obtained from drilling was either near to or greatly in excess of the theoretical: 80% of the holes yielded 78.5% of the theoretical volume, while the balance provided an average of 235% volume recovery. Amalgamation confirmed 84% of the gold weight estimated by color counting. The average sample grade was 205 mg/m^3. Three years later, Asarco drilled 500 equally successful Becker holes off the winter sea-ice in areas with an average water depth of 4.3 m.

For winter use, WestGold removed the Becker drill's nodwell, a Terra-Flex track assembly and wheels.[12] The remainder of the machine weighed 41 t and was mounted on a sled 9.1 m long and 4.0 m wide. The total weight distribution was 1104 kg/m^2 with a safe operating limit of 1-m

FIGURE 4.5   The drilling vessel *Krystal Sea* with the Becker drill deployed on the deck.

thickness of flexible sea-ice, allowing no more than 60 min at each drill site. In 1987 the whole assembly, plus the operators, was exposed to the elements and was towed by a 13.5 t, RD-85 Rolagon, all-terrain vehicle fitted with balloon tires. Ninety-one holes were completed at 4.3/d. The average penetration of 7.85 m/hole revealed a mean grade of about 500 mg/m³ for all 1.0-m length samples. The following winter, a 61.3 t sled, 13.7 m long and 6.1 m wide, reduced the distributed weight by 25%. The whole was enclosed in a tent in which the air was heated to about −10°C and which could withstand winds of over 80 km/h on ice 1 to 1.5 m thick. With the improved configuration, 528 Becker drillholes were completed by WestGold in 35 days, equivalent to 15/d, to an average penetration depth of 12 m.

Twenty years earlier, in the summer of 1967, the Becker drilling had been conducted from the 62-m-long "Virginia City."[16] The drill was deployed on the seabed by means of a hydraulically powered jacking platform to allow the hammer and pipe mechanism to operate freely, unhampered by the effects of swell. Site selection was made by divers. For its summer drilling in 1987 and 1988, WestGold deployed the Becker drill from the *Krystal Sea*, a modified landing craft with a helicopter deck, shown in Figure 4.5.[12] The working deck was 9.76 m wide and 20 m long. Four 1.5 t anchors created a 25 ha area within which 10 to 12 holes could be drilled on 100-m centers. The vessel could weigh anchor and be established at a new location with a loss of only 2 h. Four UHF transponders along the coast provided position control, and holes were drilled in water depths of 4 to 25 m.

Drilling was possible for no more than six weeks during the summer. Lost time due to sea conditions was considerable. Depending on the period of the sea swell, drilling was conducted in wave heights of up to 1.2 m with no swell compensation. Later experience revealed that the movement of the vessel accentuated other effects and introduced serious errors into the results. The Becker drill was very efficient on heavy glacial ground, successfully penetrating the indurated seabed sediments. But in unconsolidated, and especially free-running, sediments, excessive excavation and coning may occur. These consequences resulted from the vacuum effect in the drill annulus as the water is first excavated from the pipe when entering the seafloor. The effect is exaggerated when drilling is conducted from a vessel in rough seas.[5]

Some unwanted correlation between sea state, expressed as wave height, and the estimated sample grade was apparent.[18] An area could appear to possess at least twice the actual grade if drilled in waves of 1.0 m or more. The effect was particularly serious on well-washed lag gravels, within which the gold in places attained its highest concentration in soft free-running sediments in the upper 0.3 m of the vertical sequence. Surficial coning caused a serious overestimation of grade, especially on the 4.5-km moraine. Some silts and muds were slurried by the aggressive drilling system and were lost in the final cycloning process. The sample volume recovered, but not the contained gold, was reduced, and relating the two measurements further inflated the apparent gold grade.

In the summers of 1987, 1988, and 1989, WestGold completed 548, 805, and 501 drillholes, respectively. Although the drilling rate never surpassed the best achieved on the ice, it consistently improved: 6.5 holes/d in 1987; 9.9/d in 1988; and 11.1/d in 1989. Five major drilling programs were completed by WestGold from 1987 to 1989 inclusive. The 2479 drillholes, added to those undertaken by Shell and Asarco, brought the total number offshore to 3547.[12]

As a first stage of its management of the Nome project, WestGold, for four weeks in late 1985, undertook a bulk sampling program. The objective was to check on the grade indicated by Shell's and Asarco's earlier drilling in part of the West Flank. Seabed sediments were dug by a 4.6 m^3 clamshell operated from a Manitowoc 4600 dragline on the leased, 109-m-long barge, *Kokohead*. Treatment was by trommel screening and jigging. The *Kokohead* operated for about five weeks and worked for 250 h, but for only about 70 h on a continuous basis. Three separate areas were sampled with the expectation of digging and treating between 3000 and 4000 m^3/d. Various technical problems prevented more than 3390 m^3 of sediment from being processed in total to produce 65 oz. of gold, an effective grade of 595 mg/m^3. A more successful, shallow water, bulk sampling program in the nearshore part of the Central Core was undertaken by backhoe through the mid-winter ice of 1988. This approach made available large quantities of sediments for geotechnical studies. The average grade of the large samples confirmed that estimated from preexisting drillholes, but because of their larger volume, the bulk samples showed less grade variability between individual sites.

4.2.7 Reserve Estimation

WestGold's sample treatment and grade estimation procedures for individual drillholes were refined versions of those developed and followed as standard for decades by the onshore, placer gold industry.[19-22] Drill samples were washed and panned by hand, with the gold content (and thus grade) estimated by counting gold particles (or colors) according to their size, with amalgamation as a check. Shell, Asarco, and WestGold employed the same proven techniques: screening, hand panning, color counting for each 1.0 m section, and confirmation by amalgamation and parting.[3] The few logistical changes which were made, such as mechanized screening, increased the efficiency and rate of treatment. Five size classifications of gold colors, plus an oversize, were counted, and the corrected results were used to construct a typical size distribution for individual areas. The accuracy of estimation was good. The amalgamated weight divided by the counted weight averaged 0.915, varying from 1.06 (6% underestimation by counting) in higher grade areas to 0.83 (20% overestimation) in lower grade areas.[18] The difference was partly accounted for by the flatter shape of smaller gold particles, which more frequently represented lower grades.

The greatest uncertainty involved in estimating the grade of individual samples resulted from WestGold's choice of volume to which the panned gold content was attributed. It was either the theoretical or the actual volume for the 1.0-m advance, depending on volume recovery. Conventional procedure is to accept the theoretical volume if the recovered volume is the smaller of the two, a process usually regarded as being conservative since it tends to underestimate the grade. In some offshore areas the difference was serious, resulting in an understatement of the true grade. For example, in parts of Block 8, grades determined conventionally were later revealed to be merely 62% of the true grade, and required increasing by a drilling factor of 1.59 to provide the correct figure.[23] In another area, if actual volumes were used to the exclusion of theoretical ones, the gold content of the increased volume of ore reserves rose by 104%.

The problem of poor sample support resulting from a drill pipe with an internal diameter of less than about 150 mm is only partly overcome by a higher drilling density.[19] Early industrial practice in the onshore Nome area involved drilling at a preferred spacing of less than 40 ft. (12 m) and never more than 100 ft. (30 m) for dredging properties.[24] To determine the base of the reserves, churn drill samples were cut at vertical intervals of 1 ft. (305 mm) and supplemented by regular dredge bucket sampling during production.

Reconnaissance drilling undertaken by Shell and Asarco was spaced roughly 300 m east–west by 120 m north–south. The results served merely to indicate that a geological feature was mineralized where drilled. In 1985 the drilling logs were used to derive a global estimate of 61.3 Mm^3 by constructing polygons with a maximum radius of influence of 183 m. With a cutoff grade of 260 mg/m^3, the *in situ* grade was estimated to be 671 mg/m^3. The resource comprised, on average, a 5.0-m vertical thickness of sediments below 6.3 m of water, giving a mean digging limit below msl of 11.3 m. A commitment already had been made to the purchase and use of a very large bucket-ladder dredge. Practical dredging considerations, therefore, reduced the resource to 3.8 Mm^3 at 615 mg/m^3, containing 75,000 oz. of gold.

In-house reserve classifications and definitions, using the traditional, onshore, industry terms, were instituted for further exploration. Proven reserves were defined as being in those areas in which the drillhole spacing was low enough to allow short-term production planning.[25] Probable reserves were in areas in which additional drilling was required for their promotion into the proven category, perhaps resulting in significant change in volume and grade. Possible reserves comprised volumes in areas not yet drilled but for which geophysical and bathymetric data indicated sufficient geological similarities with those features already in the proven or mined category.

Probable reserves were delineated by drilling on a grid of 100 × 100 m or, rarely, 200 × 200 m. A spacing of 50 m (slightly more in Block 8) established proven reserves but was insufficiently dense to define the frequently sharp ore boundaries. Geostatistical and drilling simulation studies on Block 8 confirmed that the spacing ideally should not exceed 30 m for proven reserve estimation. Generally, to qualify for dredging, an offshore 200 × 200 m block should have contained no less than 50 drillholes. Other geologically complex areas required drillholes to be no more than 20 m apart for the same purpose. Elsewhere, reserve definition in the surficial drainages and lag gravels for very selective extraction by a suitable mining system needed a sample (not necessarily a drillhole) spacing of no more than 10 m.

However, the minimum drillhole separation achieved anywhere at Nome by WestGold was 50 m. In some exploited areas it was 100 m. The densities fell short of those necessary to delineate reserves with the necessary confidence. WestGold's inability, because of financial and time constraints, to improve the drillhole spacing contributed to the eventual termination in 1990 of the offshore mining operations. Inadequate densities resulted in an inability to identify the outlines of contiguous, payable ground. Dilution increased, through the unknowing inclusion of unpayable areas. Likewise, payable ground was excluded with only a suspicion of its existence. In some features the selection factor was adversely affected and by 1990 the recovered grade was, as a result, significantly less than expected.[23,26]

The first stage in the estimation of reserves and the formulation of a mining plan was the construction of an interpreted geological map. The predominant lithological characteristics of each of the five geological provinces were extracted from the drillhole logs. Interpretations of the sedimentology of the seabed sediments were displayed on a composite sediment map for varying depths and elevations. Within each province important geomorphological features, being areas of similar geological and sedimentological character, were delineated to determine areas potentially suitable for exploitation. The characteristics were considered in the development and interpretation of a variogram with defined parameters, and a geological model was constructed for each feature, showing its boundaries.[15] A composited grade and thickness were computed for each drillhole within, or in the relevant vicinity of, a selected area.[25] They were used to assign kriged grades and thicknesses. Polygons defining the separate reserve categories were superimposed on the geological

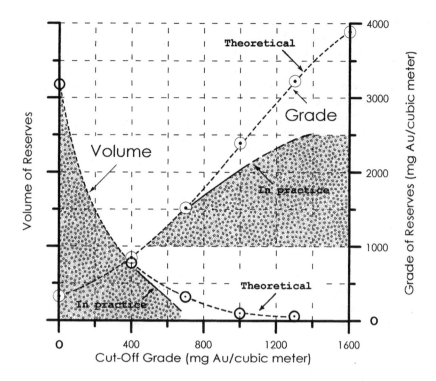

FIGURE 4.6 A grade-volume curve for the surficial sediments of the Nome marine deposit.

models. The boundaries were modified as necessary and were used to estimate the gold content and ore volume for each feature.

A range of cutoff grades, reflecting estimated costs, the gold price, and other factors, was used for planning purposes.[15,25] The effects of changes in cutoff grade were depicted on grade-volume curves of which Figure 4.6 is an example.[8] An increase in cutoff grade resulted in a theoretically higher reserve grade, accompanied by a reduction in the volume. In practice, however, the reserve areas became increasingly fragmented horizontally into many small, well-separated parts. Some became too small to justify exploitation except by a very selective mining method, further reducing the volume and causing the grade to be diluted. Areas identified as economically exploitable were divided into blocks for production planning. A block was required to encompass a sufficient volume of ground with an average estimated grade above the cutoff grade.[8] There was no standard block size, but minimum criteria were applied to the plan dimensions and orientation.

4.2.8 GEOTECHNICAL CHARACTERISTICS

Throughout all the WestGold drilling, records were maintained of blowcounts: the number of hammer blows struck to achieve each individual 1.0 m of penetration. Drilling was normally terminated at 10 m below the seabed or at a higher elevation if further advance was impossible at around 300 to 400 blows/m.[27] The greatest resistance to penetration was offered by the till in which full percussive power was used. The extent of throttle opening was not constant, however. It was decreased in very soft surface sediments in the Glacial Thrust Zone to obviate the danger of oversampling. Otherwise, a general correlation was possible, allowing estimation of the magnitude of "N," the penetration resistance, expressed in blows/ft.[28,29]

Some standard penetration tests were undertaken.[27] Representative samples were collected with a split spoon sampler. The penetration resistance indicated that the relative density of the in-place, unfrozen sediments is medium-dense to dense, with some very dense, especially in the Central

Core and Apron. The *in situ* relative density was quantified as 90 to 100%. The dry density of the *in situ* ground varies from about 1.45 to 2.35 g/cc, the highest figure being represented by till. The water content immediately below the seabed is 10 to 15%.

Triaxial tests were performed on samples and the soil shear strength was measured by means of isotrophic, consolidated, undrained tests. Silt and clay between the pebbles and cobbles of the lag gravel horizon within the first 0.5 m of seabed are generally classified as soft. As the sand and fine gravel content increases with depth the ground is firm; quickly thereafter it becomes firm to stiff. Below, it is stiff to very stiff, until by 5 m it averages very stiff to hard. It becomes consistently hard by 7.5 m, and very hard beyond 10 m. The offshore glacial sediments are overconsolidated, indicating that they were previously under greater stress than at present. The plasticity index is 3.3 for silt and 3.0 for silty sand in lag gravels. The drained angles are high: 39° for silty sand, and 46° for gravelly sand, indicating a dense packing. The allowable bearing capacity of the seafloor is about 50 kPa.

The depth-dependent cutting energy and achievable cutting rate from a unit with a power range of 300 to 500 kW in sand and gravel could be predicted from the relative density. The energy required to excavate 1 m^3 of sediment increased from 800 to 1250 kJ/m^3 as depth increased from 6 to 15 m, with a corresponding 35% reduction in cutting production measured in m^3/h. Till exposed on the seabed within 100 to 200 m of the shoreline appears to be loosened to a greater and deeper extent than elsewhere, perhaps due to winter ice gouging. None of the reserves outlined by WestGold bottomed on bedrock. Consideration of bedrock hardness and profile were therefore economically unimportant. Permafrost exists discontinuously at depth over large onshore areas immediately adjacent to the coastline. Its possible presence offshore below shallow waters, at depths not accessed by drilling, cannot be entirely discounted.

4.2.9 DREDGING

Small-scale, marine gold mining had been attempted in vain at Nome several times before West-Gold's arrival.[6] Simple, mechanized bucket-ladder dredging was tried briefly in very shallow water in 1900 until a storm wrecked the dredge a few days later. More attempts were made during the next two decades by launching machines from the low water mark. Most were more notable for their imaginative designs than for their accomplishments. The first successful marine mining activity was on the offshore extension of Daniel's Creek at Bluff. Auric Resources exploited the site, thought to be a composite of drowned colluvial, fluvial, and beach concentrations, during the winter months from 1939 to 1941 inclusive.[4] A spar set in the sea-ice allowed the gravels to be slushed to the beach for processing in the summer. Later Martin Dredging Inc. successfully recovered surficial gold from shallow waters off Nome with the Mermaid 1, a 12-inch (305-mm) hydraulic dredge. The company planned to deploy a 30-inch (762-mm) unit the following year and to use a seabed mounted system to mine under the ice in winter. These, and other imaginative plans, never fully materialized.[6]

A further, 26-hole, offshore drilling program at Bluff was completed by Aurora Mining Inc. in 1968. It revealed the existence of auriferous sediments, with some grades reported as high as 2 g/m^3. In 1976, Westfield Minerals Ltd. drilled 112 Becker holes up to 1.1 km offshore in the same paleochannel which reportedly exhibited karst structures in its limestone bedrock. Indicated reserves were estimated to be 1.05 Mm^3 at an average grade of 860 mg/m^3. Six years later, Phoenix Marine Exploration Inc. towed a cutter suction dredge to Bluff to restart offshore mining. Its suction head became entangled in old cables secured to the seabed about 100 m offshore and perhaps dating from 1939–1941.[4] A storm washed the dredge ashore. Subsequent efforts to dredge with the Teneal near Nome were equally short-lived.

A marine bucket-ladder dredge, which WestGold originally thought in 1985–6 would be the most suitable mining system for use offshore at Nome, is very similar to one designed for onshore gold dredging. But WestGold's Nome project could not withstand the prohibitively high cost of a

FIGURE 4.7 The Bima in unusually calm seas at Nome.

new unit. However, the 30 ft³ (850 l) bucket capacity Bima had been laid up in Singapore since mid-1985 with no plans for its further use. The dredge had been designed by Mining and Transport Engineering b.v., a subsidiary of IHC Holland n.v. Following its launching in 1979, the Bima was used by Billiton Marine to recover cassiterite from drowned fluvial channels in the coastal waters of Pulau Tujuh, Indonesia.[30] Local geological and sea conditions had dictated the design features. The dredge's pontoon was 110 m long by 30 m wide and 6.5 m deep. The tailings discharge launders overhung the stern by an additional 27 m. Its capacity made it the world's largest. With a maximum displacement of about 12,000 m³, the Bima's loaded draft of 3.7 m dictated that a normally safe operating water depth was 6 m, increasing to 7.5 m in rough seas. WestGold purchased the dredge and transported it to Alaska under dry tow.[7,13] Figure 4.7 shows it shortly after its arrival, and Figure 4.8 illustrates its deployment.

The Bima excavated sediments from the seabed and immediately treated them on a continuous basis at the same rate. Its diesel driven-power plant comprised five generating sets at 440 volts AC, an auxiliary set, and an emergency/harbor set. Some components were DC motor driven through thyristors from the main power system. A variable-speed, 1342 kW-drive moved an endless line of linked buckets around top and bottom tumblers mounted on the dredge ladder, which was suspended through a long well and below the pontoon. Each of the 137 rotating buckets dug into the seafloor and scooped up material which it conveyed upward on the dredge ladder, dumping its contents as it rotated on the upper tumbler. The sediment load was fed under gravity to the treatment plant mounted on the pontoon. The total 5667 kW of installed power had enabled the Bima to dig the Indonesian sediments at a maximum rate of 2300 m³/h.

The lowest elevation attained by the lips of the dredge buckets mounted on the bottom tumbler defines the digging depth.[13] This figure, less the water depth, determines the depth of cut below the seabed. The lateral extent of digging perpendicular to the direction of the dredge's advance is the face width. Bucket-ladder dredges, both on and offshore, rely for their economic success on very forceful digging power to achieve a high throughput, resulting in a low unit volume cost. A dredge should operate with the minimum downtime in easily dug ground to near full-depth capability on high volume, low grade reserves. It is very nonselective laterally in its manner of digging. Vertical selectivity decreases in shallow water with a low ladder angle, which causes the bucket line to hang in catenary. In less resistant sediments the lowest point of the catenary may be below

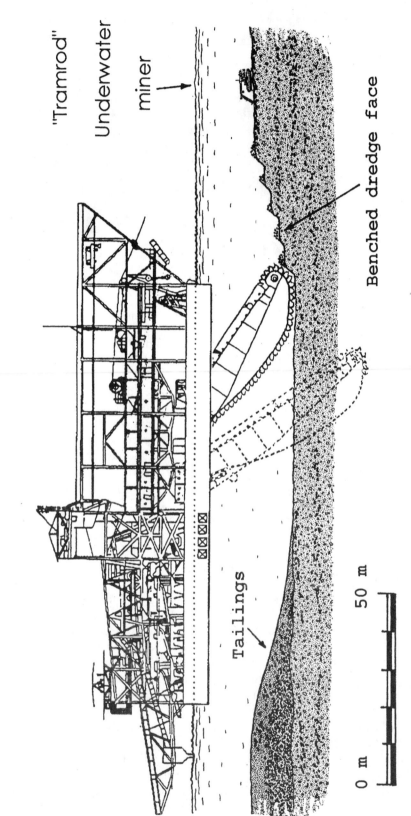

"Bima" bucket-ladder dredge

"Tramrod" Underwater miner

Benched dredge face

Tailings

0 m   50 m

FIGURE 4.8   The Bima's operational mode at Nome.

the level of the buckets on the underside of the bottom tumbler. The outcome can be overdigging and dilution.

The installed power enabled the Bima to dig in easy conditions at a rate of 30 to 40 buckets/min (bpm) at maximum power, or 0 to 30 bpm at maximum torque. Under suitable conditions, with a fully inclined ladder angle of 55° below horizontal, the dredge could attain an excavation depth of 45 m below msl with a digging face slope of 1:2. The 850 l capacity of each bucket theoretically could be achieved by digging at a depth of 26 m, defined by a ladder angle of 36.5° below the horizontal. Any lesser angle of the 88-m-long ladder reduced the capacity: at 20° it became about 500 l, giving a bucket fill factor of almost 60%. As the ladder was raised, the bucket line hung in catenary with its lowest point eventually below the elevation of the bottom tumbler. The designed, minimum, effective digging depth was advised to be 10 m below msl, with a cut depth of at least 4.5 m below the seabed to achieve any reasonable bucket fill.

WestGold had originally estimated that the Bima was capable of a sustained throughput rate of 23,000 m³/d at Nome, equivalent to 1150 m³/operating hour for, say, 20 h/d. The assumption involved a minimum 6.4 m depth of cut below the seabed in sediments expected, from the Shell and Asarco drilling, to be finely sized and averaging 98% less than 9.5 mm diameter. The dredge was expected to travel parallel to the coastline. Each anchor set up, of an anticipated 19 d duration, was estimated to allow dredging of an area measuring 182 by 365 m. The assumed recovered grade of between 510 and 725 mg/m³ suggested an annual yield of 70,000 to 90,000 oz. of gold. These incorrect assumptions rendered the forecasts illusory.

To facilitate dredging in swells, the Bima's design had incorporated a ladder buffering system to counter, within certain limits, the sea-induced natural motion of the pontoon. The limits were defined as a maximum movement of the bottom tumbler when in contact with the seabed of 1 m vertically and longitudinally or 0.4 m laterally, effectively created by a 2 m swell. The equipment was, however, inoperable and totally beyond the level of repair necessary for U.S. Coast Guard certification and essential insurance. If available it would have been of some assistance but, by itself, would not have affected the final outcome of the project.

With no motive power of its own, the Bima was maneuvered on-site by a spread of four 8 t side anchors and one 20 t headline anchor.[7,13] Operational command was effected from a control room on the starboard bow. Forward pressure and advancement was achieved by an 800-m anchored, 64-mm, headline, about which the dredge swung laterally. Sidelong traversing of the face was enabled by preferentially using the four 48-mm anchored sidelines: two forward, each of 250-m length, and two aft, each of 300-m length. A sixth stern anchor was available as necessary, intended to withstand sustained winds of up to 120 km/h. The dredge was positioned with the ladder, regarded as the bow, facing offshore into the incoming swells. At the start of dredging sediments were excavated along a newly established face. Tailings were discharged from the stern, on the shoreward side on unpayable ground some 120 to 140 m from the excavation at the bottom tumbler. Where the dredge had moved ahead by the same distance, the tailings fell into a mined area. At the end of the dredge course the most recent excavation of the same length remained empty of tailings. During dredging the swell factor averaged 12% but was as high as 20% for individual days. Figure 4.9 depicts, in plan, the Bima dredging a mining block, with a typical variation in daily face advance and accompanying recovered grade.

Depending on the expected grade, the minimum size of a block was one able to support dredging for a period of from several days, for a very high grade, to a few weeks. Nonqualifying areas owing to coastline proximity and environmental restrictions were rejected. The contained sediments were subjectively required to offer less than a specified resistance to digging such that a certain minimum throughput rate could be expected. Areas with known boulder fields, very high till contents, or to which any doubt might be attached concerning the recoverable grade, if marginal, were rejected or were awarded low priority. The water depth had to comply with or exceed that required for the defined safe dredging practice in the sea conditions expected.

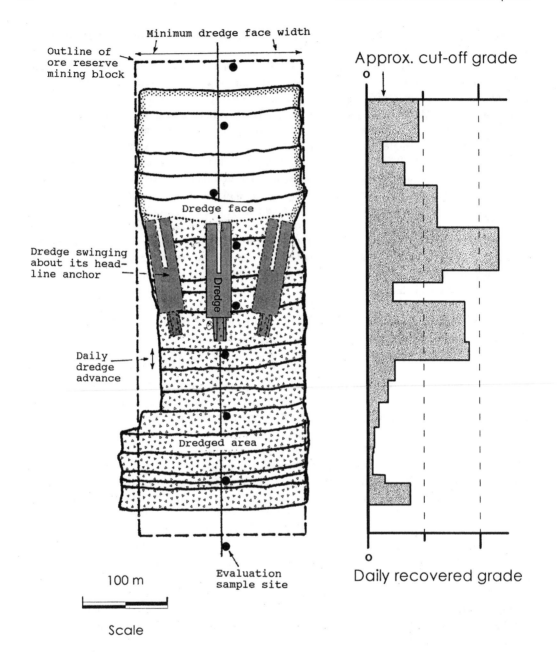

Minimum dredge face width

Outline of
ore reserve
mining block

Approx. cut-off grade

Dredge face

Dredge swinging
about its head-
line anchor

Dredge

Daily
dredge
advance

Dredged area

Evaluation
sample site

100 m

Scale

Daily recovered grade

FIGURE 4.9   The progressive dredging of a reserve block off Nome.

A number of mining blocks, preferably but not necessarily contiguous, were grouped to form a season's dredge course. With assumptions regarding the Bima's availability, the expected performance and production figures were estimated for each block: dredging time, volume throughput, recovered grade, and gold production. The relevant drillhole logs were again consulted to provide a forecast of sediment types to be excavated. The classifications of practical dredging importance were (a) silt, which created turbidity in the sea water, (b) till, which could be excavated at a low rate only with difficulty and could be passed unproductively through the trommels without its gold content being liberated, (c) clay, because it caused losses of gold from the trommels, and (d) cobbles and boulders. Oversize rocks resisted excavation and resulted in circuit blockages. They also caused severe mechanical damage and abrasion to the dredge and plant parts and resulted in shutdowns.

Continual dredging ahead caused the angles formed by the side lines relative to the line of advance to decrease continually. Usually, with every 200 m of forward movement, all the anchors had to be lifted and resited ahead. A complete anchor move could be achieved in a 12-h day shift in good wind/wave conditions. The work was undertaken by a 50-m-long, ocean-going tug, the *Aquamarine*, assisted in the hazardous shallow water operation by an anchor scow. During the very early part of the summer season, when for a short period a calm sea could occur, the Bima's attitude was reversed. With a very low ladder angle and the headline anchored on the beach, the dredge intruded ladder-first with extreme caution into very shallow waters. Extremely high-grade ground not in the reserves was recovered by this means from the northern parts of Block 8. Dredging was, in places, successfully and profitably attempted in water as shallow as 4.9 m.

When operating, the Bima's digging depth was dictated by the local water depth and the vertical, gold grade profile. Thus, the depth of cut included the payable ground with a minimum of barren dilution from below. Where the cut was several meters deep, the dredge advanced ahead for between 3.0 and 3.5 m at a time, with an incremental 0.5 m increase in digging depth. Where the removal of a single lag gravel horizon was attempted, the advance was about 6 m, taking a single 2-m cut with a lower bucket-fill factor. At the completion of the planned dredging of an area or block, the dredge could be maintained on-site, selectively and opportunistically working ahead into ground not included in the ore reserves. Alternatively, it could be towed to a new dredging site. The first course of action required constant onboard sampling to ensure that payable grades continued.

At the end of the dredging season the Bima was moored alongside a steel docking extension of the Nome stone jetty, after prior seafloor inspection by divers and the removal of any obstacles. Only smaller support and shallow-draft vessels could enter the constricted entrance to Nome's harbor. Throughout the winter the dredge was protected, together with the *Aquamarine,* from ice rafting behind a three-sided, snow and ice berm.[13] The berm was 60 m wide at its base, extended for 250 m parallel to the jetty and was 25 m high. It required two months to complete and involved the manufacture and placement by blowers of up to 250,000 m³ of snow.

All the excavated ground raised by the dredge buckets was delivered to two 3.4-m-diameter, 17.2-m-long, revolving trommels, each fitted with 10, 25, and 305 mm screens. Rotating at up to 10 rpm and fed by high-pressure water, the trommels disintegrated the feed and washed the component products. Ideally, any clay and till was completely slurried so as to liberate fine particulate gold. Oversize of +25 mm was discharged 11 m aft of the stern. Screened, 25 mm product was pumped to eight 7.6-m-diameter circular IHC jigs. These collected the gold into a heavy mineral concentrate for upgrading in rectangular jigs with a total area of 96 m². The jig concentrates were cleaned by a magnetic separator and on two large Deister shaking tables for helicopter transfer ashore. There, the remaining gangue minerals were removed to produce an almost pure gold concentrate for shipment and sale. Access to the secondary jig circuit and to the gold room was very limited, and constant monitoring was maintained by video cameras and motion detectors. Nevertheless, security was a constant problem, gold was stolen and some dismissals were involved. No chemicals or reagents were added to the feed, either to optimize the recovery of gold or for any other reason.

The main loss of gold in the dredge's recovery circuit was from the trommels. Cohesive clay and till resisted disaggregation to different degrees. Large, 0.5 m³ blocks of material proved incapable of disintegration unless retained unduly in the trommels, which would then back up and overload. Too high a digging rate exacerbated the problem and prevented the liberation of gold which remained within the discharged, non-rock oversize. With increasing amounts of clay-rich feed and/or higher digging rates, the losses of gold increased. An optimum dredge throughput rate existed, variable from one geological feature to another and depending on the local sediment types. The optimum was less than, rarely equal to, but never more than, the achievable digging rate. It represented an economic balance between the resulting unit cost/m³, adversely affected by the reduced excavation rate, and the higher metallurgical recovery. To overcome the problem, an autogenous clay scrubbing circuit was added in 1990 to treat the 25 to 305 mm fraction of the trommels' clay and stone oversize. Scrubber product of 25 mm was fed to the primary jigs in closed circuit, and the recovery improved markedly.[26]

The Bima's treatment plant had been intended, and was used originally, for the recovery of cassiterite, with a density of 6.9 (less than half that of gold). Particles of heavier gold therefore became lodged in unintentioned mechanical traps in the unsuitable circuit. When a certain loading was reached they or other particles emerged from the plant a few weeks later, as opposed to the normal half hour after having been dredged from the seabed. The phenomenon at first confused production records and reconciliation attempts. In 1987, 2.5 months of operation elapsed before the daily gold production figures reflected the excavated grade, and observed metallurgical recovery. By that time about 5500 oz. had been locked up in the circuit and was not recoverable until the end of the season. Complete, regular sampling of the operating dredge flowsheet was therefore implemented. It allowed the derivation of a full metal balance for integration into and comparison with the drillhole sample data, especially grade and gold size distribution.

Until the 1990 season, 5 to 40% of all dredge feed was discharged into the sea as trommel oversize, equivalent annually to between 85 and 235 m³/operating hour. From 1 to 5% was +305 mm rock. Jig tailings varied annually between 215 and 355 m³/h of solids. Once the optimum operating parameters of the trommels had been established, the overall metallurgical recovery improved from around 70% in 1987. Two years later it averaged 84%, and the addition of the scrubbing plant raised it to an estimated 95% in 1990.

The dredge's excavation rate was dictated by the bucket speed (bpm) and the bucket fill factor. The latter was the outcome of several varying parameters, including (a) the geotechnical characteristics of the sediments, (b) the digging depth and cut depth, (c) the performance of the trommels, (d) the weather conditions, (e) the speed at which the buckets traversed the face laterally, normally 18 m/min, and (f) the angle at which the buckets were presented to the face. Some sediments, especially till, strongly resisted any attempt to run the buckets at too high a speed and caused the bucket line to stall. A bucket speed as low as 14 bpm was required, with a major negative impact on the throughput. Under the best operating conditions any ground beyond 10 m below the seabed proved to be too resistant to dredging. The throughput rate was less than half that originally assumed, varying annually between 353 and 590 m³/operating hour, with an average of 457 m³/h. In 1987, while working in Block 8, the rate averaged 450 m³/h, with a daily figure of 100 to 700 m³/hr. The bucket speed varied from 18 to 27 and averaged 23 bpm. The following year the less consolidated lag gravels of the Glacial Thrust Zone were dredged in deeper water, allowing the average digging rate to rise to 650 m³/h at 24 bpm.

The maximum digging depth rarely exceeded 20 m of combined water and ground thickness because of the limited vertical extent of payable sediments. In shallow water of less than 10 m, with a cut depth of 2 to 9 m, the throughput rate varied from 200 to 600 m³/h. The average daily fill factor was 40% with a range from 10 to 60%, depending on the exact ladder angle and depth of cut. Where the dredge worked in deeper water exceeding 10 m on more easily excavated sediments with a cut depth of 2 to 4 m, the throughput rate was higher. It ranged from 500 to 800 m³/h, and the fill factor averaged 60%, at times attaining 80%. The Bima's ladder was too long for the required digging depths. An amputated and, therefore, steeper ladder would have allowed an increase in throughput rate, but one beyond the sensible treatment capability of the plant. In addition, even if shortening the ladder had made more efficient, deeper digging technically possible in shallow water, practical considerations would have dictated otherwise. Tailings disposal would have resulted, in visible contravention of the permit, in the creation of tailing mounds exposed well above sea level and requiring a stacker installation. The dredge would have lost all maneuverability in such a situation and would have been disabled or wrecked in the first storm.

Regardless of the digging and treating conditions, long period swells seriously limited the digging and throughput rate. They caused the Bima to "see-saw" at the same frequency, causing the bottom tumbler to temporarily lift up and lose contact with the base of the cut. With the following downward movement, the tumbler reestablished impacted contact with the non-yielding seabed. If the dredge continued operating, the bucket fill factor and throughput rate were very frequently and significantly decreased. The static and dynamic force due to deceleration could exceed 1000 t. The pontoon also surged forward, and at a speed of 0.5 m/s the horizontal force imposed on the 14,000 t Bima was of similar magnitude. The obvious precaution was taken of suspending operations and raising the ladder when conditions were regarded as being seriously detrimental. However, it did not prevent the repeated shearing of the bottom and top tumbler shafts through fatigue. Performance was also very sensitive to the wind and current. A constant 50 km/h wind against the dredge's 3200 m² sail area terminated operations because the effective force of 1.4 t for each 1.6 km/h of wind speed slowed or accelerated lateral movement to inoperable extremes. A 2.8-km/h (1.5-knot) current also created a 37.2-t pressure against the 500 m² submerged area of the pontoon. Severe summer storms each caused a complete shutdown of the dredge for some days, and the limited warning of their onset resulted in heavy equipment being damaged or irretrievably swept off the deck.

From 1987 to 1990 inclusive, the possible annual dredge operating days varied from 155 to 170.[31] The average operating time was 2700 h/year, equivalent to 16.5 h each available day, or 69%. Emergency repairs and some maintenance were undertaken during the dredging season, but as much as possible these were delayed until winter. The average throughput rate of 457 m³/h was half the 900 m³/h minimum figure achieved in Indonesia. The rate fluctuated considerably, even throughout a single day. In shallow water of less than 10 m depth, there was usually a 20 to 50% variation in total daily throughput on consecutive days. In deeper water, where the ground conditions were less severe and variable, it was 10 to 30%. The average 3.2-m depth of cut reflected attempts to dig the thinnest possible thickness of the lag gravels and involved the dredging of an average of 37 ha/year.

All WestGold's marine activities were within both Alaskan state waters and the federally mandated coastal management zone. Innumerable state and federal government agencies were therefore involved in the annual permitting of the company's exploration and mining activities. The agencies frequently could not arrive at a consensus, and the whole process was a constant source of concern.[32] Certain restrictions were applied to any offshore activity. In particular, mining could not approach any closer than 1.6 km of the mouths of the Nome, Snake, and Penny Rivers because of salmon spawning. Neither was it allowed within 30 m of the low water mark and underwater accumulations anywhere of dredge tailings within 2.1 m of the water surface were prohibited. The Bima maintained its Bureau Veritas approval and continually complied with the rules and regulations of the U.S. Coast Guard.

Considerable public interest was shown in WestGold's work, and the company established a high standard for community involvement and communications.[33] Environmental concerns centered on feared marine dispersion of trace toxic elements out of the marine sediments, sea water turbidity, and the suggested overall effect on king crabs and the benthic organisms on which they fed. The toxic elements comprised the common metals naturally occurring in the detrital sediments at ppm and ppb levels throughout the region. The greatest public interest focused on the turbidity plume produced by the tailings discharge. The calm sea and clear water at only the very start of summer made the plume briefly visible before Norton Sound became engulfed in the massive, natural discharge of turbid water from the Yukon River.

Activities under particular agency scrutiny included the dredging operation, the discharge of process water into the sea, and the return of treated sediments as tailings back onto the seabed.[32] The National Pollutant Discharge Elimination System (NPDES) permit required compliance with both federal and state water-quality criteria at the edge of a mixing zone of 500-m radius from the point of tailings discharge. State water-quality regulations stipulated that turbidity levels should not exceed 25 NTUs above background. The same criteria led to the establishment of a similar mixing zone with a radius of 100 m, at the periphery of which trace metal levels for eight potentially toxic elements were not to be exceeded for 4 d averages and 1 h maxima.

Full details of the necessary monitoring processes and of the mitigation techniques tried and proved by WestGold have been published.[32-34] The monitoring programs demonstrated that, with the benefit of the mitigation techniques developed and used, no environmental impacts were either significant or permanent.[17,34,35] In 1990 the U.S. Environmental Protection Agency modified the NPDES permit, prohibiting any dredging in less than 5 m of water. In water depths of between 5 and 7 m the permit required that judgment and mitigational procedures, such as further decrease of the already low throughput rate, should be applied.

4.2.10 UNDERWATER MINER

In 1986 the Nome offshore project was still in its exploration stage, but WestGold had committed to the Bima as the mining method. The operating conditions, however, could hardly have been more different from those for which the dredge had been designed. Within a year it was realized that a suitable mining system was required to: (a) exploit resources in both shallow and deep water, (b) mine very selectively, (c) be less susceptible to the sea swell, (d) operate in winter, through or under the ice, (e) produce gold at a lower cost/oz. than could the Bima, and (f) allow modular capacity increases. The many systems studied included hydraulic dredges, but the performance of floating bucket-wheel and cutter-suction units would, like that of the Bima, have been compromised by the swell.[36,37] Barge-mounted backhoes were similarly affected, and had operating depth limitations and low throughput capabilities. Hydraulically operated clamshells and drag-lines also lacked sufficient control. An airlift system was rejected. It lacked the aggression and control necessary for efficient clean-up of cobble-rich lag gravels against the false bedrock at a high enough rate, and the water was not sufficiently deep for the compressed air system.

WestGold's studies led to the conclusion in 1988 that a seabed mounted excavational device was necessary. Ocean trench excavators had developed into various remotely controlled, track-mounted vehicles. They carried some excavating device, usually a jet pump or chain cutter arrangement, to dig a seafloor trench. One such machine was the Tramrod, model 250, built in Britain by Alluvial Mining, a subsidiary of Royal Boskalis Westminster n.v. It was then being used for seabed landscaping in the North Sea oil industry.[6,38] WestGold leased the Tramrod, referred to locally as the "underwater miner" (UWM), and used it successfully in an experimental mode at Nome in 1989. This event was the first recorded time that such a machine had been employed for marine mineral recovery on the continental shelf. It is contrasted with the Bima in Figure 4.8.

The UWM weighed 25 t and consisted of a 5.8-m-long undercarriage on 0.76-m-wide, dual tracks with a total footprint 4.0 m long and 3.0 m across.[6] The maximum speed of travel was 500 m/h and differential track movement permitted directional changes. Mounted on the chassis was a long, hydraulically controlled dredging arm, which rotated about vertical and horizontal axes. With a maximum 50° slew from the center line, the vertical range of the arm's intake end, shown in Figure 4.10, was from 2.8 m above the seafloor to 2.0 m below. Suction-induced excavation was effected by a powerful jet pump installed at the guarded 410-mm intake and capable of pumping 250-mm solids. High-pressure water was supplied to the jet pump by a 10.55 kg/cm², surface mounted, 9690 l/min pump via two 152-mm-diameter flexible hoses. Sediment loosening was assisted by high-pressure water jets directed outward from the circumference of the intake.

The UWM was tested to determine whether it could selectively mine the lag gravels to produce an undiluted, higher grade plant feed. It was deployed in water depths of 6 to 18 m. The jet pump assembly provided merely suction and permitted the excavation of only loose sediments. Therefore, a bucket-wheel and pump assembly, suitable for later installation on the UWM, was tested separately for its cutting ability. A third, untried alternative would have been to place a centrifugal pump directly on the machine to boost the jet-pump, increasing the slurry density and excavation rate. An umbilical cable provided connections to the surface, and a multiplexed control and signal cable carried power for the electrohydraulic, submersible, power packs. Underwater monitoring facilities included video cameras for both forward and rear views; another was positioned on the digging arm 1 m from the suction face. Scanning sonar, compass, depth recorders, and heading control were also employed.

FIGURE 4.10   The suction arm of the Tramrod used at Nome in 1989.

The UWM was deployed from a 76-m-long by 23-m-wide, ocean-going, 250-1 barge equipped with a complete gold recovery plant and carrying the necessary support facilities (Figure 4.11). The barge was secured and maneuvered by means of five Danforth 7.5-t anchors, one at the bow and four laterally. Each anchor winch was equipped with 610 m of 44 mm steel cable, and the preferred vessel orientation was with the bow facing the incoming swells. The UWM was placed on, and recovered from, the seabed by a 230-t-capacity Manitowoc 4100 S-2 crane with a 60-m boom stationed at the stern. Thrusters assisted in the UWM's deployment and careful seabed positioning. Slurried sediments of 250-mm diameter were raised through a 305-mm interior diameter floating pipeline by the boosting action of a 305 × 356 mm, 11.36 m^3/min, deck-mounted gravel pump.

The conventional, gravity flowsheet for gold recovery had a nameplate capacity of 250 m^3/h. The slurried feed was screened on a 22.3 m^2 area grizzley. Washed, 200 mm oversize, usually about 10% of the feed, was discharged directly into the sea. Undersize was fed to a 20.7-m-long, 1.83-m-diameter, rotating trommel supplied with 200 m^3/h of sparge water. Oversize of 38 mm was discharged and an intermediate 13- to 38-mm product passed over a nugget trap before being discarded. Undersize of 13 mm, averaging 70 to 80% of the feed, was delivered via a dewatering cyclone as the underflow to a two-stage jig plant. Combined primary and secondary jig tailings were released into the sea, and the gold-bearing hutch concentrate was upgraded on a shaking table before despatch ashore. Effluents from the treatment plant were discharged in such a way as to minimize their settlement on virgin ground.

FIGURE 4.11   The 250-1 barge from which the Tramrod was deployed off Nome.

After being lowered to the seabed, the UWM's orientation and forward movement took it directly away from the barge on the vessel's extended center line. The UWM remained connected by its umbilical cord, the twin water hoses, and the floating pipeline. For emergency purposes the crane hook was also kept attached to the UWM's lifting strip, ensuring a quick machine recovery from the seabed if necessary. Manually operated joy-sticks in the barge-mounted control cabin governed the individual speeds of the two tracks, together with the horizontal swing and vertical movement of the dredging arm. Uneven or cobble-littered ground could be traversed easily. With the pump intake in direct contact with the lag gravels the jet pump was activated, and the operator observed the excavation process via the cameras to achieve a digging accuracy of ±100 mm.

The UWM advanced along a predetermined corridor, selectively stripping off the lag gravels from the underlying lower-grade material. A vertical cutoff horizon, the false bedrock, was determined visually and was rarely deeper than 0.5 m. It defined the road bed on which the UWM slowly advanced with a sweeping action of the arm. The shallow excavated trench was about 9 m wide. In some places, the machine was halted or reversed to ensure that loose sediments at the base of the cut were being sufficiently cleaned up. Within the limits of its anchor spread, the barge intermittently followed the UWM stern first. Its advance caused the pipelines, cord, and securing line from the barge to the UWM gradually to lose the slack required for the sea conditions. The barge was then moved by an appropriate distance, usually about 7 m, to again bring its stern to a point immediately behind the UWM. The full, uninterrupted excavation of a trench was dictated by a predetermined mining block boundary or by the barge's confinement within its anchor spread. Both events required the UWM to be hoisted on deck, as the barge's anchors were moved ahead. The barge and UWM then either returned almost to the original starting position to commence the excavation of a second corridor parallel and adjacent to the first or the barge was towed to a new mining site.

With only the jet pump installed, the mining rate in loose sand and fine gravel averaged 120 m³/h at a water depth of around 15 m. Gravels in mud and sand were mined at a maximum rate of 26 m³/h, but unwashed till could be excavated at no more than 1 m³/h. Separate tests with the bucket-wheel mounted on the boom of the crane demonstrated that it could excavate such resistant sediments if incorporated on the UWM's strengthened dredging arm. The higher mining selectivity in lag gravels yielded a recovered grade of between two and four times greater than the diluted

FIGURE 4.12 A later version of the Tramrod, fitted with a bucket-wheel as a result of the Nome experience.

grade achieved by the Bima in the same situation. The UWM worked in water as shallow as 4 m, and it proved its suitability by continuing to operate in sea conditions so unfavorable that the Bima had to be shut down. It represented the extreme opposite of the Bima. Finally, fitted with a bucket-wheel, its still comparatively low throughput would have resulted in higher unit volume costs, but it could have excavated very selectively, both laterally and vertically with minimal dilution to yield a considerably higher unit revenue. Such a machine, later constructed by Alluvial Mining for use elsewhere, is shown in Figure 4.12.

4.2.11 GOLD RECOVERY PERFORMANCE

During dredging, the Bima's position was continuously and automatically recorded, together with the digging depth and the depth of cut. The results of dredging were checked by bathymetric, side scan sonar, and underwater video surveys. With confirmation of clean-up provided by divers, the data allowed accurate estimation of the volumes dredged. The Bima's flowsheet was systematically sampled during operation to develop a metal balance. Dredge buckets were hand sampled, and an automatic device regularly and frequently collected samples of trommel undersize feeding the primary jigs. These and other routine samples yielded gold particle size distributions and grade data. The results allowed comparisons to be made with the corresponding information, especially involving particle size frequency, derived from drilling to examine the relationships with metallurgical recovery.[20] At the end of each season the entire flowsheet was cleaned out to remove locked-in gold, which was mathematically attributed to individual mining blocks.

Gold production and recovered grades from blocks within individual features were reconciled with the original estimates so as to provide feedback into the estimation methods.[3] From 1987 to 1990 inclusive, the Bima produced 118,078 fine oz. (3,672.6 kg) of gold from 4.457 Mm³ of sediment at an average recovered grade of 824 mg/m³. This included 7041 oz. from a final clean-up when the dredge was scrapped in 1991–2. The highest single day's production, in Block 8, exceeded 640 oz. (19.91 kg). Based on the drilling results, 81% of the targeted gold content, represented by 69% of the volume, was recovered. But the gold output was augmented by 14% by the dredging of ground not in reserves. The balance was sterilized as deep remnants under tailings, and as small exclusions. Dilution amounted to an additional 37% by volume at a coincidental 37% of the reserve grade. The overall recovered grade represented 106% of that estimated for all the ground dredged: an R/E factor of 1.06.[23]

An average factor of 1.06 over four years compares well with equivalent, onshore alluvial gold producers, such as Alaska Gold Company.[39] For example, dredge No. 5 over 25 years recorded an average R/E factor of 1.05, but with a range of 0.49 to 2.38 for individual summer seasons. Dredges No. 2 and 3 achieved 0.83 and 1.27 overall for five and four years, respectively.[19,23] The factor's range increases with shorter time periods, and the Bima's varied from 0.5 to 2.1 for intervals of approximately one month's dredging. For any alluvial deposit, the R/E factor compares the first and final stages of production and is the mathematical product of four others: the estimation, selection, digging (excavation recovery), and treatment (metallurgical recovery) factors.[23] Each combines its own component factors. Thus, the estimation includes the drilling factor, and the treatment factor includes the trommel recovery factor.

A best-fit curve to relate the recovered and expected grades, may be represented by:

$$\text{R/E factor} = (a) \cdot [\text{expected-grade}]^{-b} \qquad \text{(Eq. 4.1)}$$

in which a and b are constants. The relationship is expressed as:

$$\log[\text{R/E factor}] = (-b) \cdot \log[\text{expected-grade}] + \log(a) \qquad \text{(Eq. 4.2)}$$

to yield a straight line.[26] In its simplest form the graphical plot is a regression relationship which, if only apparently higher-grade areas are selected for dredging, illustrates the effect of too low a drilling density.

In different features, with dredging planned on declining drillhole density, the average R/E factors recorded for seasonal dredge courses ranged from >1.0 for 50 × 50 m, through 0.7 for 70 × 70 m, 0.6 for 70 × 100 m, to 0.5 for 100 × 200 m. A correlation also existed offshore between the Bima's R/E factors and the geological characteristics of provinces and features. Factors in the heart of Block 8 varied between 1.30 and 1.38 over three years. Exactly the same drill spacing on the Block 8 boundaries produced one of 0.75 and one of about 0.6 on bordering lag gravels.

Figures of >1.0 do not necessarily reflect efficient dredging but can be indicative of underestimation of the reserves. For example, the dredging of part of Block 8 resulted in an R/E factor of 1.31, but detailed sampling and reconciliation revealed that the 31% bonus in recovered grade was due to the following component factors: 1.87 for estimation, 0.83 for selection, 0.95 for digging, and 0.89 for treatment. The Bima had recovered 89% of the gold from the ground dug, leaving 5% spillage and remnants on the seabed. The 0.83 selection factor resulted from too few drillholes. The estimation factor reflected an underestimation of the grade, due in small part to the search radius used in kriging. It was mostly caused by the drilling and by the agreed sample volume assumptions. These had led to an undervaluation of individual samples, as indicated by a drilling factor of 1.59 as a component of 1.87. In contrast, overvaluation was a problem when drilling some lag gravels elsewhere. On part of the 4.5 km Moraine, dredging produced an R/E factor of 0.23, signifying a 77% shortfall in recovered grade. But detailed studies revealed that the combined digging and treatment factor was reasonable at 0.86. Errors caused by the time-imposed need to drill in rough seas had resulted in a combined estimation and selection factor of 0.27, creating an overvaluation of the block grades by more than 2.5 times.

During the UWM's operation, the amount of gold on the shaking table indicated the grade of the ground dug about 5 min previously. Its systematic collection at intervals, related to the depth of cut and progressive location of the UWM, provided a good depiction of the changing grade of the lag gravels along the excavated trench. The UWM's uniquely selective mining capability allowed extraction of small, high-grade patches of ore without any significant dilution. It resulted in the production of a recovered grade three to five times higher than that achieved by the Bima in the same ground. The UWM's R/E factor varied through only a small range from 1.09 to 1.22 for different features.[26] Where the Bima had achieved an average factor of 0.55 on three areas of lag gravels, the UWM performed twice as well. The UWM's mean factor was 1.15 from expected

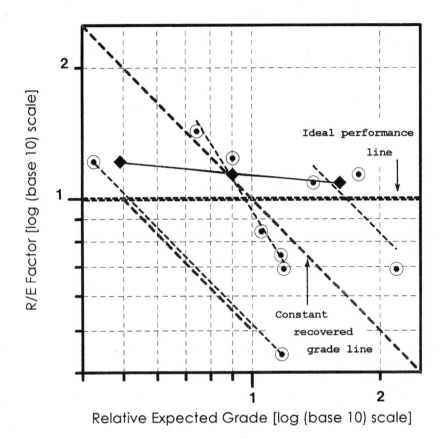

◆   "Tramrod" underwater miner

◉   "Bima" bucket-ladder dredge

FIGURE 4.13 A grade reconciliation plot showing the consistent and higher R/E factor achieved by the Tramrod, compared with the Bima.

grades varying between 700 and 3500 mg/m³. A comparative example of the achieved relationship between expected grade and R/E factor for both mining systems is shown graphically in Figure 4.13.

4.2.12 ECONOMICS

Marine exploration and mining at Nome proved to be considerably more expensive than similar activities onshore. Complete geophysical coverage of WestGold's leases during two mid-1980s surveys incurred an expenditure of US$1.28 million and was followed by a total outlay of $8.6 million on five drilling programs. The bulk sampling was almost as costly. The first 91-hole winter drilling program off the ice in 1987 averaged $4,600/hole but succeeded in proving up gold reserves at a rate of $12/oz. The costs of the next winter program in 1988 were much reduced at $2,400/hole, and those of summer drilling similarly declined, increasing experience having been gained, from $4,200 in 1987 to $3,100 in 1988 and $2,600/hole in 1989. These figures demonstrate that any future high sampling density evaluation of lag gravels at Nome requires not only a more effective, but also a cheaper method than drilling: perhaps a clamshell.

The Bima's average, annual throughput of around 1.23 Mm³ at Nome was 60% less than originally assumed. The total operating costs were about 50% higher than first anticipated, whereas the recovered grade was 6% better overall. The unit operating costs, which in Indonesia had been approximately US$1.10/m³, averaged about US$7.00/m³ on a cash basis. They were represented in a typical year by 75% for direct operating costs, 15% for exploration, and the balance on winter overhaul, repairs, freight, and ice berm construction. Of the cash costs, labor and salaries, including housing and travel, represented 37%. Supplies, including fuel, added 7%, and maintenance and repairs amounted to 13%. Reflecting the continual expenditures on environmental work, all internal and outside services totaled 21%. Insurance was expensive at 7%, and exploration constituted the remainder. The cash cost of gold production, excluding exploration expenditures, rose from US$250 to $335/oz. from 1987 to 1989, while the gold price simultaneously declined from a high of about $485/oz. to around $350/oz. The only significant, variable dredge costs were fuel, repairs, and some supplies. The fixed element consequently comprised about 90% of the cash costs. The operating profit had to be earned in the final two weeks of the typical 165-d season when the sea conditions were at their worst, employees exhausted, and the threat of storms greatest.

The four month UWM test program in 1989 cost approximately US$5 million. To place that figure in perspective, it was no more than that involved in the replacement of the Bima's top tumbler shaft, which finally broke from fatigue in 1990. The trials allowed estimates to be made of the operating costs for a machine fitted with a bucket-wheel. The UWM's low estimated throughput compared with that of the Bima's resulted in a significantly higher projected cost/m³, but it was more than offset by the substantially increased recovered grade to reduce the cash cost/oz. The relatively small estimated potential gold production per unit was balanced by numerous benefits: (a) modular capacity increases, (b) substantially lower capital costs per oz., (c) possible use through the winter sea-ice, (d) application on smaller mining blocks, (e) close geological supervision, (f) less concentrated environmental impact, and (g) ability to exploit shallow-water reserves, perhaps deployed from the beach.

WestGold's management and the company's consultants considered in 1990 that a larger version of the Tramrod would be an economically feasible UWM. If fitted with an array of digging tools and a centrifugal underwater dredge pump, the design was potentially capable of maintaining a mining rate of 150 to 300 m³/h at 20% solids up to 250 mm. A depth of cut of from 0.3 to rarely 3 m was thought necessary, with a 5% machine tilt tolerance and optional underwater screening capabilities.[40] Plans were formulated with Alluvial Mining for the establishment of a suitable corporate vehicle, Coastal Placer Mining, for strategic utilization of the unique experience. However, the Alaskan leases were surrendered in 1991, the local offices were closed, and the expertise became dispersed.

4.3   WORLDWIDE ACTIVITIES

4.3.1   PHILIPPINES

As well as off Nome, marine gold has also been mined in a bay on the east coast of Luzon Island in the Philippines.[41] The source is a drowned extension of fluvial deposits in the Paracale and Malaguit Rivers at 14°17'N and 122°48'E. The offshore geological sequence comprises auriferous gravel on bedrock, overlain in turn by sand, 6 to 10.5 m of sticky marine clay, and finally, by sand and shells. High grades of finely sized gold are concentrated in a layer from bedrock to 1.2 m above. Drilling by auger and churn drills revealed grades varying between 140 and 550 mg/m³ when the operational break-even grade was 200 mg/m³. Nearshore water depths in the bay vary from 2 to 4 m, increasing to a maximum of 14.6 m and averaging 5.5 m. Peak tide height is 1.71 m with 0.5-m storm surges. The climate is tropical with a temperature range of 20° to 34°C. A northeast monsoon blows from December to February, and typhoons occur from July to January, peaking in November and December. The bay is sheltered by mountains on three sides, but it could be dredged only from June to October.

Eight small New Zealand dredges worked in the Paracale River in 1915–1923 in grades of 2 to 6 g/m³. Coco Grove Inc. was incorporated in 1934 and started offshore work in 1936 with three old harbor construction dredges. These were replaced in 1938 by two 8 ft³ (225 l) capacity bucket-ladder dredges manufactured by Bucyrus–Erie Co. The Mary Angus and Anne Petronella had a combined capacity of 285,000 to 350,000 m³/month compared to a nameplate one of 173,000 m³ each.

The dredges dug to 19.8 m depth with a ladder angle of 45°. Each was electrically driven by a cable supplying 445 kW from the shore, and gravity recovery of gold was by tables and riffles. A 335-kW tugboat assisted with anchor moves and a 366-mm suction dredge maintained a typhoon refuge anchorage. The dredges finally operated up to 0.8 km from the shore. Historical dredging costs of US$0.155/m³ were regarded as high at the time, compared with onshore producers, because of the large inventory of supplies, the logistical support necessary, and the corrosive action of sea water. Both dredges worked until 1941 when stopped by military hostilities. Only the Mary Angus resumed work from 1951 to 1956. Operations finally ceased because of the limited season, plus the problems of mangroves, boulders, and the need to traverse barren ground. In 1987, 27 airlift holes were completed, through a 100-mm-diameter cased pipe in the seabed to a maximum depth of 4 m and an average of 3.2 m. In 1988, a 1.3-km-wide area of 1402 ha was licensed for six years, extending 11 km along the coast but with no subsequent activity.

4.3.2 NEW ZEALAND

In New Zealand waters, marine gold deposits centered at a distance of 11 km from Hokitika, off the west coast of South Island at 43°S and 171°E, were first investigated by Marine Mining Corporation in the 1960s.[42] The offshore sediments comprise sand and gravelly sand. They are commonly overlain by a gold-bearing, muddy/sandy gravel unit less than 1 m thick. The gravel is characterized by poor sorting, abundant shell fragments, and a chaotic nature. Cobbles up to 200 mm in size are usually well rounded. From 1967 to 1972 Alpine Geophysical Associates completed sparker seismic profiling, offshore sampling, and 127 vibrocore drills in water depths of 70 to 100 m.

CRA Exploration Pty. Ltd. (CRAE) undertook seabed sampling programs over 12,000 km² in 1980 and 1982. At 1-km intervals on lines 5 km apart, sediment samples of 20 to 30 kg each were collected by a Forster Anchor Dredge from the uppermost 200 to 300 mm in water as deep as 100 m. Gold, averaging 124 microns in size, was recovered by using a Wilfley table and a Super Panner to reveal an anomaly over 8 to 14 km offshore with eight samples containing 42 to 545 mg/m³ of gold. Follow-up sampling on a 1 × 1 km grid defined an area 23 km long and 6 km wide which encompassed 66 sample sites reporting over 50 mg/m³ and averaging 189 mg/m³. A gravity corer failed to penetrate more than 200 mm in 1982, whereas 34 vibrocore holes averaged 60% recovery and 1 m penetration. Two years later another vibrocore program of 39 holes at 26 sites achieved an average 47% recovery and 1.4 m penetration.

CRAE considered a minimum economic target for the project, named "Harvester," to be 150 Mm³, with payable sediments perhaps 5 m thick over 30 km², and grading 200 mg/m³. Further work commenced in 1989, with a 220 kW vibro-hammer driving 300-mm-diameter drill pipes to a depth of 10 m. The drilling, monitored by underwater video camera, progressed at a rate of 4 holes/24 h. For the 24 holes completed, the average penetration was 9.8 m, with a core recovery of 83%. The machine was deployed off the 96-m-long vessel *Stena Seahorse II*, complete with dynamic positioning, a large working deck area, a 100-t crane, and a helideck. The finely sized gold was again recovered physically in a Knelson Concentrator in order to provide a size frequency distribution, and the grade was determined for 1-m vertical intervals. In 75% of the holes, the highest grade was in the first, uppermost section. An average grade for the surface gravel less than 1 m thick was 189 mg/m³ over a 67 km² area, and work was terminated. The mineralization was interpreted as constituting a lag deposit, possibly related to a paleoshoreline, formed when the sea level was about 120 m below that of the present day.

4.3.3   SOUTH AMERICA

Parts of the southern extremities of Chile and Argentina are overlain by huge volumes of uncon-solidated glacial drift. The sediments range from lateral and terminal moraines through boulder clays and tills, to widespread fluvio-glacial outwash. They are covered by recent loess, peat, and fluvial sands and gravels.[43] Four periods of major glacial advance originated from ice fields in the mineralized Andean chain. Starting in the late Tertiary, the glaciation peaked about 1 million years ago, but the best preserved moraines remain from the last Quaternary advance. The blanket of glacial material has been modified by invading marine activity together with fluvio-glacial and fluvial action. The results demonstrate the widespread distribution and concentration of particulate gold in the clastic sediments. Evidence remains of century-old, onshore gold workings, including dredging, in the area of the Straits of Magellan, on Tierra del Fuego, and around the outer islands of Picton, Lennox, and Nueva. Gold has been concentrated by high-energy marine and storm activity on beaches and in possible lag gravels during periods of changing sea level. Coastline faulting appears to have played a role in some places.

In 1988 Geomar S.A., a subsidiary of Anglo American Corporation, successfully demonstrated the presence of gold on the seabed of the Straits and in 1990/1 conducted an offshore sampling program in the Magellan Straits and Nassau Bay concession areas.[43] To identify areas of anomalous gold mineralization, a hydraulically operated clamshell grab was deployed from the 1200-kW oil industry supply vessel, *Cruz de Froward*, 53.6 m long with a 10-m beam and 1338-t displacement. Samples were treated on board by trommels and Knelson and Knudsen concentrators to produce a concentrate which was examined onshore to determine the gold content. Work was terminated in 1992, and the Chilean marine licence is now held by Minera Mares Australes S.A.

4.3.4   OTHER AREAS

In the late 1960s an Australian company, Planet Gold Ltd., explored for a possible offshore extension of the Bermagui gold-bearing beach placer area in southern New South Wales. The company carried out a bathymetric and seismic survey, followed by a drilling program. The machine employed, the Amdrill of Alluvial Mining, penetrated the seabed to a depth of about 10 m by means of a pneumatic percussion driver. The sample was airlifted to the surface, using reverse flow flushing. Insignificant traces of gold were found.

Since the early part of this century, intermittent and economically unsuccessful marine explo-ration activity has been conducted off the coasts of Nova Scotia and East Malaysia (Sarawak). The extent to which marine placer gold deposits have been explored, and possibly exploited, in the Russian Far East is uncertain.

4.4   CONCLUSIONS

The Nome marine gold deposit owes its origin to a singular combination of (a) nearby primary mineralization, (b) glaciation, (c) recurrent coastline faulting, (d) sea level changes, and (e) a coastline fully exposed to high-energy marine conditions. Many factors, especially the declining gold price, led to the termination of the project. In hindsight, the Alaskan experience emphasized the main subject of performance risk in marine mining: ground "digability." A sufficient under-standing of system suitability, required selectivity, and mining rate prediction requires a foundation of comprehensive understanding of the geology at the outset. Selected equipment must be compat-ible with the prevailing, well-interpreted, geological, geotechnical and climatic conditions in order to achieve the desired throughput.[6] The mining rate usually constitutes the greatest variable in the medium- and long-term cash flow. The capability of selective excavation must match the *in situ* grade variability, which can be revealed only with a sufficient density of reliable sampling. The Bima was not a sufficiently selective mining system to achieve a balance between costs/m^3 and potential revenue/m^3 to produce gold at a low enough cash cost/oz.

Although the UWM was never used again at Nome, it represented an important stage in the development of similar, but improved, systems for marine mining. WestGold predicted that "once such an underwater miner has been tested and proven, a whole new chapter in (marine) mining history will open." The project was terminated, but De Beers Marine's first tracked mining machine, launched in late-1989 for diamond recovery offshore Namibia, proved its worth a year later.[44,45]

ACKNOWLEDGMENTS

The writer wishes to acknowledge the role played by all the WestGold employees and consultants involved in the collection and interpretation of data concerning the Nome project from 1986 to 1990.

REFERENCES

1. Nelson, C. H. and Hopkins, D. H., Sedimentary processes and distribution of particulate gold in the northern Bering Sea, United States Geological Survey professional paper no. 689, 1972.
2. Kaufman, D. S. and Hopkins, D. M., Late Cenozoic geologic controls on placer gold distribution in the Nome nearshore area, in *Geologic Studies in Alaska*, Dover, J. H. and Galloway, J. P. (Eds.), United States Geological Survey Bulletin, 1903, 26.
3. Bronston, M. A. and Howkins, C. A., A summary of the geology of the Nome placer deposit including methods of sediment description, WestGold staff report, Anchorage, Alaska, July 1990.
4. Barker, J. C., Robinson, M. S., and Bundtzen, T. K., Marine placer development and opportunities in Alaska, *Mining Engineering*, 42, 1, 21, January 1990.
5. Bronston, M. A., Development of proven ore reserves for the Nome expansion project, WestGold staff report, Anchorage, Alaska, 27 August 1990.
6. Garnett, R. H. T., Development of an underwater mining vehicle for the offshore placer gold deposits of Alaska, in *Alluvial Mining*, Elsevier Science Publishers Ltd., England, 1991, 157.
7. Garnett, R. H. T., Marine dredging for gold, offshore Alaska, *Offshore Technology Conference Proceedings*, Houston, Texas, paper OTC 8008, 1996, 455.
8. Garnett, R. H. T., Estimation of marine mineral reserves, *Transactions of the Society for Mining, Metallurgy, and Exploration*, 304, 69, 1998.
9. Oceanographic Services, Inc., Extreme wave analysis, Nome, Alaska, Report No. 04492, Santa Barbara, California, January 1975.
10. Daily, A. F., Technical report on off-the-ice placer prospecting project, Nome, Alaska, February–April 1964, submitted to Shell Oil Company, 23 October 1964.
11. Daily, A. F., Off-the-ice placer prospecting for gold, *Offshore Technology Conference Proceedings*, Houston, Texas, paper OTC 1029, 1969, 277.
12. Bronston, M. A., Offshore placer drilling technology: a case study from Nome, Alaska, *Offshore Technology Conference Proceedings*, Houston, Texas, paper OTC 6019, 1989, 459, and *Mining Engineering*, 42, 1, 26, January 1990.
13. Garnett, R. H. T., Problems with dredging in offshore Alaska, *Mining Engineering*, 49, 3, 27, March 1997.
14. Graul, M., Bronston, M. A., and Williams, C., High resolution seismic exploration for gold, *Offshore Technology Conference Proceedings*, Houston, Texas, paper OTC 5941, 1989, 579.
15. Nome placer project (Alaska): Bima accessible ore reserves as at 31 December 1989 — final report, WestGold staff report, Anchorage, Alaska, 30 July 1990.
16. Jenkins, R. L. and Lense, A. H., Marine heavy metals project offshore Nome, Alaska, July–August 1967, United States Department of the Interior, Bureau of Mines Heavy Metals Program technical progress report, August, 1967.
17. Ellis, D. V. and Garnett, R. H. T., Practical mitigation of the environmental effects of offshore mining, *Offshore Technology Conference Proceedings*, Houston, Texas, paper OTC 8023, 1996, 611.
18. A statistical evaluation of procedures and conditions in the Block 8 area, offshore Nome, Alaska, WestGold staff report, Anchorage, Alaska, 29 October 1990.

19. Garnett, R. H. T., Components of a recovery factor in placer gold dredging, in *Alluvial Mining*, Elsevier Science Publishers Ltd., England, 1991, 115.

20. Garnett, R. H. T., Placer evaluation: a reply, *Minerals Industry International*, Institution of Mining and Metallurgy, London, Bulletin no. 1036, 23 May 1997.

21. OCS mining program: Norton Sound lease sale, final environmental impact statement, United States Department of the Interior, Minerals Management Service (Alaska OCS Region), Report no. MMS 90-0009, March 1991.

22. Wells, J. H., Placer examination: principles and practice, United States Department of the Interior, Bureau of Land Management, technical bulletin no. 4, 1969 and 1973.

23. Garnett, R. H. T., Components of a recovery factor in gold and tin dredging, *Transactions of the Institution of Mining and Metallurgy*, section A, 100, 121, 1991.

24. Purington, C. W., Gravel and placer mining in Alaska, United States Geological Survey, bulletin no. 263, Series A (Economic Geology), 55, 1906.

25. Nome placer project ore reserves as at 31 December 1988 — explanatory report, WestGold staff report, Anchorage, Alaska, May 1989.

26. Garnett, R. H. T., Mineral recovery performance in marine mining, *Transactions of the Society for Mining, Metallurgy, and Exploration*, 304, 55, 1998.

27. R & M Consultants, Field investigation and laboratory analyses for the shallow water gold project, Nome, Alaska, Anchorage, Alaska, September 1988.

28. Standard method for penetration test and split-barrel sampling of soils, American Society for Testing and Materials, D-1586, November 1984.

29. Harder, L. F. and Seed, H. B., Determination of penetration resistance for coarse-grained soils using the Becker hammer drill, research report no. UCB/EERC-86-06 sponsored by the U.S. National Science Foundation, University of California, May 1986.

30. Dieperink, J. H. and Donkers, J. M., Offshore tin dredge for Indonesia, *Transactions of the Institution of Mining and Metallurgy*, Section A, 87, 1978.

31. Garnett, R. H. T., Offshore gold dredging in the sub-Arctic, *Mintech 1990*, Sterling Publications International Limited, London, 1990, 100.

32. Gardner, L. A., Regulatory processes associated with metal-mine development in Alaska: a case study of the WestGold Bima, United States Department of the Interior, Bureau of Mines, Open File Report, 88, 1992.

33. Rusanowski, P. C., Nome offshore placer project: a model for resource extraction projects in Alaska, in *Alluvial Mining*, Elsevier Science Publishers Ltd., England, 1991, 587.

34. Davidson, L., Ellis, D. V., Garnett, R. H. T. et al., Marine mining technologies and mitigation techniques — a detailed analysis with respect to the mining of specific offshore mineral commodities, contract report for United States Department of the Interior, Minerals Management Service (MMS 95.0003), C-CORE publication no. 96-C15, July 1996.

35. Garnett, R. H. T. and Ellis, D. V., Tailings disposal at a marine placer mining operation by WestGold, Alaska, *Marine Georesources & Geotechnology*, special issue: submarine tailings disposal, 13, nos. 1 & 2, 41, 1995.

36. Garnett, R. H. T., The availability of mining systems for exploiting offshore deposits of platinum and chromium, position paper prepared for the Marine Minerals Technology Center Workshop, Stockpile 2000, Hawaii, September 1991.

37. Garnett, R. H. T., Mining systems for exploiting nearshore marine deposits, presented at the 23rd Conference of the Underwater Mining Institute in Washington, D.C., 1992.

38. Wenlock, J., Seabed solutions to mining offshore, *Engineering and Mining Journal*, 197, 1, 16B, January 1996.

39. Garnett, R. H. T., Dredge recoveries in north American placer gold dredging, in *Alluvial Mining*, Elsevier Science Publishers Ltd., England, 1991, 77.

40. Alaska Anvil Inc., Conceptual studies of an underwater miner, Anchorage, Alaska, November 1989.

41. Johnson, G. R., Coco Grove dredges have interesting features, *Engineering and Mining Journal*, April 1942.

42. Lew, J. H. and Corner, N. G., New Zealand offshore gold exploration — the Harvester project, *Proceedings of the 1990 Annual Conference of the Australasian Institute of Mining and Metallurgy*, Rotorua, March 1990, 275.

43. Lyall, R. A., Project to explore and exploit marine placer gold deposit, Magellan Straits and Nassau Bay, southern Chile, Minera Mares Australes S.A., Santiago, Chile, April 1997.
44. Garnett, R. H. T., Offshore diamond mining in southern Africa, *Mining Engineering*, 47, 8, 738, August 1995.
45. Garnett, R. H. T., Marine diamond mining — now an established industry, presented at the *29th Conference of the Underwater Mining Institute*, Toronto, 22 October 1998.

5 Marine Placer Diamonds, with Particular Reference to Southern Africa

Richard H.T. Garnett

ABSTRACT

Diamonds were discovered on the coast of Namibia in 1908. They had been fluvially transported to the southern Atlantic after their erosion from kimberlites hundreds of kilometers inland. Long-shore currents, combined with high-energy wind and wave action during periods of considerable sea level changes, concentrated the diamonds in trap sites on paleo coastlines and other marine geological features. The host gravels generally exist as a thin veneer on an irregular bedrock. Discontinuous deposits lie on the inner and middle sections of the continental shelf along the South African and Namibian coastlines.

The same sea conditions that created these unique deposits contribute to very difficult marine operating conditions. Half a century elapsed before the first diamonds were mined from offshore. Intermittent, small-scale mining, including a spectacular success in the early 1960s, has continued to the present day in both Namibian and South African waters.

Large-scale marine diamond mining started a decade ago, the outcome of many years of exploration. Government encouragement and important technological developments, notably in positioning and geophysical exploration, contributed. Operations in waters as deep as 200 m have necessitated the successful development of new sampling and mining systems. A recent trend has been toward the deployment of seabed-mounted equipment. Higher production, lower grades, and increased throughput with new mining systems are expected in future.

The principal marine diamond producer is Namdeb, for which De Beers Marine acts as a contractor. Four smaller but significant public companies are in various stages of development, ranging from sustained, profitable production to trial mining. Marine operations are more capital intensive, with higher unit costs, than equivalent onshore activities, but the grades are higher with little overburden.

Total southern African, marine diamond production for 1998 is expected to be 0.7 to 0.8 Mct. Namibian production represents about 80% of the total, worth around US$170 million, and equals that from the country's onshore sources. Exploration has also been conducted offshore West Africa, Australia, and Indonesia, but without economic success.

5.1 SOUTHERN AFRICA

5.1.1 MARINE DIAMOND INDUSTRY

Summaries of the early history of the southern African marine diamond mining industry have been provided by Corbett and Williams.[1,2] The first offshore concession was granted in 1957. Shortly afterward, attempts commenced to confirm that the terrace deposits so profitably worked onshore in Namibia and South Africa since 1908 extended beneath the sea.[3] Around 1.5 Mct were produced as a result of ensuing shallow-water operations between 1961 and 1970. However, relatively low diamond prices at the time, inadequate technology, and lack of familiarity with all the geological factors thwarted efforts to achieve sustained financial success. By 1971 annual production had declined to 40,800 ct.

However, the potential rewards to be won from the ocean had been revealed. De Beers Marine (Pty) Ltd. (DBM) quickly acquired almost all of the offshore concessions, some of which were later consolidated, and began an extensive 15 year exploration program.[3-6] The search concentrated increasingly on the middle section of the continental shelf, and an area as large as Holland was surveyed and sampled.[7,8] DBM recognized that new mining techniques would be needed to exploit the marine deposits in Namibian open waters to depths of 150 m.[1]

In 1989 DBM declared a production from their first, new mining vessel: 21,545 ct, increasing to 29,195 in 1990. As the organization matured from an exploration into a mining company, production rose to 170,744 ct in 1991 and 260,298 ct in 1992. DBM had finally reached the stage of trial and operational mining of proven areas after 20 years of continuously escalating exploration. Marine diamond mining had come of age. Encouraged by this success, other companies increased their combined output: 10,476 in 1990, 42,120 in 1991, and 59,599 ct in 1992. Meanwhile, from South African harbors some tens of small, converted fishing vessels were operated in shallow waters by various owners, using both divers and airlifts. Production from this source averaged about 7000 ct/year from 1971 to 1977 inclusive, and about 40,000 ct/year to 1989. Thereafter, it increased to a peak of 170,000 ct in 1991 from which it has since declined.

The principal marine diamond mining areas of both countries are shown in Figure 5.1. Namibia awards marine mining grants and exclusive prospecting licence areas. Shallow-water grants extend from the shoreline to about 5 km offshore. There they are bordered by the mid-water licence areas, with western boundaries up to 12 km from the irregular coastline by which they are defined. The rest of the middle shelf has been subdivided as far as the 200-m isobath, with one grant covering more than 80% of the total possible area.[9] In places, prospecting licences have been awarded in recent years out to the 500-m isobath and beyond.[1] South African marine concessions are also defined by distance and by water depth. Inshore "a" concessions extend from 33 m seaward of the mean low water mark to 1000 m seaward of the mean high water mark and are each about 30 km² in area. Seaward, the contiguous "b" concessions are each bounded on the west by a straight line, between defined latitude and longitude coordinates, 5 km west of the mean high water mark. They are 50 to 120 km² in area and extend out to an average depth of 80 m. Further westward, the "c" concessions, varying in area from 500 to 6500 km², extend to the 200-m isobath. They are followed by the "d" concessions covering the outer shelf from 200 to 500 m water depth.

Mining is now taking place on the inner and middle shelves of Namibia and on the inner shelf in South African waters. Requirements for a mining permit in South Africa are detailed in the Minerals Act of 1991. Minimum work requirements are stipulated. Licence holders must report their diamond production but are not required to publish their resources and reserves in a uniform manner, nor to advise the efficiency with which the same are exploited. The Namibian government has a direct interest in the industry through the Namibian Diamond Corporation (Pty) Limited (Namdeb), an equal partnership with De Beers.[10] The South African government's interest is relatively small. Other significant existing and potential operators comprise public companies. Their production and reasonable aspirations are limited partially by public funding constraints but also because the majority of the preferred, deeper-water licence areas are held by Namdeb or De Beers. The remaining areas are presently under licence to more than 30 companies engaged in exploration and some mining.

DBM acts as contractor to Namdeb in deep waters and also independently conducts marine exploration and mining offshore South Africa and internationally.[4] De Beers' experience in mining the onshore, diamondiferous terraces has facilitated the laudable success of its enormous investment in a marine industry of which it is the leader. DBM has five vessels dedicated to mining: the *Louis*

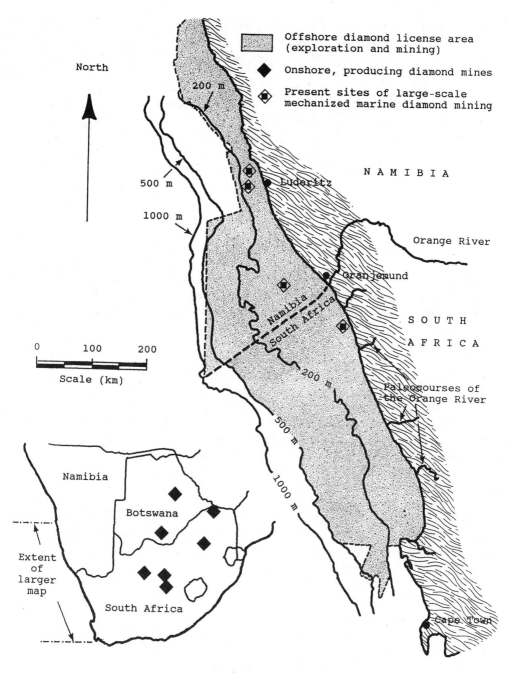

North

200 m

500 m

1000 m

Offshore diamond license area
(exploration and mining)

Onshore, producing diamond mines

Present sites of large-scale
mechanized marine diamond mining

N A M I B I A

Lüderitz

Orange River

Oranjemund

Namibia

South Africa

S O U T H

A F R I C A

200 m

Paleocourses of
the Orange River

500 m

1000 m

0   100   200

Scale (km)

Namibia

Botswana

Extent
of
larger
map

South Africa

Cape Town

FIGURE 5.1 Marine diamond exploration and mining licence areas of southern Africa, showing bathymetry and the main operational sites.

G. Murray, Grand Banks, Debmar Atlantic, and *Debmar Pacific.* The *Voyager* has recently joined the production fleet,[11] and another is reportedly scheduled for a later date. As a result, DBM's output has risen rapidly: 302,754 ct in 1993, 406,925 in 1994, 457,397 in 1995, 470,892 in 1996, and 481,000 in 1997.[4]

Ocean Diamond Mining Holdings Limited (ODM) has been involved in exploration and production since its incorporation in 1983.[12] The company has successfully developed its own expertise and technology for operations in shallow and mid-depth waters.[13] ODM uses the *Oceandia* for

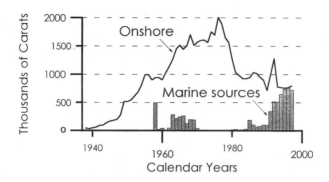

FIGURE 5.2   Historical annual diamond production from Namibian onshore and marine sources. (From Louw A. C., Ocean Diamond Mining's quest for more efficient diamond recovery, presented at the 29th. conference of the Underwater Mining Institute, Toronto, 22 October, 1998, Ocean Diamond Mining Holdings Ltd., Internet home page, <www.odm.za>. With permission.)

exploration work. In 1995 a new airlift mining vessel, the *Namibian Gem*, profitably boosted annual production, which is now maintained at nearly 60,000 ct/year. A third vessel, the *Ivan Prinsep*, has recently been converted for mining.[14] Benguela Concessions Limited (Benco) has undertaken marine exploration since 1989.[15] Its airlift vessel, the *Moonstar*, was commissioned in 1997 and quickly undertook evaluation sampling and trial mining before commencing productive operations.[15] The Namibian holdings of Diamond Fields International Ltd. (DFI) were sampled in 1995, using the vessel *Geomaster* under a joint venture agreement with BHP Minerals and Benco. Confirmatory and bulk sampling, followed by trial mining using DBM's *Coral Sea* on contract, was commenced in late 1998. Namibian Minerals Corporation (Namco) started exploration in 1993 using the *Fox*, followed in 1995 with a program of evaluation and bulk sampling with the *Sprut*.[16] Production was initiated in 1998.[17]

The industry has experienced a protracted gestation period, from the late 1950s to about 1990, for mechanized mining. Not only has the learning curve been prolonged, but it has also been capital intensive.[18] For ODM, 10 years elapsed before the company became an independent diamond producer. It took Benco eight years from incorporation to achieve its first fully mechanized production in 1997. Four or five years have elapsed since Namco and DFI began exploration. Several reasons exist in addition to the time-consuming surveys. These investors have been awaiting the confirmed success of the industry leader, DBM, and the development of appropriate technology. Improved geophysical techniques, accurate position fixing, more reliable sampling techniques, and controllable mining systems are now available.

In 1995 1.34 Mct were produced by Namdeb out of a total Namibian yield of 1.38 Mct, valued at about US$288/ct and worth US$400 million. Over 40% were marine diamonds.[4] DBM contributed 35% of Namdeb's total output in 1996 by producing 470,892 ct.[4,10] As illustrated by Figure 5.2, marine diamonds now contribute 48% of Namibia's diamond production. South African marine output in 1996 was around 0.1 Mct.[19] By comparison, world diamond production in 1995 totaled 104.5 Mct. The average world value was slightly over US$70/ct, giving a total market value of US$7,300 million, and increasing by about 3% annually.[20]

Producers of rough stones either sell privately by means of their own sightings or through the Central Selling Organisation. The market value of a rough diamond, quoted in US$/ct, depends on several features. Size is one, with which the value increases almost exponentially. The result, as pointed out by De Decker and Woodbourne,[9] is that smaller concentrations of larger diamonds may be more important economically than large concentrations of small ones. Average values of southern African marine diamonds are among the highest in the world. Depending on the sales route and the prevailing market conditions, they range upward from about US$120/ct for stones of 0.25 ct. Parcels of rough stones with an average size of 1 ct command prices of more than US$500/ct.

The typical size distribution is skewed. A parcel averaging 0.25 ct may comprise 20% exceeding 0.5 ct. It will contain occasional stones of 2.0 ct, with, very rarely, the largest being about 5 ct. Very selective diving operations have yielded individual diamonds exceeding 10 ct, and with an average size of over 0.5 ct. Those stones presently being recovered from the Orange River pale-odeltaic environment on the Namibian middle shelf are thought to average 0.7 to 0.8 ct.[5,21] Diamonds being mined on the inner shelf and, to a lesser extent, recovered by sampling elsewhere on the middle shelf, are smaller. They vary between 0.3 and 0.4 ct, with a value approaching US$200/ct. At one Namibian inner shelf site the mean diamond size is 0.34, with a mode of 0.22, a standard deviation of 0.24, and an upper limit of 2.24 ct. The population is a composite of two or three component ones, each reflecting a stage in the marine deposit's formational history.

5.1.2 GEOLOGICAL SETTING

The marine geological setting has been described by Murray et al. and others.[1,9,22] The west coast continental shelf extends up to 250 km offshore and is divisible into inner, middle, and outer shelves, each with important differences. As shown in Figure 5.3, the separation of the middle from the inner shelf is at a water depth of between 100 and 110 m. In many places the shelf change lies beneath a 20 to 30 m thick layer of young, unconsolidated sediments known as the "mudbelt." With a width of about 8 km, the inner shelf is covered by an average 30 m of water. Precambrian, crystalline, basement bedrock of gneiss and metamorphic rocks is well exposed and sparsely covered by clastic sediments. Diamondiferous deposits of the inner shelf are similar to those onshore.[1]

Late Cretaceous and Tertiary age sedimentary rocks form the bedrock on the middle shelf, which has very low relief and a seafloor gradient of 1.5%. Water depths vary between 100 and 250 m, where the outer shelf begins. In Namibian waters the bedrock comprises a very hard sandstone exhibiting 2 to 3 m of relief in seafloor outcrop. Frequent, large slabs and blocks of rock were released during regressive and transgressive shoreface erosion.[1] To the south a predominantly clay pavement constitutes bedrock. It is grey to black in color and in places is interbedded with fine-grained, grey-green sandstone. The bedrock topography is subdued and the sediment cover is generally thin. Some bedrock outcrops as low discontinuous ridges parallel to the regional strike.

With a density of 3.5 g/cm³, a diamond is the lightest of the economic placer minerals. It travels farther and is concentrated in some significantly different ways. Over a period of 100 million years, diamonds were liberated by erosion from numerous Cretaceous kimberlites in the interior of southern Africa (see Figure 5.1). Many were fluvially transported westward to the South Atlantic coast via a system of rivers, generally referred to as the Great Karoo River.[1,9] The system is now evident as paleoriver courses and as the present-day Orange River, which forms the boundary between South Africa and Namibia. Between 1.5 and 3 billion carats are thought to have made the journey of hundreds of kilometers to the ocean.[23] During that time, alternating wet and dry climates caused the sea level to vary through +300 m to −500 m relative to today's msl.

From paleoriver mouths, the diamonds were redistributed along and on ancient beaches that developed in response to the sea level changes. Powerful southwesterly winds, combined with high-energy swells, created mostly northward-moving, longshore drift currents which continue to this day. These moved sediments, including the diamonds, in water less than about 35 m deep. Coastal configurations, such as headlands, islands, indentations and reefs, affected the local marine energy level and thus the diamond distribution which extended from each river mouth. Sand and diamonds on suitably oriented beaches were also blown northward by the wind along shallow, eroded valleys (aeolian transport corridors) and across desert deflation surfaces in topographical depressions.[1] When sea levels were lower, the beaches were wider and more extensive than they are now and the effect was more widespread. In places where the coastline veered back to the east, some diamonds eventually reentered the sea. Such inner shelf features, both exposed and inundated by the sea, are illustrated in Figure 5.4.

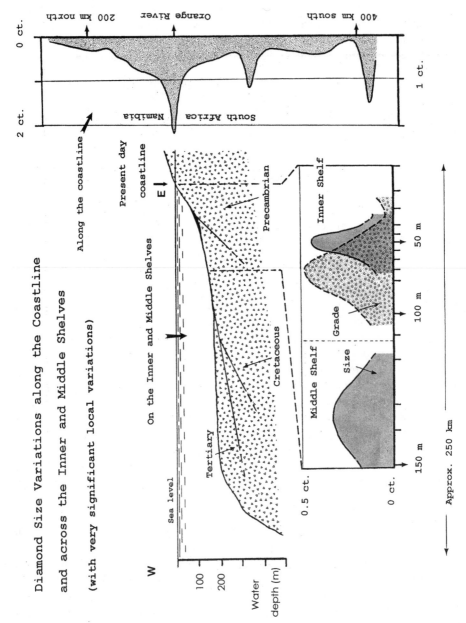

FIGURE 5.3 Inner and middle shelves of Namibia and South Africa, showing the general variation in diamond size.

TARGET PALEO FEATURES

A....Coastline

B....Embayment   C....Depression   D....River

E....Corridor   F....Valley   G....Major gullies

H....Area of energy change   I....Sheet gravels

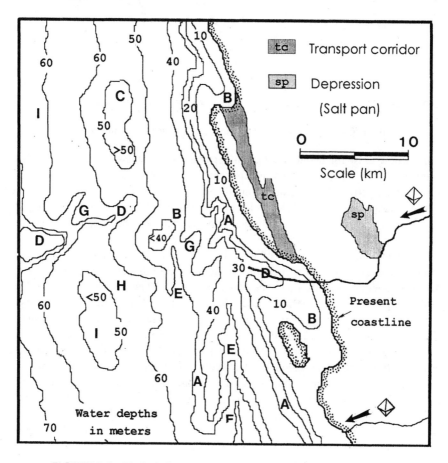

FIGURE 5.4 Typical diamond accumulation features of the inner shelf.

   The resulting diamond-bearing deposits have been continually upgraded by vigorous marine action. According to De Decker and Woodborne,[9] the winnowing of the host sediments in the shallow marine deposits by waves on the coast was the single most important process that contributed to the enrichment. Present-day diamond concentrations have resulted from the continual reworking of older mineralized deposits and the subsequent redeposition of diamonds in preferred trap sites. Wave-induced sea bottom currents persisted throughout periods of sea level change. Deposits in the surf zone were subjected to continual sorting, which concentrated the diamonds closer to the bedrock. They were trapped in gravels which collected in the gullies, pot-holes, and channels eroded into the wave-cut terraces. High energy caused their dispersal, but their final site of entrapment was more usually in a middle-energy environment. Fluvial and aeolian erosional processes acted on emerged marine deposits and vice versa.[9]

Typical Bedrock Profiles

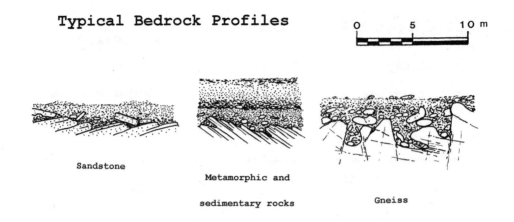

Gravel thicknesses and profile irregularities are variable in scale

FIGURE 5.5   Typical bedrock profiles from the inner and middle shelves.

Diamonds now exist for about 1400 km along the continental shelf, extending from 100 m above to at least 200 m below msl. They are concentrated between 450 km south and 300 km north of the present Orange River. The marine diamond deposits of Namibia and South Africa are unique. Onshore paleobeaches and terraces have yielded more than 100 Mct of high-quality gem diamonds. Although not extracted from the sea, they were concentrated by the ocean's action and may be regarded, geologically as opposed to statistically, as marine diamonds. Less overburden is encountered offshore where diamond grades tend to be higher than those now available onshore, but both the thickness of overburden and the location of marine diamonds may change with time.

On the inner shelf, diamonds are concentrated within drowned paleoriver channels, and in the vicinity of their paleomouths. The most frequently occurring diamondiferous deposits are those which formed in shallow, higher-energy environments on submerged terraces at a variety of water depths. To the south of the Orange River, diamonds exist in drowned paleobeaches at seven elevations ranging from 9 to 84 m. To the north the 20-m beach is the best defined, and is characterized by thin gravels and boulders. Host sediments border the edges of offshore ridges and headlands, and have collected in off-lying bars and south-facing paleobays. Diamonds are trapped in a variety of features on submerged marine terraces, including south–southwesterly trending bedrock valleys and gullies, shallow basins, and north–south seafloor depressions (see Figure 5.4).

On a smaller scale, the distribution of diamonds demonstrates a fractal relationship with the geological features. The rough gems' presence depends on the availability of favorable areas of concentration, or trap sites, in which they gather. Such sites are related to the seabed and bedrock morphology, ranging from shallow depressions to highly irregular surfaces. Diamonds invariably are associated with gravels. They collect among rounded rocks wedged in gullies and potholes and are caught up in bedrock cracks. Secondary channels containing fluvial and other terrestrial deposits exist as complex, drowned remnants at all elevations. Some examples of typical bedrock profiles are shown in Figure 5.5.

One quarter to one half of the seabed on the inner shelf is bare bedrock or is rock covered by only a thin veneer of sediments. The balance comprises unconsolidated sediments, often filling slight depressions, and usually less than 5 m thick. Thicknesses of 7 m and more exist in places where a coastal embayment is open to the north. Thick overburden may also conceal a paleoriver channel, but preservation of the underlying fluvial sediments is variable.[9] The overall, average sediment thickness is around 1 to 2 m. The poorly sorted sediments range from soft muds, through silts and sands, to gravels, coarse cobbles, and boulders up to 1 m^3 in size. Some areas contain beds of seashells and a tenacious, sticky, green clay. A free-running, rounded, siliceous sand with an average 2-mm diameter is known as "bird seed." Paleobeach deposits include lag gravels, in places no more than 1 m thick and covered by a few meters of sand. They lie with frequent boulders against the base of paleocliffs carved by waves into the bedrock.

On the middle shelf, gullies and potholes are less frequently formed. Instead, resistant ridges up to 8 m high trend generally north–northwest and provide a rough bedrock on which diamonds have been trapped. Erosion of the existing shorefaces during transgressive and regressive events released large portions of sandstone bedrock which provided the physical catalyst for diamond trap sites.[1] Cemented, diamond-bearing gravels on the dip slope ridges were flooded during marine transgressive and erosive events. Their consequent reworking produced a lag gravel from the original fluvial and deltaic sediments, increasing the diamond grade by a factor of 4 to 20.[4] Further south, where dip slope ridges controlled the ancient shorelines, diamonds occur in South African waters mostly on or in front of west-facing slopes. Behind the landward eastern ridge any gravels that may be present rarely host diamonds, unless there is a second dip slope immediately behind the first.[24] The geomorphology is complicated by multiple, subaerial, erosional events, and numerous paleoriver channels have penetrated the ridges.

Exposed bedrock and clastic sediment veneers less than 2 m thick generally occupy a smaller percentage of the seafloor on the middle shelf. In places, sandstone bedrock is exposed over half of the area.[1] Sheet gravels are widespread and in some sites cover 15% of the ocean floor. They are rarely thicker than 1.0 m. In Namibian waters boulders occur in localized areas, especially as sandstone slabs. At around 100 m elevation, sand, mud, and gravels up to 5 m thick in some places cover 70 to 80% of the seabed. Detrital sediments, with diamonds, in the 120 to 140 m range may include material washed down from around –60 m. Evidence of the inland origin of the gravels is provided by well-rounded jasper, agates, and banded ironstone typical of known terrestrial deposits.

The regional, marine distribution of diamonds is a result of subtle interactions between their general movement northward and their entrapment in local features of macro and micro scale. Bedrock morphology and sediment type are the most important influences. Diamonds are dispersed neither uniformly nor totally at random. Irregularly shaped, well-mineralized areas exist in places as widely separated "islands" within extensive tracts which are either low grade or barren. Inside such areas, the diamonds exhibit very localized distribution characteristics with a high nugget effect and variable grades over short distances. On the inner, and parts of the middle, shelf, reconnaissance sampling may be justified over less than 20% of a concession. Payable ground eventually may comprise less than half the area sampled.

Diamondiferous gravels vary in thickness and continuity. They may be overlain by, or inter-bedded with, low-grade or barren sediments. Suspended gravel layers may exist anywhere in the vertical sequence. Exposed, onshore deposits provide a model which is generally applicable to those offshore. Isolated diamonds occur almost anywhere in the vertical sequence, but the typical basal gravel unit shown in Figure 5.6 forms the main ore horizon in most areas. The sequence may be partially or repetitively represented. Gullies and potholes are well developed in the more resistant quartzitic or gneissic bedrock.[25] They contain gravel of much higher grade than that of sediments lying above the general bedrock level.[26] Unless the marine deposit is aeolian in origin, three quarters of the diamonds may be in the basal section of any gravels. The total diamond content is generally related more to areal extent than to volume and can be correlated with the sediment type and the bedrock features.[25,26]

The overall dimensions of a rough diamond, loosely termed its size, are quantified as its mass measured in carats, with 1.0 ct equalling 0.2 g. Average sizes of rough stones decline north of the river mouths.[27,28] At the Orange River mouth, for example, the mean size is 1.5 to 2 ct, but, as shown in Figure 5.3, some 200 km to the north the average diamond weighs only 0.1 to 0.2 ct. Ore grades show a tendency to increase in the same direction, perhaps reflecting the easier transport of more numerous smaller stones. Paleochannels usually contain the largest diamonds. Sizes and grades exhibit a similar pattern with increasing distance offshore. Along the offshore, submerged

Incremental volumetric grade

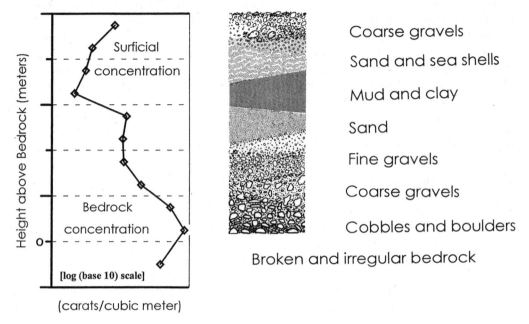

Coarse gravels

Sand and sea shells

Mud and clay

Sand

Fine gravels

Coarse gravels

Cobbles and boulders

Broken and irregular bedrock

FIGURE 5.6 Typical grade and lithological profile of an inner shelf, diamond-bearing feature.

continuation of the present Orange River and its paleodelta, however, localized increases in grade and, especially, size mirror the coastline trend. Approximately 95% of the marine diamonds are of gem quality because the industrial and less perfect stones have been destroyed during transport.

In each diamond-bearing geological feature a sympathetic relationship usually exists between grade and average diamond size. Beyond a certain threshold, however, more elevated grades are increasingly represented by larger numbers of smaller stones.[29] As a mining industry commodity, a diamond grade quoted in an economic context should always be qualified by its mean rough stone price.

The grade of a thin, diamondiferous gravel horizon, expressed volumetrically as ct/m^3, is reduced by the diluting effect of any significant thickness of overburden. Volumetric quantification does not then properly reflect the higher grade of the mineralized gravels. The grade is therefore best expressed in terms of the accumulation of carats in a planar or areal sense, as ct/m^2. The exception is where the ore thickness is known to exceed a few meters and the volumetric expression can be used.[30] However, the complication of varying ore and total sediment thickness must be taken into consideration during mine planning and in feasibility studies. The volumetric grade is then also used, and for marine sediments a specific gravity of 1.5 to 1.7 is applied.

5.1.3 OPERATING CONDITIONS

The offshore environment is not ideal for mining. Fog and gale force winds are frequent, and wave heights may exceed 5 m in all seasons. From September through May high-intensity swells occur with significant wave heights of 1 to 3 m and a period of 6 to 8 s. During the winter months of June to August the figures increase to 5 to 7 m and at least 10 to 15 s.[31] However, waves with 20 s periods have been recorded in every season. Wave conditions for 95% of the time are capable of transporting very coarse sand at a depth of 30 m. Storm waves occurring for 5% of the time can move medium pebbles (10 mm diameter) at 30 m depth and small (100 mm diameter) cobbles at 15 m.[9]

The geotechnical properties of the bedrock and its sediment cover influence the choice, and determine the effectiveness, of seabed sampling and mining systems. Physical stability and mining efficiency are governed by the strength characteristics and bearing capacity of the sediments.[32] In general, deposits of the middle shelf are more amenable to mechanized mining than those of the inner shelf. Sand overburden presents problems of removal and disposal because it exists through the full spectrum from free running to fully cemented. DBM has recently dedicated the 2324 gross tonnage *Zealous,* commissioned in January 1997, to geotechnical surveys.[20] The twin-screw vessel is 79 m long overall with a total installed power of 5680 kW, and is equipped with bow and stern thrusters.[8]

5.1.4 EXPLORATION

The technical approach to exploration is dictated by the physical conditions, the nature of the target, and the informational requirements for the mining stage. An established sequence of tasks is similar to that employed for onshore prospecting: geophysical surveys, ground truthing, interpretation, reconnaissance and evaluation sampling, geotechnical studies, bulk sampling, and trial mining. Finally, a feasibility study may lead to commercial mining. The sequence is followed by most operators but with significant differences in emphasis on the component stages.

A program starts with the identification, by means of bathymetric and geophysical surveys, of seabed features favorable for the accumulation of diamondiferous sediments. Vibrocoring and grab sampling, typically with a 350-l, hydraulic clam shell grab for water depths of up to 80 m, assist in ground truthing the interpretations. Remote and direct observations, the latter by small submarine, are increasingly being made. The results are used to produce maps of the seafloor at an appropriate scale such as 1:5,000. These show the distribution and possible lithology of any exposed bedrock, the sediment types and their estimated thicknesses. Figure 5.7 is an example.

On a macroscale, favorable features include reefs and changes in slope, coastline embayments, wave cut terraces, paleobeaches and submarine cliffs, fluvial paleochannels, bedrock basins and depressions conforming to paleopans, major gullies, and lag gravel sequences. For some companies a study of sediment dynamics on paleoshorelines is an integral part of the work.[9] Wave refraction diagrams, involving estimations of the wave energies and velocities for the old coastlines and water depths are produced. An assessment is made of bedrock roughness, and the frequency of gullies and fractures is quantified.

Recognized features of interest are studied further with more detailed mapping and evaluation. Reconnaissance sampling is used to identify areas of diamondiferous sediments and to confirm a feature's prospectivity. In the event of positive results, a zone of mineralization is delineated by evaluation sampling on a 50 × 50 m grid. Figure 5.8 provides an example from Namibian waters of this progression. Using a variety of methods, the grade is estimated and a resource may be determined to exist.[29] Ideally, any resource estimation is followed in selected sites by confirmatory and bulk sampling. These stages increase confidence in the grade estimation, yield sufficient diamonds for carat valuation, and provide data for downstream planning. Different sampling systems may be used for grade confirmation and to measure the geomorphological characteristics for mining system consideration.[29]

Geotechnical studies are essential in order to proceed to a feasibility study stage which should include trial mining before a reserve may be declared. Only then may detailed mine planning, production scheduling, and hence full-scale operations commence. DBM, for example, now deploys three vessels for seabed characterization but has published no results. The sparsity of generally available geotechnical information may result less from lack of publication than from nonacquisition of data by other industry participants. There may be insufficient definition by some potential producers of the objectives for each stage of a program. The important distinction between evaluation and bulk sampling and between trial mining and production is often blurred. Definitive tests and decision stages are thereby omitted. Planning and scheduling are then based on insufficient

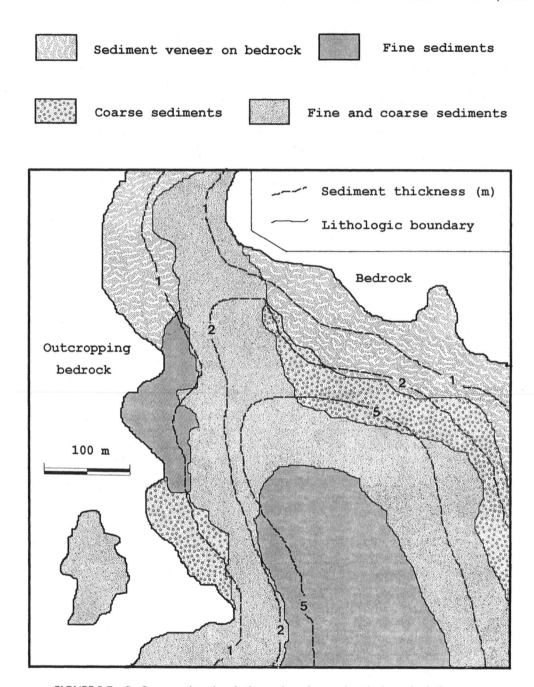

FIGURE 5.7   Seafloor map based on bathymetric and ground-truthed geophysical surveys.

data with which to estimate properly the expected mining rates. A resource with inadequate, geological description and no geotechnical information is incorrectly regarded as being ready for mining.[29] Without adequate test mining using the selected system, the first indication of the problems lying ahead is a serious inability to mine in production mode at the planned rate.

Accurate positioning is usually achieved by using a differential global positioning system. It involves the tracking and positioning of survey, sampling, and mining vessels together with towfish and excavation tools. A combination of surface and subsurface techniques is often involved and accuracies of better than 5 m are now expected. The ability to relocate exactly on a remotely

◇ -- Reconnaissance samples   ◯ .... Confirmation samples

● --- Evaluation samples   ▨ ----- Bulk samples

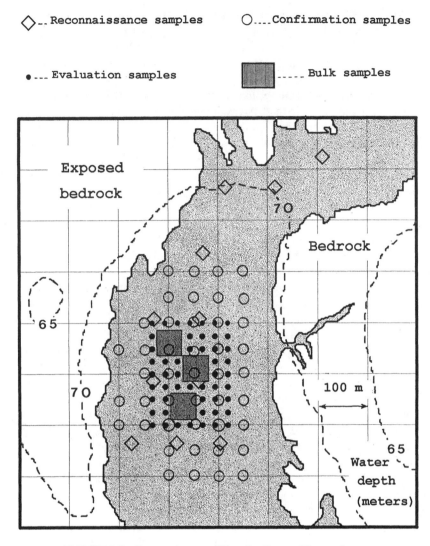

FIGURE 5.8 Progressive sampling of a diamondiferous feature.

identified seabed feature is essential for the mining of microfeatures and for efficient diamond extraction. DBM crews reportedly can relocate a vessel to within 1.5 m and apparently place a drill head to within 0.5 m of a targeted seabed area.[11] Accurate depiction of an area, before and after excavation, is essential, and computerized geographic information systems are widely used.

Once on station, a vessel is maneuvered accurately with an anchor spread or dynamic positioning system, which is then used to maintain precisely the vessel's location. Computers on board DBM's *Coral Sea* and *Grand Banks* respond at 2-min intervals to directional fixes from satellite and shore beacons. Each vessel's position is corrected by adjustment of the four anchor chains. Means of positioning and tracking seabed equipment are described by Donavan and Norman.[33] A short baseline acoustic navigation system consists of a hull-mounted transducer array, which both interrogates and receives signals from a transponder mounted on the seabed equipment. Remotely operated systems on the seafloor are positioned and tracked by means of a baseline acoustic navigation system: an array of seabed transponders with a baseline several kilometers long.

Companies other than DBM have contracted out their geophysical survey work, but the trend is for them to become increasingly self-sufficient with their own vessels.[13] Bathymetric, side scan sonar, and seismic sub-bottom profiler techniques are used with towfish tracking equipment. For regional side scan surveys, 100% coverage is usually sought, whereas reconnaissance work may commence at a 200-m line spacing.[34] For detailed surveys of geological features, the line spacing may close to 50 m.[9,35] ODM and DBM have completed surveys at 10- and 12.5-m spacing, respectively.[13]

Pinger or chirp, high-resolution seismic methods provide a 0.50-m resolution and a 5-m penetration, with 15 m maximum in mud. Areas of thick sediment accumulations are surveyed with deeper penetrating sparker or boomer systems.[9] A 2-m resolution and 25-m maximum penetration in unconsolidated sediments is achievable. Chirp seabed profilers are used for mapping any pale-ochannel extensions on the inner shelf.[1] The survey line spacing depends on the objectives and local conditions. An initial reconnaissance survey may involve 2- to 5-km line spacing north–south and east–west. Infill surveys at a few hundred meters spacing in selected areas confirm geological features which are examined progressively in greater detail. A final 5-m spacing is now being considered by some operators.

Alluvial diamond deposits constitute one of the most difficult sampling problems in the international mining industry. Diamonds are extremely irregularly distributed and the ratio of waste materials to gemstone is incredibly high, requiring very large samples for the results to have any validity. Finally, the grade must be qualified by the $/ct value. The problem is compounded offshore by being in a marine environment.

The primary objective of sampling is to establish whether or not the combined diamond grade and value is sufficiently high to allow possible commercial exploitation of a deposit. If so, the purpose becomes the delineation of a resource.[29] Sampling requires the extraction and recovery at surface of a volume of ground which matches the measurable dimensions of the seabed excavation created by its collection. For a sample to be meaningful, all the diamonds within the excavated area down to, and on, the bedrock should be recovered. Each sample should be as representative as possible of the surrounding ground, and the program must yield a sufficient number of rough diamonds to allow their reliable $/ct valuation. Each and every sample has the potential to increase the geological understanding, and it is under-utilized if not designed to do so. An important, but often less well recognized, objective is to gather quantifiable information concerning the lithology, stratigraphy, and geotechnical characteristics of the sediments. These data are used to determine the optimum mining method. Without them a realistic estimation of a mining rate and expected recovery of volume and grade cannot be achieved.

Even in a relatively high-grade deposit of 0.5 ct/m^3, the diamond content is only about 0.15 ppm. If the average diamond exceeds 1.0 ct, then each is contained in several m^3 of sample. The ratio of valuable weight to that of waste is about 1:10 M. A less extreme example would be a grade of 0.2 ct/m^3 and diamonds of 0.50 ct in size. Then, on average, 2.5 m^3 of gravel must be treated to find one stone. Any sample of 1 m^3 volume has a 67% chance of containing no diamonds. The chance of there being one diamond is 27%, reducing to 1% for three diamonds. The difficulty of obtaining a representative sample intensifies with decreasing stone frequency and with increasing stone size.[36] In higher grade and geologically homogeneous areas, the sample size can be reduced. The parameters to be determined during the planning of a sampling pattern are the required volume or seabed coverage of each sample and their total number.[30] These may be expressed as the sample spacing and, more completely, as the sampling density, which relates, as a ratio, the individual sample volume to that volume of the ground each is required to represent.

Experience gained from onshore diamond investigations has provided a basis for the approach to marine sampling. De Beers Consolidated Mines Ltd. commences onshore with a reconnaissance phase of probe drilling on a 200 × 200 m grid, sampling at 1.0-m vertical intervals, and penetrating 3 m into bedrock.[24] All the geological features are recorded and plotted, and drilling may continue down to a spacing of 50 m. A second phase involves taking larger samples in defined areas of revealed potential. Overburden is removed and a discontinuous, 1-m-wide trench is dug in 10-m lengths each 10 m apart. Alternate trenches are divided into two horizontal 5-m sections, each separated into 0.5-m vertical zones. In soft or semiconsolidated sediments cased, 1.6 m diameter, auger holes yield samples at 1.0-m vertical intervals. One operator uses a grabbing rig to cut a slot 2 m long and 1 m wide with the ground supported by bentonite mud.

Uniformity, or otherwise, of grade and the desired confidence level of the grade estimate influence the choice of sample size.[24] In common with those onshore, the marine diamond deposits display a cluster effect. The trap sites in which the rough gems usually gather may measure only a few meters in size.[9] If the sample is smaller and the sample spacing greater than the average trap site's dimensions, then the probability of collecting a representative selection of all the diamond sizes is reduced. Large, higher $/ct stones tend to be under-represented. Geomorphological interpretations are inhibited, and the trap site size is not revealed.[30] Sample size is also governed by the purpose of the program, whether for reconnaissance, evaluation, or confirmatory bulk sampling. Each stage requires progressively larger samples and some sampling systems are suitably used for more than one stage.

Specification of the minimum sample size necessitates definition of the required probability, at the marginal grade, of recovering a predetermined number of diamonds.[36] The acceptable percentage of samples which by chance will contain none is decided. Too small a sample results in barren or low-grade ones being over-represented and in some being seriously over-valued.[29,30] The sample's volume should be large enough to suggest that on average at least one diamond per sample is recovered, but its size attainment may be inhibited by practical limitations. Instead, groups of subsamples may be collected by cluster drilling, providing in total a sufficient bulk volume, or support.

Sample sites should be spaced closely enough to allow not only the establishment of grade but also of ore continuity at the boundaries of features.[29] If the spacing exceeds the average trap site's dimensions, a variogram fails to depict the spatial clustering, and kriging ceases to be the best unbiased estimator. The spacing may allow for anisotropy, and both ODM and Benco have used a separation of about 50 to 90 m for resource delineation. Grade distribution in an area encompassing numerous trap sites approximates to log-normality but can be fitted only when a constant is added.[29] As the sample size, or support, increases, the histogram becomes less skewed, more normal, and the confidence limits decrease.

Reserves generally can be delineated with sampling ratios ranging from 1:500 to 1:20,000, depending on the type of feature. No fixed rules exist, but if enough samples are taken, the mean of the grades remains approximately the same with increasing sampling density. In the 1960s DBM's reconnaissance sampling on the inner shelf was conducted on a grid of 100 × 2000 m, closing up to 10 × 400 to 500 m.[3,6] Continued sampling for evaluation was at 10 × 50 m, an effective density of 1:1500. In similar areas, ODM has employed a sampling density in shallow water of between 1:17,000 and 1:36,000. BHP Minerals and Benco used one which ranged from 1:18,000 for reconnaissance sampling to 1:625 for evaluation sampling.[37]

5.1.5 SAMPLING SYSTEMS

The process of sampling involves three main procedures: excavation of the sediments down to a cleaned up bedrock, the delivery of them all, or a diamondiferous portion, to the surface, and the metallurgical recovery of the diamonds. At the seabed the sediments are loosened as necessary and gathered, and the excavation must be stabilized.[31] Penetration should be possible with an ease which allows excavation to a sufficient depth and at a high enough speed.[33] Exploration is cost intensive, and sample integrity must be maintained. It should never be compromised by allowing equipment failures or procedural errors to cast doubt on the allocation of recovered diamonds between consecutive samples. Bedrock irregularity must be recognizable and, ideally, measurable. The extracted sediments should be attributable as closely as possible to the theoretical volume of the excavation as based on the footprint area and average penetration depth. Preferably, but within practical limits, if more than one sediment type, such as gravel and sand overburden, exists each should be identifiable and individually collectable for study. A sampling system must be economically efficient and yield results in which there is confidence. It must be suitable for the marine conditions, the water depths, and the positioning requirements. The degree of engineering sophistication influences the cost and equipment availability.[33]

There are three categories of systems. Those used in the worldwide, onshore diamond industry and deployed offshore without any modification include Banka, churn, rotary, and percussion drills. Reverse circulation drills and airlift systems are also available. Each has limitations and inherent sources of error which may be magnified under certain marine conditions. Second, specialized equipment from other industries onshore, such as the Wirth drill and Bauer cutter, have been specially modified for marine use. Finally, greater acceptance of remotely controlled mining equipment has triggered the replacement of earlier methods, especially the airlift system if used alone, by seafloor-mounted systems and mobile vehicles. In addition to allowing controlled, selective excavation, such seabed units have an important advantage. The same design criteria may be employed on machines intended specifically for sampling or mining. Alternatively, one machine may be used for both tasks. The compatible, or multi-duty, equipment approach ensures reliable estimates of the recoverable diamond content and greater confidence in predictions of mining viability. Sampling and production units are almost interchangeable, and the operator, not the process or system, determines when bedrock clean-up is complete.[32]

Some mining systems are incapable of completely excavating all the sediments, and thus every diamond, underlying a sample area. Use of a sampling system with an equal weakness provides an estimation of the recoverable diamond content of the area by the specified mining system. However, such an approach must be used with care, and preferably only if there is no doubt about the mining system later to be used. The least desirable approach is to employ a sampling system that imitates an inefficient mining approach which may not eventually be the chosen tool for production. Sampling may so significantly under-value the sediments as to lead to the assumption that mining is unjustified. Alternatively, the economic advantages of potentially higher diamond production from using a more efficient mining system may be disguised. No amount of mathematics, statistics, or geostatistics can overcome any such deficiencies in the sampling procedure.

All the systems used in southern African waters are deployed from ocean-going vessels, the dimensions of which are dictated partly by the sampling systems deployed from them. Bow thrusters and powerful main engines ensure that a vessel is capable of sufficient maneuverability and of cruising at a reasonable speed.[31] Ones not equipped with dynamic positioning rely on a spread of anchors, which preferably can be set and recovered without assistance. Vessels are normally stationed with the bow heading into the oncoming swell, minimizing the movement as the prevailing wind varies in direction. Launch and recovery of the sampling tool is either through a moonpool from a drill tower, or via an A-frame amidships or at the stern. Donovan and Norman describe the launch and recovery of the sampling or mining equipment from a vessel to the seabed and its return onboard.[33] Equipment design requires that due consideration be given to the various stages involved. Swell compensation is necessary for the operation of some systems. It is either active, being adjustable during operation, or passive, pre-set, and nonadjustable during use.

For onboard diamond recovery, the larger operators now use dense media separation (DMS) plants. Primary screening and storage, ferro-silicon separation, and X-ray recovery, are followed by hand sorting. The screen product destined for heavy media treatment by cyclone is usually about 1.5 to 20 mm, reflecting the expected size range of the diamonds, and is commonly 5 to 40% of the feed. The denser cyclone underflow of concentrates containing diamonds is screened, washed, and fed to the X-ray recovery plant. Modular plants of up to 150 t/h capacity, and containing predesigned DMS and X-ray modules, have been successfully integrated into sampling and mining operations at sea.[19] They generally yield 1 t or less of diamondiferous concentrate from 80 to 100 t of screen product. On DBM's vessels the final concentrate, 20 to 30% of which usually comprises diamonds, is less than 0.01% of the original feed.[26] It is automatically sealed into 1.0-l tins, each of which is bar-coded to indicate the source coordinates, for secure transport ashore.[11] For all the operators, any theft of diamonds from exploration and evaluation samples is more critical than from production because the resulting undervaluation may adversely affect major investment decisions.[18]

The vibrocore drill is used extensively for ground truthing any geophysical interpretations. The vibrator and core pipe are supported by a tower integrated into a base which is set on the seabed. The drill produces a 6-m-long by 84-mm-wide, relatively undisturbed, sample of the sediments penetrated and has been successfully used in water depths of 350 m. A conventional Banka drill, mounted on a 97-t barge, has been employed for ground truthing and offshore target identification in Indonesian waters. Larger drills use the elevating power of compressed air, instead of mechanical means, to raise the sample to the surface. Nesbitt and Murray have described some of the large-diameter drilling systems employing airlift retrieval and used offshore in the 1970s.[3,6] The term *airlift* refers to the method of upward transport of a sediment slurry impelled by expanding air in a pipe. The term is also used to mean a complete sediment recovery system for sampling and/or mining and which depends almost entirely on compressed air for loosening, collecting, and transporting.

The performance of any drill varies according to the geological environment in which it is used. In particular, the thickness, degree of unconsolidation, and range of particle sizes have a major influence. Where onshore drills are used at sea, the prevailing hydrostatic conditions and the generally looser sediments complicate their behavior. Sediments can be pushed aside, flushed out of the bottom of the drill, or subjected to a jigging effect which may be accentuated by movement of the vessel. Sloughing of the side walls may result in "coning" and "belling," with either barren dilution or diamond salting as a result.[6] Sediments may be disturbed excessively and rapidly, confusing or invalidating geological observations.

DBM has made extensive use of a 0.96-m-diameter airlift device known as the "Mega-drill," which was originally developed for use in shallow waters. Each 7-t drill is fitted with a penetrometer which continually indicates the nature and depth of strata changes. The *Douglas Bay*, is equipped with two Mega-drills, one each to port and starboard. The 84-m-long, 2172-gross-tonnage vessel has a total installed power of 4006 kW and is refueled at sea.[8] She is equipped with three mooring winches and sets her own 5.5-t anchors. The Mega-drill's footprint is 0.72 m^2 in area, and the drill spacing may be either a few meters or multiples of 20 m. Holes are normally completed in clusters of three or six to provide the necessary sample support. The maximum seabed penetration is 8 m. The time required for a 2- to 3-m thickness of sediments is generally 10 min, and about 45 min for 7-m penetration. Despite the attachment of 75- or 100-mm boulder guards, large flat stones can turn lengthwise and enter the drill head. Recovered samples are treated on board through a 20-t/h DMS plant, X-ray machine, and heavy minerals laboratory. Sampling operations can be delayed by sea conditions, and all attempts to drill are halted if the swells exceed 4 to 5 m. The Mega-drill is now recognized to be capable of overvaluing sediments susceptible to slumping and excessive volume recovery, and, in places, to undervalue thin veneers of gravels in deeper waters of the middle shelf.

BHP Minerals and partner (Benco) selected trench-cutting equipment manufactured by Bauer Spezialtiefbau GmbH, for the development of a marine sampling tool depicted in Figure 5.9. The chosen machine incorporated vertically aligned cutter wheels which counter rotated. Mounted on drives, they continuously loosened and reduced the size of any material being impacted.[31] The Bauer cutter was 12 m high and 2.8 m long by 1.2 m wide in plan. It was housed within an outer, 5 × 5 m guide frame with four adjustable legs by means of which the wheels were lowered to impact with the seabed. The 60-t cutter was deployed from the *Geomaster* (Figure 5.10), a specially converted vessel of 1566 gross tonnage and 123 m in overall length. She was dynamically positioned, and the cutter was deployed through a 6 × 6 m moon pool. Samples were fed to an onboard, 150-t/h screening plant; the 2 to 16 mm sized product then went to a 10-t/h DMS plant. Concentrates were hand sorted on board.

De Beers Marine's Rotary ("Wirth")

Drill System

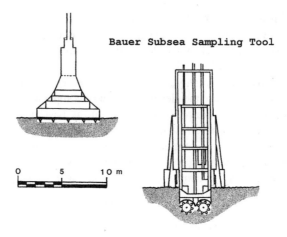

Bauer Subsea Sampling Tool

FIGURE 5.9   Large drilling systems modified for marine use.

FIGURE 5.10   The Geomaster leaving Cape Town for Namibia.

The cutter was suspended by two 43-mm-diameter non-spin wire ropes on a hydraulically driven, 300 kW, double drum winch with line pulls of up to 80 t. It was lowered, landed and recovered by means of a 23-m-high drill tower and active heave compensation system.[31] Diesel generated hydraulic power was supplied to the cutter via a band of 10 hoses. The cutter wheels, a hydraulic cylinder on the sampling tool, and the hose storage drums were driven by a 430-kW power pack. The whole assembly was lowered to the seabed at 25 m/min, and the cutter descended vertically into the sediments to excavate a trench theoretically equal in horizontal dimensions to those of its own 4m² footprint. The penetration rate averaged 4 m/h in hard soil conditions; otherwise it was higher. Slurried sediments were pumped to the surface through a 200-mm rubber hose by a

150-mm centrifugal pump above a suction box in the cutter frame. During demonstrations, 600 m^3/h of slurry could be lifted at a velocity of 7m/s with a maximum particle size of 90 mm. Larger cobbles were either broken by the wheels or remained on the seabed.

Several limitations were imposed by BHP Minerals on the Bauer sampling program. Targeted geological features were each required to be at least 5 ha in area with a minimum plan dimension of 100 m and in water deeper than 30 m. Samples were taken only if the location exhibited sand at surface and was no closer than 25 m to any known bedrock or coarse sediment outcrop. Drilling to a maximum depth of 6 m was conducted on 50-m centers.[23] For each location the depth of penetration, estimated stratigraphy, amount of concentrate recovered, and the number and description of all sorted diamonds were recorded.[37] From more than 4000 sample sites, 80,000 t of sediment were recovered in one year at an overall rate of 11 sites/d. The weekly performance ranged from 10 to 210 samples collected in all types of features and in water as deep as 129 m.

On parts of the South African inner shelf, the Bauer cutter revealed an estimated average grade of 0.145 ct/m^2 in areas totaling about 1.8 km^2.[37] In individual features the volumetric grade ranged from 0.12 to 0.44 ct/m^3. Further north, 882 samples were collected in three months, totaling 9400 m^3 and equivalent to 10.7 m^3/sample. The numerous geological features, with a combined area of 23.9 km^2, were subjected to a sampling density of one sample site per 2.7 ha of seabed.

The Bauer cutter penetrated sand, gravel, calcarenite, clay and cobbles. Where excessive quantities of clay, sand, and shell existed, the sample size was restricted to avoid problems in the treatment plant. Volume recovery was low in areas with boulders and where bedrock topography variations exceeded 0.5 m within the cutter footprint. The tool was not completely effective in clean-up unless bedrock was fully penetrated over the entire 4m^2. Some material usually remained at the bottom of the hole, piled between the rotating cutter wheels, and perhaps resulting in underestimation of the grade.[37] Where penetration of the sediments was easy, it was accompanied by slumping and excessive volume recovery. In places, holes were terminated prematurely because the cutter wheels continually jammed when boulders, particularly crystalline, rounded ones, were encountered. Some areas were therefore avoided, and large gullies were sampled only if thought to be more than 70 m wide. Details of the Bauer cutter's behavior and limitations nevertheless reflect the welcome public availability of information rather than any negative performance relative to other less well described systems. The Bauer cutter continues to be used for drilling deep, 1.0 by 1.8 m holes onshore for De Beers.

DBM has perfected, and continues to employ, two different mining methods in deep water: a remotely operated, tracked vehicle and a large-diameter reverse circulation drill.[4] The latter, shown in Figure 5.9, is variously referred to as a "type of rotary maul" or as the Wirth drill because the original was manufactured by Wirth Maschinen-und-Bohrgerate-Fabrik GmbH of Germany. A vertically suspended string of 500-mm-diameter steel pipes, with a combined weight of about 130 t, is connected at its seabed end to a large-diameter, circular steel table. Several tungsten carbide-tipped drill heads or cutting wheels around the underside circumference bite into the seabed when the drill is rotated on its vertical axis at about 5 rpm. The stirred-up sediments are ingested through 300-mm intakes in the table and transported by airlift up the pipe to the treatment plant at a velocity of about 14 m/s.[11] Brute force applied to the table causes it to descend through the sediments and to grind into the bedrock for full diamond recovery.[33] Three table sizes are used: 3.5 m diameter for sampling and 5 or 7 m for production.

The Wirth drill was first employed as a mining tool in 1991 on DBM's *Coral Sea*. The vessel, of 6054 gross tonnage and 122 m in overall length, is shown in Figure 5.11.[26] Since 1997 she has operated as an evaluation sampling vessel, also equipped with the smaller, 20-m^2, drill suspended from a derrick tower through a moonpool.[8] Although the drill in places tends to track a preexisting circular excavation, the system nevertheless should provide the best estimate of the future grade to be recovered by the larger production unit. The *Coral Sea* operates in swells of up to 4 m, and work is halted when winds exceed 55 km/h (30 knots).[11,28]

FIGURE 5.11   The *Coral Sea* of De Beers Marine undergoing a refit in Cape Town harbor.

The smaller companies' workhorse for sampling and mining has been the simple airlift system, using compressed air assisted physically by only the action of the intake being dragged across the seabed. The system comprises a steel suction pipe, about 20 m long, terminating with the suction head at the lower end. A simple mining head is illustrated by Figure 5.12. Others may be rectangular. The pipe is suspended from and recovered by two side-mounted A-frame davits, which hold it in an inclined position parallel to the length of the vessel from which it is deployed. Compressed air, supplied by hoses, is injected close to the suction head opening which rests on the seabed. Slurried sediments captured by the intake are transported up the pipe and through a flexible hose from its upper end to the onboard plant. A boulder guard limits the maximum particle size entering the head to about half the pipe diameter, or less. The system is simple, robust, reliable, and easily maintained. Its inherent buoyancy assists in the pipe's handling, but digging efficiency is low and can be controlled to only a limited extent. The airlift is unsuitable for use in water of less than 30 m, but it has been useful in mid-depth waters of the inner shelf in the absence, until recently, of any alternative system for the smaller operator.

ODM, with its experience in airlift operations, dedicated the 56-m-long, 860-gross-tonnage *Oceandia* to such sampling. The vessel is equipped with a 355-mm dredging head and a 10-t/h diamond recovery, DMS, plant. Approximately 10 m^2 of seabed are sampled at each site, and the recovered sediments are evaluated on board for diamond content.[13] In 1995 ODM sampled a total inner shelf area of some 11 km^2 in order to delineate reserves in advance of a mining program. Around 2200 samples were collected on a 50-m grid at an average rate of 10/d to yield 2580 ct.[35] In 1997 Benco used the *Moonstar* in airlift sampling mode on the inner shelf in slightly more than 30 m of water. Sample trenches, theoretically 5 m long and 1 m wide, were excavated to give a 0.9% sample coverage of the ground. The vessel completed 154 samples in one month, 74% of which were diamondiferous. A total of eight features, having an average estimated grade of 0.17 ct/m^2, amounted to 1.3 km^2 in combined area.[37]

The simplicity of design also has its drawbacks. The airlift system is disadvantaged by the imprecise control of the suction head, and it lacks any easily directed, aggressive excavation. Bedrock attainment and clean-up efficiency range from poor to good, depending on the prevailing conditions, but can be monitored only after the event. Cleaning may be limited to a distance no greater than the diameter of the suction head and is reduced by the presence of cobbles and

FIGURE 5.12   An airlift mining head, fitted with a boulder guard.

boulders.[32] Oversize gravels tend to form an armoring over the underlying sediments, preventing their full penetration. Blockages occur at the intake, restricting throughput. They are cleared by intermittent operation but the intake can become rock bound. In areas of any significant depth of loose overburden and running sands the airlift digs a cone-shaped pit or a trench with collapsed sides. Attainment of a depth of 5 m below seabed in such conditions has been known to generate a 30-m-diameter pit. Up to about 7-m depth of sediments can be removed, but a small vessel may take 3 or 4 days to open up an underwater mining face.

Diamonds recovered by airlift trench sampling are not easily related to a very specific area and sediment depth. Excessive volumes may not be measurable, depending on the plant layout, and may result in overvaluation. Equally, insufficiently cleaned bedrock prevents all the diamonds from being reported and therefore causes underestimation of the grade. The system may illustrate the diamond content recoverable from subsequent airlift mining but can seriously underestimate that extractable by a more efficient mining system. ODM reports that, in its own operations using this method, as much as 30% of the sediments were not being mined, presumably involving a larger percentage of the contained diamonds.[14] The airlift has been used extensively for evaluation, bulk sampling, and mining, but recently available alternatives may now be preferable.

Sampling systems deployed on the seabed, and positioned and operated by remote control with underwater monitoring, have the potential to overcome the recognized inadequacies of the airlift system. Royal Boskalis Westminster n.v. and its subsidiaries have designed, sometimes with partners, various such underwater excavators. The units have nameplate outputs of up to 200 m³/h, and are capable of operating to 600-m water depth.[38] The first to be used for mining was the tracked Tramrod, employed in gold mining offshore Alaska in 1989.[39] A hybrid, static version, the Namrod, was subsequently designed. It was used for sampling by Namco.

Namrod, the subject of Figure 5.13, is a complete, self-contained, workstation for sampling and is intended to operate in water depths of up to 150 m.[16,19,40] It is supported by landing pads which exert a very low ground pressure and can be remotely adjusted from the surface to ensure leveling and optimum operation of the component systems.[38] Two recovery booms gather the sediments which are delivered to a built-in, underwater screening plant. A full instrumentation package enables the operator during sampling to monitor the screening, the operational parameters, and the machine's attitude, which should not exceed a 10° slope. Terrain variations are scanned and logged by computer, and clean-up performance is recorded on video.

FIGURE 5.13   The Namrod being lowered into harbor waters for testing. (Photograph supplied by J. Wenlock.)

The static, 35-t machine is 7.8 m long overall by 7.8 m wide and stands 5.4 m high.[40] An excavation boom, fitted with either a backhoe bucket or clam shell grab attachment, can reach 2 m below the seafloor. It removes the seabed material and transfers it to a grizzley bar separator, after which a rotating trommel separates the finer sediments. A second boom with a 200-mm, centrifugal pump suction unit handles cobbles up to 120 mm in diameter and cleans up the bedrock, assisted by a highly powered jetting ring. A 10-t force is exerted on the suction head which may incorporate a short impact pick and can reach into fissures and gullies to a depth of 4 m. The underwater screening is an integral, power-saving, part of the system. It retains only the –50 or –75 mm fraction, which is transferred by belt conveyor into a collection hopper for pumping to the surface.[16] Oversize is returned directly to the seabed.

The remotely controlled Namrod was deployed by Namco using a passive compensation system from the dynamically positioned support vessel, *Sprut*.[19] It reportedly cleared down to bedrock areas of up to 10 m^2 each, and can handle the full range of mud, sand, gravel, cobbles, and boulders up to 0.5 m.[38,40] During Namco's program volumes of up to 400 m^3/h at 400 kPa were pumped, and sediment throughput varied from 25 to 50 m^3/h. Up to 40 m^3 of sediments were excavated at each sample site. The greatest penetration achieved below seabed was reportedly 7 m,[19] and trenches as long as 100 m were dug. In one year, 638 samples were collected in water depths of between 30 and 125 m. The average rate was 1.5 samples/d, with five being the maximum. The Namrod's design expectations in terms of bedrock clean-up of the diamondiferous sediments were reportedly fulfilled. One disadvantage, however, was the slow rate of operation, aggravated by temporary launch and recovery problems.

During Namco's survey, 2956 diamonds were recovered by Namrod from 184 samples, each of 10 m^2 area, indicating an average estimated grade of 0.62 ct/m^2. A single, albeit unusually large, sample yielded 219 diamonds and the most prospective feature reportedly averaged 2.1 ct/m^2.[41] Subsequent bulk sampling at three sites, each representing approximately 130 m^2 over a distance of 0.75 km yielded 3161 diamonds, indicating an average grade of 0.76 ct/m^2 at 0.37 ct/stone.[16] From another parcel of nearly 300 diamonds, the average weight was 0.29 ct, with the largest stone being 1.3 ct.[18]

More recently, Soil Machine Dynamics Ltd. (SMD) and Paragon International b.v. (PI) designed a 15-t, seabed drilling system, called Subsam II.[40] The hydraulically powered, skid-mounted drill, is simple and reportedly capable of quick deployment and collection of large samples of defined volume.[32] It comprises a base frame supported on three adjustable legs for leveling on an uneven seabed. Two supports guide the drive platform and a vertically mounted suction pipe to which a downward force of 40 kN can be applied. The pipe is equipped with a rotating head of 0.8- to 2.0-m diameter for mechanical cutting, suction through the head, and optional jetting to a depth of 6 m. The machine can operate to a water depth of 200 m at an estimated rate of up to 20 samples/d.[32] Built-in instrumentation allows accurate logging of the penetration depth and monitoring of all operating variables. Slurried sediment removal is by airlift, advantageously allowing Subsam II to be deployed from an existing airlift mining vessel with an installed crane, or A-frame, and stabilizing system.

The same two companies have also designed and constructed an articulated dredge arm (ADA) bulk sampler to complement their new SDM 50 mining system.[32,40] ADA was supplied to ODM in 1998 for installation on the *Oceania*. Weighing 10 t, the electrohydraulic system allows the remote and accurate guidance of an inclined suction arm with airlift head on the seabed. Manual heading control uses a thruster during deployment, and the front pair of skids are adjustable through 0 to 750 mm for operation on an uneven or a sloping seafloor. The suction arm is hydraulically rotated through 90° and can be lifted, tilted, and extended with a horizontal and vertical force of 10 and 40 kN, respectively. With a position control and instrument system, the 1.2-m-diameter suction has optional cutting facilities. The head can be positioned at any point within the sample area of up to 4 m² to excavate to a depth of 2.5 m below the seafloor. Samples of 1.5 m² extent can be taken to a depth of 4.5 m, and of 1.13 m² to one of 6.0 m. In water depths of up to 200 m, loosened sediments are lifted to the surface via a 350-mm-diameter hose with two 0.375 m³/s air compressors, or by a dredge pump.[40] Full instrumentation includes cameras, transducers and sensors, and on-screen simulation of the operation. Immediate data plotting is possible. ADA is designed to be attached to an existing vessel-mounted airlift system.

5.1.6 MINING SYSTEMS

The selection of an appropriate mining system requires full knowledge of the seabed geomorphology, with a quantified understanding of the nature and composition of the sediments. The seabed must be characterized in order to estimate and, preferably test, how the sediments react during each stage of the mining process: loosening or disaggregation, collection, bedrock clean-up, and transport to the surface.[32] The selection of a mining system considers the same factors, and more, as a sampling system.[31,33] The design alternatives are either a mining head connected mechanically and directly to a vessel, such as the Wirth drill or an airlift, or an increasing choice of remotely controlled systems placed on the seabed. The areal extent of a deposit, the overall grade, horizontal continuity, and vertical variability dictate the required excavational selectivity. The desired technological sophistication, the supporting engineering requirements, and the cost are all considered.[33] A proposed system is then judged in terms of expected availability, excavation efficiency, degree of selective extraction realistically provided, mining rate, and overall economics.

In the most selective of all systems, divers operating from small boats less than 30 m in length manipulate suction hoses in water depths of up to 25 m (Figure 5.14). To offset the extremely high unit costs/m³ they extract small, high-grade pockets of gravel from very irregular bedrock trap sites. Sediments are raised to surface by the suction of a deck-mounted gravel pump through 100- to 200-mm, reinforced, flexible pipes. Oversize and undersize are removed by screening, and the intermediate fraction, representing 5 to 25%, is usually within the size range of 1.5 to 16 mm. Referred to as "plant feed," it is either brought ashore or is treated on board with a Pleitz jig, which is said to give a 65 to 75% diamond recovery with one pass, and 90 to 95% with multiple passes.

FIGURE 5.14 Small diving vessels awaiting improved sea conditions in Port Nolloth harbor, South Africa.

Vessels typically are maintained on site by 70- to 110-kg, five-anchor spreads, and a vessel with five divers may stay at sea for a week.[5] Larger, steel-hulled boats have a crew of nearly 20, of which half may be divers. Two men usually work together, remain underwater for 2 h at a time, and decompress at surface in a chamber within 3 to 5 min of ascending. Sea temperatures range from 7 to 16°C, with an average of 10 to 12°C. Bad weather and heavy ocean swells prevent diving for about 250 to 275 d/year, allowing work for only 8 to 10 d/month for small vessels. The largest may achieve about 160 d/year. Production has been limited by restricted financial resources and the simple technology. Some small companies have lacked management control and diamond security at sea.

A small operator may progress from a diving boat to an airlift mining system. The smallest scale is represented by vessels of less than 1000 t and equipped with a 255-mm airlift and recovery plant. Slightly larger ones may have 305- to 455-mm airlifts. For example, ODM's, 79-m-long, 2555-t *Namibian Gem,* shown in Figure 5.15, has operated two 405-mm airlift heads, while staffed by a crew of 40.[13] A more recent addition to the industry is Benco's *Moonstar:* 80 m long, of steel construction, and 2330 gross tonnage.[15,37] Two 510-mm, 30-m-long airlifts are powered by about 5 m³/s of compressed air. The slurry speed in the airlift pipe is 4.5 m/s and 90 m is the maximum mining depth. Each airlift feeds a dedicated 50-t surge bin ahead of a 150-t/h screening facility. Gravels sized 1.5 to 16 mm are supplied to a 50-t/h DMS plant, and concentrates are routed through an X-ray unit.[19] The elapsed time from seabed excavation to recovery of a diamond on board is less than 1 h.

In mining mode the *Moonstar* is aligned roughly north–south on a four-point mooring, dragging the twin mining heads for the length of the mining block. At the completion of each traverse, the vessel is moved onto a parallel course. Without any wave compensation system *Moonstar* can operate in swells of up to 4 or 5 m. The agitative action of the swell transferred to the heads via the airlift pipes may be beneficial to volume extraction in the absence of any aggressive assistance to the suction. The targeted efficiency is reported to be 75 to 95% recovery of sediments.[37] The vessel's theoretical, maximum mining rate is about 250 m³/h of sediments, with an average coverage of 300,000 m²/year.[19] Fuel consumption of 9 to 10 t/d allows for 55 d on site. During 1997, in waters slightly deeper than 30 m, the vessel reportedly averaged 130 m²/d of seabed, compared with ODM's vessel at 260 m²/d on ground of different character and sediment thickness.

FIGURE 5.15 An airlift vessel, *Namibian Gem*, at anchor off the southern African coast. (Photograph courtesy of Ocean Diamond Mining Holdings Limited and Colin Sunkel Photography, Cape Town.)

Clay, cemented sediments, and any pronounced irregularities in the bedrock profile retard an airlift system's m^3/h throughput rate, the m^2/d rate of advance, and the percentage recovery of diamonds off the bedrock. Two or three airlift passes ensure better diamond recovery but reduce the overall mining rate. Divers have reported that one typical airlift operation on the inner shelf generally achieved a 70 to 80% bedrock clean-up. Elsewhere, the figure has been as low as 60%. Airlift systems can operate without the advantage of swell compensators in moderate seas, depending on the size of the vessel, but their production is adversely affected by abnormally unseasonal weather. Winter storms in April and May 1997, for example, caused a 40% reduction in *Moonstone*'s availability and contributed to a 50% drop in diamond output.

The Wirth drill is the mining system which has probably been contributing the greatest diamond production from the Namibian marine grants. DBM's *Grand Banks*, with an overall length of 122 m and 6054 gross tonnage, was commissioned in late-1992 as a production vessel equipped with such a mining device.[26] The Wirth drill was also deployed from the *Debmar Atlantic* in 1994, followed by the *Debmar Pacific*, which began operating in March 1997.[4,11] All DBM's vessels carrying the Wirth drill are fitted with swell compensators. Movement due to the swell is offset by a system which exerts a constant downward pressure on the rotating table. The limiting swell for operations varies with each vessel at between 6 and 8 m and exceeds the 4 to 5 m usually experienced. The *Grand Banks* has a 9-m compensator, allowing the vessel to continue operating even in 7-m swells.

Seabed gravels are mined by drilling overlapping holes as the vessel gradually moves forward. The precise operation is closely monitored by appropriate instrumentation and by a manned submarine, the *Jago*. Mining reportedly covers 96% of each mining block.[11] DBM's production data indicate that each 50 × 50 m block of 120 Wirth holes typically yields 2000 ct in a 24-h operating period.[28] The Wirth drill is regarded by DBM as a mode of vertical attack, able to work on rocky ground with boulders.[1] It is preferred for use in extreme situations in water depths to 200 m. Most existing surface reliefs and sediment thicknesses reportedly can be tolerated, but any significant slope to the bedrock presents a problem. In places, sandstone slab boulders up to 2 m in size are encountered, and some can be pushed aside. A retractable, central drill head prevents the table from spinning on top of a large rock. The drill has been used for only a short time on a large gully feature typical of the inner shelf.

Original De Beers Marine "crawler"

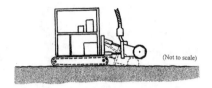

(Not to scale)

"NamSSol" seabed "crawler"

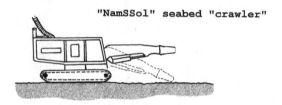

Subsea Diamond Miner "SDM 50"

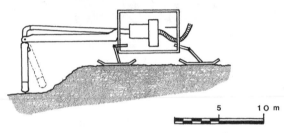

5   1 0 m

FIGURE 5.16   Seabed mounted mining machines now in use.

DBM was the first operator to use a remotely controlled mining system in southern African waters. Referred to as the Crawler, it was delivered to the company in August 1987. The machine was installed on DBM's first production vessel, the *Louis G. Murray*, which was commissioned in late-1989. By coincidence a related company had just completed the first trial gold mining offshore Alaska, of the Tramrod.[39] After several modifications the Crawler continues to operate in Namibian waters. The 50-t, track-mounted, machine is electronically operated and hydraulically powered via an umbilical cable from the 77-m-long and 3413-gross tonnage vessel.[7,26] As depicted in Figure 5.16, it was originally fitted with a 1.25-m-diameter bucket-wheel with a power input of 63 kW. This was later removed. Two 250-mm hoses bring the airlifted sediment to the surface. An acoustic array system allows for very accurate positioning of the Crawler, which mines in lanes, cutting a furrow in the 1- to 3-m-thick gravels.[11] A balance exists between the area of the intake footprint and the pump capacity. Past practice has been to run the machine for 3.5 d on the seabed and then to bring it to the surface for 2.5 h of maintenance. However, it can remain submerged for several days at a stretch and is able to operate at depths of 200 m.

The Crawler is used in a middle shelf environment similar to that of the Wirth drill but is regarded as providing horizontal, rather than vertical, attack.[1] It is employed on smoother, relatively flat areas with low bedrock irregularity and on more finely sized sediments than is the Wirth drill. A Crawler continues to be operated from the *Louis G. Murray*, and has been subjected to various design modifications as a result of experience gained. In 1993, a second Crawler was commissioned to act as a back-up and to improve production continuity. It also served as a test platform for the perfection and development of DBM's mining techniques.[4] The *Voyager* is to be equipped with a new, larger generation of Crawler, which was subjected in 1998 to exhaustive onshore testing. Trials have involved the use of full-scale concrete models of the seabed where the machine is to be operated.

Increasing awareness of the capabilities of seabed mounted equipment is reflected in Namco's commissioning of SubSea Offshore Ltd. a Scottish subsidiary of Dresser Industries, to build a mining machine. The resulting NamSSol started production in 1998.[20,42] It is a 120-t, 2-MW vehicle supported on, and driven by, wide tracks. It is 8 m long, 6-m high, and 5 m wide, and its appearance in Figure 5.16 is reminiscent of the Tramrod used in Alaska nine years previously.[39] An articulated suction boom is fitted with a Warman, centrifugal pumping system and can generate a 20-t downward force. Jetting water is provided at the intake to agitate, loosen, and capture the sediments.

The machine is deployed on the *Kovambo,* the renamed vessel used for Namco's earlier bulk sampling program: 95 m long with a gross registered tonnage of 2687. NamSSol moves independently of the vessel on a flexible riser and operates on lines according to a predetermined pattern. A 50-t/h DMS plant is on board.[17] Namco forecasted an average availability of 20 h/day for a minimum of 300 d/year in sea swells of up to 6 m. Attainable throughput was designed to be 1 Mm^3/year of solids, involving the clean-up of between 30 and 40 ha of seabed with 90% efficiency. The machine moves at 30 m/h, with its excavation performance and assessment thereof assisted by underwater cameras and a scanning-monitoring system.

Meanwhile, a company of the Royal Boskalis Westminster group has combined with SMD and PI to produce a skid-mounted, seabed, mining machine, known as the Subsea Diamond Miner, SDM 50, also shown in Figure 5.16.[40] The unit is also remotely controlled and monitored from the surface. Its design benefits from the practical experience gained during the use of the Namrod sampler. Reflecting the preference of some operators for pad- or skid-mounted support, the machine is a production version of the ADA bulk sampler.[32] The whole system comprises the 75-t, seabed mining machine, umbilical winch, slurry transport hose, and a hose-handling device. The mining component is 22.4 m long by 9 m wide and 6 m high. It incorporates a chassis mounted on four walking skids, which can be adjusted by as much as 750 mm for leveling. The minimal ground pressure allows operation on overburden with a compressive strength of as low as 5 kPa. Other practical advantages include the usual provision of a stable operating platform and mobility over the roughest terrain. The SDM 50 reportedly can negotiate its way over boulders up to 1 m in diameter, and steering is by differential skid movement.

A cantilevered suction boom with a 400-mm-diameter intake is mounted on a rotating turret with the supporting hydraulic and electrical control system. Combined vertical and horizontal movement of the boom along the excavation face allows separate removal of overburden and gravels.[32] To dislodge material and to ensure bedrock clean-up, the boom has a penetration force of 75 kN, and the suction intake carries a jet agitation ring and an optional mechanical cutter or rock breaker. A submersible dredge pump, or existing airlift system, transports the slurried sediments to surface through a 400-mm-diameter, heavy-duty hose. The mining unit is designed to be operated from a vessel equipped with a 100-t A-frame and swing gantry with a machine stabilizing system. Total power requirements amount to 900 kW.

In free flowing, thinner sediments the SDM 50's design capacity allows a seabed coverage of 1000 m^2/20 h day. Removal of sediments thicker than 4 to 5 m may involve stripping successive layers by multiple passes. The machine is designed to penetrate any ground with a compressive strength of less than 5 MPa.[40] The planned operating depth, based on hose length, is 150 m, but the maximum capability of the SDM 50 is 500 m. Assisted by underwater cameras, the system is fully instrumented, allowing complete operational monitoring and on-screen simulation. Sonar allows profiling and the avoidance of seabed obstacles, and an acoustic positioning system enables all relevant data to be plotted directly as part of a GIS.

5.1.7 Production Planning

For a new operator, production planning is an iterative process, undertaken in conjunction with the selection of the most appropriate mining system. Planning cannot be completed until the test mining and system choice has been finalized. The benefit of a relatively thin mineralized sediment cover is that it allows a marine mining system to be relocated easily from one mining site to another. Sediments can be selectively removed from defined areas which are of a sufficient size to be accepted for mining or be rejected in their entirety.[29] The degree to which selectivity is possible depends on the means of excavation and the minimum size of an area or parcel for which the grade and its continuity can be defined by the sampling results. Recent increases in the variety of mining systems has made a wider choice available, allowing a balance to be struck between selectivity and mining rate. The extent to which such mechanical advances are equalled by improvements in ore parcel delineation are revealed to individual operators by their production results and reserve reconciliations.

Worldwide, two opposite extremes of marine mining selectivity are represented by the diving vessel and the bucket-ladder dredge. A diver may excavate single trap sites with a high level of control and avoidance of dilution. Single cobbles can be rejected, but a day's volume output represents no more than a few seconds of a dredge's throughput. The diver's unit mining costs are extremely high and require very elevated grades. Individual diamond trap sites may be discovered by extremely dense, trial and error gravel recovery, equivalent to a very high sampling density with relatively small samples. The bucket-ladder dredge, in contrast, sacrifices selectivity in order to achieve an extremely high volume throughput with high dilution at the lowest possible unit cost to render economic very low grades. More widely spaced, larger samples may be acceptable, unless the average grade is close to the cut-off, because the overall dredge length dictates one minimum dimension of the mining block in plan. Other mining systems lie between the two extremes, with the extent of control being an important variable.

International standards of resource and reserve definition are intended to encompass all mineral deposits,[29] but they do not easily accommodate those of marine diamonds. No standards are available from either an industry body or regulatory authorities. Some producers apply variations of existing onshore definitions and methods. Certainly DBM, and probably ODM, now have their own tried and proven standards related to substantial production experience, but some estimates originating in the industry until recently have lacked the rigor normally expected.[29] Estimating and reporting will doubtless benefit in future from increased production experience and should refer to the tested, proven mining system.

Widely spaced reconnaissance samples in places provide an estimated, global, mean diamond grade and value for a resource with an acceptable degree of confidence for further exploration planning, but the low sampling density is insufficient for the assignment of a reliable grade and value estimate to component areas for mine planning purposes. Such a resource is upgraded to the necessary measured resource, and subsequently reserve, status only by more detailed evaluation sampling, confirmatory high-volume bulk sampling, trial mining, and completion of a positive economic study.[29] Practical considerations also influence the most dense sampling to be attempted. No firm rules exist because of the variety of diamondiferous, geological features, but a minimum of a 50×50 m grid and a sampling density of 1:625 is one company's requirement in order to outline an indicated resource on the inner shelf.[37] A lesser density, where there is reasonable evidence of geological continuity, yields a lower status inferred resource.

Resource and reserve estimation requires consideration of available mining systems and their related capabilities, especially selective extraction.[29] A sampling density sufficient to delineate a reserve for one mining system may be inadequate for another, more selective one. An expected mining rate is influenced by the geomorphology and dictates the unit cost. That cost determines the cut-off grade by which a production choice is made between two different courses of action. A parcel of ore with an estimated grade higher than the cut-off is considered economic, to be mined and treated. One with a grade below the cut-off is regarded as unpayable waste not to be mined. Figures in the range of 0.05 to 0.1 ct/m^2 are frequently quoted for a cut-off grade, but without clear explanation of their derivation and assumed equal applicability to different mining scenarios.

Various established procedures are used in the partly iterative estimation process, depending on the geology and proposed mining system.[29] First, distinct features are defined and classified according to their geomorphological characteristics and their apparent grade continuity. The areal and volumetric diamond grades from individual sample sites are then gridded at a spacing of 50 m using an inverse-distance-squared algorithm with search radii of, say, 100 and 150 m. The resulting polygons are vetted to ensure their compliance with the feature's interpreted geological boundaries. Only those with estimated grades higher than the selected cut-off are included in the final indicated resource. Some operators employ sophisticated methodologies on progressively more closely spaced sampling to outline areas of economic potential.[26] Kriging techniques provide the grade estimates, using search distances compatible with the geology and grade continuity of individual features. Semi-variograms reflect the local grade trends which may be related to diamond transport directions and bedrock structure. Different distribution models, depending on the frequency with which samples may contain no diamonds, assist in estimating diamond size and grade.

A parcel is the smallest area of ground to which a mine or no-mine decision is applied when a mining system is on station.[29] A number of contiguous parcels above the cut-off grade are combined into a mining block or sub-block. A block is the minimum area of regular outline into which movement of a mining vessel for exploitation is economically justified. It is defined in terms of its dimensions, area, volume, or the minimum number of sample sites it should encompass. A feature may contain several contiguous blocks of equal or similar size. On the deep-water middle shelf, DBM uses blocks which are typically 2500 m^2 in area. Half a dozen such blocks, each 50 × 50 m, are exploited in one anchor setup by a mining vessel. Benco also uses a block size of approximately 50 by 50 m, and which, if contiguous, may be amalgamated into a larger mining area. ODM's practice, once minable ore reserves have been delineated, is to divide them into rectangular mining blocks of 10 × 20 m.[13] Figure 5.17 shows a typical inner shelf feature with mining blocks outlined and ready for production.

A mining program is drafted and finally constructed at block size, not at sample support size. The course, or order, to be followed during mining may be modified at both block and parcel size during production. The grade histogram for each area is different. With increasing sample support, the average sample grade and estimated block grade decrease, but the diamond content and ore volume increase. For example, 31 Bauer cutter samples, each of 4 m^2, on an area of the inner shelf yielded for Benco an estimated average grade of 0.80 ct/m^2. Subsequently, 5 m^2 trench samples, 154 in number, dug by airlift, reduced the estimated mean grade to 0.67 ct/m^2. But the total contained diamonds in the resource increased by 80% from 36,000 to 65,000 ct in more than double the original area.[15] The relationship between cut-off grade, cumulative ore volume, average grade, and total contained diamonds is expressed graphically by a grade-volume curve.[29] Any increase in the cut-off grade in response to increasing costs causes the economically minable fraction of a feature to decrease. Although areas of payable ore improve in estimated grade, they also become smaller, increasingly fragmented, and isolated. Additional sampling becomes necessary to reestablish confidence in ore continuity and to define boundaries. Some areas may be so reduced in dimensions or volume that they fail to qualify as ore blocks, and the contained diamonds cease to be part of the resource.

Block boundaries are plotted on maps which depict the parameters to be considered in production and which form the basis for the preparation of a mining plan. Some sediments may be prohibitively thick for the available mining system if the contained diamonds are concentrated on bedrock. Those areas are excluded. Any places into which a vessel's entry may be considered imprudent and its safety compromised are also omitted. The mining plan stipulates the order in which blocks are mined and highlights their salient characteristics, including the estimated grade and the expected diamond production from each. However, such detailed planning procedures are not universally adopted. One contractor has produced 2000 to 3000 ct/month at inner shelf localities simply by working on the fringes of areas mined by previous operators. Mining invariably is commenced in the highest grade areas so as to achieve capital payback as rapidly as possible. This policy, although having financial merit, incurs the risk that the better reserves are exploited with a

Estimated block grade

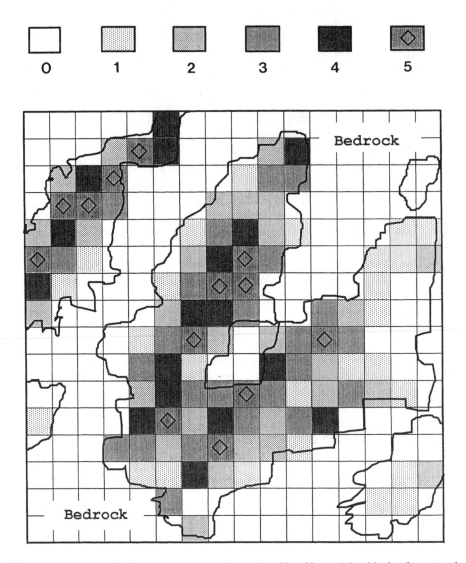

FIGURE 5.17 An inner shelf feature with grades allocated to 50 × 50 m mining blocks, from very low (0) to very high (5).

low mining efficiency during the learning process, before sufficient operational experience of the area has been gained. Each locality is unique in subtle ways.

No resource or reserve figures are quoted by DBM, but smaller operators advise their shareholders of any exploration successes.[15,16] The Bauer cutter program determined for DFI that several features within its 660-km² Namibian licence area hosted a composite 8.44 km² resource containing 1.15 Mct at a cut-off grade of 0.1 ct/m² and a stone value of US$165/ct. Similarly, a low-grade resource containing 0.72 Mct of small diamonds was revealed in a 3646 km² concession explored by Benco on the South African middle shelf. Bulk sampling by Benco's airlift system confirmed another resource estimated to contain about 0.325 Mct, 40% of them indicated and the balance inferred, with an average stone size of about 0.30 ct and a value of US$200/ct.[43] ODM reported its combined reserves to contain some 0.25 Mct, equivalent to four years of production at current rates.[13] Use of the Namrod allowed Namco to announce success within 3.85 km^2 of its 400-km^2 licence area. Based on a 100 × 200 m grid, the reported diamond contents in two features were 0.26 Mct in a measured resource and 2.4 Mct indicated and inferred.[16]

5.1.8 Mining Performance

Mining performance is measured in many ways. They include output, availability, throughput rate, and grade and carat recovery vs. the reserve estimates.[44] In common with all forms of alluvial mining, marine achievements are enhanced by experience which is gained over months and years. For example, DBM's equipment reliability has continually improved and now allows production during some 80% of the available time at sea, with the average vessel working for 2.5 months from a single anchor spread.[8] From 1965 to 1969 the company's forerunner recovered 150,000 ct/year from sampling and experimental mining. By 1992 DBM was effectively operating two and a half mining production vessels. The following year, the improved availability and increased fleet size allowed a 41% expansion in the area of seabed mined so that the average diamond output was probably around 100,000 ct/vessel.[4,5] Approximately 0.75 km^2 of ocean floor was mined, suggesting a recovered rough stone grade of about 0.33 ct/m^2. Two years later the mining rate had doubled with the increased Wirth drill fleet, each vessel being capable of covering between 0.25 and 0.4 km^2/year.[19] DBM probably has been mining 1500 to 3000 m^2/d from its reserves recently by means of all its vessels combined. In 1998 production by DBM out of Namibian waters was expected to be around 0.5 Mct, generated by the Wirth drills and Crawler.

Also in Namibia, ODM's annual diamond output by using the airlift system has grown steadily since 1993 to reach 60,000 ct/year.[13] Some 5% of that is recovered during sampling operations, with the balance from mining production. In 1994 the *Oceania* was achieving a time breakdown of 55% for mining, 28% for sampling, and 17% for port calls and repairs.[34] Most operators have now followed DBM's example, whereby their crews work a cycle of 28 d on and 28 d off, in 12-h shifts. Adverse weather conditions can cause production to decline through increased downtime, especially if it develops without much warning, and a reduced mining rate while still operating.

The recovered grades reported by the different mining systems reflect both their cut-off grades and the reserves available to each. Diving boats locate and selectively extract very high sustained grades of up to about 7 or 10 ct/m^3, although the industry average is around 5 ct/m^3. Recovered grades of 1.0 to 1.50 ct/m^2 have historically been achieved by airlift systems, but are now between 0.15 and, for short periods, 1.0 ct/m^2.[3] Some commentators suggest, in the absence of any firm corporate statements, that DBM recovers an average grade of close to 0.35 ct/m^2. Where the company's vessels operate on Namdeb's middle shelf grant, most gravels are said to be between 1 and 3 m thick, inferring an excavated grade of between 0.125 and 0.35 ct/m^3. Confirmatory sources hint at one of 0.25 ct/m^3 in sheet gravels averaging 1 m thick, equivalent to 0.25 ct/m^2.

On DBM's production vessels all diamond-bearing concentrates are processed and hand sorted as discrete batches, each representing a mining block. The results are compared with the original block forecasts and any significant discrepancy is closely examined.[11] Such comparisons of diamond output and recovered grade with the expectations based on sampling are important, but they can be deceptive. None of the operators claims that complete mining recovery of diamonds from a block is always achieved. The fact that an operator's vessel can return to a mined area and recover another 25% of the original expected grade is sufficient evidence of incomplete excavation recovery in some parts of the industry. In other places an unexpectedly high production may reflect inefficient sampling or an underestimation error rather than mining efficiency.

Performance may be expressed as the ratio of the recovered grade or output over that estimated. The ratio is referred to as the R/E factor or mine call factor (MCF).[44,45] If neither the sampling nor the later mining recovered diamonds efficiently, the long-term R/E factor may merely suggest which process has been more (in)efficient. Mining output and the sampling based prediction are each used to judge the verity of the other process without knowing which, if either, correctly reflects the true grade and diamond content. If both performed equally, then variations in the factor, especially short-term ones, display a regression effect, which may simply reflect an inadequate sampling density.[44]

During the early 1970s, for example, the estimated block grade from some inner shelf airlift sampling varied between 0.14 and 0.55 ct/m³, but the recovered grade was less, between 0.11 and 0.27 ct/m³. The R/E factor therefore ranged from 0.46 to 0.90, perhaps due to the mining efficiency being lower than that of the earlier sampling. In the late 1980s some similar sampling offshore Namibia indicated an average grade of about 0.7 ct/m³, but subsequent airlift mining was completed with an R/E factor of 2.0. The outcome suggests that the original sampling had seriously under-estimated the grade. In both areas the true, *in situ* grade was, and remains, unknown. The range in R/E factor could have resulted entirely from insufficient sampling. Only the diamond production is beyond doubt. More recently, an area of the inner shelf which had been sampled by the Bauer cutter was further tested by Benco using an airlift system. Mining by airlift achieved a recovered grade of 0.61 ct/m² for a period, vs. one estimated from sampling of 0.67 ct/m², an R/E factor of 0.91. The airlift reportedly performed with an efficiency similar to that of the Bauer tool, allowing Benco to conclude that "actual recoveries match the predicted quantities of diamonds present in the ore blocks."[15] In 1998 DFI contracted DBM's *Coral Sea* to undertake confirmatory sampling with the Wirth drill and allow a comparison with earlier Bauer sampling results.

Neither Namdeb nor DBM publishes details of performance against expected grade. However, one source reported that the *Coral Sea*, while still a Wirth drill production vessel, mined 15,000 ct from seven sub-blocks, and cited the output as a positive discrepancy.[28] If each sub-block was 50 by 50 m in area, the result indicates a very high recovered grade of 0.86 ct/m². If such a performance were maintained, the higher revenue would be pleasing. However, an unusually high R/E factor of perhaps 2.5 to 3.5, if such was recorded, would raise serious questions concerning the validity of the original sampling technique, perhaps used some years earlier with less efficient equipment. Predictions are fallible, but operators generally experience a high R/E factor variability in areas of rugged bedrock and a lesser variation over a smooth one. Experience elsewhere suggests that the use of seabed mounted systems should ensure better diamond recovery in both sampling and mining modes because of the superior mining head control.[39]

5.1.9 ECONOMICS

Marine exploration and diamond mining costs off the southern African coast are high compared to those of an alluvial operation onshore, but the rewards are commensurate. DBM, for example, is said to have spent more than US$50 million from 1991 to mid-1994 on exploration alone.[7] In joint venture with Benco, BHP Minerals expended over US$20 million on a geophysical survey and Bauer cutter sampling program.[37] Each Bauer sample cost US$4,000 to acquire, but in one area Namibian diamonds were brought into the resource category for a sampling expenditure of US$4.35/ct — less than 3% of their value. Following commencement of a program of bathymetric and geophysical surveys in 1993, Namco announced an annual exploration expenditure of US$3.5 million in 1994/5, increasing to $9.5 million the following year.[16] By July 1997 a total of US$18.8 million had been expended by the company over three years.[44]

In 1996 a fully equipped and staffed geophysical survey vessel could be obtained for US$12,600/d. A larger vessel with full ground truthing facilities could incur twice that expense. In 1993/4 the full cost of contracting an available airlift sampling vessel for one year was about US$15 million. A fully crewed airlift vessel with a capacity similar to that of the *Moonstar* would now probably command a charter fee of around US$0.5 million/month, plus taxes.[43] The full charter cost of a vessel of the tonnage and capabilities of, say, the Coral Sea is probably about US$50,000/d. Trans Hex International announced that its joint venture program of offshore, bulk sampling in Indonesia cost US$6 million. In northwestern Australia one company, until a budget reorganization, was expected to spend at least A$10 million in 1996 on marine diamond exploration, including the charter of a dredge for bulk sampling of selected sites. The cost of Namco's NamSSol project, based on long-term charter of one vessel, including commissioning and sea trial costs, is variously estimated to be between US$16 and 20 million, up from US$12.5 million in 1994.[16,17]

By 1996 DBM's total development costs had probably exceeded US$500 million.[20] Of this amount some US$200 million is said to have been expended by 1994 on the purchase and conversion of the vessels then comprising its marine fleet.[7] On a more modest scale the overall cost of transforming a 2500-ton oil rig supply vessel into the *Namibian Gem* in 1993 cost ODM about US$12 million.[2] Similarly, a 30-year-old vessel was purchased by Benco, overhauled and rebuilt as the *Moonstone* for a total outlay by late-1996 of US$10 million.[37] About US$3.2 million is the reported expenditure incurred on the conversion of the *Kovambo*, to accommodate NamSSol, and which was purchased for around US$7.2 million. A processing plant increases the investment by approximately US$2.5 million.

Estimates of the annual operating costs of each of DBM's mining vessels ranged from about US$6 to 9.5 million in 1992/3.[7] In 1994/5 the *Coral Sea*'s average operating charge was estimated at between US$10.0 and 12.5 million/year.[46] A bucket-ladder dredge, typified by one used in Indonesian waters, provides the lowest unit cash cost: as inexpensive as about US$1.00/m^3 under perfect operating conditions and if digging soft overburden. Typical contract costs for overburden removal by other conventional dredges in the same, easy conditions could range between US$1.25 and 1.50/m^3. An analyst has estimated the unit mining cost of NamSSol to be US$30/m^2. Operational, cash costs of mining are reflected in breakeven diamond output. The *Trident Cape,* a dredge pump vessel operated by Dawn Diamonds needed a production of 1500 ct/month in 1987 to meet all operating costs. The figure probably remains valid now for a similar operation because rough diamond prices have, until late 1997, generally kept pace with cost inflation. The *Moonstone*'s breakeven production, for example, is said to be 2000 ct/month at US$200/ct, before interest, tax, and depreciation. Namco is reported to have assumed a breakeven grade of 0.16 ct/m^2 in some situations.[41] Other companies apply 0.1 ct/m^2 as a minimum economic grade to different exploration targets.

Estimates of the profitability of diamond mining vary widely. In 1992/3 it was suggested that a DBM vessel could generate an annual operating profit of some US$10 million.[5] Two years later the *Coral Sea,* was thought to be returning a profit of between US$20 and 50 million annually in a mining mode.[18] Of the other major producers, only ODM has graduated to a dividend declaration level, to the benefit of its shareholders. It is not expected to remain alone in the next few years. Financial prudence has proved to be a recipe for success in the industry because risk is always present on the inhospitable coast. Wrecks of vessels such as Dawn Diamonds' *Poseiden Cape* are still visible on the rocky shoreline. Usually the risk results in loss of investment rather than human life, although that has not unfortunately always been so.[2] Risk most often derives from technical misjudgments: the choice of mining system, the forecast mining rate, and the estimation of recoverable grade.[47]

Grade estimation is often perceived to hold the greatest dangers because of the complexities of diamond distribution and the difficulties of sampling. An understanding of the geomorphology is an essential part of successful reserve estimation, but it is also the foundation on which the design and/or choice of the mining system and, most important, the forecast mining rate is based. Neglected acquisition of geological and geotechnical data for meaningful analysis and application to engineering details is manifested in marine mining as an inability to excavate and recover diamonds at the expected rate. In the planned mining of any sediments, other than fine gravels on an ideal, smooth bedrock, the greatest risk lies in overestimation of the attainable excavation rate and of a realistic bedrock clean-up capability.[47]

Despite recent major technological improvements, the need for experienced technical and financial management remains paramount in this high-risk industry. In earlier years possible ignorance of, or disregard for, the risks, combined with over-optimism, was a dangerous cocktail. Some organizations have attempted to enter the industry without realizing that the lead time can be so long. The enthusiastic support of financial analysts is a prerequisite for initial funding by private placement or public offering. In consequence, some smaller, public companies may be under pressure to disclose over-optimistic production forecasts. Joint ventures have also constituted a successful financing route.

5.1.10 INDUSTRY TRENDS

Discernible technical trends in the industry include the continuing development of new and more efficient geophysical, survey and marine equipment. Higher equipment availability, greater extractive selectivity and control of the mining head, and higher-volume mining systems are being recognized as the key to long-term success. The potential of seabed mounted sampling and mining systems is increased when they are allied with advances in underwater sonar and mapping devices.[38] ODM, using a real-time sonar visualization system developed by Sonar Research & Development to detect seabed remnants of unmined ore and uncleaned bedrock, has raised the ct/m^2 yield by 60%.[14] The more carefully executed mining, despite being at a 34% lower m^2/h rate, has increased the hourly diamond production by 5% and substantially extended the life of the reserves.

Evaluation methodologies are being strengthened to refine reserve definition. The monitoring and reconciliation of a mining system's recovery performance is improving. Advances should eventually permit the compilation of a mining vessel's full diamond flowsheet balance, from seabed to sale. Improvements in operational efficiency are necessary as the industry matures and available grades inevitably decline. Namdeb, for example, reported that the seabed area mined in 1996 by DBM's vessels increased by 15% over that for 1995, whereas diamond production rose by only 3%, indicative of a 10% fall in areal grade.[28] Maximizing the time at sea has been a first step by all operators toward improving seabed coverage rates. Lowering the unit costs and cut-off grade by higher mining rates and improved control is recognized to be the necessary second one.[2,10,28]

Mining systems being deployed on the seabed, such as the SDM 50, NamSSol, and DBM's Crawler, continue to benefit from accomplishments in other marine industries and are having an accelerating, beneficial impact on offshore diamond mining.[31] They provide a controlled platform on which to mount a submersible dredge pump, together with equipment with which to monitor and direct the digging performance in real time. Seabed vehicles are capable of positioning a variety of remotely controlled drills, sampling systems, and excavational devices. Dramatic weight savings have already been achieved through the use of very high strength steel for structural components.[32] Perhaps combined with underwater screening, such machines will augment mining rates by a modular approach. Not only may the number of vessels be increased, but more than one crawler can be operated from a single vessel or platform.[33]

Lower grade deposits and those beneath several meters of overburden are also becoming targets. The necessary, higher volume throughput rates required from existing mining systems will have an impact on the necessary, onboard, metallurgical treatment capacities. Excessive clay in future mining areas may encourage consideration of scrubbing and jigging to be incorporated in the diamond recovery flowsheet. A new 40-t model of the Bauer cutter for deep sampling reportedly can penetrate 20 m below the seabed with a 4.2 m^2 footprint and a capacity of 400 m^3/h. The diamondiferous gravels of such target areas eventually may be exposed and mined economically by very high-volume, less selective, methods such as trailing suction hopper (TSH) dredges. Several companies have researched their application for both overburden removal and diamondiferous gravel excavation, because previous depth capabilities have been substantially expanded. Such dredges have been designed to work in water depths of between 55 and 80 m, with the capability of extending to 120 m if required. An average size TSH dredge moving 20 to 25 Mm^3 of sediments annually would probably yield a sufficient return on the capital investment at a recovered grade of 0.05 ct/m^3 and a stone value of US$200/ct. The problems of disposal of saline tailings, the result of necessary batch treatment of a hopper load, could probably be overcome, with prior environmental permitting, by anchoring a floating treatment plant offshore.

Tailings effluents from existing mining vessels are directed and monitored to prevent the over-sanding and dilution of unmined areas. Mining courses and block dimensions are influenced and chosen accordingly. Remedial measures to reduce tailings plumes and to diminish other consequences of seabed excavation and effluent disposal are receiving increasing attention. The authorities now require an environmental impact study (EIS) and an environmental management plan report (EMPR) to be completed for offshore mining. A scoping exercise includes studies of the social impacts of marine mining and ensures baseline studies are undertaken.[11,16]

Diamond production from mining on the inner and middle shelves of Namibia and South Africa continues to grow, but no regular production yet occurs in any deep water areas of the latter country.[19,20] South African middle shelf diamonds are generally significantly smaller than those exploited to date in Namibian waters of the same depth. But their economic recovery may soon become possible. Whereas diving operations contribute a declining proportion of production, mechanized mining vessels have been increasing in size, sophistication, and numbers during the last decade. DBM's fleet comprises eight vessels. Five of them, including the new *Voyager*, are dedicated to production.[8] The recent addition has been described as a second-generation mining unit capable of adding 100,000 to 120,000 ct/year to DBM's production. From 1999 onward it will add to a 1998 output estimated to be about 500,000 ct.[20] Namibian marine output from all sources now exceeds 0.6 Mct annually, with a market value of some US$170 million. A production of 0.7 to 0.8 Mct was expected in 1998 from offshore southern Africa. Some optimistic forecasts for future years range as high as 2.0 Mct/year as existing and potential producers launch new systems and vessels. To achieve their long term objectives, the smaller companies may have to consider merger and acquisition as a strategy in order to guarantee the necessary financial stamina. The efforts to locate and recover increasing quantities of diamonds appear well justified because the southern African continental shelf is one of the largest diamond exploration targets in the world.

5.2   WORLDWIDE ACTIVITIES

5.2.1   Australia

In northwestern Australia's Joseph Bonaparte Gulf, at 14°S and 128°E, tenements were granted in the last few years to several companies for diamond exploration. Along more than 300 km of coastline, the grants traced the offshore extensions of major rivers draining the Kimberley district which hosts the Argyle diamond mine. Some, such as the River Ord, are known to host diamondiferous gravels onshore. At sea, fluvial gravels are overlain by a sequence dominated in places by carbonate clasts, clays, gravelly clays, and some lithified horizons. They, in turn, are overlain by sands and clays possibly deposited in a shallow marine or estuarine environment. The whole package is a few tens of meters thick. The extreme environment, particularly the strong tidal currents, which can attain 13 km/h (7 knots), impose restrictions on the equipment and drilling methodology applicable. September to November are the calm months, otherwise the area is very exposed to the monsoon northwesterlies during the wet season. The summer cyclones present an additional hazard.

The most active exploration company was Cambridge Gulf Exploration N.L. (CGE). A report that diamonds were recovered in 1992 during CGE's program of airlift bulk sampling by the *Lady S* was not supported by further sampling, which yielded none. Neither is there any firm evidence to date of any offshore sorting or upgrading of sediments by present-day or previous marine or wind action. CGE undertook a follow-up program of reverse circulation drilling in the Gulf during 1996–7. The drill was operated on the *Gulf Explorer*, a 90-m-long barge, with a 67-m-high derrick, and an operating displacement of 7918 t. The vessel is shown in Figure 5.18. Her position was maintained by a computer-controlled, eight-point anchoring system. Each 9-t anchor was deployed and recovered by an oil rig supply vessel. With 400 m of wire on each anchor winch, the derrick could be positioned accurately in 30 min over a 100-m radius within the anchor spread. Very time-consuming moves

FIGURE 5.18   The Gulf Explorer being prepared for drilling in Joseph Bonaparte Gulf, Australia.

between drill sites were restricted to daylight hours. A 140-mm reverse-circulation drill was used for ground truthing but even a light swell, insufficient to trigger the swell compensator, forced sediment into the drill string during the barge's downward movement. Uncased drills suffered from caving and produced excessive plant feed with little penetration. Despite the problem of higher skin friction, larger cased holes were needed to ensure volume control, to prevent contamination and recirculation, and to avoid hole collapse. A 406-mm bit welded to the same diameter casing, with cuttings rising through the 152-mm dual-wall drill pipe, succeeded in penetrating 51 m of sediments, but not without problems caused by blockages. A 1016-mm, cased system was generally far more successful. The average hole penetration achieved varied inversely with the drill diameter: 20.5 m for the 406 mm, 17.5 m for the 1016 mm, and 9.1 m for larger diameters. All work was discontinued in 1997.

5.2.2   INDONESIA

In southeast Kalimantan, Indonesia, at 3°S and 115°E artisanal miners have for decades washed 0.3 to 1.5 ct diamonds from the Maluka, Danau Seran, and Cempaka paleochannels. The 2-km-wide Maluka fluvial gravels have been traced downstream some 20 to 30 km to the coast near the city of Banjarmasin.[48] A short distance offshore they average 3 m thick on a bedrock of weathered metasediments and are covered by about 30 m of clay overburden. Water depths of 1 to 2 m increase very slowly to the west. The climate is tropical and monsoon winds gust up to 140 km/h. The clay varies from very soft to firm and stiff with a maximum estimated shear strength of 100 kPa. In 1996 Trans Hex International Ltd. entered into a joint venture agreement with Ocean Resources N.L., which had been awarded a sixth-generation Contract of Work for the 1450 km² Sunda Shelf project area.[48] The objective was to continue the exploration of the offshore Maluka gravels. These had been geophysically surveyed and Banka-drilled in 1994 by P.T. Indo Mineratama, a third partner. The 89-mm-I.D. machine completed a 100 × 500 m grid to a depth of nearly 40 m, followed by infill drilling at 10 × 30 m. Each meter's penetration produced 22 kg of material.

On the Kalimantan coast the calmer sea conditions, unlike those of southern Africa, allow either mechanical or hydraulic dredging. Ocean Resources leased a seagoing, 14-ft³ (400-l) capacity, bucket-ladder dredge with which to undertake the joint venture's bulk sampling.[48] The 82-m-long Aokam No. 3 dredge, built in 1966–68, had previously been employed in offshore tin dredging on the coast of Thailand. It was suitable for later mining in the event that payable, diamond-bearing ground could be identified. The dredge has a theoretical capacity of 420,000 m³/month with a maximum digging depth of 35.5 m at a ladder angle of 54°.

In 1996 the dredge dug its way for 3.4 km through 2-m-deep water.[32] On site it was maneuvered by means of a five-anchor spread. Access to the 2- to 5-m-thick gravels lying on bedrock at between –27 m and –35 m was gained by digging down at an angle of 22°. Overburden down to –13 m was pumped for 250 m via a floating pipeline to a discharge point. Deeper overburden was discharged directly from the stern of the dredge into a preexcavated tailings disposal pond. Two bulk samples were processed by digging trenches 25 m wide at bedrock and perpendicular to the paleochannel. Loss of availability, and thus higher unit costs, were caused by some mechanical breakdowns and major slippage of the digging face. During the monsoon season, high winds impacting the dredge's large sail area caused the anchors to drag. Considerable difficulties were also encountered with slumping clay overburden, solved only by laying back the trench's side walls to a low angle.

Excavated gravels were screened and treated on board. The 2 to 14 mm screen fraction was processed through a 45-t/h jig plant using 12.7- and 25.4-mm ceramic balls for ragging. The 90-m-long first trench, dug in two sections, provided 6750 m³ of gravels from beneath 175,680 m³ of overburden: a strip ratio of 26:1. The second 85-m trench involved the removal of 191,500 m³ of overburden to yield 2900 m³ of gravel. Bulk sampling continued until late-1997 when work was discontinued due to the negative results, and the dredge has since been returned to Thailand.

5.2.3 WEST AFRICA

Off the west African coast of Sierra Leone a 15,867 km² prospecting licence was awarded to De Beers in 1994, but the exploration results were disappointing and work has since ceased.[4,18] Malaysia Mining Corporation also undertook some work to the east offshore from the Mano and Moa Rivers. Other areas considered prospective for marine diamonds, and which have been subject to different levels of exploration activity, include offshore Angola, Liberia, Ghana, Uruguay, and the White Sea.

5.3 CONCLUSIONS

The successful establishment of the marine diamond mining industry of southern Africa is attributable to many factors. The most important has been the unique set of natural conditions creating the marine concentration of diamonds. Declining sources of good quality stones from onshore resources necessitated the move offshore in order to maintain supply. This action was facilitated by governmental encouragement, coupled with the relative price stability of rough stones. The long lead time from discovery to sustained, large scale, commercial production has amounted to decades and has required significant corporate and personal dedication. Final achievement has been realized as a result of technological developments, some from other industries, and considerable financial resources. Especially in recent years there has been a necessary emphasis on geological understanding of the marine deposits, and the experience gained from adjacent, onshore, alluvial diamond mining has been invaluable. Risks nevertheless remain for medium-sized operators. Correct selection of appropriate sampling and mining systems for different environments is critical. The learning curve can be prolonged and financially challenging. Forecasting the realistic excavational performance in terms of attainable mining rate and bedrock clean-up are as important as grade estimation.

REFERENCES

1. Corbett, I. B., A review of diamondiferous marine deposits of western southern africa, *Africa Geoscience Review*, 3, 2, 157, 1996.
2. Williams, R., *King of Sea Diamonds: The Saga of Sam Collins*, W.J. Flesch & Partners (Pty.) Ltd., Cape Town, 1996.
3. Nesbitt, A. C., Diamond mining at sea, Diamond Research Laboratory, Mines Division, Anglo American Corporation of S.A. Ltd., 1967.
4. Annual reports of De Beers Centenary AG and De Beers Consolidated Mines Ltd.
5. Yorkton Natural Resources, *The Revitalised Diamond Market,* 1993.
6. Murray, L. G., Exploration and sampling methods employed in the offshore diamond industry, *Ninth Commonwealth Mining and Metallurgical Congress*, South Africa, 1969, 714.
7. Gooding, K., De Beers sucks up fortune in gems from its floating mine, *Financial Times*, London, 29 August 1994.
8. De Beers Marine (Pty) Ltd., *Company Overview*, 1997.
9. De Decker, R. H. and Woodborne, M. W., Geological and technical aspects of marine diamond exploration in southern Africa, *Offshore Technology Conference Proceedings*, Houston, Texas, paper OTC 8018, 1996, 561.
10. Thompson, J. O., Namdeb — a new era of partnership, *Optima*, 40, 2, 23, 1994.
11. Wannenburgh, A., De Beers, diamonds and the deep blue sea — pioneering marine diamond mining off the west coast of southern Africa, *Optima*, 41, 2, 24, 1995.
12. Annual reports of Ocean Diamond Mining Holdings Ltd., Cape Town.
13. Ocean Diamond Mining Holdings Limited, investor information, March 1997.
14. Louw, A. C., Ocean Diamond Mining's quest for more efficient diamond recovery, presented at the 29th. conference of the Underwater Mining Institute, Toronto, 22 October, 1998, Ocean Diamond Mining Holdings Ltd., Internet home page, <www.odm.za>.
15. Annual reports of Benguela Concessions Limited, Cape Town.
16. Annual reports of Namibian Minerals Corporation, London.
17. Media releases of Namibian Minerals Corporation, London.
18. Duval, D., Green, T., and Louthean, R., *New Frontiers in Diamonds — The Mining Revolution*, Rosendale Press Ltd., London, 1996.
19. Chadwick, J., Dredging, *Mining Magazine*, 176, 3, 145, March 1997.
20. Diamond Supplement, *Mining Journal*, London, 22 November 1996.
21. De Beers v De Beers, *Diamond International,* September/October 1995, 45.
22. Murray, L. G., Joynt, R. H., O'Shea, D., Foster, R. W., and Kleinjan, L., The geological environment of some diamond deposits off the coast of South West Africa, Proceedings of ICSU/SCOR Working Party 31 Symposium, Cambridge, 1970, in *The Geology of the East Atlantic Continental Margin,* Delaney, F. M. (Ed.), Inst. of Geological Sciences Report no. 70/13, London, 1970, 119.
23. Gurney, J. J., Levinson, A. A., and Smith, H. S., Marine mining of diamonds off the west coast of southern Africa, *Gems & Gemnology*, winter volume, 206, 1991.
24. Kuhns, R., African continental shelf and its control and distribution of alluvial/fluvial and marine diamonds, BHP Minerals, 1996.
25. Notes on the geology of the Buffels Marine Complex and prospecting activities in this area, staff report, geology department, De Beers Consolidated Mines Ltd., Namaqualand Mines, September, 1986.
26. Oosterveld, M. M., Ore reserve estimation and depletion planning for a beach diamond deposit, Proceedings of the Tenth International Symposium on the Application of Computer Methods in the Mineral Industry, South African Institution of Mining and Metallurgy, Johannesburg, 1972.
27. Schneider, G. I. C. and Miller, R. McG., Diamonds, in *The Mineral Resources of Namibia*, Geological Survey, Ministry of Mines and Energy, Republic of Namibia, 1992.
28. Sutherland, D. G., The transport and sorting of diamonds by fluvial and marine processes, *Economic Geology*, 77, 7, 1613, 1982.
29. Garnett, R. H. T., Estimation of marine mineral reserves, *Transactions of the Society for Mining, Metallurgy, and Exploration*, 304, 69, 1998.
30. Rombouts, L., Ore reserve calculations on alluvial diamond deposits, in *Alluvial Mining*, Elsevier Science Publishers Ltd., England, 1991, 443.

31. Schwank, S. K., Large size sampling tools for offshore exploration, *Offshore Technology Conference Proceedings*, Houston, Texas, paper OTC 8283, 1996, 91.
32. Bedford, S. and Van der Steen, A., Extending the limits of subsea earthmoving, presented at the World Diamond Conference, Perth, Australia, 7-8 October 1997.
33. Donovan, R. C. and Norman, R. V., Engineering philosophies associated with subsea diamond sampling and mining, *Offshore Technology Conference*, Houston, Texas, paper OTC 8021, 1996, 589.
34. Annual reports of Ocean Diamond Mining Holdings Limited, Cape Town.
35. New vessel boosts ODM's diamond production, in Panorama, *Mining Magazine*, London, p. 4, July 1996.
36. Sutherland, D. G. and Dale, M. L., A method of establishing the minimum sample size for sampling alluvial diamond deposits, *Transactions of the Institution of Mining and Metallurgy*, Section B, 93, B55, 1984.
37. Moonstone Diamond Corporation NL., information memorandum, Perth, Australia, July 1996.
38. Wenlock, J., Seabed solutions to mining offshore, *Engineering & Mining Journal*, 197, 1, 16B, January, 1996.
39. Garnett, R. H. T., Development of an underwater mining vehicle for the offshore placer gold deposits of Alaska, in *Alluvial Mining*, Elsevier Science Publishers Ltd., England, 1991, 157.
40. Brochures published by Royal Boskalis Westminster n.v., Boskalis International b.v., Paragon International b.v., and Soil Machine Dynamics, Netherlands and United Kingdom.
41. Diamond update: seabed bonanza, *African Review of Business and Technology*, April 1996.
42. Marine diamond mining comes of age, *Mining Magazine*, 177, 6, 337, 1997.
43. Annual report of Moonstone Diamond Corporation N.L., Perth, Australia, 1996.
44. Garnett, R. H. T., Mineral recovery performance in marine mining, *Transactions of the Society for Mining, Metallurgy, and Exploration*, 304, 55, 1998.
45. Garnett, R. H. T., Components of a recovery factor in gold and tin dredging, *Transactions of the Institution of Mining and Metallurgy*, Section A, 100, A121, 1991.
46. Miller, P., *Diamonds — Commencing the Countdown to Market Renaissance*, Yorkton Securities, London, February, 1995.
47. Garnett, R. H. T., Risks in marine diamond mining — lessons from the past and present, presented at the *29th Conference of the Underwater Mining Institute*, Toronto, 22 October 1998.
48. Annual reports of Ocean Resources N.L., Perth, Australia.

Part II

Ferromanganese Oxide Minerals

6 Resource Estimates of the Clarion-Clipperton Manganese Nodule Deposits

Charles L. Morgan

ABSTRACT

Manganese nodules are concretions of manganese and other transition metal oxides which form in marine and freshwater sedimentary environments worldwide. The Clarion–Clipperton zone (CCZ) deposits are on the surficial sediments of the deep seabed in the northeastern tropical Pacific Ocean. In general, manganese nodules are believed to form in aqueous, unconsolidated sedimentary environments which include a flux of dissolved or chemically adsorbed manganese and iron through a gradient between relatively reducing conditions where Mn(II) is stable in solution (or as an adsorbed species on particles) and oxidizing conditions where the manganese will precipitate as basically tetravalent oxides.

Two sources have been proposed for the metals in the CCZ deposits: hydrothermal sources from seabed volcanoes, and terrestrial sources from North and Central American rivers and airborne particles. Active vulcanism is significant only in the extreme eastern part of the CCZ, so the metals in almost all of the CCZ deposits must involve some sort of lateral transport from the east for either primary source. Research completed to date suggests that metals in the CCZ surface waters consist mostly of adsorbed species on fine particles concentrated at the top of the thermocline, moving with the surface currents to the west. Plankton concentrate these adsorbed and fine-grained particulate metals into body parts and fecal pellets which are large enough to settle through the water column to the seabed. Thus accumulations of manganese nodules can be expected to be proportional to the productivity of the overlying surface waters. Growth rates of the nodules appear to vary between about 3 to perhaps 20 mm/My. Deposits are patchy, exhibiting wide variations of abundance in apparent response to local geographic and oceanographic factors.

Available data for the CCZ cover an interpolated area of 9 million km^2, which hold 34 billion metric tons of manganese nodules. Specific estimates of metal resources include the following (units, millions of metric tons): Mn — 7500, Ni — 340, Cu — 265, Co — 78.

6.1 INTRODUCTION

Manganese nodules are concretions of manganese and other transition metal oxides which form in marine and freshwater sedimentary environments worldwide. The Clarion–Clipperton deposits are on the surficial sediments of the deep seabed in the northeastern tropical Pacific Ocean, approximately between Baja California and the Hawaiian Islands to the east and west, and between the Clarion and Clipperton tectonic fracture zones (abbreviated here as CCZ) to the north and south (see Figure 6.1). These concretions are of particular interest because of the large extent and abundance of the deposits and because of their relatively high contents of the base metals nickel, copper, and cobalt. Since the early 1960s the CCZ deposits have been examined in some detail by

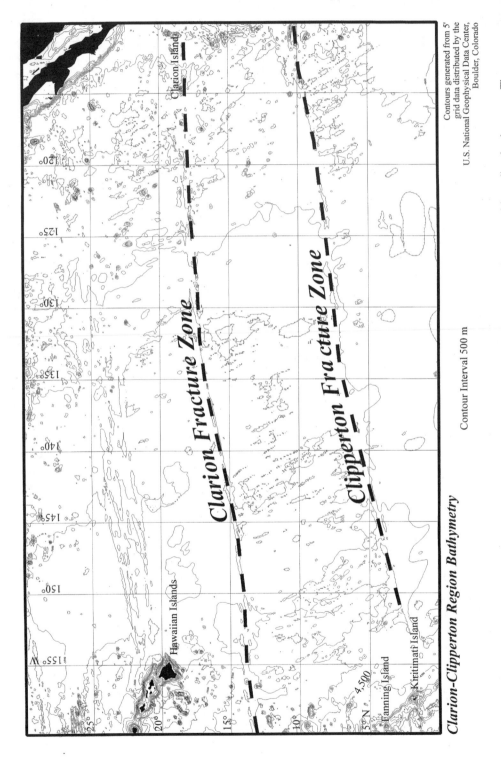

Clarion-Clipperton Region Bathymetry

Contour Interval 500 m

Contours generated from 5'
grid data distributed by the
U.S. National Geophysical Data Center,
Boulder, Colorado

FIGURE 6.1 Clarion–Clipperton Fracture Zone Region (CCZ), showing the major bathymetric features and bounding land masses. The manganese nodules with the highest concentrations of nickel and copper and with the highest densities (kg/m^2) are found between these two fracture zones.

more than a dozen international consortia with aims at commercial exploitation. Several estimates of resource potential have been published to date (e.g., References 1 through 3).

The purposes of this chapter are (1) to provide an update of these resource estimates using a relatively large database consisting of the available published data as well as some unpublished data; and (2) to review the basic factors known about how these deposits form in the light of these data and the current understanding of the tropical North Pacific environment. More comprehensive reviews of the CCZ and other deposits and their associated geological and oceanographic influences are presented elsewhere (e.g., References 4 and 5).

First, a brief overview is presented of the basic factors believed to be related to the formation of the CCZ deposits, including the formation processes believed to be important, the growth rates of the deposits, and the key local factors which may influence the deposit characteristics. This is followed by a description of the deposit data and methods used in the resource assessment, and a presentation of the results obtained. Finally, some comments are included which will hopefully lead to an improved understanding of how the observed distribution of the high-grade, high-abundance CCZ deposits are related to the formation processes involved.

6.2 FORMATION PROCESSES

In general, manganese nodules are believed to form in aqueous, unconsolidated sedimentary environments which include a flux of dissolved or chemically adsorbed manganese and iron through a gradient between relatively reducing conditions where Mn(II) is stable in solution (or as an adsorbed species on particles) and oxidizing conditions where the manganese will precipitate as basically tetravalent oxides. A key prerequisite for significant deposit formation includes a stable environment for the precipitation in which the flux of Mn(II) is not overwhelmed by other sedimenting materials and in which the oxidizing conditions prevail throughout the formation process. Formation rates vary widely, depending mostly on the flux of Mn(II) to the deposit site. Some freshwater deposits are known to accumulate at rates greater than a few millimeters of new oxide precipitates per year.[6] As discussed below, the deep seabed deposits probably accumulate at rates several orders of magnitude slower than this.

The following sections outline the key factors of interest for the CCZ deposits, including the peculiar chemistry and mineralogy of the manganese oxides which are ultimately responsible for the existence and morphology of the deposits, the key oceanographic and geological factors relevant to their formation, the specific sources for the manganese and other metals, and what is known of the growth rates of these deposits.

6.2.1 CHEMICAL AND MINERALOGICAL FACTORS

As mentioned above, the oxidation/reduction chemistry of manganese in natural aqueous systems appears to be the primary factor which leads to the accumulation of the deposits. In its simplest form, dissolved or adsorbed manganese in the presence of dissolved oxygen precipitates as manganese oxide, or:

$$2Mn^{+2}_{(aq)} + O_{2(aq)} + 2H_2O \Rightarrow 2MnO_{2(ppt)} + 4H^+$$

The equilibrium constraints for this reaction have been examined for the different possible structures of manganese oxide and different reaction mechanisms[7,8] and provide a solid conceptual basis for the process. Two mineral forms of manganese oxide, birnessite and todorokite, are the dominant species in the CCZ deposits,[9,10] although microcrystalline and apparently amorphous iron and manganese oxides and oxide-hydroxides may constitute the bulk of the material.[11]

Crerar and Barnes[12] showed how these minerals act to catalyze oxidation of further manganese oxides and also enhance adsorption of many metal species, including the economically interesting nickel, copper, and cobalt. Burns and Burns,[13] using electron micro-probe techniques and basic arguments of mineral structure, suggested basic relationships between the manganese and iron phases in the nucleation process which initiates formation of manganese nodules on silicate, phosphate, and other substrates.

In general, the first transition-metal oxide phase to precipitate is iron oxide-hydroxide ($FeOOH \cdot xH_2O$, goethite or amorphous iron oxide), which forms a surface particularly conducive to manganese oxide precipitation. Once in place, the manganese oxide surface is itself autocatalytic and effective in attracting subsequent manganese into the growth structure and also highly effective in attracting and sequestering other transition metals such as nickel and copper. Glasby and Thijssen[14] showed how the loose, hydrated todorokite structure characteristic of the manganese phase in the CCZ deposits could be stabilized by the addition of nickel, copper, zinc, and other divalent transition metals, leading to enhanced incorporation of these metals.

Li[15] completed a statistical analysis of manganese nodule compositions from several marine and freshwater environments which suggests that nodules consist of three different components, including detrital aluminosilicates, iron oxides, and manganese oxides, with the nickel and copper included in the manganese-oxide component and the cobalt included in the iron-oxide component. Dymond et al.,[16] using factor analysis and other statistical methods, showed how the compositions of manganese nodules from the CCZ, as well as from sites to the east and north of the CCZ, could all be explained by a model with three distinct precipitation processes: hydrogenous precipitation, oxic diagenesis, and suboxic diagenesis. Hydrogenous precipitation occurs as metals accumulate into manganese nodules directly from sea water. The metals are either in solution or adsorbed onto fine-grained particles, which confront the nodule surfaces directly and are incorporated into the nodules without significant interaction with seabed sediments.

Oxic diagenesis involves a surficial sediment intermediary. The metals are initially deposited on the seabed as adsorbed cations, primarily on fecal matter from surface waters, and are subsequently liberated as divalent cations after ingestion and excretion by benthic organisms, to be finally incorporated into the nodules through the adsorption/oxidation processes outlined above. Suboxic diagenesis involves this same biological intermediary but also includes some surficial burial of the manganese nodules.

Recently Knoop, Owen, and Morgan,[17] using similar statistical techniques, documented the expression of all three of these growth processes in the compositions of more than 5,000 manganese nodule samples from the CCZ and showed how the relative dominance of suboxic diagenesis increases from west to east at the expense of oxic diagenesis, consistent with the relatively higher sedimentation rates to be expected eastward. Subsequent diagenesis, which implies significant chemical reduction and remobilization of manganese after burial,[18] does not appear to be a significant contributor to the CCZ nodules, except in the extreme eastern portion of the area.[19,20]

At this time the above mechanisms appear to provide a reasonable description of the general chemical and mineralogical controls for the CCZ deposits. The following sections outline what is currently known to provide the environmental context that results in the observed distributions of composition and abundance (i.e., kg nodules per m^2 of seabed) for these deposits.

6.2.2 METAL SOURCES AND PATHWAYS

Since the CCZ deposits were discovered during the classic *Challenger* cruises,[21] two ultimate sources have been proposed for the metals in the manganese nodule deposits, hydrothermal sources from seabed volcanoes and basically terrestrial sources from freshwater discharges into the ocean or through airborne particles. Through the subsequent studies of the deposits, it became apparent that both sources are likely to be operative throughout the region. Elderfield[22] showed how the overall accumulation rates of manganese in marine sediments, including nodules, are consistent with known and estimated worldwide fluxes from these general sources.

However, since active volcanic processes appear to be significant only in the extreme eastern part of the CCZ, and since the predominant surface ocean currents move from east to west, the metals in almost all of the CCZ deposits must involve some sort of lateral transport whether they originate from terrestrial or volcanic discharges or emissions. Indeed, some of the most interesting scientific questions related to the CCZ deposits focus on the natural pathways taken by the metals to their final deposit sites and how these pathways may have changed over geological time scales.

Because the CCZ deposits reside in the center of the largest of the oceans, they represent the ultimate receptacle of metals transported through the planet's marine environment. More simply, the CCZ manganese nodules represent the final ocean dumping site for transition metals and other materials. In this capacity, it is possible that the deposits record the geochemical history of the northeastern Pacific Ocean and thus might hold important clues to how the global climate has changed through time. During the past 25 years, much progress has been made in the determination of the important aspects of the natural pathways followed by transition metals in the Pacific on their journeys to the seabed. Some understanding of these pathways, outlined in the following passages, is a necessary prerequisite to interpretation of the potential stratigraphic messages contained in the deposits and also provides an essential context for the observed distributions of the economically interesting deposits.

6.2.2.1 Lateral Transport in Surface Waters

The North Equatorial Current dominates the surface waters in the CCZ, except in its southern extreme (Figure 6.2), between the equator and approximately 10° N latitude at average speeds of about 10 to 20 cm/sec. South of this latitude the equatorial countercurrent transports water in a generally eastward flow.[23] Moored current-meter arrays and free-drifting drogues deployed in the CCZ during the U.S. government-sponsored Deep Ocean Mining Environmental Study (DOMES, Sites A, B, and C in Figure 6.2) partially confirmed this general picture, but also showed significant variations in the surface currents and mid-water currents.[24]

Manganese is essentially insoluble in sea water, and careful sampling of sea water and sea water filtrates taken in the CCZ and other northern Pacific sites indicates that this element exists primarily in oxide phases or as species adsorbed onto particles, with low-level maxima (<1nmol/kg) of dissolved concentrations being found in surface waters and at depth in the oxygen-minimum zone.[25-28] In the CCZ only very fine-grained, clay-sized particles of terrigenous origin can be found in the water column, since all particles with significant settling velocities in water will settle out much closer to land.

To investigate the behavior of manganese and other metals in the ocean waters overlying the areas of commercial interest, Martin[29] deployed an array of particle interceptor traps in a free-floating array at 14°17.28'N 129°37.74'W on April 7, 1980, and recovered the array at 14°13.26'W, 131°46.56'W on May 7, 1980. While the particle traps were deployed, numerous water samples were also collected.

Labile Mn fluxes, measured with particle interceptor traps at depths of 125, 275, 525, and 900 m, slowly increased with depth from 270 at 125 m to 450 μg-Mn/cm^2/1000 yr at 900 m. Amounts of dissolved Mn slowly decreased with depth, and little, if any, relationship with oxygen was found. In contrast, suspended particulate Mn values increased markedly at the top of the oxygen minimum. Concentrations of up to 880 μg of weakly leachable Mn per gram of dry suspended particles were measured and are probably the result of microbial oxidation. These rates are on the same order of magnitude as those estimated for global excess manganese accumulating in open-ocean sediments,[22] and they strongly suggest that manganese transport in the CCZ surface waters mostly consists of adsorbed manganese on fine particles concentrated at the top of the thermocline, moving with the surface currents to the west.

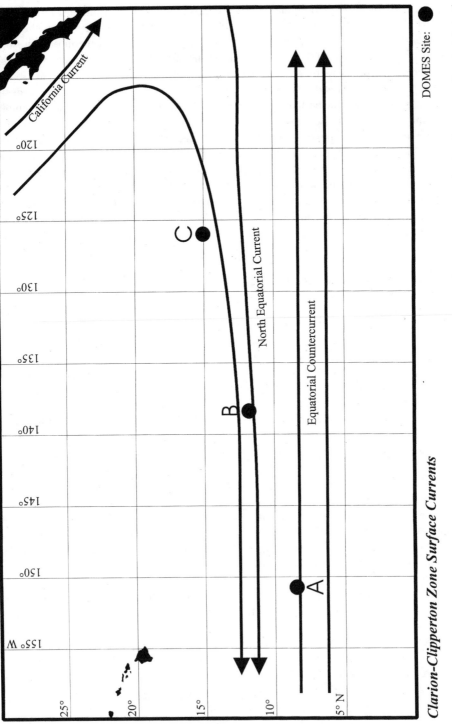

Clarion–Clipperton Zone Surface Currents

FIGURE 6.2 Dominant surface currents in the CCZ. The northern equatorial current may be an important mechanism for transporting metals to the CCZ deposits. Fine particles and adsorbed metals discharged from North American freshwater sources travel south with the California Current to join terrestrial materials from Central America, and possibly volcanically derived metals upwelled from the East Pacific Rise, in the general flow to the west. Sites of study by the Deep Ocean Mining Environmental Study (DOMES Sites A, B, and C) are indicated by filled circles.

6.2.2.2 Vertical Transport

Though the above studies show how it is possible to move sufficient metals through the surface waters to the prime sites of CCZ deposit accumulation, they do not explain how these metals manage to find their way to the seabed. Since most of the metals are closely associated with very fine-grained particles, they have essentially no settling velocities and are moved passively along with the currents. Studies of zooplankton and deep-sea sediments by several investigators (e.g., References 30 through 33) have made it clear that marine plankton are extremely efficient scavengers of dissolved and fine-grained metals in the oceans. In general, phytoplankton act to concentrate dissolved metals in their skeletons during their normal metabolic activities, including photosynthesis. Zooplankton further concentrate dissolved and fine-grained particulate metals into fecal pellets which are large enough to settle through the water column to the seabed.

The best available metric for plankton activity in the CCZ comes from the U.S. National Aeronautics and Space Administration (NASA) Coastal Zone Color Scanner (CZCS) program. The Coastal Zone Color Scanner (CZCS) on the Nimbus-7 satellite was designed for the quantitative remote sensing of the bio-optical properties of marine waters by measuring the visible spectral signatures of the sea. The CZCS had six spectral channels and five visible channels sensing solar back-scattered radiance and one infrared channel sensing emitted thermal radiance. The CZCS data are used to map chlorophyll concentration and other parameters in the world's oceans. The reduced data derived from the scanner observations were produced by the Nimbus Project Office for general, unrestricted distribution in collaboration with the NASA Goddard Space Flight Center (GSFC) Space Data and Computing Division, the NASA GSFC Laboratory for Oceans, and the University of Miami Rosenstiel School of Marine and Atmospheric Science. This global processing effort was initiated in 1985 and completed in early 1990. Composite data products were generated as $1° \times 1°$ averages.

CZCS data for the CCZ were used to generate contoured values for surface-water chlorophyll concentration for this study. Since chlorophyll is directly related to photosynthesis, it can be used as a proxy for primary productivity. The contoured chlorophyll values thus obtained are presented in Figure 6.3. Generally, primary productivity, and the directly proportional chlorophyll concentrations, increase from north to south in response to increasing inputs of solar energy to surface waters and equatorial upwelling of nutrient-rich deep water. The clearly apparent east–west trend shown here is closely related to the Northern Equatorial Current, which brings nutrients in from the North and Central American freshwater stream discharges, leading to enhanced growth at all trophic levels in the surface waters.

Thus, based on the above considerations, we would expect metal fluxes to the seabed, and presumably also the manganese nodule deposits, to be relatively concentrated beneath the areas of high productivity. This is in fact the primary mechanism for manganese nodule growth proposed here for the relatively high-grade, high-abundance CCZ deposits, with some modification due to the tectonic movement of the Pacific Plate during the past 10 million years or so. The matter of growth rates is considered in the next section. The possible relationship to plate movements is considered below, after the resource calculations are presented.

6.3 NODULE GROWTH RATES

Researchers have used a variety of techniques to date the CCZ deposits, including radiometric methods, a growth model based on nodule size distributions, stratigraphic constraints imposed by non-oxide nodule inclusions and geological settings, and others. Ku[34] and Krishnaswami and Cochran[35] used ^{238}U and ^{235}U decay series to obtain consistent recent (~50,000 years) growth rates of a <5-mm radial growth per million years (mm/My). Krishnaswami et al.[36] used ^9Be, ^{10}Be, ^{230}Th,

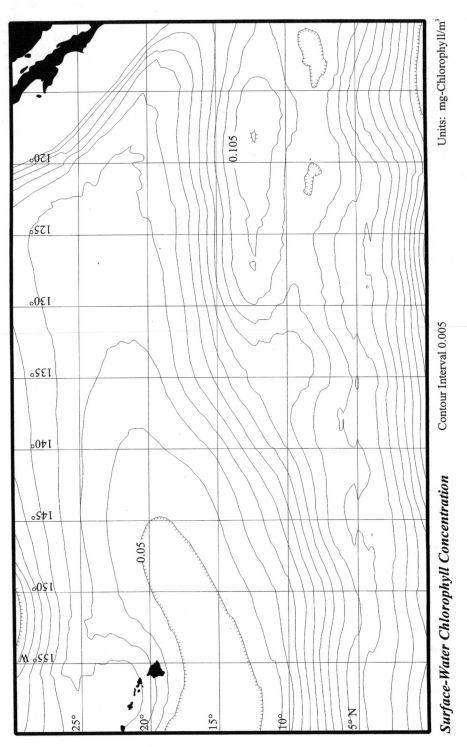

Surface-Water Chlorophyll Concentration   Contour Interval 0.005

Units: mg-Chlorophyll/m³

FIGURE 6.3   Surface-water chlorophyll concentration in the CCZ. Chlorophyll concentration is directly related to primary productivity and generally increases from north to south in the world's oceans, responding to increasing solar input. The east–west trend clearly apparent in this figure is closely associated with the northern equatorial current, which brings nutrients into the oligotrophic Pacific Basin waters, enhancing primary productivity.

[231]Pa, and [238]U to investigate both recent growth and, with the beryllium isotopes, growth histories of a few million years. This latter study of a nodule in the western part of the CCZ (DOMES Site A; see Figure 6.2) suggested that the nodule had been growing at approximately 5 mm/My for the past 80,000 years and at an overall rate of approximately 2 mm/My for the past 3 million years. Using similar methods with nodules from a site in the middle of the CCZ (11° N 140° W), Huh and Ku[37] derived growth rates of 3 to 8 mm/My.

A key limitation of all the above estimates based on radiometric methods is that only the outer few mm of nodules are amenable to such dating because of the relatively short half-lives of the isotopes available. Thus, even with the longest-lived species, [10]Be, only the outer 1 to 2 cm of the CCZ nodules can be examined. These studies also showed that, though the average rates appear to be comparable from study to study, there are large variations observed within single nodules. It is therefore likely that growth rates have varied significantly over time, and so the estimates based only on the outer rims of the nodules cannot be safely extrapolated throughout the entire history of the deposits.

An important problem posed by these growth rates is the reconciliation with observed sedimentation in the area, which can be orders of magnitude higher, on the order of mm of sediments per thousand years (e.g., Reference 30). Why aren't these nodules buried? Though a clear and quantitative explanation for this apparent contradiction has not been obtained to date, there is some evidence, from basic considerations and observations of benthic photographs from the area,[38] and from long-term repetitive photographs of nodule fields, that detritus-feeding benthic organisms are quite effective in removing accumulated sediments from nodule surfaces and even in depositing sediments under them.[39]

Another method of estimating manganese nodule growth rates was proposed almost 20 years ago by Heath[40] and further developed by Finney et al.[41] These workers reasoned that, if burial of nodules is independent of the size of the nodules and if they grow more or less continuously, then their size distribution should follow a log-linear slope, or:

$$\ln\left(N/N_0\right) = -\frac{B}{2G}\left(D - D_0\right)$$

where N/N_0 is the ratio of the number/area of nodules of a certain size class to the numbers/area of some initial, smallest size; B is the burial rate for nodules in the area; G is the growth rate; D is the size class with N nodules; and D_0 is the size of the initial, smallest size class.

Using two samples from the CCZ deposits (a box core from 14°15'N, 124°59'W and a free-fall grab from 9°3'N, 146°29'W), Heath showed how the size distributions of the recovered nodules do follow the predicted log-linear trend. He estimated the burial rates of nodules from the frequencies and depths of nodules and measurements of sedimentation rates found in core samples from the same general area (estimate 0.96 nodules/My-m²). He assigned values for D_0 and N_0 based on the smallest nodules found in the samples (6 mm), and deduced a growth rate of 4 mm/My for the free-fall grab sample, which is consistent with the radiometric dates determined before and since this study. For the box core, he used radiometric dates determined for the recovered nodules (7 mm/My) to calculate a burial rate of 2.9 nodules/My-m².

Sizes of 6,706 nodules collected in free-fall grab samplers from this same general area (between 8°41.84'N, 143°18.31'W and 15°49.21'N, 125°54.40'W) were measured by the Ocean Minerals Company.[42] Each nodule was characterized by three orthogonal measurements: maximum, middle, and minimum dimensions. The distributions of these measurements are presented in Figure 6.4. Median sizes for the three dimensions are, respectively, 8, 6, and 4 cm. It should be noted that these sizes are artificially limited on both upper and lower bounds. Nodules smaller than 2 cm maximum dimension were not included in the tally. Because the free-fall grab sampler had a sampling size of 25 × 25 cm, larger nodules can be expected to be preferentially excluded in the sampling process.

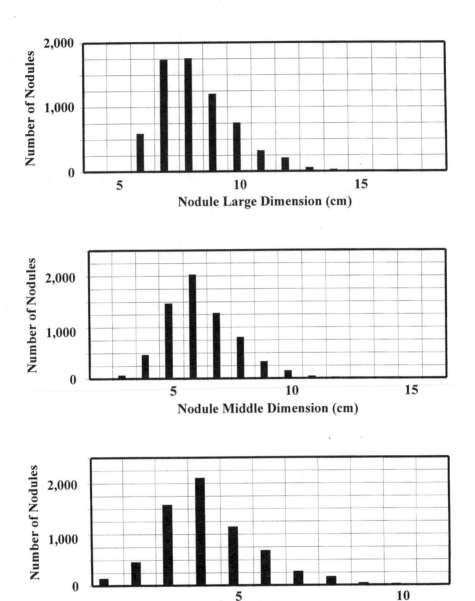

FIGURE 6.4 Manganese nodule sizes in the central part of the CCZ. Nodules collected in free-fall grab samples in the area between 8°41.84'N, 143°18.31'W and 15°49.21'N, 125°54.40'W were measured for maximum, middle (orthogonal to the other two measurements), and minimum size. 6,706 nodules were collected and measured. Nodules with maximum dimensions less than 2 cm were not included in the analysis, and nodules approaching the size of the free-fall grab sampler used (25 × 25 cm) can be expected to be increasingly excluded with increasing size.

If we ignore the small and large extremes for these size distributions, the expected log-linear behavior is clearly in evidence (Figure 6.5). Using Heath's estimated burial rates, (~1 to 3 nodules per My-m²) and using the above relationship, the following growth rates are implied: for a burial rate of 1 nodule/My, the large, middle, and small dimension distributions indicate growth rates of

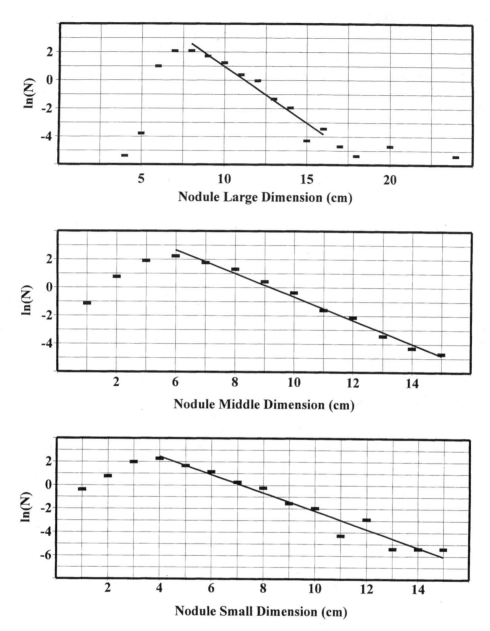

FIGURE 6.5 Log-linear plots of the measured nodule size distributions. N = number of nodules in the respective size class. If the large and small extremes are ignored, excellent fits are obtained for all three dimensions. Correlation coefficients are −0.97, −0.99, and −0.98 for the large, middle, and small dimensions, respectively. Slopes for the regression lines are, respectively, −0.804, −0.832, and −0.775 cm^{-1}.

6.2, 6.0, and 6.5 mm/My, respectively; for a burial rate of 3 nodules/My-m^2, these rates become 19, 18, and 19 mm/My. The first set of growth rates is quite consistent with the radiometric measurements, while the second set is considerably higher. The median nodule size of 8 cm thus implies an age of about 13 My, based on a burial rate of 1 nodule/My-m^2, and 4.4 My, based on the burial rate of 3 nodules/My-m^2.

6.4   LOCAL FACTORS

One of the most important characteristics of the CCZ deposits, relevant both to their geological significance and to their commercial potential, is the local variability of the deposits, particularly with respect to abundance. Even in the areas that exhibit the highest average abundance levels in the region, nodules occur in patches with highly variable abundance. Thousands of benthic photographs in these areas, taken along multiple azimuths with continuous coverage for several kilometers, clearly show that the typical pattern of occurrence consists of areas with high abundance alternating with areas with few or no nodules.

Piper and Blueford,[43] working in the DOMES Site A vicinity, note the apparent relationship between the flow of Antarctic Bottom Water through an abyssal valley and a corresponding accumulation of high-abundance deposits. Skornyakova et al.,[44] working in the same area, show some evidence, from acoustic measurements, sampling, and examination of photographs, that the highest abundances occur on the topographic highs. However, they suggest that high abundance is essentially independent of bottom relief. Demidova et al.[45] present evidence for significant redistribution of recent sediments due to benthic current activity and suggest that the most significant accumulations of nodule deposits occur in intermediate zones between maximum erosion and maximum deposition. It is possible that different factors are important in different areas, which might account for these apparent discrepancies. Because the accumulation of high abundances is highly dependent on the immediate geochemical environment of the sediment–sea water interface, it is perhaps not too surprising that local variability remains largely unexplained. Hopefully, when more of the extensive efforts of commercial exploration efforts work their way into open literature, the local constraints on deposit formation for the CCZ deposits will be much better understood.

6.5   RESOURCE ASSESSMENT

This resource assessment uses a database which includes the publicly available data assembled by the U.S. National Geophysical Data Center (NGDC),[46] as well as the controlled-access database collected by one of the commercial explorers for the CCZ deposits, Ocean Minerals Company (OMCO). This latter database was donated to the author for scientific study (though not for commercial exchange) in 1991 and has been described in Morgan et al.[42] and used in other scientific examination of the CCZ deposits.[17] More than 8,000 individual sample collection sites are included in the aggregate database.

Some modifications of the original data were necessary to derive a data set that provided some basis for uniform treatment of each datum. First, only those samples from whole sample recoveries of the NGDC data were used. Analytical results from isolated single nodules or portions of nodules were not included. Second, samples recovered within a few kilometers of each other were averaged to constitute a single site. This was necessary because much of the data consist of multiple free-fall grab sample stations collected at each sample site. Initial inclusion of the individual free-fall grab recoveries introduced a local spatial variance in resultant contour maps, which greatly impaired the determination of the regional trends. This averaging of individual free-fall sample recovery data into grouped stations resulted in between 2,000 and 3,000 (depending on the variable) separate station locations within the bounds selected for the analysis, i.e., 160° to 110° W longitude and 0° to 30° N latitude.

The methods used in the resource analysis and the results obtained are presented in the following sections.

6.5.1   KRIGING

During the past 50 years or so much progress has been made in the quantitative estimation of ore deposits on land, and a branch of applied statistics, termed *geostatistics,* has been developed to formalize and organize this progress. D. G. Krigue, working with data from South African gold mines, showed that simple techniques of averaging and interpolating between samples of commercial deposits to generate uniform grid block values usually do not give very accurate estimates of the amounts and grades of ore which are actually present.[47,48] He found that by using the actual spatial covariance to define the weighting of samples used to determine estimates, much better results could be obtained. Since that time, this basic idea has been developed by several industrial and academic professionals into a fairly well-defined discipline. The primary set of techniques used in geostatistics is termed "kriging" in honor of this pioneer. Several textbooks exist that give comprehensive treatments of kriging (e.g., References 49 and 50).

The primary tool used for the examination of spatial covariance of ore-grade variables (e.g., ore abundance and ore grade) is the variogram. Variograms are generated by summing the squared differences between all possible pairs of data values from sites that are similar distances from each other. Half the mean value for this squared difference is plotted as a function of distance between sample pairs.

Variograms are defined as the quantity γ plotted against the distance between sample points. The value for γ is defined as follows for any deposit variable x with N pairs of samples (x representing the difference for each pair) at distance h from each other:

$$\gamma(h) = \frac{\left[\sum_{i=1}^{N} \Delta x_i^2\right]}{2N}$$

This formulation is useful, because, with a few assumptions about the data set, the following relationship holds:

$$\sigma(h) = V - \gamma(h)$$

where $\gamma(h)$ is the same as defined above for a particular distance between samples (in one, two, or three dimensions), V is the overall sample variance, and $\sigma(h)$ is the spatial covariance at distance h. The assumptions necessary for this relationship to hold include the existence of the variance for the variable population as well as the local "stationarity" of the variogram. This latter assumption implies that the variogram calculated from points in any sub-area of the region of interest will be essentially the same. The primary consequence of this relationship is that, as h increases, the spatial covariance, or autocorrelation, between samples decreases to zero. Thus, given these assumptions, γ should increase from an initial value (termed the *nugget*, and denoted by C_0) to a plateau (termed the *sill* and denoted by C), which equals the overall sample variance. The distance at which this plateau is reached is termed the range, and denoted a.

Distributions of the variables iron, manganese, nickel, copper, and abundance were made using this kriging technique. The nugget, range, and sill for the variograms generated by these variables are presented in Table 6.1. Cobalt proved unstable in the analysis. Its overall variance is very small and exhibits essentially no spatial covariance that can be confidently used to generate grid blocks or contour plots. For the resource estimates generated here, a simple linear weighting was used for cobalt to generate concentration estimates at each grid block. Representative variograms are presented in Figure 6.6. Grid blocks with dimensions of 0.25 × 0.25 degrees were used in the kriging analysis performed with the aggregate manganese nodule data.

6.5.2 GENERAL RESULTS

Frequency distributions for the variables abundance, cobalt, copper, iron, manganese, and nickel are presented in Figure 6.7. Except for cobalt, the distributions of the derived grid blocks are used instead of the original data to eliminate the effects of the highly irregular spatial distribution of the

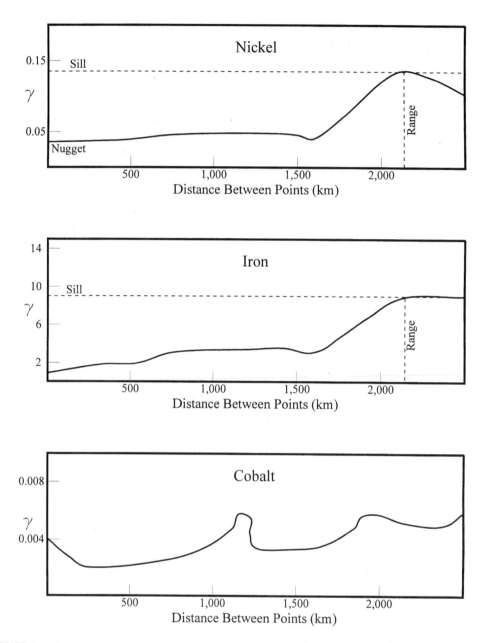

FIGURE 6.6 Representative variograms for CCZ manganese nodule variables. The ordinate, γ, represents the spatial variation of the variable in question, and has units of the variable squared. Definition of γ and the other terms used are standard for kriging analysis and are found in the text.

sample sites. All variables except cobalt are bimodal or multi-modal. Cobalt appears to represent a single population. Table 6.2 presents the basic descriptive statistics for these data.

The variable cobalt is the exception that proves the rule and provides an illustration of the weaknesses and strengths of the kriging technique. Kriging does not work for cobalt because there is no apparent relationship between the distance between sample pairs and the difference between their cobalt concentrations. That is, cobalt exhibits no coherent spatial covariance. Thus kriging provides a useful screen in the generation of contour maps for sparse data in general. If the data do not have usable variograms, then it is probably a waste of time to generate a contour map. The

TABLE 6.1
Variograms of Key Manganese Nodule Variables in the CCZ Deposits

Variable	Units	Nugget	Sill	Range (km)
Abundance	$(kg/m^2)^2$	16	30	1600
Manganese	$(Dry\ Wt.\ \%)^2$	4.8	19	1300
Iron	$(Dry\ Wt.\ \%)^2$	0.9	8.3	2300
Nickel	$(Dry\ Wt.\ \%)^2$	0.04	0.13	2300
Copper	$(Dry\ Wt.\ \%)^2$	0.04	0.14	2700

Variograms are defined as the quantity γ, which is plotted against the distance between sample points. The value for γ is analogous to and has the units of variance for any deposit variable. It is inversely related to the spatial covariance of the variable. The nugget is the γ value at zero distance between sample points; the sill is the γ value at the range, the distance at which the spatial covariance is apparently zero, and γ is at its maximum value.

TABLE 6.2
Descriptive Statistics for Manganese Nodule Variables

Variable	Units	# Grid Blocks	Min.	Max.	Median	Mean	Std. Dev.
Abundance	kg/m^2	11,907	0.9	11.6	3.5	3.8	1.7
Cobalt*	Wt. %	1354*	0.01	0.63	0.15	0.16	0.06
Copper	Wt. %	19,733	0.27	1.31	0.67	0.74	0.3
Iron	Wt. %	19,273	4.8	13.2	9.2	9.1	2.1
Manganese	Wt. %	19,733	11.2	31.9	21.6	21.6	4.7
Nickel	Wt. %	19,733	0.49	1.48	1.02	0.96	0.28

Except for cobalt, all data are from krigued, grid block interpolations derived from the original data sets.

*Original grouped station data used due to lack of coherent spatial covariance.

negative aspects of kriging include its relatively high requirement of computing time and its relatively complex interpretation when compared with other contouring algorithms. Also, some professionals involved with mining resource estimation note that the technique has been in some ways abused, since it is common practice to produce error estimates, or variances, from the derived grid-block data and to base confidence limits on such estimates [e.g., Reference 49, p. 55-71]. Any variance estimate based on grid-block data is suspect, since it is derived from data that have already been smoothed and averaged.[51]

Figures 6.8 through 6.12 present the contoured grid blocks for the variables abundance, copper, iron, manganese, and nickel. The similarities among these figures are striking and clearly show the well-established positive correlations among Mn, Cu, and Ni and the negative correlation between each of these components and Fe. The figures also share general features with the map of chlorophyll concentrations (Figure 6.3) presented earlier, including clear maxima (or a minimum for iron) between 10° and 15° N. The possible significance of these common features is considered in Section 6.6. The results of the resource assessment are described in the next section.

6.5.3 RESOURCE ASSESSMENT RESULTS

Table 6.3 presents the overall resource calculations produced from the kriging analysis. For comparison, the first, well-publicized estimate made by John Mero for the entire Pacific Basin was 1700×10^9 tons of nodules,[1] and a somewhat earlier estimate by the Russians Zenkevitch and

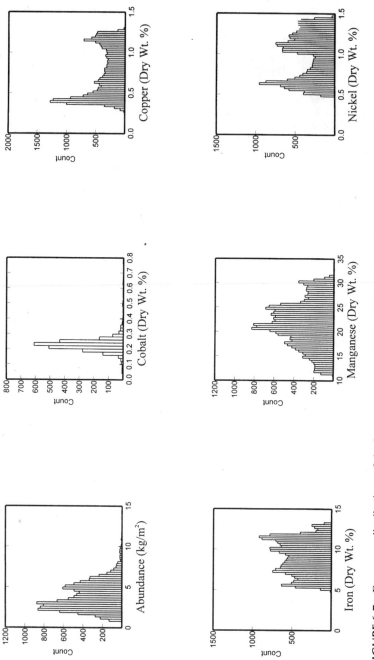

FIGURE 6.7 Frequency distributions of the key manganese nodule variables. Except for cobalt, these data are the distributions of the calculated grid block interpolations generated by a kriging analysis. Cobalt shows no coherent spatial covariance, so the original data are used. All variables except cobalt exhibit bimodal or multi-modal distributions.

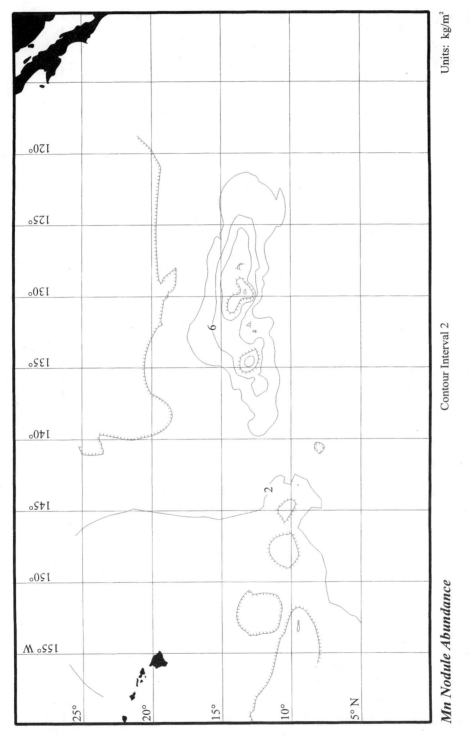

Units: kg/m²

Contour Interval 2

FIGURE 6.8 Manganese nodule abundance in the Clarion–Clipperton Zone.

Mn Nodule Abundance

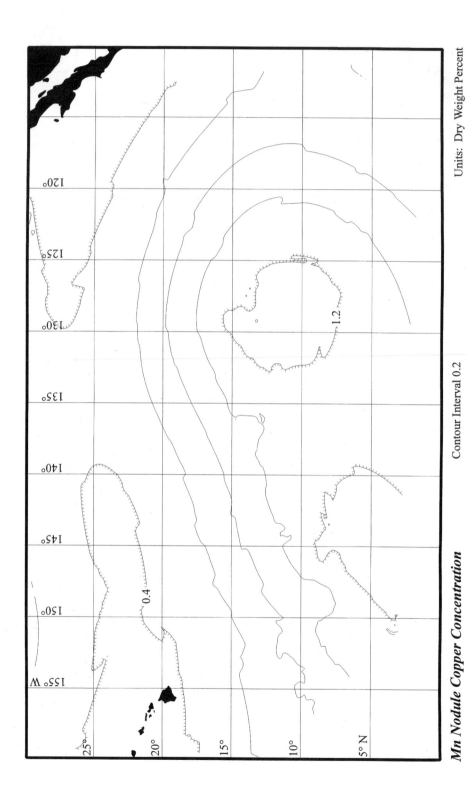

Mn Nodule Copper Concentration   Contour Interval 0.2   Units: Dry Weight Percent

FIGURE 6.9   Manganese nodule copper concentration in the Clarion–Clipperton Zone.

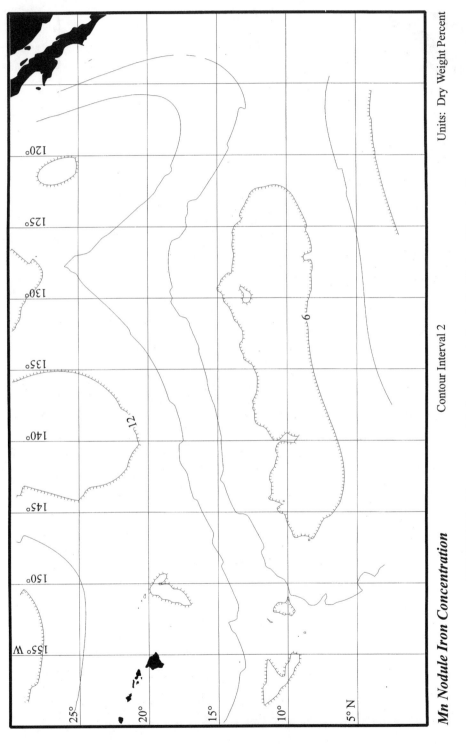

Mn Nodule Iron Concentration   Contour Interval 2   Units: Dry Weight Percent

FIGURE 6.10 Manganese nodule iron concentration in the Clarion–Clipperton Zone.

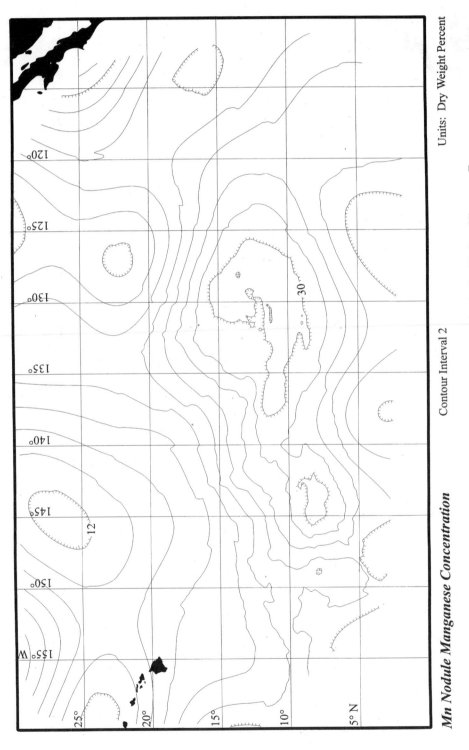

FIGURE 6.11   Manganese nodule manganese concentration in the Clarion–Clipperton Zone.

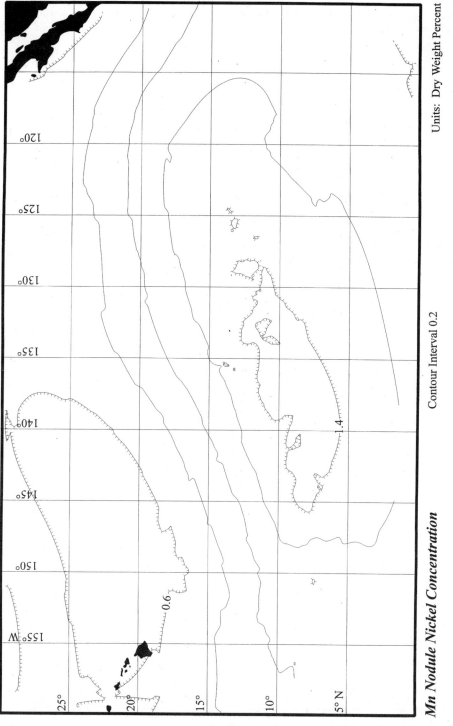

FIGURE 6.12 Manganese Nodule Nickel Concentration in the Clarion–Clipperton Zone.

TABLE 6.3
Resource Estimates for the CCZ Deposits, Based on Krigue-Block Calculations

Component	Total Area (km²)	Total (Metric Tons × 10⁶)	Area: Ab >8	Tons: Ab >8 (Metric Tons × 10⁶)
Nodules	8,956,000	34,390	153,400	1419
Manganese	"	7478	"	413
Nickel	"	340	"	18.9
Copper	"	265	"	16.2
Cobalt	6,542,000	78	150,400	3.4

Skornyakova for the Pacific Basin was 90×10^9 tons.[52] Though all data available to the author were used, the resultant areas represented by the kriging interpolation (maximum 8.9 million km²) only represents about 50% of the CCZ (defined here as between 0 and 30° N and 110 and 160° W) overall. The area within the CCZ included in this krigue-block network represents about 15% of the Pacific Basin (between depths of 4000 and 5000 m[53]), and holds about 34×10^9 tons of manganese nodules (Table 6.3). Thus, if the Russian estimate is accurate, about 38% of the Pacific manganese nodule resource occurs within 15% of the area.

The nodule abundance of 8 kg/m² was selected for separate consideration from the rest of the resource based on the apparent separate population of nodules above this abundance level, indicated by the abundance histogram (Figure 6.7). It has no fundamental operational or economic implications. It is important to note, however, that this high-abundance subgroup of the grid blocks is surprisingly contiguous and, not surprisingly, lies in the approximate center of the existing international exploration areas (Figure 6.8). As shown in the table, these high-abundance blocks represent 1.7% of the grid area and contain 5.5% of the Mn, 5.6% of the Ni, 6.0% of the Cu, and 4% of the Co.

The current (estimated 1998) annual world production rate for nickel is 900,000 metric tons/year.[54] Nickel is the primary metal of commercial interest overall in deep seabed manganese nodules, and so a future manganese nodule mining industry must be judged by its nickel production. The resources identified here could clearly provide a significant proportion of potential world production. The high-abundance blocks alone contain enough nickel to satisfy the entire world, at this current production rate, for 21 years. The entire resource includes enough nickel to provide more than 377 years' production at this rate.

The above resource estimates show that the CCZ manganese nodule deposits constitute a world-class nickel resource, with significant quantities of manganese, copper, and cobalt. Whether or not the exploitation of this resource occurs within any particular time frame is a question of politics, economics, and technology and lies well beyond the scope of the present study.

6.6   ENVIRONMENTAL IMPLICATIONS

The relevance of the CCZ deposits to several important scientific and environmental questions cannot be denied. As outlined above in the sections on formation processes, there are compelling links between the theories of manganese nodule formation and primary productivity in the northeastern tropical Pacific Ocean and between the occurrence of nodule deposits and the natural scavenging mechanisms that continually act to remove fine-grained solids and metals from the oceans worldwide.

One addition to this debate is presented in Figure 6.13. When the surface generated by the CCZ nodule abundance data is offset to the east and south, the resultant pattern of linear correlation between the corresponding values of nodule abundance and chlorophyll concentration line up remarkably well with the azimuth of the relative motion between the North American and Pacific

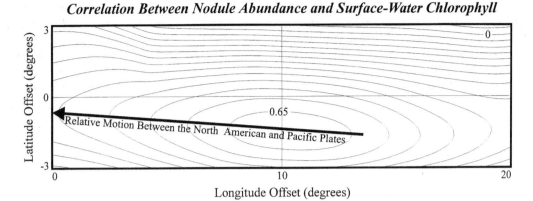

FIGURE 6.13   Plot of Pearson correlation coefficients with varying offsets between the surfaces representing chlorophyll concentrations in surface waters and manganese nodule abundance. Figure 6.3 shows the long-term average of chlorophyll concentration in Pacific surface waters, the best available indicator of primary productivity for this area. Figure 6.8 shows the abundance of manganese nodules in the region. This plot shows how the two surfaces represented by these two variables are correlated as a function of spatial offset. The azimuth of the tectonic motion of the Pacific plate is well aligned with this correlation, suggesting a strong relationship between nodule abundance and primary productivity.

Plates.[55] The age for the high-abundance nodule deposits implied by the offset representing the best correlation obtained (r = 0.682; longitude offset 10.25 to 10.5 degrees; latitude offset –1.5 degrees) and the spreading rate of the Pacific plate (0.885° westward/million years; 0.071° north-ward/million years) is 12 million years for the east–west component, and 21 million years for the north–south component. The age estimate from the east–west component is quite consistent with the ages calculated from the higher radiometric growth rates and also from the nodule size distri-butions, using a burial rate of 1 nodule/My-m^2.

Of course, this intriguing spatial correlation pattern does not clearly explain the growth histories of these deposits if growth has been uniform over time. If the ages are approximately correct, then uniform growth would place the high-abundance deposits along the plate azimuth, approximately half way between the site of the existing high-abundance deposits and the current productivity maximum. Also, the north–south offset of the highest correlation is about twice what would be predicted if the median nodule age is 12 My. Another possibility, more consistent with these data and the growth model, would be that the deposits grew much more rapidly during their early stages, while they were closer to the productivity maximum, and that, before the well-documented Miocene hiatus, which occurred 6 to 7 million years ago, the productivity maximum occurred at a slightly higher latitude.[56-58]

These considerations relating to the links between the CCZ deposits and primary productivity in surface waters are intriguing and suggestive of a host of further research efforts on the CCZ deposits to determine the relationships more precisely. If the stratigraphic record contained in these deposits can be properly interpreted, it may hold important clues to the basic processes which determine the overall health of Pacific ecosystems in general.

6.7   CONCLUSION

The extensive manganese nodule deposit data collected over the past 30 years confirm John Mero's and others' early speculations and prove that the CCZ deposits contain very significant amounts of manganese, nickel, copper, and cobalt. Marine technologies related to oil exploration and exploitation have dramatically improved during the past five or six years. Many of these technologies are directly relevant to exploration and commercial recovery of the CCZ deposits and are likely to become even more applicable to deep seabed mining as the oil industry moves farther and farther offshore. As land-use problems and environmental impact concerns become more constraining for land-based mineral developments, particularly in the tropical rainforest environments where new prospects are currently common, the CCZ deposits will become relatively more attractive. If the international regulation of seabed mining can be accomplished without imposing untenable restrictions on the development of the deep seabed, commercial development of these deposits will surely occur.

ACKNOWLEDGMENTS

The author wishes to thank Mr. Conrad G. Welling for providing the original data for this study; the Distributed Active Archive Center (Code 902.2) at the Goddard Space Flight Center, Greenbelt, MD, 20771, for producing the chlorophyll data and making them available for general distribution over the Internet; and Professor Cronan for his constructive and insightful editorial critique.

REFERENCES

1. Mero, J. L., *The Mineral Resources of the Sea*, Elsevier, Amsterdam, 1965, Chapter 3.
2. Archer, A. A., The deep seabed and its mineral resources, *Proceedings of the Third International Ocean Symposium*, Tokyo, 54, 1978.
3. U.N. Ocean Economics and Technology Branch, *Assessment of Manganese Nodule Resources: The Data and Methodologies*, Graham & Trotman, Ltd., London, 1982.
4. Glasby, G. P. (Ed.), *Marine Manganese Deposits*, Elsevier, Amsterdam, 1977.
5. Bischoff, J. L. and Piper, D. Z., *Marine Geology and Oceanography of the Pacific Manganese Nodule Province*, Marine Science Series, Volume 9, Plenum Press, New York, 1979.
6. Morgan, C. L., Moore, J. R., and Meyer, R. P., Investigations of the sediments and potential manganese nodule resources of Green Bay, Wisconsin, *University of Wisconsin Sea Grant Technical Report* WIS-SG-73-21, 1973, 144.
7. Stumm, W. and Morgan, J. J., *Aquatic Chemistry,* 2nd Edition, John Wiley & Sons, New York, 1981, 465.
8. Cotton, F. A. and Wilkinson, G., *Advanced Inorganic Chemistry,* 3rd Edition, John Wiley & Sons, New York, 1972, 846.
9. Sorem, R. K., Mineral recognition and nomenclature in marine manganese nodules, *Acta Mineral. Petrogr.* 20, 383, 1972.
10. Frondel, C., Marvin, U., and Ito, J., New occurrences of todorokite, *American Mineralogist*, 45, 1167, 1960.
11. Cronan, D. S. and Tooms, J. S., The geochemistry of manganese nodules and associated deposits from the Pacific and Indian Oceans, *Deep-Sea Research*, 16, 335, 1969.
12. Crerar, D. A. and Barnes, H. L., Deposition of deep-sea manganese nodules, *Geochimica et Cosmochimica Acta*, 38, 279, 1974.
13. Burns, R. G. and Burns, V. M., Mechanism for nucleation and growth of manganese nodules, *Nature*, 255, 130, 1975.
14. Glasby, G. P. and Thijssen, T., Control of the mineralogy and composition of marine manganese nodules by the supply of divalent transition metal ions, *N. Jb. Miner. Abh.*, 145, 291, 1982.
15. Li, Y. L. Interelement relationship in abyssal Pacific ferromanganese nodules and associated abyssal sediments, *Geochimica et Cosmochimica Acta*, 46, 1053, 1982.
16. Dymond, J., Lyle, M., Finney, B., Piper, D. Z., Murphy, K., Conrad, R., and Pisias, N., Ferromanganese nodules from MANOP sites H, S, and R — control of mineralogical and chemical composition by multiple accretionary processes, *Geochimica et Cosmochimica Acta*, 48, 931, 1984.
17. Knoop, P. A., Owen, R. M., and Morgan, C. L., Geochemical analysis of manganese nodules from the Clarion–Clipperton Zone of the Northeastern Tropical Pacific Ocean, *Marine Geology*, 147, 1-12, 1998.

18. Calvert, S. E. and Price, N. B., Diffusion and reaction profiles of dissolved manganese in pore waters of marine sediments, *Earth and Planetary Science Letters*, 16, 245, 1972

19. Calvert, S. E. and Piper, D. Z., Geochemistry of ferromanganese nodules from DOMES Site A, northern equatorial Pacific: multiple diagenetic metal sources in the deep sea, *Geochimica et Cosmochimica Acta*, 48, 1913, 1984.

20. Boudreau, B. P. and Scott, M. R., A model for the diffusion-controlled growth of deep-sea manganese nodules, *American Journal of Science*, 278, 903, 1978.

21. Murray, J. and Renard, A. F., *Deep Sea Deposits*, Report of the scientific results of the *H.M.S. Challenger*, 1873-76, Her Majesty's Stationery Office, London, 1891, as described in Cronan, D. S. Manganese nodules: controversy upon controversy, *Endeavor, New Series*, 2(2), 80, 1978.

22. Elderfield, H., Manganese fluxes to the oceans, *Marine Chemistry*, 4, 103, 1976.

23. Wyrtki, K., Surface currents of the eastern tropical Pacific Ocean, *Inter-American Tropical Tuna Commission Bulletin*, 9, 271, 1965.

24. Halpern, D., Observations of upper ocean currents at DOMES Sites A, B, and C in the tropical central North Pacific Ocean during 1975 and 1976, in *Marine Geology and Oceanography of the Pacific Manganese Nodule Province*, Bischoff, J. L. and Piper, D. Z. (Eds.), Plenum Press, New York, 1979, 43.

25. Klinkhammer, G. P. and Bender, M. L., The distribution of manganese in the Pacific Ocean, *Earth and Planetary Science Letters*, 46, 361, 1980.

26. Landing, W. M. and Bruland, K. W., Manganese in the North Pacific, *Earth and Planetary Science Letters*, 49, 45, 1980.

27. Martin, J. H. and Knauer, G. A., Manganese cycling in the northeast Pacific waters, *Earth and Planetary Science Letters*, 51, 266, 1980.

28. Martin, J. H. and Knauer, G. A., Lateral transport of Mn in the north-east Pacific gyre oxygen minimum, *Nature*, 314, 524, 1985.

29. Martin, J. H., Transition metal distributions in the water column in the OMCO exploration area, unpublished report to Ocean Minerals Company, 1980.

30. Arrhenius, G. O. S., Pelagic sediments, in *The Sea*, Volume 3, Hill, M. N. (Ed.), 655, 1963.

31. Osterberg, C. L., Carey, A. G., Jr., and Curl, H., Jr., Acceleration of sinking rates of radionuclides in the ocean, *Nature*, 200, 1276, 1963.

32. Martin, J. H. and Knauer, G. A., The elemental composition of plankton, *Geochimica et Cosmochimica Acta*, 37, 1639, 1973.

33. Fowler, S. W., Trace elements in zooplankton particulate products, *Nature*, 269, 51, 1977.

34. Ku, T. L., Rates of accretion, in *Marine Manganese Deposits*, Glasby, G. P. (Ed.), 1977, Chapter 8.

35. Krishnaswami, S. and Cochran, J. K., Uranium and thorium series nuclides in oriented ferromanganese nodules: growth rates, turnover times, and nuclide behavior, *Earth and Planetary Science Letters*, 40, 45, 1978.

36. Krishnaswami, S., Mangini, A., Thomas, J. H., Sharma, P., Cochran, J. K., Turekian, K. K., and Parker, P. D., [10]Be and Th isotopes in manganese nodules and adjacent sediments: nodule growth histories and nuclide behavior, *Earth and Planetary Science Letters*, 59, 217, 1982.

37. Huh, C. A. and Ku, T. L., Radiochemical observations on manganese nodules from three sedimentary environments in the north Pacific, *Geochimica et Cosmochimica Acta*, 48, 951, 1984.

38. Piper, D. Z. and Fowler, B., New constraint on the maintenance of Mn nodules at the sediment surface, *Nature*, 286, 880, 1980.

39. Gardner, W. D., Sullivan, L. G., and Thorndike, E. M., Long-term photographic, current, and nephelometer observations of manganese nodule environments in the Pacific, *Earth and Planetary Science Letters*, 70, 95, 1984.

40. Heath, G. R., Burial rates, growth rates, and size distributions of deep-sea manganese nodules, *Science*, 205, 903, 1979.

41. Finney, B., Heath, G. R., and Lyle, M., Growth rates of manganese-rich nodules at MANOP Site H (East North Pacific), *Geochimica et Cosmochimica Acta*, 48, 911, 1984.

42. Morgan, C. L., Nichols, J. A., Selk, B. E., Toth, J., and Wallin, C., Preliminary analysis of exploration data from Pacific deposits of manganese nodules, *Marine Georesources and Geotechnology*, 11, 1, 1993.

43. Piper, D. Z. and Blueford, J. R., Distribution, mineralogy, and texture of manganese nodules and their relation to sedimentation at DOMES Site A in the equatorial North Pacific, *Deep Sea Research*, 29(8A), 927, 1982.

44. Skornyakova, N. S., Gordeyev, V. V., Anikeyeva, L. I., Chudayev, O. V., and Kholodkevich, I. V., Local variations in nodules of the Clarion–Clipperton ore province, *Oceanology*, 25, 488, 1985.

45. Demidova, T. A., Kontar, E. A., and Yubko, V. M., Benthic current dynamics and some features of manganese nodule location in the Clarion–Clipperton province, *Oceanology*, 36, 94, 1996.

46. Moore, C. J., National Geophysical Data Center Marine Geology and Geophysics data archives, URL: http://www.ngdc.noaa.gov/mgg/, 1997.

47. Krigue, D. G., A statistical analysis of some of the borehole values in the Orange Free State goldfield, *J. Chem Metall. Min. Soc. S. Afr.,* 53(3), 47, 1952.

48. Krigue, D. G., A statistical approach to some basic mine valuation problems on the Witwatersrand, *J. Chem. Metall. Min. Soc. S. Afr.,* 52(6), 119, 1951.

49. David, M., *Geostatistical Ore Reserve Estimation*, Elsevier Scientific Publishing Co., Amsterdam, 1977, Chapter 4.

50. Journel, A. G. and Huijbregts, Ch. J., *Mining Geostatistics*, Academic Press, London, 1978.

51. Merks, J. W., *Metrology in Mining and Metallurgy*, Matrix Consultants Limited, Vancouver, British Columbia, Canada, 1992, 3-11.

52. Zenkevitch, N. and Skornyakova, N. S., Iron and manganese on the ocean bottom, *Natura (USSR)*, 3, 47, 1961.

53. Sverdrup, H. U., Johnson, M. W., and Fleming, R. H., *The Oceans, their Physics, Chemistry, and General Biology*, Prentice-Hall, Englewood Cliffs, New Jersey, 1942, 15, 21.

54. Nickel Producers Environmental Research Association, Internet Unrestricted Resource Location: http://www.nipera.org/, 1998.

55. Minster, J. B. and Jordan, T. H., Present-day plate motions, *Journal of Geophysical Research*, 83(B11), 5531, 1978.

56. Keigwin, L. D., Jr., Late Cenozoic stable isotope stratigraphy and paleoceanography of DSDP sites from the east equatorial and central North Pacific Ocean, *Earth and Planetary Science Letters*, 45, 361, 1979.

57. Sierro, F. J., Flores, J. A., Civis, J., González Delgado, A. J., and Francés, G., Late Miocene globorotaliid event-strategy and biogeography in the N.E.-Atlantic and Mediterranean, *Marine Micropaleontology*, 21, 143, 1993.

58. Roberts, A. P., Turner, G. M., and Vella, P. P., Magnetostratigraphic chronology of late Miocene to Pliocene biostratigraphic and oceanographic events in New Zealand, *Geological Society of America Bulletin*, 106, 665, 1994.

7 Ferromanganese Nodules from the Central Indian Ocean Basin

Pratima Jauhari and J.N. Pattan

I do not know what I may appear to the world, but to myself I seem to have been only a boy playing on the seashore, and diverting myself in now and then finding a smoother pebble or a prettier shell than ordinary, while the great ocean of truth lay all undiscovered before me.

Isaac Newton

ABSTRACT

The Central Indian Ocean Basin (CIOB), located between the Central Indian Ridge and the Ninety East Ridge receives sediments from the Ganges and Brahmaputra runoff, dispersed far south of the equator by turbidity currents. In order to delineate a mine site for ferromanganese nodules, extensive surveys were conducted in this area, which led to India being allocated a 150, 000 sq km area mine site for further developments related to ferromanganese nodule mining.

Mapping of the basin by multibeam swath bathymetry (Hydrosweep) has revealed many new bottom relief features. These features have a strong influence on the movement and distribution of water masses, sediment transport, chemical activity, and enrichment of metals in the nodules and their distribution on the seafloor. Size of the nodules ranges from less than 2 cm to above 10 cm in more than 13 morphological facies in the basin. The highest abundance of nodules is generally associated with high relief areas like valleys, followed by hilltops and slopes. Abundance is least in abyssal plains. Nodules from the plains have the highest contents of Mn, Cu, and Ni, while those from hilltops have the lowest concentrations of these metals. There exists an inverse relationship between nodule abundance and their grade. Nodules are associated with practically all types of sediment, and their surface density varies from traces to more than 20 kg/m^2 in the basin.

Genetically, three types of nodules are discernible in the basin — Type A, enriched in Mn, Cu, Ni, and Zn are concentrated in siliceous sediments. They get their supply of metals through early diagenetic processes occurring in the sediments. These nodules are small in size, rough surfaced, and contain todorokite as the dominant mineral phase. Type B, enriched in Fe and Co, are normally associated with red pelagic clays and are formed by accumulations mainly from sea water. These are represented by larger-sized nodules with smooth surfaces and containing ∂-MnO$_2$ as the principal mineral phase. The third, Type AB, are mixed phase and represent both processes. Rare earth elements in the nodules are contained in a single authigenic phase comprised of Fe, P, and Ti. ^{230}Th and ^{230}Th$_{exe}$/^{232}Th measurements on nodules from the siliceous ooze region show an accumulation rate of 1.2 mm/my for the nodule tops and 1.9 to 3.2 mm/my for the bottoms. Dendritic, collomorphic, laminated, and well-developed internal structures comprising alternating light-colored Mn-rich and dark-colored Fe-rich bands in sections of nodules have been observed under the ore microscope and scanning electron microscope.

Organisms belonging to seven different phyla and also a variety of biogenic traces were observed on various geological features in the basin. These features highlight the importance of further ecological studies pertaining to the Environmental Impact Assessment prior to mining.

7.1   INTRODUCTION

Until two decades ago there was very limited information available on Indian Ocean nodules as compared to those in the Pacific Ocean. Cronan and Moorby[1] made a comprehensive study of the Indian Ocean nodules and encrustations based on all the published data available plus about 100 new analyses and stated the CIOB to be the most promising for ore-grade nodules. Siddiquie et al.[2] evaluated all the published and unpublished data (from the databanks) — over 900 chemical analyses from 350 stations in the Indian Ocean — and noticed that most of the basins contain submarginal nodules, except for the CIOB where paramarginal (Ni+Cu+Co = 2.47% and abundance of 5 kg/m^2) nodules are present. Since India was importing 100% of her Ni and Co and 60% of her Cu requirement at the time,[3] preparations for the exploration of ferromanganese nodules in the Indian Ocean began in 1977[4] to meet the country's growing demand for metals. The first Indian cruise to locate potential-nodule bearing areas in the Indian Ocean started in December 1980, and on January 26th, 1981, the Indian scientists, while working onboard *RV Gaveshani,* dredged their first haul of nodules from the Arabian Sea. This discovery led the research program into an exploratory program with the aim of identifying a mine site in the Indian Ocean. An enormous amount of data was collected in the Central Indian Ocean Basin over the subsequent years using five different ships, but much of the information remains unpublished due to the classified nature of the project.

Over 4 million sq km of the seafloor were surveyed, the "prime areas" containing nodules were identified, and an application was submitted in 1982 to the United Nations Convention on the Law of the Sea (UNCLOS) for the registration of the mine site. India was recognized as a "Pioneer Investor" in deep seabed mining by the UNCLOS in 1982, followed by France, Japan, and the USSR. The Indian mine site was registered in 1987 based on a revised application containing updated information. More than 140 research papers have been published during the past decade on nodules and related aspects from the Central Indian Ocean Basin.[5] In the present chapter we broadly review some of these aspects.

7.2   CENTRAL INDIAN OCEAN BASIN

The CIOB is located (Figure 7.1) between the Central Indian Ridge and the Ninety East Ridge and is bounded in the south by the southwest Indian Ridge. It receives sediments mainly from the Ganges and Brahmaputra runoff, dispersed far south of the equator by turbidity currents.[6] The waters of the CIOB indicate the presence of the northward spread of Antarctic Bottom Water (AABW) from the Weddell Sea across low latitude saddles over the Ninety East Ridge.[7,8] In the middle portion of the basin, the absence of an index radiolarian species of Neogene Zone I, coupled with the erosion of the chronological record in the top several cms of sediments, indicates the presence of turbulent Antarctic Bottom water entering this region through northern saddles of the Ninety East Ridge.[9]

7.2.1   Bathymetry and Structure

Based on over 420,000 line kilometers of single beam echosounding data, three regions of different relief have been identified in the basin — high relief areas, medium relief areas, and plain areas represented by the western, eastern, and central portions, respectively.[10] The northern part, above 5° S latitude is occupied by terrigenous sediments, between 5° S and 15° S latitude by siliceous sediments with isolated calcareous patches in between, and by pelagic red clay sediments south of

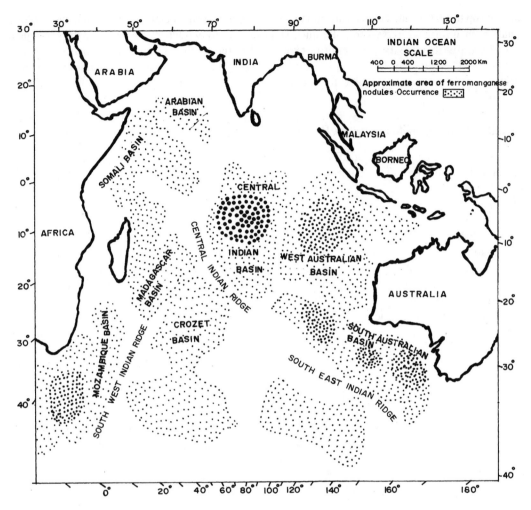

FIGURE 7.1 Location of Central Indian Ocean Basin and other physiography showing occurrence of ferromanganese nodules.

15° S latitude.[11] Topographic features have a strong influence on the movement and distribution of water masses, on sediment transport, chemical activity, and on ocean dynamics.[12] The water depth in the basin ranges from 4,500 m at 5° S to 5600 m at 10° S.[13] In this area the carbonate lysocline associated with North Indian Deep Water occurs at around 4000 m depth, while the CCD exists at around 5000 m.[14] These features have a substantial impact on the metal enrichment in nodules and their distribution and abundance on the seafloor.[15,16]

Part of the CIOB area was subjected to detailed bathymetric analyses by the high resolution multibeam swath bathymetry system, Hydrosweep, installed on *ORV Sagar - Kanya*. This mapping has brought to light many geological features in greater detail. For example, the area south of 10°55'S and 79°20'E longitudes (Figure 7.2) revealed the extension of the 79° E Fracture Zone with a ridge and trough topography and an elevation difference of over 300 m[17]; prominent east–west lineations, which appear to bend toward the south against the 79° E Fracture Zone; and several seamounts. Mukhopadhyay and Khadge[18] speculated that the CIOB was subjected to tectonic reactivation due to wrench faulting during late Miocene time, resulting in deformation of the central part of the basin. There are several guyots in this central part, and the seamounts have a much lower ratio of height to basal width as compared to Pacific seamounts.[18] Three distinct NNE–SSW trending seamount chains, along 75°23', 75°44' and 78°55' East have been identified. These seamounts have heights up to 1300 m and cover an area up to 300 km[2].[19] Studies on their distribution and morphology indicate their formation either from Reunion Hotspot or from two separate hotspots in the geological past[20] and that they are the product of fast spreading at the southeast Indian Ridge between 50 and 60 Ma.[21,22] Generation of fracture zones, differential spreading rates, localized vulcanism, and other tectonic activities are responsible for the development of secondary bathymetric features in the CIOB.[23]

7.2.2 NODULES IN RELATION TO BATHYMETRY

Nodule distribution in terms of kilograms per square meter (abundance) varies according to the topography of the area. The highest abundance of nodules is generally associated with the high-relief areas, such as the valleys followed by hilltops and slopes. Abundance is lowest on the plains.[24,25] However, nodules from the plains have the highest content of Mn, Ni, Cu, and Zn, while those from the valleys have the lowest. Similar distribution patterns were observed by Kodagali and Sudhakar,[23] using a database from 479 locations spread over an area bounded by 9°30'S and 16°S latitudes and 72°30'E to 82°E longitude, and also from seabed photographs.[26] Average abundance, obtained mainly by boomerang grabs, in this area shows highest concentration in the valleys (6.94 kg/m[2]) followed by hilltops (5.81 kg/m[2]) and slopes (4.30 kg/m[2]). Plain areas have the least abundance (2.72 kg/m[2]).

Quantification of nodule occurrence by the seabed photographs shows similar results,[26-28] though in places nodules seem to be partially buried due to partial sediment cover.[29] Higher abundance on slopes, hilltops, and valleys is an indication of availability of more nucleating material originating from volcanic activity around abyssal hills and seamounts or breaking down of basaltic rock fragments by weathering processes.[30,31] Exposure of fresh basaltic rock as well as volcanic flows could produce a new supply of nuclei associated with each tectonic episode. Rolling down of nodules from slopes and hilltops into neighboring low-lying areas, such as valleys, increases nodule abundance in this domain.[23] The wide range of nodule sizes and morphologies typical of such areas can be explained by multiple episodes generating different nuclei.[32] Thickness of the sediments in an area is also responsible for diverse sizes of nodules in it.[33] The hilltops and flanks normally have a very low sediment cover, while the plains may have a very thick sediment cover which tends to result in a more uniform, albeit low, distribution of nodules.[23]

7.3 DISTRIBUTION OF FERROMANGANESE DEPOSITS

Since the Central Indian Ocean Basin is most promising from the point of view of ore-grade nodules,[1,2] efforts to locate a mine site were concentrated in this area. Over 4.24 million sq km were surveyed over about three years using six different ships, including two chartered from Britain (*M.V. Farnella* and *G.A. Reay*), and one from Norway (*M.V. Skandi Surveyor*) in addition to the Indian Research Vessels *Gaveshani, ORV Sagar Kanya,* and *Nand Rachit.* Samples were obtained mainly by free-fall grabs, Pettersson and Van veen grabs, box, spade, gravity corers, and dredges. Cameras, deep-towed cameras, and single beam and multibeam echosounders were also used in addition to gravity and magnetic observations.

During the reconnaissance sampling, the basin was subdivided into a one-degree-square grid. The nodules were found associated with each sedimentary facies with varying occurrence and composition. Locally, they are even associated with terrigenous sediments in the northern basin, though are absent or present in traces in most of the places there. The high terrigenous influx from the Ganges and Brahmaputra Rivers into the northern part of the basin to 8° S,[34] inhibits the growth and abundance of nodules in this area. Ore-grade nodules occur only in the central part of the basin, which is occupied mainly by the siliceous sediments. Maximum occurrence of nodules occurs on the red pelagic clays. Based on the results obtained from the regional surveys, detailed surveys were conducted in the "prime areas" of nodule occurrence.[35] Sampling was subsequently narrowed

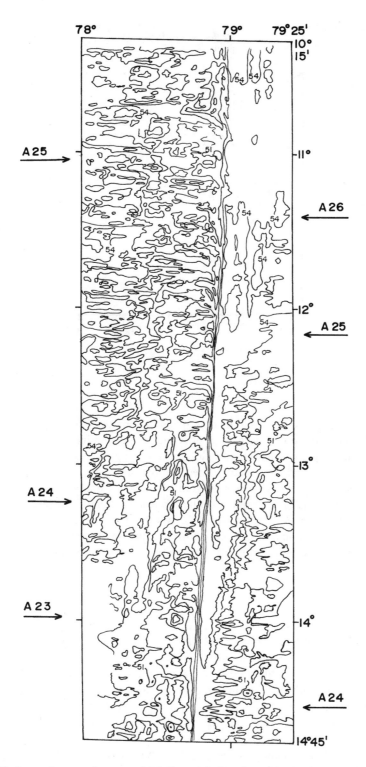

FIGURE 7.2 Bathymetric map of a part of the Central Indian Ocean Basin showing 79°E fracture zone. (From Kamesh Raju, K., Ramprasad, T., Kodagali, V.N., and Nair, R.R., Multibeam bathymetry, gravity and magnetic studies over 79°E fracture zone, Central Indian Basin, *Jour. Geophy. Res.*, 98, B6, 9605, 1993. With permission.)

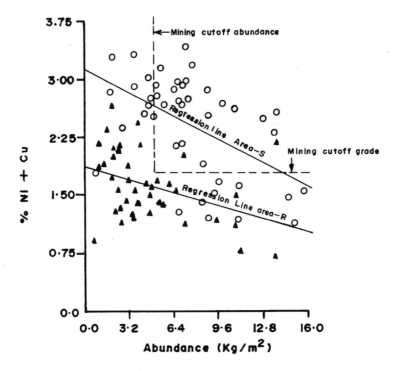

FIGURE 7.3   Inverse relationship between nodule abundance and their grade. (From Sudhakar, M., Ore grade manganese nodules from the Central Indian Basin, *Mar. Mining,* 8, 201, 1989. With permission.)

down to 50 km and 25 km spacing, eliminating mainly the northern and other parts of the CIOB with low nodule concentration. Surveys to identify the candidate mine site formed the final phase of exploration. These were carried out at a close grid spacing to detail the information acquired during previous surveys.[35] At every station three sets of data were collected to calculate the wet abundance of nodules in kg/m², the percentage grade (Ni + Cu + Co) of nodules, and the topography of the seabed. By extrapolating and interpolating the results after statistically analyzing the data obtained, potential areas were identified where the weight of nodules per square meter was above 5 kg, the enrichment of Ni and Cu in them above 2%, and the topography of the area not very rugged. There is an inverse relationship between grade and abundance (Figure 7.3).

7.3.1   THE PIONEER AREA

Out of over 4 million sq km surveyed; two areas of 150,000 sq km each containing nodules of equal commercial value were identified in accordance with Resolution II of the United Nations Convention on the Law of the Sea (UNCLOS) (Figure 7.4). An application for its registration was made with the Preparatory Commission in 1984. A revised application was filed with the commission on July 20, 1987, after modifying the database with further investigations. Since India's case was free from overlap with other claims, the Preparatory Commission accepted India's claim favorably and registered her application on August 17, 1987. India thus obtained its registration as the first Pioneer Investor and a mine site of 150,000 sq km in the Central Indian Ocean.[4] The certificate of registration was handed over to the Indian delegation in New York.

Of two areas containing nodules of equal commercial value, one, Area A, was allocated to India, whereas the other, Area B, was surrendered to the International Seabed Authority as a "reserved area" for conduct of activities by the Authority in the area through the Enterprise or in association with developing states (Figure 7.4). Average nodule abundance in area A is 4.39 kg/m²

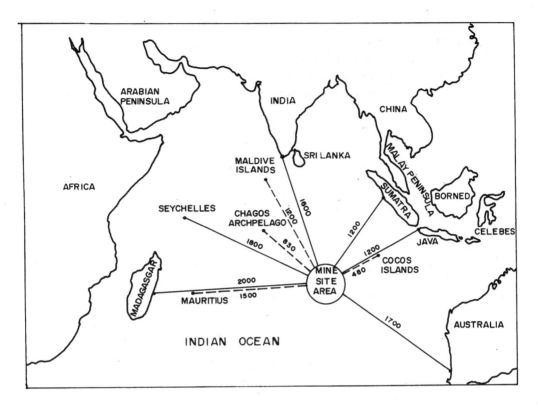

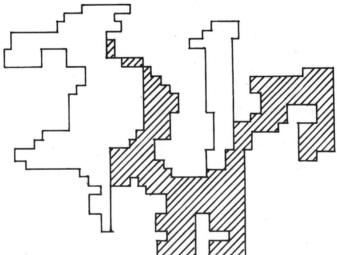

FIGURE 7.4 Mine site allocated to India (Area A – not hatch marked) in the Central Indian Ocean Basin.

(Table 7.1) and grade 2.31% (1.12% Ni, 1.05% Cu, and 0.14% Co.[35] Around 676 million tonnes of nodules are available in area A.

7.4 PHYSICAL PROPERTIES OF NODULES

The various physical properties of nodules, such as wet and dry densities, porosity, water content, P and S velocities of nodules from siliceous, calcareous, pelagic, and transition zones, are presented in Table 7.2.[36,37]

TABLE 7.1
Abundance, Grade, and Resources
of Manganese Nodules in the Pioneer Area
of the Central Indian Ocean Basin

	Total Application Area	Area A	Area B
Area (sq km)	300,000	150,000	150,000
Average abundance (kg/m^2)	4.51	4.39	4.45
Total nodules (million tonnes)	1335.0	676.50	658.50
Combined metal Values (Ni + Cu + Co) million tonnes	21.84	10.88	10.96

Source: Mudholkar, A.V., Pattan, J.N., and Sudhakar, M., Techniques and results, in *From the First Nodule to the First Mine Site: An Account of the Polymetallic Nodule Project*, Qasim, S.Z. and Nair, R.R. (Eds.), Department of Ocean Development and NIO, Goa, 29-36. 1988. With permission.

TABLE 7.2
Variations in Physical Properties of Ferromanganese Nodules
from Different Sediment Types

Parameters	Siliceous Ooze	Calcareous Ooze	Pelagic Clay	Trans. Zone
Wet den. (gcm^{-3})	1.98	1.88	1.95	2.10
Dry den. (gcm^{-3})	1.35	1.29	1.36	1.47
Porosity (%)	62.9	58.9	59.0	63.1
Water cont. (%)	47.9	45.5	43.5	42.7
P.vel.dry (ms^{-1})	2123	1926	2044	2229
P.vel.wet (ms^{-1})	2464	2957	2630	2800
S.vel.dry (ms^{-1})	1083	952	1091	1254
Atten (dBcm^{-1})	22.1	28.4	22.7	21.6
Coef. anisotropy	1.17	—	—	—
Den.incre.sat. (%)	.32	31	30.3	29.9
P.vel.incr.sat (%)	15.8	34.9	16.7	12.0

Atten (dBcm^{-1}) = Attenuation (dBcm^{-1}), Coef. anisotropy = Coefficient of anisotropy, Den. incre. sat(%) = Density increase(%) on saturation, P. vel. incr.sat(%) = P. velocity increase(%) on saturation.

Source: Mukhopadhyaya, R. and Ramana, Y.V., Acoustic properties of Indian Ocean manganese nodules in relation to physical constitution and chemical composition, *Deep Sea Res.*, 37(2), 337, 1990. With permission from Elsevier Science Ltd, The Boulevard, Langford Lane, Kidlington, U.K.

7.4.1 MORPHOLOGY

Based on size, shape, and surface texture the CIOB nodules exhibit a wide range of morphologies, and this is reflected in their composition.[38] Their size in the basin ranges from less than 2 cm to more than 10 cm. Spheroidal and ellipsoidal nodules with rough surfaces are more common among

TABLE 7.3
Average Concentration of Mn, Fe, Cu, Ni, and Co (%) in Ferromanganese Nodules of Different Sizes from the Central Indian Basin (n = 891)

Size →	<2 cm n = 52	2–4 cm n = 410	4–6 cm n = 309	6–8 cm n = 80	8–10 cm n = 17	>10 cm n = 23
Mn	27.70	25.00	23.80	21.90	20.90	21.70
Fe	6.60	6.30	6.90	7.60	9.40	9.50
Cu	1.31	1.16	1.04	0.91	0.68	0.68
Ni	1.14	1.19	1.13	0.95	0.77	0.76
Co	0.11	0.11	0.10	0.12	0.11	0.10

n is number of samples analyzed.

Sources: Compiled from Valsangkar and Khadge,[39] Valsangkar et al.,[41] Valsangkar et al.,[42] Banakar et al.,[61] Valsangkar and Ambre,[115] Valsangkar and Karisiddaiah.[116]

smaller sizes.[39] The shape changes to discoidal and elongated with increase in size, the largest becoming polynucleate.[40] In a specific size group, nodules have common properties, such as morphology, chemistry, and mineralogy, and they all appear to have been formed by one process. The different size groups, therefore, at least reflect different environmental conditions where nodule forming processes act differently.[41,42] Concentrations of Mn, Fe, Cu, Ni, and Co in different sized nodules show that smaller nodules are more enriched in Mn, Cu, and Ni than larger nodules (Table 7.3). This could be due to burial of smaller nodules where Mn, Cu, and Ni will be supplied by diagenesis. Basin depth also influences nodules to a certain extent. Nodules from relatively shallower depth are larger in size, contain rock fragments as nuclei, and are rich in Co and Fe in contrast to nodules from deeper depths, which are mostly medium to small in size with consolidated mud as a nucleus.[43] A study conducted in an area of 71,424 km[2] by Valsangkar and Khadge[39] revealed 13 different shapes of nodules, including the common ones like spheroidal, ellipsoidal, discoidal, intergrown, polynucleate with smooth and rough textures, which were grouped into six size classes. Spheroidal to ellipsoidal nodules (up to 4 cm) are the most abundant. Valsangkar and Khadge[39] concluded that the smaller nodules are diagenetic, and the larger are hydrogenetic. Acoustic properties of nodules from different types of sediments in the basin suggest that smaller nodules are more anisotropic than the larger ones, which have a low density and low porosity.[36,37]

The surface textures exhibited by nodules in the basin are smooth, rough, granular, and mixed — smooth mainly on the top and granular on the bottom.[38] The rough and granular texture generally is associated with nodules formed diagenetically, whereas the smooth texture is the product of hydrogenous deposition. Valsangkar et al.[42] believe that, compared to their larger counterparts, the smaller nodules have a greater ability to roll, which allows them to remain in contact with the sediment/water interface for longer periods. Smooth texture is common in tabular, elongated, irregular discoidal and polynucleate nodules.[42] Compositional differences with respect to morphology has been well documented in other parts of the oceans.[44-46]

7.4.2 NUCLEUS

Accretion of nodules is initially dependent on the existence of a solid substrate (seed) in an oxidized environment.[47] The distribution and morphology of nodules are controlled by the availability of potential agents on which manganese oxide material can accrete.[31] Horn et al.[30] found that the "seeds," or nuclei, are essential for the formation of nodules and that their distribution pattern is influenced by seamounts.

In the CIOB, nodules have been found to accumulate around shark teeth,[48] the tympanic bulla of the Minke Whale,[49] large phillipsite crystals,[50] zeolites,[51] pumice,[52] rock fragments, and volcanic material.[53] Some of the nuclei contain hard grounds, ferromanganese micronodules, volcanic glass, ichthyoliths, and phillipsites.[54]

Iyer and Sharma[53] studied the correlation between the occurrence of ferromanganese nodules and rocks in a part of the CIOB and discovered three rock types — fresh basalts, weathered basalts, and pumice providing nuclei. The rock fragments found on abyssal hill summits are fresh basalts with very thin veneers of oxide coating. Those occurring along slopes and flanks are partly weathered, with substantial oxide deposition, and those transported to the abyss form highly weathered nodule cores. The availability of nucleus material directly affects the abundance of nodules. The availability of nucleus material and the occurrence of thin sediment cover have resulted in a high abundance of nodules along hill slopes and flanks.[53]

Martin-Barajas et al.[55] studied nodules from the CIOB and suggested that their nuclei material (mostly altered ash layers and rhyolitic pumice) have possibly been derived since Miocene time from Indonesian arc vulcanism. Based on geochemical analyses of the pumice, they attributed the volcanic products to distal fallout from the Indonesian volcanic arc.[56] Iyer and Sudhakar[52] studied 1,925 data points in the CIOB and observed that the ferromanganese nodule fields coexist with the distribution of pumice. They noted the trends of seamounts and abyssal hills and the near north — south orientation of the fracture zones coinciding with the distribution of nodules rich in Ni and Cu. This observation led them to speculate on the presence of *in situ* silicic vulcanism in the basin as a source for this pumice[52] However, based on modal, mineralogical, and geochemical studies, Mudholkar and Fujii[57] ruled out the possibility of any recent intraplate vulcanism in the CIOB and concluded that the fresh CIOB pumice originated from the Krakatau eruption in 1883.

7.5   INTERNAL STRUCTURES

Ore microscopic studies reveal the presence of well-developed internal structures comprising dendritic, collomorphic, laminated, and alternating light (Mn-rich) and dark (Fe-rich) bands.[58,59] The layer closer to the nucleus contains more Fe and Co, phillipsite, quartz, and feldspars and is hydrogenetic in origin; whereas the reverse is true for the outer layer containing Mn, Ni, Cu, and Zn, todorokite, δ-MnO_2, phillipsite, and quartz which represents early diagenetic growth. A mixing model showed that the contribution of interstitial water to the outer diagenetic layers is about 75% and that of sea water is about 25%.[59]

7.6   MINERALOGY

Todorokite is the dominant mineral phase in the nodules of the northern CIOB. These nodules are associated with siliceous sediments rich in montmorillonite, chlorite, and illite. δ-MnO_2 is dominant in nodules from the southern CIOB. They are associated with pelagic clay sediments containing keels of foraminifera, zeolites, chlorite, and illite.[60] The intensity of the X-ray peak of todorokite changes with nodule size class within the same sediments.[61] The abundance of todorokite decreases with increasing size class, whereas δ-MnO_2 behaves independently.

7.7   GROWTH RATES

Growth rates have been determined in two nodules from the basin by Banakar.[62] Based on the exponential depth decay profile of $^{230}Th_{exe}$ activity and the $^{230}Th_{exe}/^{232}Th$ activity ratio, the accretion rates obtained for nodule tops were 1.2 to 1.3 mm/my, and for the bottom layers, 1.9 to 3.2 mm/my.[62] The nearly three times higher accretion rates on the nodule undersides compared to their tops reflects the fixed position of the nodules throughout their growth history.[63] The accumulation rates of sediments in the basin averages 2 mm/ka.[9] This suggests that nodules overlying the sediments accrete nearly 1,000 times slower than the sediments.

7.8   ELEMENTAL VARIATIONS

The range of metal concentration in different domains of the CIOB are comparable to ranges reported from similar domains in other basins of the world oceans.[23,64] Iron content is reported highest in nodules from hilltops (8.9%) and lowest in plains (6.96%), in contrast to Mn, which is highest in plains (25.6%) and lowest on hilltops (22.4.%). Cobalt does not show much variation in different domain nodules (0.12 to 0.15%), whereas Ni and Cu are highest in plains and lowest in valleys and on hilltops.[23] An inverse relationship between nodule abundance and grade based on 600 analyses has been observed[15,65] on a database from over 1,000 locations in the CIOB (Figure 7.3), confirming earlier observations in the Pacific Ocean.[66]

The maximum, minimum, and mean concentration of major, trace, and rare earth elements in the CIOB nodules is presented in Table 7.4. The elements show variations in concentration with the sediment type (Table 7.5) and geographical location. Manganese is generally high to 15° S and west of 79° E, where mainly siliceous–diatomaceous sediments occur, but its concentration decreases southward in the areas of red pelagic clays. Fe shows a reverse trend to that of Mn. Copper and Ni cluster toward Mn, and Co toward Fe, and the two groups are negatively interrelated.[67] Scatter diagrams (Figure 7.5) show that the Ni and Cu contents consistently increase with an increase in the Mn/Fe ratio, whereas Co decreases.[15] R-mode factor analysis carried out on ferromanganese nodule data comprising major, trace, and REE (Pattan, unpublished data) shows (Figure 7.6) three factors — Factor-1 represents 55.1% of the variance and has strong negative loadings of Fe, Ti, P, Pb, Y, Sr, and all the REE, suggesting a hydrogenous phase supplied from sea water. Factor-2, has strong loadings of Mn, Ca, Mg, Mo, Zn, Ni, Cu, Ba, and V, representing 28.4% of the variance, suggesting a diagenetic phase where metals have been supplied from the underlying sediment; these cations mostly stabilize the todorokite structure. Factor –3, (8.5%) has loadings of Al, K, and P (partially), suggesting a mostly aluminosilicate phase.

Aluminum concentration is known to be a measure of lithogenous content in nodules.[68] The CIOB nodules show a positive correlation between Al and Si (r = 0.71), the latter being another indicator of mainly detrital source (Figure 7.7). However, the mean Si/Al ratio in nodules (3.3) is more than in pelagic clays (2.97), suggesting that excess biogenic Si in the form of siliceous skeletal material may also be present. Phosphorus and Fe in the nodules show a positive correlation (r = 0.6) (Figure 7.7), perhaps due to the adsorption of P by hydrous ferric oxides or to the formation of a ferric-phosphate phase.[31,69] There appears to be a colloidal Ti-hydrate present in the Fe-rich hydrogenous phase of the nodules, as evidenced by a strong positive association (r = 0.8) between Fe and Ti (Figure 7.7).[70,71] A positive correlation of Mn with Mo (r = 0.8), Ba (r = 0.6), Ca (r = 0.4), and Mg (r = 0.3) (Figure 7.8) may be attributed to the stabilization of the todorokite structure by divalent cations.[31,69,72-74] The composition of todorokite, the major mineral phase in the nodules, includes Ca, Mg, and K.

Concentrations of REE in nodules from the CIOB show a wide variation, from 398 to 2020 ppm (Table 7.4)[71,75-77]; however, their mean concentration is similar to that in Pacific Ocean nodules.[78] When the total REE content of nodules is compared with the REE in the surrounding sediment and average shale (North American Shale Composite), nodules are found to be enriched up to fourfold and 2 to 10 times, respectively. Higher and lower REE contents in the top and bottom portions, respectively, of oriented nodules was observed by Nath et al.[75] and is similar to an earlier observation made on Pacific Ocean nodules.[79] There are two opinions regarding the REE carrier phases in the nodules. Elderfield et al.[78] and Nath et al.[75] are of the opinion that REE in the nodules are carried by two independent phases comprised of Fe oxyhydroxide and phosphate. On the other

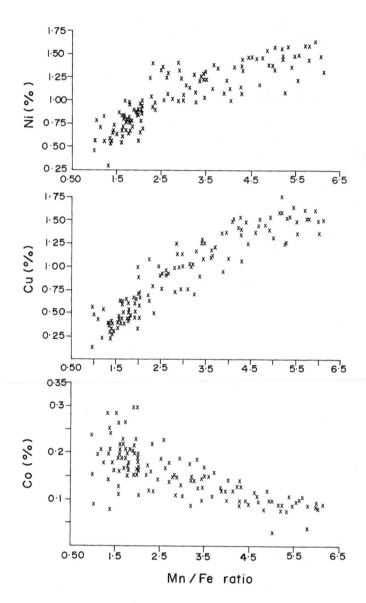

FIGURE 7.5   Scatter plots showing positive relation between Mn/Fe ratio versus Ni and Cu, and inverse relation between Mn/Fe ratio and Co. (From Jauhari, P., Classification and inter-element relationships of ferromanganese nodules from the Central Indian Ocean Basin, *Mar. Ming*, 6, 419, 1987. With permission.)

hand, Glasby et al.[80] and Pattan and Banakar[71] believe REE in the nodules are carried by a single authigenic phase comprising Fe, P, and Ti. All existing data on REE, and combined data of Fe, P, and Ti from the CIOB (Figure 7.9), shows very strong positive interelemental correlation (r = 0.8). A reexamination of the data of Elderfield et al.[78] (r = +0.87) and Nath et al.[75] (their Table 7.6, r = +0.85) shows a positive correlation between Fe and P in the nodules, suggesting the presence of a single phase containing Fe and P. Further, no apatite phase is recorded in the nodules by X-ray diffraction. The mean Fe/P values of CIOB nodules is 42, suggesting Fe is the main authigenic carrier phase. REE in the nodules are not associated with aluminosilicate or diagenetic phases.[71,75] Elderfield et al.[78] noticed that the highest REE content of nodules in the Pacific Ocean is associated

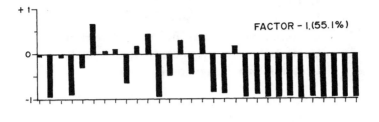

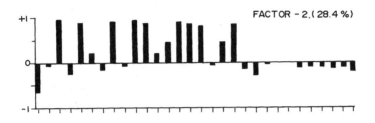

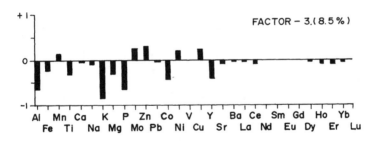

FIGURE 7.6   R-mode factor analysis of major, trace and rare earth elements depicting hydrogenous (Factor-1), diagenetic (Factor-2), and aluminosilicate phases (Factor-3).

with the lowest REE content in associated sediments and vice versa, suggesting a competitive scavenging between them. Glasby et al.[80] and Courtois and Clauer[81] did not report any such relation in SW and SE Pacific Ocean nodule–sediment pairs. Pattan et al. (unpublished data) showed that the higher REE content of nodules from the CIOB is associated with higher REE content of associated sediment, suggesting a common REE source, most probably from sea water.

The behavior of cerium is quite different from that of its neighboring REE in the nodules.[82,83] Cerium is highly enriched compared to other REE members in the nodules because in an oxic environment Ce^{2+} is oxidized to insoluble Ce^{4+}. In the CIOB nodules, positive Ce-anomalies have been reported,[71,75-77,84] suggesting an oxic environment. There are a few reports of negative Ce-anomalies in Pacific Ocean nodules,[80,81,85,86] but none have been reported in the CIOB. Cerium-anomalies show an inverse relationship with Mn/Fe ratio (r = –0.7) (Figure 7.9), suggesting that nodules with highest diagenetic signature have lowest Ce-anomalies. Shale-normalized REE patterns of nodules exhibit a convex pattern depicting enrichment of middle REE (MREE) over heavy REE (HREE) and light REE (LREE) (Figure 7.10).

TABLE 7.4
Maximum, Minimum, and Mean
Concentration of Major, Trace, and Rare Earth
Elements in Ferromanganese Nodules from the
Central Indian Ocean Basin

Element	Max	Min	Mean	No. of Samples Analyzed
Si (%)	18.8	5.90	9.20	23
Al	4.9	1.40	2.80	33
Fe	20.5	2.40	7.10	1119
Mn	48.6	6.50	24.4	1119
Ti	0.83	0.18	0.43	37
Ca	2.10	0.74	1.63	37
Mg	2.46	0.98	1.90	37
Na	2.64	0.51	1.80	22
K	2.27	0.32	1.10	33
P	0.29	0.07	0.17	33
Cu	2.73	0.13	1.04	1108
Ni	2.21	0.18	1.10	1108
Zn	1.14	0.02	0.12	676
Co	0.43	0.07	0.11	1108
Pb (ppm)	1090	382	712	22
Mo	871	160	570	22
Li	198	12	97	22
Ba	2854	434	1570	26
Y	173	54	102	22
Sr	1032	232	679	37
La	240	59	132	37
Ce	1296	178	528	11
Pr	49	12	33	37
Nd	257	49	147	37
Sm	52	10	33	37
Eu	13	3	8	33
Gd	63	12	34	11
Tb	8	2	5	37
Dy	46	10	27	33
Ho	9	2	5	33
Er	30	5	13	11
Tm	6	1	2	11
Yb	21	4	12	33
Lu	8	1	2	37

Source: Compiled from Jauhari,[38] Valsangkar and Khadge,[39] Valsangkar et al.,[41] Valsangkar et al.,[42] Banakar et al.,[61] Pattan and Banakar,[71] Nath et al.,[75] Banakar and Jauhari,[76] Pattan et al.,[77] Valsangkar and Ambre,[115] Valsangkar and Karisiddaiah,[116] Sudhakar,[117] Pattan et al., unpubl. data.

TABLE 7.5
Comparison of Mn, Fe, Cu, Ni, and Co (%) Concentration in Ferromanganese Nodules from Siliceous Ooze and Red Clay from the Central Indian Ocean Basin

Element	Siliceous Ooze (n = 75)			Red Clay (n = 74)		
	Maximum	Minimum	Mean	Maximum	Minimum	Mean
Mn	34.60	18.10	26.50	29.10	13.20	21.70
Fe	13.60	2.40	6.60	18.70	4.80	9.90
Cu	1.86	0.36	1.12	1.30	0.23	0.52
Ni	1.60	0.64	1.14	1.54	0.32	0.66
Co	0.33	0.06	0.11	0.43	0.12	0.18

n is number of samples analyzed.

Source: Jauhari, P., Relationship between morphology and composition of manganese nodules from the Central Indian Ocean, *Mar. Geol.*, 92, 115, 1990. Sudhakar, M., Ore grade manganese nodules from the Central Indian Basin: an evaluation, *Mar. Ming.*, 8(2), 201, 1989. With permission.

7.9   FACTORS AFFECTING THE COMPOSITIONAL VARIABILITY OF CIOB NODULES

7.9.1   OXIDATION STATE OF MANGANESE AND IRON

The incorporation of trace metals and the mineralogy of ferromanganese nodules is influenced by the oxidation state of Mn and Fe.[87] Manganese occurs in Mn(II), Mn(III), and Mn(IV) valency states in naturally occurring minerals and surface waters. Mn(II) is thermodynamically unstable in the oxic environment and normally is oxidized to Mn(IV).[88] Nodules from the different sedimentary environments in the CIOB have an O:Mn ratio of 1.73 to 1.75, indicating that around 73 to 75% of the Mn is Mn (IV).[89] CIOB nodules are less oxidized than the Pacific Ocean nodules (Mn-IV — 98%).[90,91]

However, variation in the Mn/Fe ratio of nodules does not show any clear relation with the oxidation state of Mn.[89,91] Electron spin resonance (ESR) is very sensitive in detecting and measuring Mn(II) in a diamagnetic system. Its sensitivity decreases in the higher paramagnetic matrix of manganese nodules.[92] The ESR spectra of CIOB nodules shows a qualitative presence of Mn(II),[89] that can accommodate a high percentage of cations in the nodules. Identification of the Fe oxide phase in the nodules by X-ray diffraction has been unsuccessful due to its fine grain size and poor crystallinity.[93] However, Mossbauer spectroscopy can distinguish the relative amounts of ferric and ferrous iron present. The results on the CIOB nodules are very similar to those observed by Carpenter and Wakham.[94] All the spectra exhibit a well-resolved doublet, which is characteristic of paramagnetic Fe(III).[95-98] Fe(III) in ferromanganese nodules can be present in a number of minerals, including ferric manganite, δ-MnO_2, and colloidal hydrated iron oxides.[96] None of the ferromanganese nodules studied showed the presence of Fe(II) despite considerable variation in iron content, associated sediment, and water depth. Smaller quadruple splitting values obtained for the CIOB nodules from siliceous ooze reflect a smaller particle size of iron compared to that in those from pelagic and seamount deposits.[99]

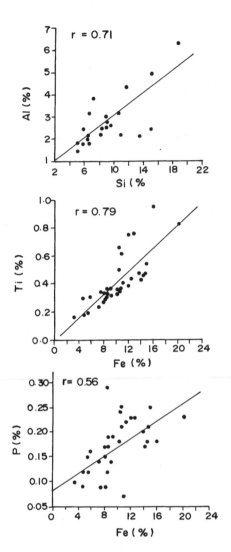

FIGURE 7.7   Scatter plots showing positive correlation between Si and Al, Fe and Ti, and Fe and P.

7.9.2   METAL SOURCES

To account for the compositional differences in nodules various sources of metals have been suggested by various workers.[15,41,42,89,100-102] Variations in the Mn/Fe ratio (0.96 to 8.76) and the trace element content of the nodules reflect the variable contribution of distinct accretionary processes acting on the seafloor.[15] Upon plotting Mn, Fe, and Ni + Cu + Co on a triangular diagram, three distinct types of nodules are discernible — Type A, B, and AB (Figure 7.11). The Type A nodules have Mn/Fe ratios of >5 and a high Ni and Cu content. The high Mn/Fe ratio reflects the increasing influence of early diagenetic activity. The Type B nodules have Mn/Fe ratios up to 2.5. These deposits get their metal content as a result of slow precipitation from the sea water and show the influence of hydrogenous activity. The third subgroup contains a mixture of Type A and Type B nodules — a mixture of hydrogenetically and early diagenetically grown phases.[15]

   Type A nodules, show a high Mn/Fe ratio (>5.0) and high Ni and Cu contents and reflect early diagenetic activity. Their association with siliceous ooze reinforces the influence of diagenesis by biogenic components on the genesis of these nodules. The siliceous ooze sediment area of the CIOB could reflect divergence and upwelling due to the equatorial currents[1] where there would be

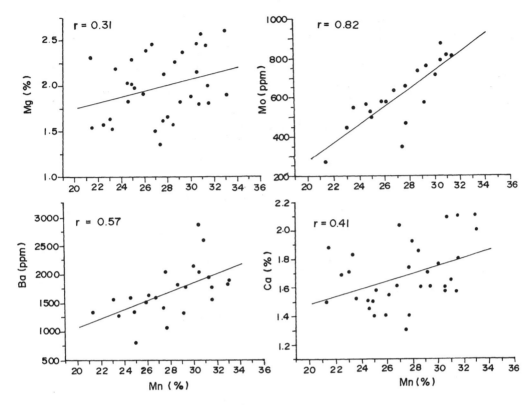

FIGURE 7.8 Positive correlation between Mn against Mg, Ba, Mo, and Ca, depicting stabilization of Mn phase by divalent cations.

greater surface productivity and attendant flux of organisms to the seafloor. The siliceous sediment accumulates under regions of high biological productivity rather than other sediment types.[103] The productivity pattern influences the geochemical characteristics of nodules.[104] Beneath the zones of productive waters, nodules generally grow fast and are rich in Ni and Cu contents, indicating the importance of organic matter in the incorporation of trace metals into them.[105] The high biological productivity in the surface water leads to an enhanced supply of organic matter to the bottom sediments.[14] Siliceous sediments are important carriers of Cu and Ni in the deep sea environment. Oxic diagenetic reactions within the sediments favor metal remobilization and reprecipitation from pore waters on the nodule surfaces. The Mn, Ni, and Cu are derived from the pore waters of sediments in which they are enriched by early diagenetic remobilization processes. The varying concentrations of Mn and associated elements illustrate a variable intensity of supply of diagenetic Mn and its remobilization toward the growing nodule.[15]

Dymond et al.[87] and Aplin and Cronan[102] observed that oxic diagenesis in areas of oxic siliceous sediments was caused by the decomposition and oxidation of organic matter and dissolution of biogenic components, which releases biologically bound Mn, Cu, and Ni for incorporation into the nodules. Pattan et al.[106] reported 30 to 35% biogenic silica in the CIOB surface sediments from 11° to 13° S, due to high productivity and good preservation of siliceous tests. The areas under high biological productivity waters show high concentrations of Ni and Cu in the nodules due to the oxidation of organic matter, leading to diagenetic reactions which can reduce and remobilize these metals in the uppermost sediments. Hartmann and Muller[107] showed that in the interstitial water of the upper cm of Pacific siliceous ooze sediments, Cu, Ni, and Mn are enriched compared with the near bottom sea water. The inter-element relationships between Ni and Cu and Mn reflect a balance between the release of Ni and Cu from biogenic phases and the reductive remobilization

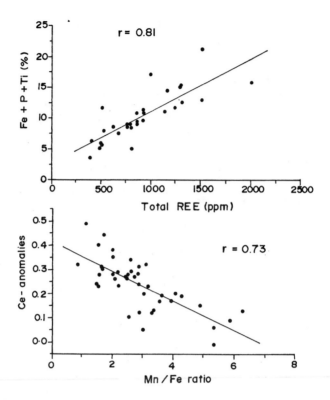

FIGURE 7.9 Positive correlation between Total REE and Fe+P+Ti and inverse relationship between Mn/Fe ratio and Ce-anomalies.

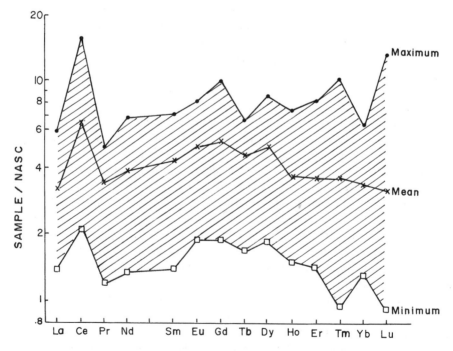

FIGURE 7.10 North American Shale Composite (NASC) — normalized REE pattern of ferromanganese nodules.

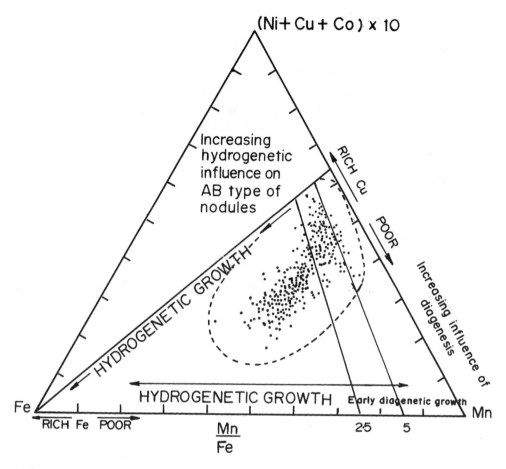

FIGURE 7.11 Triangular diagram showing nodules occupying hydrogenous and early diagenetic fields. (From Jauhari, P., Classification and inter-element relationships of ferromanganese nodules from the Central Indian Ocean Basin, *Mar. Ming*, 6, 419, 1987. With permission.)

of Mn oxide.[15] The nodules resulting from early diagenetic processes are small to medium in size and are spherical to ellipsoidal in shape; their surface texture is rough and granular.

Type B nodules are the most common in the CIOB. They show a greater influence of hydrogenetic activity rather than diagenetic activity and are characterized by low Mn/Fe ratios and Ni and Cu contents. By contrast, Fe and Co contents are comparatively higher in them than in other types of nodules. Type B nodules are associated with red pelagic clay sediments,[11,67] above which the primary productivity is low and whose total biogenic content is less than in sediments associated with Type A nodules.[103] The concentration of Mn and Fe in these nodules indicates that the probable source of these elements is largely sea water. The metals in the bottom waters may have been derived from the continents, or the weathering of submerged rocks, or from active spreading centers. Much of the dissolved Mn, Fe, and Co may have originated from shallow waters near the continents and was transported great distances by advective–diffusion processes.[108] Diffusion loss of Mn from nearby highly reduced sediments may also have contributed Mn to the bottom waters.[87] Oceanic circulation and mixing is the most obvious mechanism of transportation of these metals to reaction sites at the sediment–water interface where they are incorporated into the framework of ferromanganese oxides.[67] The removal of Fe and Mn from sea water depends on the form in which the metals exist.[109] Both elements can occur as colloids or be associated with fine-grained clay minerals. Over long periods they agglomerate into larger particles or take part in biological processes.

Adsorption by organic matter can influence the stability of colloidal Fe and Mn.[67] The concentration of Mn, Fe, and other transition metals in nodules reflects their affinity for nodule surfaces.[110] The hydrogenous Mn and Fe oxides, being characterized by high specific areas, interact with other cations in the sea water and precipitate along with these ions. By the process of adsorption, cations get incorporated on the surfaces of the nuclei of nodules. This, being a cyclic process, keeps the nodule growing.[67] The Co also represents the input of hydrogenous oxides, because it is diagenetically very weakly mobile. Cobalt correlation with Fe in the nodules is a result of adsorption and oxidation of Co^{+2} to Co^{+3} on the ferric hydroxide silicate particles.[102,111,112] The Co enrichment in hydrogenous nodules may also be attributed to its ability to substitute for Mn^{+4} in its higher oxidation state. On the other hand, negative correlation between Co and Mn may be the result of hydrogenetic material mixing with the early diagenetic substance. Variability in the composition of hydrogenous nodules reflects variability in the composition of bottom water.

AB nodules are a mixed type. The metal enrichment in these nodules could be attributed to hydrogenous or early diagenetic processes or both. In the latter case, the upper part of a nodule exposed to sea water is likely to get enriched in metals by precipitation from sea water, while the part buried within the sediments may be enriched in metals by precipitation from interstitial pore waters.[67] A mixing model is proposed by Valsangkar et al.,[42] who suggest that small nodules formed at or near the CCD are transported by gravity down valleys where they come into contact with larger nodules formed by hydrogenous precipitation.

7.10   ENVIRONMENT IMPACT STUDY

Though the commercial future of marine minerals has improved since the UNCLOS came into force in 1994, the environmental impact of nodule mining remains an important unsettled factor. Due to the nature and the distribution pattern of the nodules on the seafloor, the area affected by mining is going to be very large. Therefore, to avoid any adverse consequence, a complete knowledge of the biogeochemical cycles within the oceans on a global scale is a prerequisite for mining operations. A preliminary study in the CIOB[113] concluded that because of its location, and other factors, the impact of mining on the environment in this area is likely to be minimal. This speculation needs confirmation by actual field experiments. Seabed photographs obtained using boomerang cameras and deep-sea towed cameras indicate the presence of benthic activities in the sub-surface. Organisms belonging to seven different phyla are present, and a variety of biogenic traces, such as fecal casts, body tubes, fecal pellets, burrow openings, feeding tracks, locomotion trails, and mounds and cones, were observed on various geological features in the CIOB.[114] Nodule deposits appeared fairly hospitable to abyssal life because many benthic organisms, along with approximately 4 to 8 biogenic traces per square meter, were observed.[114] The above features highlight the need for and importance of further ecological studies in the basin. An environmental impact study program commenced in 1996, aiming toward collection of baseline data pertaining to oceanographic parameters in the CIOB at the surface, in the water column, and on the seabed. A five-year plan started in 1997 to carry out benthic disturbance experiments by artificially generating benthic plumes and then carrying out post-impact studies every six months. Predictive models for large-scale mining are to be developed.

ACKNOWLEDGMENTS

We thank Prof. D.S. Cronan for giving us the opportunity to write this review and for his constructive suggestions, which helped immensely in improving the manuscript. Thanks are also due to Mr. R.R. Nair for his comments, Mr. G. Parthiban for computer plots, and Mr. Shyam Akerkar for the line diagrams. This is N.I.O. contribution No. 2624.

REFERENCES

1. Cronan, D.S. and Moorby, S.A., Manganese nodules and other ferromanganese oxide deposits from the Indian Ocean, *J. Geol. Soc. London*, 138, 527, 1981.

2. Siddiquie, H.N., Gujar, A.R., Hashimi, N.H., and Valsangkar, A.B., Superficial mineral resources of the Indian Ocean, *Deep Sea Res.,* 31 (6-8), 763, 1983.

3. Siddiquie, H.N. and Rao, P.S., Notes and comments on exploration for polymetallic nodules in the Indian Ocean, *Ocean Dev. Int. Law*, 19, 323, 1988.

4. Qasim, S.Z. Retrospect and prospect, in *From the First Nodule to First Mine Site*, Qasim, S.Z., and Nair, R.R. (Eds.), The Offsetters, New Delhi, 1988, 1.

5. Anon., *Central Indian Ocean Basin: A Bibliography of NIO's Contributions with Abstracts.* Library, NIO, Dona Paula, Goa, India 1996.

6. Kolla, V. and Kidd, R.B., Sedimentation and sedimentary processes in the Indian Ocean, in *The Ocean Basins and Margins, 6, The Indian Ocean,* Nairn, A.E.M. and Stehli, F.G. (Eds.), Plenum Press, New York, 1982, 1.

7. Warren, B.A., The deep water of the Central Indian basin, *Jour. Mar. Res.,* 40, 823, 1982.

8. Johnson, D.A. and Nigrini, C., Radiolarian biogeography in surface sediments of the Eastern Indian Ocean, *Mar. Micropal.*, 7, 237, 1982.

9. Banakar, V.K., Gupta, S.M., and Padmavathi, V.K., Abyssal sediment erosion in the Central Indian Basin: evidence from radiochemical and radiolarian studies, *Mar. Geol.*, 96, 167, 1991.

10. Kodagali, V.N., Kamesh Raju, K.A., Ramprasad, T., George, P., and Jaishankar, S., Detailed bathymetric surveys in the Central Indian Basin, *Intnat. Hydrogra. Review*, LXIX(1) 143, 1992.

11. Udintsev, G.B., *Geological and Geophysical Atlas of the Indian Ocean*, Academy of Sciences of USSR., Moscow, 1977, 23.

12. Cronan, D.S., *Under water Minerals.* Academic Press, London, 1980, 61.

13. Banerjee, R. and Mukhopadhyaya, R., Characteristics of manganese nodules from sub-equatorial Indian Ocean between 4° 30'S latitudes, *Indian J. Mar. Sciences*, 9(1), 17, 1990.

14. Belyaeva, V.N. and Burmistrova, I.I., Critical carbonate levels in the Indian Ocean, *Foraminiferal Res.*, 15, 337, 1985.

15. Jauhari, P., Classification and inter-element relationships of ferromanganese nodules from the Central Indian Ocean Basin, *Mar. Ming*, 6, 419, 1987.

16. Nair, R.R. and Jauhari, P., Polymetallic nodule resources of the Indian Ocean, in *Science and Quality of Life,* Qasim, S.Z. (Ed.), The Offsetters, New Delhi, 1993, 393.

17. Kamesh Raju, K., Ramprasad, T., Kodagali, V.N., and Nair, R.R., Multibeam bathymetry, gravity and magnetic studies over 79° E fracture zone, Central Indian Basin, *Jour. Geophy. Res.*, 98, B6, 9605, 1993.

18. Mukhopadhyay, R. and Khadge, N.H., Tectonic reactivation in the Indian Ocean: evidence from seamount morphology and manganese nodule characteristics, *J. Geol. Soc. India,* 40(5), 443, 1992.

19. Kodagali, V.N., Morphologic investigation of unchartered seamount from Central Indian Basin revisited with multibeam sonar system, *Mar. Geod.*, 15, 47, 1991.

20. Mukhopadhyaya, R. and Khadge, N.H., Seamounts in the Central Indian Ocean Basin: indicators of the Indian plate movement, *Proc. Indian Acad. Sci. (Earth Planet. Sci.)*, 99 (3), 357, 1990.

21. Mukhopadhyaya, R. and Khadge, N.H., Growth constraints of the Indian Ocean seamounts, in *Proceedings of the Seminar on Space Applications in Earth System Science*, Indian Geophysical Union, Hyderabad, 100, 1993.

22. Mukhopadhyaya, R. and Batiza, R., Basinal seamounts and seamount chains of the Central Indian Ocean: probable near axis origin from a fast spreading ridge, *Mar. Geophys. Res.*, 16, 303, 1994.

23. Kodagali, V.N. and Sudhakar, M., Manganese nodule distribution in different topographic domains of the Central Indian Basin, *Mar. Georesour. Geotechnol.*, 11(4), 293, 1993.

24. Pattan, J.N. and Kodagali, V.N., Seabed topography and distribution of manganese nodules in the central Indian Ocean, *Mahasagar,* 21(1), 7, 1988.

25. Kodagali, V.N., Influence of regional and local topography on the distribution of polymetallic nodules in Central Indian Ocean Basin, *Geo-Mar Letts.*, 8, 173, 1988.

26. Sharma, R. and Kodagali, V.N., Influence of seabed topography on the distribution of manganese nodules and associated features in the Central Indian Basin: a study based on photographic observations, *Mar. Geol.*, 110 (1-2), 153, 1993.

27. Sharma, R., Quantitative estimation of seafloor features from photographs and their application to nodule mining, *Mar. Georesour. Geotechnol.*, 11(4), 311, 1993.

28. Sharma, R., Distribution of nodules and associated features in the Central Indian Basin: observations from seabed photographs, in *Oceanography of the Indian Ocean*, Desai, B.N. (Ed.), Oxford and IBH, New Delhi, 1992, 549.

29. Sharma, R., Effect of sediment-water interface "boundary layer" on exposure of nodules and their abundance: a case study from seabed photos, *J. Geol. Soc. India*, 34 (3), 310, 1989.

30. Horn, D.R., Horn, B.M., and Delach, M.N., Factors which control the distribution of ferromanganese nodules, and proposed research vessels tracks in the North Pacific, Lamont Doherty Geological Observatory Tech. Report, 8, 1-20, 1973.

31. Glasby, G.P., Role of submarine volcanism in controlling the genesis of marine manganese nodules, *Oceangr. Mar. Biol. Ann. Rev.*, 11, 27, 1973.

32. Craig, J.D., The relationship between bathymetry and ferromanganese deposits in the north equatorial Pacific, *Mar. Geol.*, 29, 135, 1979.

33. Moore, T.C. and Heath, G.R., Manganese nodules, topography and thickness of quaternary sediments, *Nature*, 212, 983, 1966.

34. Nath, B.N., Rao, V.P.C., Becker, K.P., Geochemical evidence of terrigenous influence in deep — sea sediments up to 8° S in the Central Indian Basin, *Mar. Geol.*, 87, 301, 1989.

35. Mudholkar, A.V., Pattan, J.N., and Sudhakar, M., Techniques and results, in *From the First Nodule to the First Mine Site: An Account of the Polymetallic Nodule Project*, Qasim, S.Z. and Nair, R.R. (Eds.), Department of Ocean Development and NIO, Goa, 29-36. 1988.

36. Mukhopadhyaya, R. and Ramana, Y.V., Acoustic properties of Indian Ocean manganese nodules in relation to physical constitution and chemical composition, *Deep Sea Res.*, 37(2), 337, 1990.

37. Ghosh, A.K. and Mukhopadhyay, R., Internal constitution of ferromanganese nodules from the Central Indian Ocean Basin, *Ind. J. Mar. Sci.*, 20(1), 17, 1991.

38. Jauhari, P., Relationship between morphology and composition of manganese nodules from the Central Indian Ocean, *Mar. Geol.*, 92, 115, 1990.

39. Valsangkar, A.B. and Khadge, N.H., Size analysis and geochemistry of ferromanganese nodules from the Central Indian Ocean Basin, *Mar. Ming.*, 8, 325, 1989.

40. Mukhopadhyaya, R., Morphological variations in the polymetallic nodules from selected stations in the Central Indian Ocean, *Geo-Mar. Letts.*, 7(1), 45, 1987.

41. Valsangkar, A.B., Khadge, N.H., and Desa Erwin, J.A., Geochemistry of polymetallic nodules from the Central Indian Ocean Basin, *Mar. Geol.*, 103, 61, 1992.

42. Valsangkar, A.B., Karisiddaiah, S.M., and Parthiban, G., Variation in size, morphology and chemical composition of polymetallic nodules from the Central Indian Ocean Basin, in *Oceanography of the Indian Ocean*, Desai, B.N. (Ed.), Oxford & IBH, New Delhi, 1992, 561.

43. Mukhopadhyaya, R. and Banerjee, G., Underwater geomorphology as a function of the variation in ferromanganese nodule characters in the Indian Ocean, *Curr. Sci.*, 58 (3), 115, 1989.

44. Glasby, G.P. and Thijssen, T., Control of the mineralogy and composition of marine manganese nodules by the supply of divalent transition metal ions, *Neus. Jahrb. Mineral*, 145, 92, 1982.

45. Usui, A., Nishimura, A., Tanahashi, M., and Tereshima, S., Local variability of manganese nodules facies on small abyssal hills of the Central Pacific Basin, *Mar. Geol.*, 74, 237, 1987.

46. Raab, W., Physical and chemical features of Pacific deep sea manganese nodules and their implications to the genesis of nodules, Tech. rep. *Int. Decade Ocean Explor.*, 1, 31, 1972.

47. Cronan, D.S., Regional geochemistry of ferromanganese nodules in the World ocean, in *Ferromanganese Deposits on the Ocean Floor*, Horn, H.D. (Ed.), Washington, DC, NSF, 19, 1972.

48. Banakar, V.K. and Sudhakar, M., Ferro-manganese oxide growth on shark teeth from Central Indian Ocean Basin, *Indian Jour. Mar. Sci.*, 17 (4), 265, 1988.

49. Banakar, V.K., Ferro-manganese encrustation on the tympanic bulla of a minke whale from Central Indian Ocean Basin, *Indian Jour. Mar. Sci.*, 16 (4), 261, 1987.

50. Ghosh, A.K. and Mukhopadhyay, R., Large philipsite crystal as ferromanganese nodule nucleus, *Geo-Mar. Letts*, 15, 59, 1995.

51. Iyer, S.D. and Sudhakar, M., A new report on the occurrence of zeolites in the abyssal depths of the Central Indian Basin, *Sed. Geol.*, 84, 169, 1993.

52. Iyer, S.D. and Sudhakar, M., Coexistence of pumice and manganese nodules fields-evidence for submarine silicic volcanism in the Central Indian Basin, *Deep-Sea Res.*, 40(5), 1123, 1993.

53. Iyer, S.D. and Sharma, R., Correlation between occurrence of manganese nodules and rocks in a part of the Central Indian Ocean, *Mar. Geol.*, 92, 127, 1990.

54. Gupta, S.M., Paleogene hardgrounds and associated intraclast lag deposits as the substrates of ferromanganese crusts and nuclei of nodules: Inference of abyssal current in the Central Indian Ocean, *J. Paleont. Soc. India*, 40, 11, 1995.

55. Martin-Barajas, A., Lallier-Verges, E., and Leclaire, L., Characteristics of manganese nodules from the Central Indian Basin: Relationship with the sedimentary environment, *Mar. Geol.*, 101, 249, 1991.

56. Martin-Barajas, A. and Lallier-Verges, E., Ash layers and pumice in the Central Indian Basin: relationship to the formation of manganese nodules, *Mar. Geol.*, 115, 307, 1993.

57. Mudholkar, A. and Fujji, T., Fresh pumice from the Central Indian Basin: a Krakatau 1883 signature, *Mar. Geol.*, 125, 143, 1995.

58. Banerjee, R., Iyer, S.D., and Dutta, P., Buried nodules and associated sediments from the Central Indian Basin, *Geo-Mar. Letts.*, 11, 103, 1991.

59. Pattan, J.N., Internal microfeatures of manganese nodules from the Central Indian Ocean, *Mar. Geol.*, 81, 215, 1988.

60. Rao, V.P., Mineralogy of polymetallic nodules and associated sediments from the Central Indian Ocean Basin, *Mar. Geol.*, 74, 151, 1987.

61. Banakar, V.K., Pattan, J.N., and Jauhari, P., Size, surface texture, chemical composition and mineralogy interrelations in ferromanganese nodules of Central Indian Ocean, *Indian Jour. Mar. Sci.*, 18(3), 201, 1989.

62. Banakar, V.K., Uranium–thorium isotopes and transition metal fluxes in two oriented manganese nodules from the Central Indian Basin: implications for nodule turnover. *Mar. Geol.*, 95, 71, 1990.

63. Moore, W.S., Thorium and radium isotopic relation in manganese nodules and sediments at MANOP Site S, *Geochim. Cosmochim. Acta*, 48, 987, 1984.

64. Frazer, J.Z. and Fisk, M.B., Geological factors related to characteristics of seafloor manganese nodule deposits, *Deep Sea Res.*, 28, 1533, 1981.

65. Sudhakar, M., Relation between grade and abundance of manganese nodules, *Curr. Sci.*, 57(12), 662, 1988.

66. Menard, H.W. and Frazer, J., Manganese nodules of seafloor, inverse correlation between grade and abundance, *Science*, 199, 869, 1978.

67. Jauhari, P., Variability of Mn, Fe, Ni, Cu and Co in manganese nodules from the Central Indian Ocean Basin, *Mar. Geol.*, 86, 237, 1989.

68. Goldberg, E.D. and Arrhenius, G.O.S., Chemistry of Pacific pelagic sediments, *Geochim. Cosmochim. Acta.* 13, 153, 1958.

69. Calvert, S.E. and Price, N.B., Geochemical variation in ferromanganese nodules and associated sediments from the Pacific Ocean, *Mar. Chem.*, 5, 43, 1977.

70. Halbach, P., Scherhag, C., Hebisch, U., and Marchig, V., Geochemical and mineralogical control of different genetic types of deep-sea nodules from the Pacific Ocean, *Minerala Deposita*, 16, 59, 1981.

71. Pattan, J.N. and Banakar, V.K., Rare earth element distribution and behaviour in buried manganese nodules from the Central Indian Basin, *Mar. Geol.*, 112, 303, 1993.

72. Burns, V.M. and Burns, R.G., Mineralogy, in *Marine Manganese Deposits,* Glasby, G.P. (Ed.), Elsevier, 1977, 185.

73. Piper, D.Z., Leong, K., and Cannon, W.F., Manganese nodule and surface sediment compositions: Domes Sites A, B and C., in *Marine Geology and Oceanography of the Manganese Nodule Province*, Bischoff, J.L. and Piper, D.Z. (Eds.), Plenum Press, New York, 1979, 437.

74. Bischoff, J.C., Piper, D.Z., and Leong, K., The aluminosilicate fraction of North Pacific manganese nodules, *Geochim. Cosmochim. Acta*, 45, 2047, 1981.

75. Nath, B.N., Balaram, V., Sudhakar, M., and Pluger, W.L., Rare earth element geochemistry of ferromanganese deposits from the Indian Ocean, *Mar. Chem.*, 38, 185, 1992.

76. Banakar, V.K. and Jauhari, P., Geochemical trends in the sediments and manganese nodules from a part of the Central Indian Basin, *J. Geol. Soc. India*, 43 (5), 591, 1994.

77. Pattan, J.N., Colley, S.C., and Higgs, N.C., Behaviour of rare earth elements in coexisting manganese macronodules, micronodules and sediments from the Central Indian Basin, *Mar. Geores. Geotechnol.*, 12(4), 283, 1994.

78. Elderfield, H., Hawkesworth, C.J., Greaves, M.J., and Calvert, S.E., Rare earth element geochemistry of oceanic ferromanganese nodules and associated sediments, *Geochim. Cosmochim. Acta,* 45, 513, 1981.

79. Elderfield, H., Hawkesworth, C.J., Greaves, M.J., and Calvert, S.E., Rare earth element zonation in Pacific ferromanganese nodules, *Geochim. Cosmochim. Acta,* 45, 1231, 1981.

80. Glasby, G.P., Gwozdz, R., Kunzendorf, H., Friedrich, G., and Thijssen, T., The distribution of rare earth and minor elements in manganese nodules and sediments from the equatorial and S.W. Pacific, *Lithos,* 20, 97, 1987.

81. Courtois, C. and Clauer, N., Rare earth elements and strontium isotopes of polymetallic nodules from southeastern Pacific Ocean, *Sedimentology,* 27, 687, 1980.

82. Goldberg, E.D., Koide, M., Schmitt, R.A., and Smith, R.H., Rare earth distribution in the marine environment, *J. Geophy. Res.,* 68, 4209, 1963.

83. Piper, D.Z., Rare earth elements in ferromanganese nodules and other marine phases, *Geochim. Cosmochim. Acta,* 38, 1007, 1974.

84. Nath, B.N., Roelandts, I., Sudhakar, M., Balaram, V., and Pluger, W.L., Cerium anomaly variations in ferromanganese nodules and crusts from the Indian Ocean., *Mar. Geol.,* 120, 385, 1994.

85. Elderfield, H. and Greaves, M.J., Negative cerium anomalies in the rare earth element patterns of oceanic ferromanganese nodules, *Earth Planet. Sci. Lett.,* 55, 163, 1981.

86. Calvert, S.E., Piper, D.Z., and Baedecker, P.A., Geochemistry of rare earth elements in ferromanganese nodules from DOMES site A: Northern Equatorial Pacific, *Geochim. Cosmochim. Acta.,* 51, 597, 1987.

87. Dymond, J., Lyle, M., Finney, B., Piper, D.Z, Murphy, K., Conrad, R., and Pisias, N., Ferromanganese nodules from MANOP sites H.S. and R-Control of mineralogical and chemical composition by multiple accretionary processes, *Geochim. Cosmochim. Acta,* 48, 931, 1984.

88. Calvert, S.E. and Pederson, T.F., Geochemistry of recent oxic and anoxic marine sediments: implications for the geological record, *Mar. Geol.,* 113, 67, 1993.

89. Pattan, J.N. and Mudholkar, A.V., Oxidation state of manganese in ferromanganese nodules and deep-sea sediments from the central Indian Ocean, *Chem. Geol.,* 85, 171, 1990.

90. Murray, J.W., Balistrieri, L.S., and Paul, B., The oxidation state of manganese in marine sediments and ferromanganese nodules, *Geochim. Cosmochim. Acta,* 48, 1237, 1984.

91. Piper, D.Z., Basler, J.R., and Bischoff, J.L., Oxidation state of marine manganese nodules, *Geochim. Cosmochim. Acta,* 48, 2347, 1984.

92. Carpenter, R., Quantitative electron spin resonance (ESR) determinations of forms of total amounts of Mn in aqueous environmental samples. *Geochim. Cosmochim. Acta,* 47, 875, 1983.

93. Murray, J.W., Iron oxides, in *Marine Minerals,* Burns, R.G. (Ed.), Mineralogical Society of America. Short Course Notes, Washington, DC, 47, 1979.

94. Carpenter, R. and Wakham, S., Mossbauer studies of marine and fresh water manganese nodules. *Chem. Geol.,* 11, 109, 1973.

95. Gager, A.M., Mossbauer spectra of deep sea iron manganese nodules, *Nature,* 220, 1021, 1969.

96. Heizenberg, C.L. and Reily, D.C., Interpretation of the Mossbauer spectra of marine manganese nodules. *Nature,* 224, 259, 1969.

97. Johnson, C.E. and Glasby, G.P., Mossbauer effect determination of particle size in microcrystalline iron manganese nodules, *Nature,* 222, 376, 1969.

98. Johnson, C.E. and Glasby, G.P., The secondary iron oxyhydroxides mineralogy of some deep sea and fossil manganese nodules. A Mossbauer and X-ray study, *Geochem. Jour.,* 12, 153,1978.

99. Pattan, J.N. and Mudholkar, A.V., Mossbauer studies and oxidised manganese ratio in ferromanganese nodules and crusts from the Central Indian Ocean, *Geo-Mar. Letts.,* 11(1), 51, 1991.

100. Calvert, S.E., Price, N.B., Heath, G.R., and Moore, T.C., Relationship between ferromanganese nodule composition and sedimentation in a small survey area of the equatorial Pacific, *Jour. Mar. Res.,* 36, 161, 1978.

101. Usui, A., Regional variation of manganese nodule facies on the Wake–Tahiti transect: morphological, chemical and mineralogical study, *Mar. Geol.,* 54, 27, 1983.

102. Aplin, A.C. and Cronan, D.S., Ferromanganese oxide deposits from the Central Pacific Ocean, II. Nodules and associated sediments, *Geochim. Cosmochim. Acta,* 49, 437, 1985.

103. Parsons, T.R.M., Takahashi, M., and Hargrove, B., Biological Oceanographic Processes, 2nd edition, Pergamon, New York, 1977.

104. Andrews, J., Friedrich, C., Pautot, G., Pluger, W., Renard, V., Melguen, M., Cronan, D., Craig, J., Hoffert, M., Stoffers, P., Shearme, S., Thijssen, T., Glasby, G., Le Notre, N., and Saget, P., The Hawaii – Tahiti transect: the oceanographic environments of manganese nodule deposits in the Central Pacific, *Mar. Geol.*, 54 (1/2), 109, 1983.

105. Platt, T., Jauhari, P., and Nath, S.S., The importance and measurement of new production, in *Primary Productivity and Biogeochemical Cycles in the Sea,* Falkowski, P.G. and Woodhed, A.D. (Eds.), Plenum Press, New York, 43, 1992, 272.

106. Pattan, J.N., Gupta, S.M., Mudholkar, A.V., and Parthiban, G., Biogenic silica in space and time in sediments of Central Indian Ocean, *Ind. J. Mar. Sci.*, 21 (2), 116, 1992.

107. Hartmann, M. and Muller, P.J., Trace metals in interstitial waters from Central Pacific Ocean Sediments, *Proc. Int. Oceanographic Assembly,* Edinburgh, 1976.

108. Martin, J.H. and Knauer, G.A., The lateral transport of manganese within the north–east Pacific Gyre oxygen minimum, *Nature*, 314, 524, 1985.

109. Murray, J.W. and Brewer, P.G., Mechanisms of removal of manganese, iron and other trace metals from sea water, in *Marine Manganese Deposits,* Glasby, G.P. (Ed.), Elsevier, Amsterdam, 1977, 292.

110. Murray, J.W., The interaction of ions at the manganese dioxide – solution interface, *Geochim. Cosmochim. Acta,* 39, 505, 1975.

111. Knauer, G.A., Martin, J.H., and Gordon, R.M., Cobalt in northeast Pacific waters, *Nature*, 297, 49, 1982.

112. Halbach, P., Giovanoli, R., and Borstel, D., Geochemical processes controlling the relationship between Co, Mn and Fe in early diagenetic deep sea nodules, *Earth Planet Sci. Lett.,* 60, 226, 1982.

113. Sharma, R. and Rao, A., Environmental considerations of nodule mining in Central Indian Basin, in 23rd *Annual Offshore Technology Conference.* Proceedings vol.1: *Geology, Earth Sciences and Environment,* OTC, Houston, U.S.A., 1991, 481.

114. Sharma, R. and Rao, A., Geological factors associated with megabenthic activity in the Central Indian Basin, *Deep-Sea Res.,* 39(3-4), 705, 1992.

115. Valsangkar, A.B. and Ambre, N.V., Relationship between size and geochemistry of polymetallic nodules from the Central Indian Ocean Basin: significance in selection of high grade nodules, in *Ocean Technology Perspective,* Kumar, S., Agadi, V.V., Das, V.K., and Desai, B.N. (Eds.), Publ. and Infor. Directorate, New Delhi, 1994, 827.

116. Valsangkar, A.B. and Karisiddiah, S.M., Evidence for the formation of different sized nodules by different accretionary processes, *Mar. Geores. Geotechnol.*, 11(10), 87, 1993.

117. Sudhakar, M., Ore grade manganese nodules from the Central Indian Basin: an evaluation, *Mar. Mining*, 8(2), 201, 1989.

8 Manganese Nodules of the Peru Basin

Ulrich von Stackelberg

ABSTRACT

Results from *R.V. Sonne* cruises SO-79 (1992) and SO-106 (1996) are presented together with a review of older data from the Peru Basin. The extended nodule field of the Peru Basin is situated at the southern margin of the equatorial maximum of bioproduction where accumulation rates of organic carbon are relatively high, which enhances diagenetic growth of nodules. The maximum abundance of nodules (kg/m^2) occurs at around 4,150 m water depth. Here a maximum of Corg in surface sediments is responsible for high nodule growth rates up to 250 mm/Ma, while mainly hydrogenetic growth in more elevated positions is characterized by accretion rates up to 50 times lower. However, due to a steep geochemical gradient between the sediment surface and a redox boundary at 5 to 10 cm depth, accretion rates of large nodules (up to 21 cm maximum diameter) show great differences between their bottoms and tops which is not observed in small nodules. Nodule size strongly determines the accretion rate, the type of internal growth, and the composition due to the different depth of immersion in sediment of large and small nodules. The layered character of nodules is mainly due to repeated biogenic lifting. Pulses of these movements are much more frequent than climate-induced oscillations, which may be superimposed on the bioturbation-induced microlayered growth pattern. Large nodules in basins tend to grow asymmetrically, and as soon as they become stuck in the stiff sediment below the redox boundary, they become buried. Parts of the nodules buried in the suboxic sediment below the redox boundary dissolve.

8.1 INTRODUCTION

The occurrence of manganese nodules in the Peru Basin was first mentioned by Agassis in 1902.[1] Glasby, in a review paper, points out that the highest combined Mn–Cu–Ni concentration of nodules from the South Pacific is found in the Peru Basin, which is influenced by a tongue of high-productivity water extending from the coastal upwelling zone off Peru (Figure 8.1).[2] Piper and Williamson, and Skornyakova present nodule data from the Peru Basin showing high Mn/Fe ratios, Ni concentrations, and coverage.[3,4]

During research cruise SO-04 of *R.V. Sonne* in 1978, a reconnaissance of this manganese nodule occurrence was made in the northern part of the Peru Basin (Figure 8.2).[5] The average size of nodules was found to be much larger than in other regions of the Pacific. Abundance and grade of the nodules was considered to be of economic value. Therefore, in 1979, the area was surveyed by *R.V. Sonne* again, on cruise SO-11, which was subdivided into two consecutive legs. During leg SO-11/1 the western sector of the northern Peru Basin, and during leg SO-11/2 the eastern sector, were surveyed in detail.[6,7]

During another *Sonne* cruise, SO-13 in 1980, a selected area was surveyed to assess the economic value of the nodule field.[8] The *Sonne* cruises mentioned before were carried out by the German industrial consortium AMR (Arbeitsgemeinschaft meerestechnisch gewinnbare Rohstoffe),

0-8493-8429-X/00/$0.00+$.50

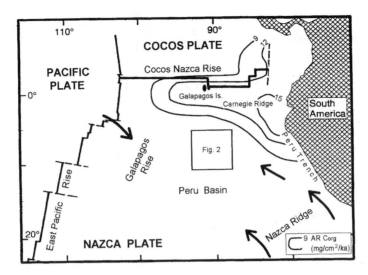

FIGURE 8.1   Location of survey area (Figure 8.2) in the Peru Basin. AR Corg (accumulation rate of organic carbon) according to Lyle.[74] Arrows = bottom currents according to Lonsdale.[22]

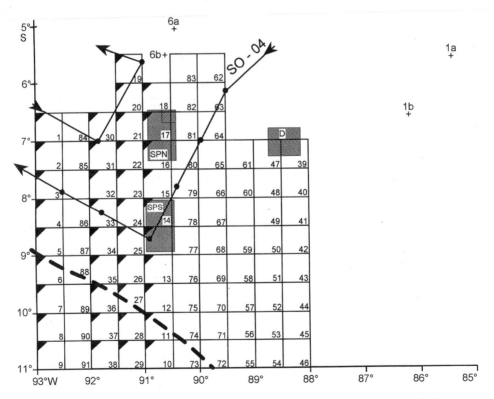

FIGURE 8.2   Location of areas in the Peru Basin surveyed by *R.V. Sonne*. SO-04 = dots on track line mark sampling locations. SO-11/1 = locations 1-38 marked with black triangles; SO-11/2 = locations 39-91. SO-79 = SPN (Sediperu North) and SPS (Sediperu South). SO-106 = D (DISCOL), NE part of SPN, and locations 1a, 1b, 6a, and 6b. Dashed line = fault zone.

which filed a manganese nodule mine site claim in the Peru Basin, and therefore, the coordinates of sampling locations and of track lines were not published. Also, during a Russian research cruise in 1980 nodules and sediments were recovered from the southern Peru Basin (about 13°15'S; 86°00'W) and the geochemistry, especially the rare earth element concentrations, were studied.[9]

Due to low metal prices and uncertainties arising from the new Law of the Sea Convention, western industrial interest in manganese nodules has faded since the mid-1980s. On the other hand, aspects of the environmental impact of possible future deep-sea mining has initiated various research activities in a number of nodule fields.

A summary of the German activities in manganese nodule research in the Pacific Ocean, including the Peru Basin, was given by von Stackelberg.[10] A program aimed at protecting the deep-sea environment was launched in 1989 with two cruises in the Peru Basin (SO-61 and SO-64) by *R.V. Sonne*. A long-term, large-scale, disturbance–recolonization experiment, DISCOL, was carried out and a specially designed "plow-harrow" was used to disturb the seafloor at the NE margin of this nodule field (Figure 8.2).[11,12] This experiment was designed to produce information on the potential impact of a deep-sea mining operation. During cruises SO-61 and SO-64, biologists investigated the disturbance of the benthic community. Bluhm presented a photographic assessment of megabenthic communities in the Peru Basin which was compared with that of the Clarion–Clipperton Zone (CCZ).[13] An additional *Sonne* cruise (SO-77) in 1992 aimed at studying the recolonization of the ecosystem in the DISCOL area.[14,15] During *Sonne* cruise SO-79 (1992) the areas Sediperu North (SPN; Figure 8.3) and Sediperu South (SPS) were surveyed, and the sediments including manganese nodules were investigated.[16-19] In 1996 a joint biological–geological *Sonne* cruise (SO-106) carried out detailed multidisciplinary studies in the DISCOL area (D; Figure 8.4), the northeastern part of the SPN area, and at four locations at the eastern and northern margins of the Peru Basin (1a, 1b, 6a, and 6b; Figure 8.2).[20]

8.2 HYDROGRAPHY, BATHYMETRY, AND GEOLOGICAL SETTING

The Peru Basin is situated between the Galapagos Rise to the West, the Peru Trench to the East, the Galapagos Islands and the Carnegie Ridge to the North, and the Nazca Ridge to the South (Figure 8.1). It is relatively strongly influenced by the equatorial maximum of bioproduction (Figure 8.1).[21] The southern margin of this maximum approaches the northern part of the Peru Basin. Antarctic bottom water passes the Nazca Ridge via transform faults and enters the Peru Basin in a NW-ward direction.[22] This NW-ward direction of bottom currents was observed during moored near-bottom current measurements in the DISCOL area carried out by *R.V. Sonne* from 1989 to 1992.[23] During *Sonne* cruise SO-106 (1996) moorings of only two-months' duration revealed the same NW-ward direction of bottom currents in area D, and a weak indication of SSE-ward directed bottom currents in area SPN.[24] By contrast, the eastern part of the Peru Basin is weakly influenced by coastal upwelling and high biological productivity off Peru, as well as by supply of detrital material derived from South America and transported by the Peru Current to the NNW.[25]

The survey areas of the *Sonne* cruises are situated on the northern Nazca Plate (Figure 8.1 and 8.2). The oceanic basement originated at the Galapagos Rise about 18 to 20 Ma.[26,27] The sediment thickness increases from tens of meters in the south at 11°S to 150 m in the north at 6°S due to the approach to the zone of equatorial high productivity.[7,28,29] In the southern part of the survey area south of a NW–SE running fault zone there is a mountainous region with water depths between 3,500 m and 4,500 m (Figure 8.2). The central and northern part deepens from regional water depths of 3,950 m in the north to 4,200 m in the south. A smooth abyssal hill topography with prevailing N–S orientation characterizes the seafloor of the survey area (Figure 8.3).[6] Less pronounced E–W lineaments are due to offsets by fracture zones.[17] In the northern part, individual seamounts rising to a depth of 2,000 m were observed. Wiedicke and Weber assume an age of some of the volcanic cones well after the time of crustal accretion, possibly connected to the jump of spreading activity from the Galapagos Rise to the EPR about 6 Ma ago.[17]

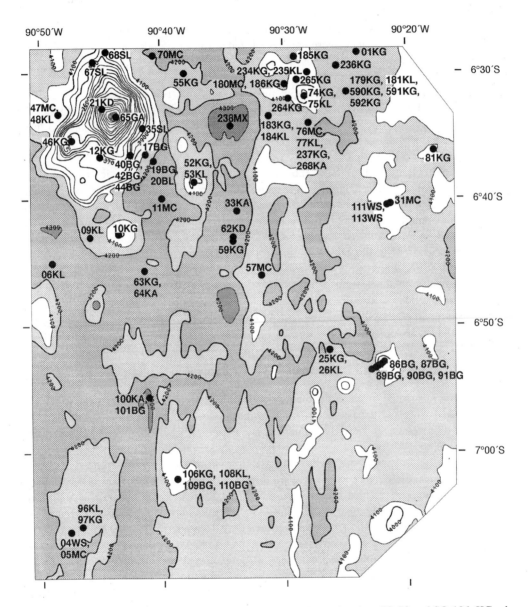

Area SPN (SO 79 + SO 106)

FIGURE 8.3   Area Sediperu North (SPN) with sampling locations of cruises SO-79 and SO-106. KG = box grab, BG = free-fall grab, GA = TV-grab, KD = dredge, MC = multi corer, KL = piston corer, SL = gravity corer, BL = free-fall corer, MX = maxi corer, WS = water sampler.

8.3   SAMPLING AND ANALYTICAL METHODS

Seventy-five free-fall grab samples and three piston cores were taken at 14 locations during cruise SO-04 (Figure 8.2). Within the study area of cruises SO-11/1 and 11/2, seven to eight free-fall grab samples were recovered at each of 91 locations — in total 680 free-fall grabs. Sampling at each location took place at station intervals of approximately 0.5 nm along a profile. Additionally, 18

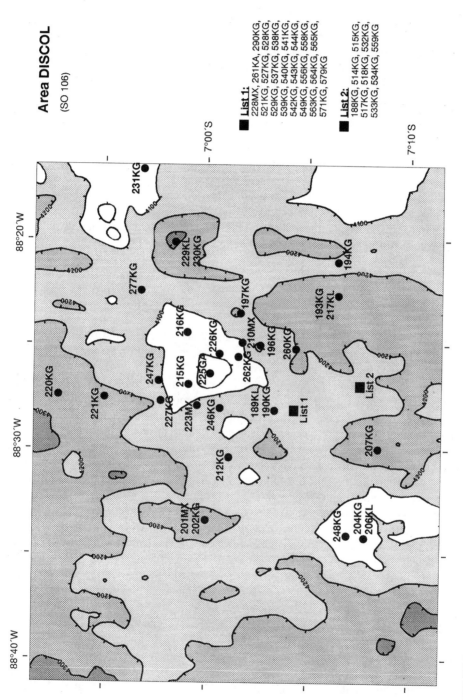

FIGURE 8.4 DISCOL area with sampling locations of cruise SO-106. Lists 1 and 2 represent two areas with very dense sampling. For explanation of abbreviation of names for sampling devices see Figure 8.3.

box grabs, four long box cores and four piston cores were taken at selected stations. The position of the locations of SO-11 was arranged along a grid with a 30 nm spacing (Figure 8.2); 6,408 manganese nodules were recovered from 420 stations during cruise SO-11/2.

During cruise SO-79 manganese nodules and crusts were recovered by the following devices: box grab (KG), free-fall grab (BG), TV-grab (GA), dredge (KD), multi corer (MC), piston corer (KL), gravity corer (SL), free-fall corer (BL), and even by accident with a water sampler (WS). On cruise SO-106 a maxi corer (MX) was deployed able to recover almost undisturbed surface samples, including nodules, in three plastic liners 80 cm long with an inner diameter of 30 cm. A TV-guided photo sled was also used 23 times.

During cruises SO-79 and SO-106 manganese nodules were recovered at 172 stations and crusts at 16 stations (Figure 8.3 and 8.4). About 3,500 manganese nodules were described and classified following the schemes of Meylan, Moritani et al., Usui, and von Stackelberg and Marchig.[30-33] Four hundred and twenty-five nodules were cut, and polished sections were microscopically studied. This number includes 177 nodules buried >20 cm.

The upper surface of nodules, which appeared to be in an undisturbed position, was marked with a red dot immediately after recovery. The mineralogical and chemical composition of individual layers of nodules and crusts was determined by X-ray diffraction analysis, XRF, AAS, and ICP-MS methods. Grain-size distribution was measured for nodule assemblages from 88 stations and nodule abundance (kg/m^2) was calculated at 74 stations.

8.4   RESULTS

8.4.1   Sediments

Surface sediments in water depths above 3,800 m are calcareous oozes. Below 4,100 m the carbonate content is <10%. This indicates that the depth of the survey area is within the range of the calcite compensation depth (CCD).[6]

Stoffers et al. studied sediment cores up to 5 m long from the SO-04 and SO-11/2 survey areas.[34,35] In general, the upper 5 to 15 cm is a semiliquid homogeneous brown mud with variable amounts of calcareous and siliceous organic remains and abundant manganese micronodules. It overlies a layer of highly bioturbated calcareous–siliceous yellowish-brown mud, which is underlain by a consolidated dark-brown clay, which in turn is followed by a white calcareous ooze (Figure 8.5). The calcareous–siliceous mud is characterized by a mix of Plio-Pleistocene coccoliths. No coccoliths were found in the dark-brown clay, while they are abundant in the calcareous ooze where they indicate an upper middle Miocene age.

During the eastward drift of the Nazca Plate the survey area subsided below the CCD; hence Miocene carbonate sediments were replaced by dark-brown clays. A hiatus marks the boundary between the white calcareous ooze and the dark-brown clay. It is assumed that nodule growth has started here, as was observed in the CCZ by von Stackelberg and Marchig.[33]

Micronodule abundance is high in the brown surface sediments, low in the yellowish-brown muds and very high again in the dark-brown clay.[35,36] The micronodules from the brown surface sediments and the dark-brown clay are characterized by high Mn/Fe ratios and high Ni and Cu concentrations, while in the yellowish-brown mud the Mn/Fe ratio and the Ni and Cu concentrations are low. This indicates strong diagenetic remobilization of transition elements in the yellowish-brown sediments due to suboxic conditions, while there is no diagenetic remobilization in brown surface sediments and the dark-brown clay due to their well-oxidized sedimentary environment. Compared to other regions in the Pacific, micronodules of the Peru Basin show higher Mn/Fe ratios (up to 22), which reflects the very strong diagenetic supply of Mn^{2+} from deeper sediments and its precipitation as micronodules in the near-surface sediments. Micronodules and macronodules from the surface sediments show similar mineralogy and chemical composition except for higher Mn/Fe and Si/Al ratios and slightly higher Cu and Zn concentrations in micronodules. The change from

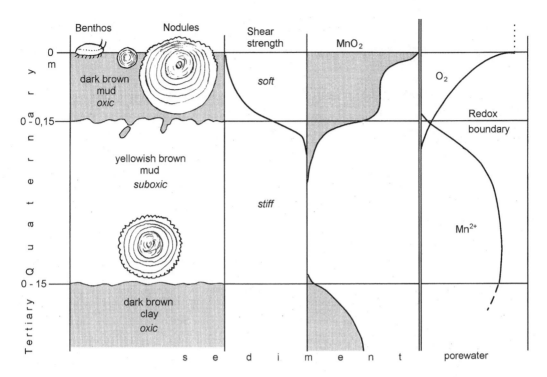

FIGURE 8.5 Schematic diagram of sediment properties with relative values of shear strength and MnO_2 concentrations and with relative concentrations of O_2 and Mn^{2+} in pore water.

dark-brown Mn-rich clays to yellowish-brown muds is explained by a significant redox shift at the end of the Pliocene.[37]

The concentration of organic carbon in surface sediments varies from 0.97 down to 0.10% and decreases from north to south with increasing water depths.

According to Th-isotope dating, sedimentation rates for the uppermost 30 cm of the core sections are in the range of 3 to 5 mm/ka.[35] Hence, the age of the base of the brown surface sediments is between 30 and 50 ka.

Marchig and Reyss have calculated the flux of diagenetically remobilized Mn in the sedimentary column.[38] They found that in a core from the northern part of the SO-11 survey area (6°13,9'S; 91°31,2'W), where large rapidly grown, cauliflower-shaped nodules occur at the sediment surface, the Mn flux is 15 times higher than in a core from the southern part (10°14,0'S; 91°59,2'W), where nodules are smaller and grow much more slowly. The authors conclude that the large northern nodules are mainly grown by the diagenetic supply of metals. An additional hydrothermal supply is not needed to explain the amount of Mn in those rapidly growing nodules.

Mangini et al. observed in a sediment core from the central part of the SO-11 survey area, (8°40'S; 90°30'W; 4,180-m water depth) within the yellowish-brown siliceous muds of the Late Quaternary, a normal sedimentation rate of about 2 mm/ka with intervals of carbonate-rich sediments having an accumulation rate one order of magnitude higher.[39] The authors assume that this material, rich in reworked Miocene nannofossils, was eroded and transported from the southern regions of the area during periods of global climatic cooling when bottom currents were enhanced.

Weber et al. describe the carbonate preservation history in eight sediment cores up to 18 m long from the survey areas SPN and SPS of cruise SO-79.[16] The Pliocene is characterized by carbonate variations showing 400 ka cycles produced by the pulsation of the Antarctic ice sheet. At 2.4 Ma B.P. an abrupt collapse of the carbonate system occurred. During the last 800 ka, 100 ka eccentricity cycles determine the carbonate variability. During glacials, higher productivity was

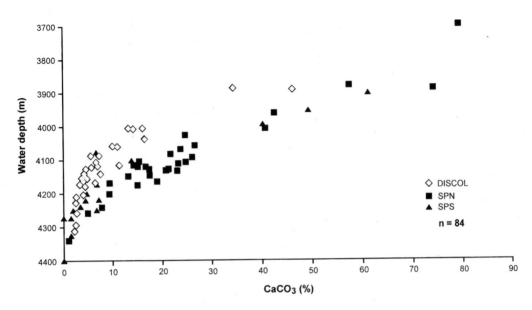

FIGURE 8.6 $CaCO_3$ concentration in surface sediments of the main SO-79 and SO-106 survey areas vs. water depth.

observed with increased flux of biogenic material and good preservation of planktonic foraminifera, while during interglacials, dissolution was greatest.

Sediment cores from the SO-79 survey area reveal a wide range of average sedimentation rates between 1.6 and 11.6 mm/ka.[16] A number of hiatuses were observed, mainly at the Pleistocene–Pliocene and the Pliocene–Miocene boundaries. There is a great variability in the depth of the Pleistocene–Pliocene hiatus between >13 m and less than 1 m. This variability in the sedimentary sequence strongly influences the nodule type.

Mass accumulation rates (MAR) calculated for the Quaternary in SO-79 and SO-106 cores indicate a tenfold increase from south to north, from 40 mg/cm²/ka in area SPS up to 430 mg/cm²/ka at location 6a, and a decrease from W to E from 400 mg/cm²/ka in area SPN to 120 mg/cm²/ka in area D.[40]

In the NE part of area SPN an asymmetric distribution of sediments and nodules was observed around a seamount. There are higher sedimentation rates, higher nodule number (n/m²), and lower abundances (kg/m²) on the western side than on the eastern side at similar water depths.

The $CaCO_3$ concentration of surface sediments in the SO-79 and SO-106 areas decreases dramatically from 80% to 5% between the lysocline near 3,700-m water depth and the CCD at 4,200 to 4,250-m water depth (Figure 8.6). In area D the CCD lies at about 4,200-m water depth; in areas SPN and SPS at about 4,250 m.

Organic carbon content increases from 0.35 to 0.75% in the transitional zone between the lysocline and CCD due to increasing dissolution of $CaCO_3$ (Figure 8.7). The highest Corg concentrations were found at locations 1a and 1b at the NE margin of the Peru Basin. There is a weak indication of decreasing Corg concentration below the CCD, which may be due to increased O_2 concentration of bottom water. A similar observation was described by Cronan and Hodkinson from the central South Pacific.[41]

Pore water profiles show an abrupt disappearance of Mn^{2+} at a sediment depth of 5 to 20 cm marking the redox boundary, although O_2 penetrates somewhat deeper (Figure 8.5).[42] This boundary more or less coincides with the lower boundary of the dark-brown surface layer.

Shear strength measurements clearly reflect the thickness of the semiliquid top layer. This thickness increases from 7 to 17 cm with increasing water depth in all SO-79 and SO-106 areas.[39]

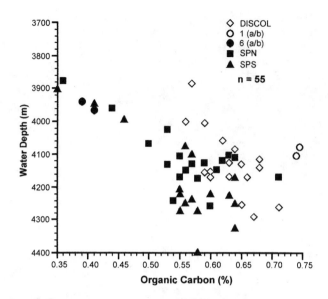

FIGURE 8.7 Organic carbon concentration in surface sediments of the SO-79 and SO-106 survey areas vs. water depth.

8.4.2 NODULES

8.4.2.1 Distribution on the Seafloor

Bottom photographs show that sediment coverage of nodules is more common in the Peru Basin than in other nodule areas of the Pacific.[7]

Thijssen et al. studied 1,617 nodules from 75 stations at 14 locations of cruise SO-04. The average abundance was 8.5 kg/m² with a maximum of 35.1 kg/m².[5]

From cruise SO-11/2, the average nodule abundance, based on 420 surface samples for which data were available, was 11.2 kg/m² with a range of 0 to 66.8 kg/m².[7] This compilation includes, however, data from box grabs, some of them with buried nodules. The maximum nodule abundance from free-fall grabs was 52.6 kg/m². Water depths at sampling stations along location profiles normally show a range of only a few tens of meters. Along some profiles, however, the range was up to a few hundred meters (maximum 465 m). A distinct variation in nodule abundance was found along many of such profiles. A high average nodule abundance of >10 kg/m² is found below 4,100-m water depth, which is at or below the presumed CCD. Maximum abundance (66.8 kg/m²) occurred at 4,235-m water depth. This result is an average based on nodule samples from all over the large SO-11 area. In each of the sub-areas the position of the maximum, however, is slightly different as the position of the CCD and of the tops of the seamounts varies. The AMR mining claim covers parts of the SO-11 survey area with >10 kg/m² average nodule abundance.

During cruises SO-79 and SO-106, nodules were collected in areas SPS, SPN, and D from water depths between 3,726 and 4,404 m. The upper part of the large seamount in SPN was nodule free. Manganese crusts were observed and recovered mainly on seamounts, abyssal hills, and ridges. Elongated patches of distinct low acoustic reflectivity on ridges, less distinct in basins, were observed during side-scan sonar surveys in areas SPN and SPS.[17] Erosional bottom current activity is assumed to be responsible for these patches which, by TV-observation, are seen to be more or less nodule-free. A great variability of nodule size and coverage was observed on bottom photographs (Figures 8.8A and 8.8B). Furthermore, there was a great number of plow marks and burrows as well as rings and double rows of nodules produced by benthic organisms.

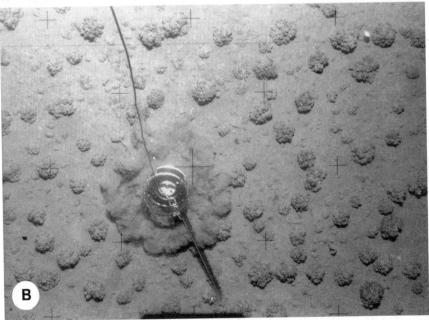

FIGURE 8.8 Bottom photographs. (A) dense coverage of relatively small nodules (median diameter: 5 cm). Area SPN, water depth: 4,030 m. Outer ring of compass: 25 cm. (B) Less dense coverage of large nodules (median diameter: 15 cm). Area SPS, water depth: 4,090 m.

Maximum numbers of SO-79 and SO-106 nodules (n/m^2) occur on slopes at about 4,000-m water depth in area D, while nodule abundance (kg/m^2) shows two lobes, one on slopes near the maximum number, the other in basins at about 4150-m water depth near the maximum of the ratio abundance/number (Figures 8.9 and 8.10). Although this observation is based on few data only, the same phenomenon of two lobes of nodule abundance can be observed in the SPN and SPS areas. The lower lobe of abundance in SPN is also at about 4150 m, while in SPS it occurs at about 4250 m. Two stations in areas SPN and SPS with high values of abundance do not fit this model.

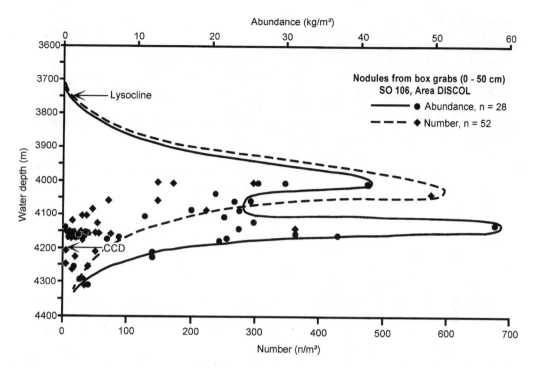

FIGURE 8.9   Abundance (kg/m²) and number (n/m²) of nodules from box grabs of area DISCOL vs. water depth.

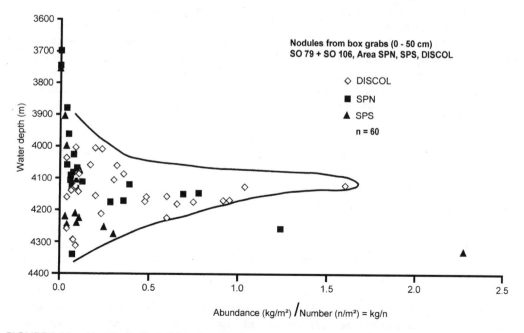

FIGURE 8.10   Abundance (kg/m²)/number (n/m²) ratio of nodules from box grabs of the main SO-79 and SO-106 survey areas vs. water depth.

8.4.2.2   Morphology

The morphology of nodules from the SO-04 and SO-11 areas was described by Halbach et al. and Thijssen et al.[5-7,28] It was determined by shape and surface texture. Nodules greater than 6 cm were found to be highly mammillated with rough botryoidal surface texture on the upper surface and smooth to microbotryoidal surface texture on the lower surface (Figure 8.11). These cauliflower-shaped nodules are mostly spheroidal to irregularly spheroidal, sometimes discoidal. Nodules of the size class 2 to 6 cm are mostly ellipsoidal to discoidal with a mammillated and microbotryoidal upper surface and a mammillated smoother lower surface. Small nodules (<2 cm) show a botryoidal surface texture and are mostly irregularly shaped due to fragments of larger nodules forming the nucleus. It was found that Peru Basin nodules with increasing size become more regularly shaped, i.e., spheroidal and discoidal. Similar observations were reported by Huh and Ku for nodules of so-called Site H in the Guatemala Basin.[43]

In the southwest of the SO-11 area, near the fracture zone, a mixture of normal Peru Basin nodules together with highly irregular ferromanganese material was observed. This material has a pahoehoe-like structure and a nucleus of palagonite often replaced by ferromanganese oxides and overlain by a layer of ferromanganese oxides. The palagonite nuclei indicate volcanic activity in the vicinity of the fracture zone.

During the study of SO-79 and SO-106 material, nine different nodule shapes were distinguished, ranging from irregular types such as I (irregular) and T (tabular) to regular ones such as S (spheroidal) and D (discoidal). Cylindrical-shaped nodules (C) develop from indurated burrow tubes. Ellipsoidal-shaped nodules (E) are frequently fragmented, forming halves (E2) of ellipsoidal nodules, or are more or less irregularly shaped (EI). Polylobate nodules (P) consist of two or several small agglutinated nodules.

S- and D-shaped nodules are mainly found in basins at water depths >4,200 m while E2, P, C, and T-type nodules occur in shallower water on or near ridges and hills. E, EI, and I-type nodules are found everywhere (Figure 8.12).

To classify the morphology of nodules, the surface texture was also studied, and three different types were distinguished: r (rough), s (smooth), and r.s (intermediate texture between rough and smooth). The most frequent type of nodule surface texture in the Peru Basin is the r-type (56.7%), with a rough surface on both sides, followed by the r.s-type (15.3%)(Table 8.1). Nodules with an s-type surface texture on both sides are only found in shallower water depths, while r-type nodules predominate in deeper water. Nodules, the top surface of which is smoother than the bottom, are more frequent (12%) than nodules with a smoother bottom surface (6%). The latter nodule type is found mainly in large cauliflower-shaped nodules of up to 21-cm maximum diameter. The nodule type with a smoother bottom side shows the maximum frequency of occurrence in area D (13.4%) and the minimum in area SPS (1.4%). This nodule type was also found in the Guatemala Basin, while it is not known in the CCZ where large nodules show a smooth top and a rough bottom surface.[33,44]

8.4.2.3   Size Distribution

The size class and the weight were determined on the basis of 6,408 nodules from 420 stations in the SO-11/2 area.[7] The highest number of nodules (50%) occur in the size class 2 to 4 cm, while 70% of the nodule weight is found in nodules larger than 6 cm. The largest nodule found measured 19 × 17.5 × 18 cm and weighed 4.8 kg.

Nodules collected from areas SPN, SPS, and D showed a size distribution similar to that from areas of SO-11/2. The histogram for all surface nodules in area SPN is characterized by a distinct maximum at 4 cm diameter (Figure 8.13). The regularly shaped nodules of S- and D-type predominate in the size class >7 cm, with an increasing D/S ratio in larger size classes. The histograms for areas

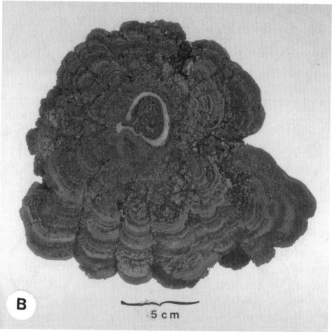

FIGURE 8.11   (A) Large cauliflower-shaped nodule with smooth surface texture (s) on the bottom side and rough surface texture (r) on the top side. 25 KG from area SPN at 4170 m water depth. (B) Polished section of a nodule with a nucleus of a cetacean earbone showing asymmetric growth of pale-grey dense, laminated h-type layers at the lower part. 63 KG from area SPN at 4257 m water depth.

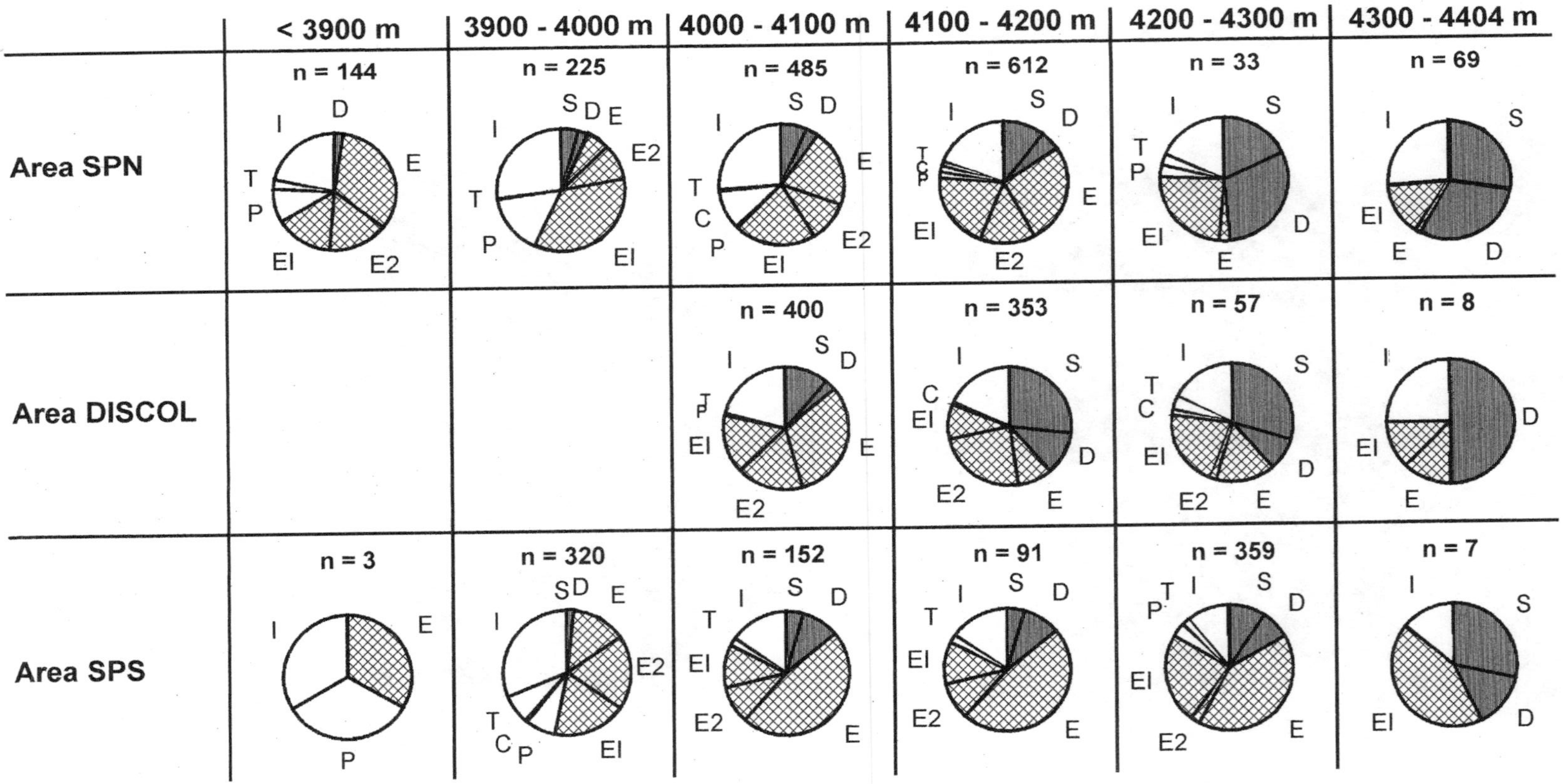

FIGURE 8.12 Distribution of nodule shapes at different water depths in areas SPN, DISCOL, and SPS. S = spheroidal, D = discoidal, E = ellipsoidal, E2 = half ellipsoidal, EI = irregularly ellipsoidal, P = polylobate, C = cylindrical, T = tabular, I = irregular.

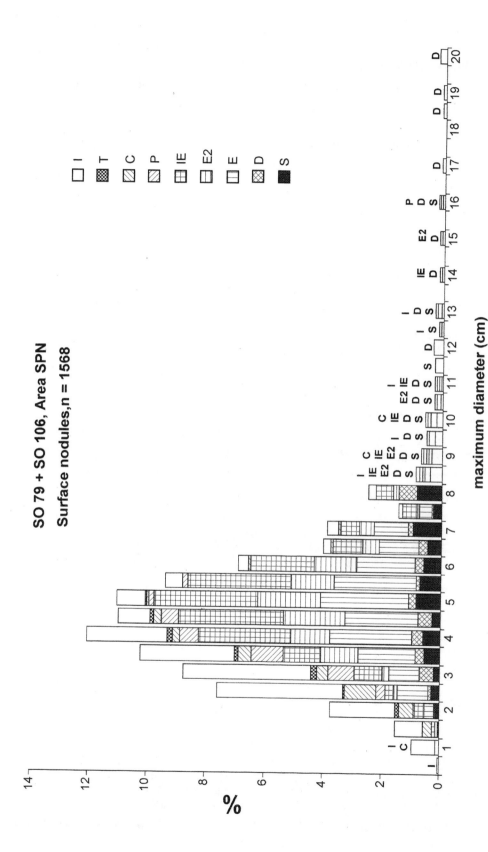

FIGURE 8.13   Size distribution of all surface nodules in area SPN with indication of their shape. For explanation of symbols for nodule shape see Figure 8.12.

TABLE 8.1
Surface Texture of Manganese Nodules (incl. buried nodules)

Area	Top/Bottom		Both Sides			Top Not Identified		Total
	r-r.s/s-r.s	r.s-s/r-r.s	r	r.s	s	r + r.s	r-r.s + s	
	n (%)	n (%)	n (%)	n (%)	n (%)	n (%)	n (%)	n
D	118 (13.4)	168 (19.1)	431 (49.1)	129 (14.7)	5 (0.6)	23 (2.6)	3 (0.3)	877
SPN	78 (4.8)	109 (6.7)	1034 (63.5)	195 (12.0)	156 (9.6)	13 (0.8)	43 (2.6)	1628
SPS	14 (1.4)	142 (14.6)	507 (52.0)	209 (21.4)	30 (3.1)	60 (6.2)	12 (1.2)	974
All areas	210 (6.0)	419 (12.0)	1972 (56.7)	533 (15.3)	191 (5.5)	96 (2.5)	58 (1.7)	3479

r = rough, s = smooth, r.s = intermediate between rough and smooth

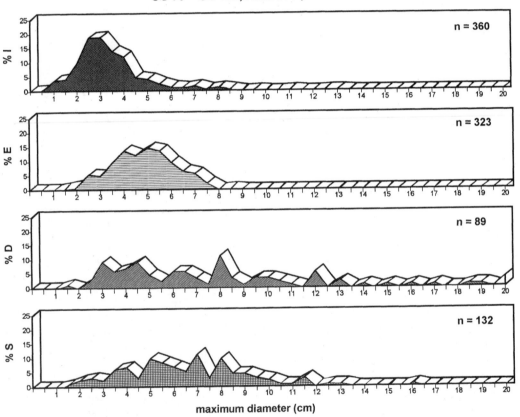

SO 79 + SO 106, Area SPN, surface nodules

FIGURE 8.14   Size distribution of different types of nodule shape in area SPN.

SPS and D are only slightly different. Figure 8.14 shows that the size distribution of S- and D-types is poorly sorted, while E- and I-types show a well-sorted size distribution. A comparison of nodule types, e.g., IE, within different areas reveals a great similarity of size distribution (Figure 8.15).

In addition to collective histograms, 88 histograms of individual nodule samples from box cores (KG), maxi cores (MX), and free-fall grabs (BG) were plotted. We may distinguish two main types of histograms: Type A is poorly sorted with a low number (n) of nodules (Figure 8.16 A). Type B is well sorted with a distinct maximum of diameter between 2 and 4 cm and a high nodule number

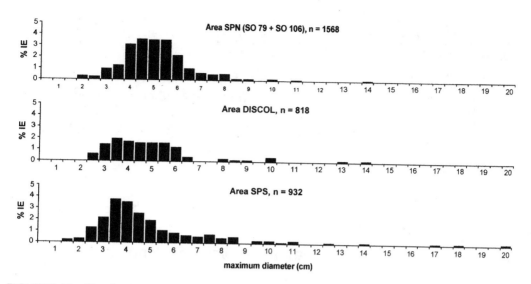

FIGURE 8.15 Size distribution of nodules with shape type IE (irregularly ellipsoidal) in areas SPN, DISCOL, and SPS.

(Figure 8.16 B). Additionally, of course, intermediate types of histograms were found. Type A is restricted to wide basins and type B mainly occurs on slopes, tops of seamounts and ridges, and in narrow basins. A-type histograms are characterized by a predominance of S and D nodules, while in B-type histograms E, E2, IE, and I-shaped nodules prevail.

8.4.2.4 Internal Growth Structures

Internal growth structures were studied in polished sections of nodules from areas SPN, SPS, and D, and five different types were identified: laminated to columnar = b, dendritic = d, massive, nondirectional = e, pillar = f, and dense, laminated = h. Structures of the b-type grow at the top of nodules representing hydrogenetic growth from the sea water, while d- and h-growth occurs at the bottom side of nodules due to diagenetic processes within the sediment. Small nodules and the inner part of large nodules show a narrow sequence of b and d layers. Growth of e- and f-type is restricted to crusts representing pure hydrogenetic growth conditions.

Large nodules always show a pale-grey h-growth with submetallic luster at the bottom side where smooth s-type surface texture was observed. Smooth s-type surface texture respectively h-growth may cover up to 70% of the lower surface of nodules. Looking upward, s-type surface texture is followed by r-type texture, representing an internal growth of d-type. The boundary between s- and r-surface texture is sharp and horizontal. A lateral wedging and interfingering of h-growth layers and d-growth layers is observed in nodule sections at this boundary. This h-growth forms dense, massive to thinly laminated layers up to 10 mm thick. The h-layers contain many inclusions of siliceous microfossils, such as diatoms and radiolarians which mostly are dissolved and only imprints and casts are preserved (Figure 8.17 A). The h-growth layers are composed of very pure todorokite or birnessite. The crystals of those minerals grow more or less perpendicular to the outer nodule surface with a length up to 15 μm (Figure 8.17 B).

Looking down the lower part of a nodule section, the transition from d- to h-growth is gradational, while h-growth changes to d-growth at a sharp boundary where d-growth starts with a thin layer of dark-grey laminated to columnar Mn oxide similar to b-growth (Figure 8.17 C). Apparently, there is a sawtooth-shaped development of nodule growth.

The h-layers are mainly found in the outer part of nodules. They are thicker and more frequent in large nodules than in small ones.

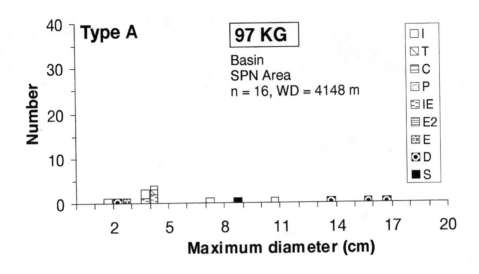

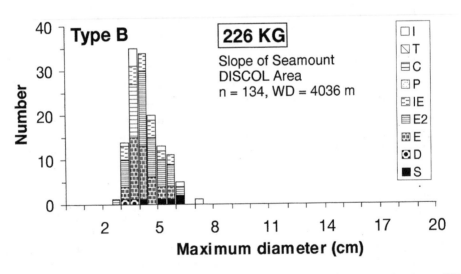

FIGURE 8.16 Size distribution of nodules from two box grabs at different water depths in areas SPN and DISCOL. For explanation of symbols of nodule shape see Figure 8.12.

Large nodules often show an asymmetric growth, while smaller ones are more symmetrically grown. Out of 35 surface nodules with maximum diameter >9 cm, 12 nodules had grown asymmetrically and 13 symmetrically. Nodules from area SPS revealed a less frequent occurrence of asymmetric growth than nodules from the D and SPN survey areas.

The h-growth, typical of the Peru Basin, had already been observed by Halbach et al. and Thijssen et al., and may be compared to the "massive zone" of nodules from the CCZ;[5,28,45] h-growth was also described for nodules from the Guatemala Basin.[46]

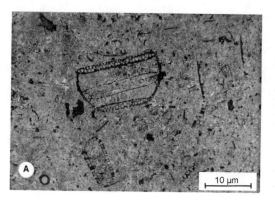

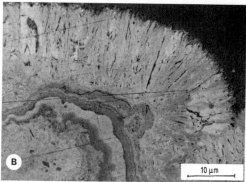

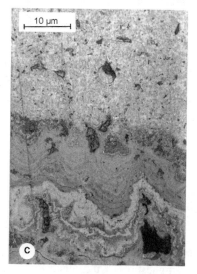

FIGURE 8.17   Photographs of polished section of nodules. (A) h-type growth layer with inclusions of diatoms. (B) h-type growth layer with todorokite crystals up to 15 μm long. (C) h-growth in upper half changes to d-growth at a sharp boundary starting with dark-grey laminated to columnar Mn oxide similar to b-growth.

8.4.2.5   Nuclei

According to previous workers, the nuclei of nodules from area SO-04 and SO-11/2 mostly consist of broken fragments of an older nodule or a rock fragment almost entirely replaced by ferromanganese oxides.[5,47]

From areas SO-79 and SO-106 a total of 407 polished nodule sections were studied, and 499 nuclei were identified. Some nodules contained more than one nucleus. The following types of nuclei were distinguished: N = nodule fragments, S = sediment, SI = manganese impregnated sediment, O = organic remains (Figure 8.11B), B = basalt or palagonite, M = mineral grains, and Y = hydrothermal material. The numerical distribution of nucleus types is presented in Table 8.2. A distinct predominance of N-nuclei (51.1%) followed by SI + S (25.6%), and B (11%) is observed. Hydrothermal material (Y) is represented by Mn oxide, Fe silicate and Fe oxhydroxide, goethite, and quartz.[48] The nucleus sizes range from 1 mm to 7 cm maximum diameter. N-nuclei are mainly found in basins, while B, M, and Y-nuclei predominate near seamounts and ridges, indicating the proximity of outcropping volcanic rocks. SI and S-nuclei were supplied from indurated horizons within the sediment column.

TABLE 8.2
Frequency of Occurrence of Nucleus Types in Manganese Nodules
from Areas of Cruises SO-79 + SO-106 and SO-25

Area	N n (%)	S n (%)	SI n (%)	I n (%)	O n (%)	B n (%)	M n (%)	Y n (%)	P n (%)	Total n
SO79 +106	255 (51.1)	20 (4.0)	108 (21.6)	—	17 (3.4)	55 (11.0)	16 (3.2)	28 (5.6)	—	499
SO25	559 (26.6)	1112 (52.9)	—	22 (1.0)	106 (5.0)	167 (7.9)	44 (2.1)	—	91 (4.9)	2101

N = nodule fragments, S = sediment, SI = manganese impregnated sediment, I = impregnated material of unknown provenance, O = organic remains, B = basalt or palagonite, M = mineral grains, Y = hydrothermal material, P = pumice, n = number.

Most N-nuclei are composed of Mn-oxide material with d- and h-growth, fragmented from relatively rapidly grown basin nodules. This material is friable, and fragments may easily be detached if nodules bump due to movement by benthic organisms or bottom currents. Fragments may also be detached by organisms feeding on other organisms attached to the nodule surface or living in crevices of the nodules.[49] A few N-nuclei from shallower water consist of densely grown Mn oxide with b-growth or sometimes with e- or f-growth. In this case fragmentation of slowly grown nodules or crusts takes place due to shrinkage on aging.

8.4.2.6 Bulk Composition

Based on bulk analysis of 171 nodules selected from samples from about 300 stations and averaged at 35 locations, Halbach et al. divided the SO-11/1 survey area into three different ranges of geochemical facies from north to south. The north facies is characterized by large spheroidal, cauliflower-shaped nodules (maximum 25 cm) with a Mn/Fe ratio >7. With an increasing Mn/Fe ratio the Ni + Cu content decreases.[6] The main mineral phase is todorokite with some admixture of birnessite. The mid facies shows smaller nodules (3 to 8 cm) of ellipsoidal to discoidal shape with Mn/Fe ratios of 4 to 7. Here, highest Ni + Cu contents up to 2.2% are observed. In the southern part of the mid facies, discoidal shaped nodules show a finely porous to smooth surface texture on their upper sides. This material contains lower contents of Ni and Cu and is composed of δ-MnO_2 intimately intergrown with amorphous $FeOOH \times H_2O$. The south facies is characterized by a Mn/Fe ratio <4. The size of the mostly discoidal nodules ranges between 2 and 7 cm. Shape, surface texture, and composition are similar to that of the nodules from the southern part of the mid facies. Thus, the Mn/Fe ratio increases from south to north in the SO-11/1 area. The Ni + Cu content reaches maximum concentration in the mid facies associated with Mn/Fe ratios of about 5. Due to the covariance of Ni and Cu with the Mn/Fe ratio in nodules from the SO-11/1 area, they could be divided into two groups with different hyperbolic regression curves separated at a Mn/Fe ratio of 5, called the "point of reversal" because of the culmination of the Ni + Cu values there.[6,28,50]

Bulk analyses of 67 nodule samples from 46 stations at 12 locations in the SO-04 survey area were presented by Thijssen et al.[5] They show high Mn/Fe ratios (average 6.8), intermediate concentrations of Ni and Cu (1.14% and 0.62%), and extremely low concentrations of Co (0.05%). The bulk composition of 225 nodules of different sizes from 107 stations in the SO-11/2 area have also been chemically analyzed.[7] In the northern part of this area, Mn/Fe ratios are high (>10; max. 53.5) and Ni + Cu contents low (<2%), while Mn/Fe ratios are low in the southwest of the area, south of the fault zone. Other elements do not show well-defined trends. Compared to nodules from the CCZ, the Peru Basin nodules have high average Mn/Fe ratios (4.7), Ni/Cu (2.03), and Mn/Co (>300). One hundred and seventy-five nodules of five size classes between 0 and 20 and >80 mm maximum diameter showed an increase of average Mn/Fe ratio from 4.1 to 6.4 and a slight decrease of Ni + Cu contents from 2.1 to 1.8% with increasing nodule size. The composition of nodules within an individual size class is highly variable. Therefore, Thijssen et al. deduced that nodule size is only a secondary parameter controlling nodule composition.[7]

Nodules from location 62 (station 517 GB) in the northern part of the SO-11/2 area, show a very low content (8%) of acid-insoluble residue and a high Mn/Fe ratio (31.3) in the hydrolysate fraction.[51] This is assumed to be due to diagenetic remobilization of manganese in the sediment column.

8.4.2.7   Composition of Individual Layers

The internal growth structures of 25 nodules from the SO-04 survey area were studied microscopically in 44 polished sections.[5] The predominant features are "massive zones," which occur especially in the outer part of large nodules. They are up to 3 mm thick and consist of fine-grained, dense anisotropic ferromanganese material with high reflectivity. A high abundance of microfossils is observed in them. The massive zone is mainly composed of todorokite with lesser amounts of birnessite. It shows a sharp outer contact, while the inner contact is diffuse, grading into columnar structures. Electron microprobe analysis of the massive zone shows high Mn and low Fe contents (28 to 45% Mn; 0.5 to 3.7% Fe) and low Ni and Cu contents (0.1 to 0.5 Ni; 0.2 to 0.4 Cu).

The composition of microlayers of large, cauliflower-shaped nodules from the SO-11/1 area was investigated by Halbach et al.[28] The upper part of the nodules consists of concentrically banded zones composed of two types of layers. The first type (A1) show thicknesses up to 3 mm and are composed of lamellar particles of nearly pure todorokite with a grain size of about 3×8 μm. The second type (A2), only 1 to 20 μm thick, contain X-ray amorphous Mn and Fe oxide hydroxide phases and aluminosilicate. The lower part of the nodules with a smooth surface texture consists of concentrically parallel layers mainly of well crystallized todorokite with some admixture of birnessite and inclusions of sedimentary debris and planktonic fragments. In this part, A1-type layers predominate, while the portion of A2-type microlayers is very low. A third type of microlayer predominates in smaller nodules. This so-called *B-substance* consists of an intimate mixture of δ-MnO_2, X-ray amorphous $FeOOH \times H_2O$, and aluminosilicate. A1 and A2 microlayers represent an early diagenetic environment within the near-surface sediment, while B layers indicate hydrogenetic growth conditions from sea water.

The compositional variations within a single large nodule (10 cm maximum diameter, 4.8 kg) from location 81 in the northern part of the SO-11/2 survey area were studied by Thijssen et al.[7] Nineteen samples were taken across the section of the nodule, five of them from the nucleus, which proved to be a mammalian bone more or less impregnated by ferromanganese oxide. A decrease of Mn/Fe ratio and increase of Ni + Cu content toward the center (excluding the nucleus) were observed. The outer layer of the assumed bottom side of the nodule shows the highest Mn/Fe ratio (73.4) and the lowest Ni + Cu content (0.55%), which indicates an increasing diagenetic contribution in that direction. The internal growth structures of the lower and upper layers of this nodule are distinctly different. Therefore, the authors deduce that the nodule was not rotated during its period of growth. All samples, excluding the nucleus, show a mineral assemblage consisting of todorokite, δ-MnO_2, birnessite, and quartz. There is a tendency for the ratio of birnessite/todorokite to decrease toward the center. This might indicate increased stability of todorokite during aging (see Section 8.4.2.9.).

Microprobe point analyses of microlayers of nodules from the Peru Basin indicated different compositions for A1 and A2 layers: A1 = 44.7% Mn, 1.2% Fe (Mn/Fe ratio = 37.2), 1.9% Ni, 0.7% Cu, <0.1% Co, 1.0% SiO_2; A2 = 19.8% Mn, 9.6% Fe (Mn/Fe ratio, 2.1), 0.4% Ni, 0.2% Cu, 0.3% Co, 9.5% SiO_2.[52]

Following Halbach et al., precipitation of A1 substance with todorokite presumably leads to pH depression in the microenvironment of the nodule surface, which decreases the mobility of silicate resulting in the formation of A2 substance characterized by X-ray amorphous silicate phases in an intimate mixture with hydrated iron hydroxide and manganese oxide compounds.[52] After

TABLE 8.3

Mean Values of Chemical Composition of Different Growth Structures of Manganese Nodules and Crusts from SO-79 Areas

Growth Structure	Number	Fe$_2$O$_3$ %	MnO %	Mn/Fe	Cu ppm	Ni ppm	Cu + Ni ppm
h	10	0.77	54.87	80.00	1384	2724	4108
d + h	9	4.48	41.87	20.51	6680	8445	15,125
d	8						
d + b	+ 3	4.36	42.78	15.69	4966	10,461	15,427
b	7						
f	+ 2	16.56	26.77	2.13	3870	8357	12,227
e	3	19.69	21.51	1.50	3735	4840	8575

h = dense, laminated; d = dendritic; b = laminated to columnar; f = pillar; e = massive, nondirectional.

restoration of the previous pH level, todorokite precipitation occurs again. Thus, the authors explain the rhythmic sequence of A1 and A2 microlayers by physicochemical changes in the nodule microenvironment.

The chemical and mineral composition of 42 individual growth layers of nodules and crusts from the SPN and SPS survey areas have been analyzed by von Stackelberg.[19] Based on these data, the average concentrations of Fe$_2$O$_3$, Cu, Ni, and the Mn/Fe ratio of different individual layers as well as of mixed growth types are presented in Table 8.3. A distinct decrease in the Mn/Fe ratio from 80 to 1.5 from h- to e-growth structures is observed, reflecting the range from extreme diagenetic to extreme hydrogenetic growth conditions. The lowest values for Ni + Cu (4108 ppm) are found in h-growth structures, and the highest (15,427 ppm) in d-plus d+b-structures.

Additionally, layers of some nodules from areas SPN and SPS were examined in more detail. The chemical composition of the different growth structures is presented in Tables 8.4 and 8.5. Figure 8.18 shows a plot of Mn/Fe vs. Ni + Cu in some of these nodules. In general, there is a similar point of reversal at a Mn/Fe ratio of 5, as was observed by Halbach et al.[6] Analysis of individual layers, however, reveals a much wider range of Mn/Fe ratios and Ni + Cu contents than for bulk analyses of nodules. Identical types of growth structures cluster distinctly within the diagram. Diagenetic h-growth shows the highest Mn/Fe ratios up to 134.4 and lowest Ni + Cu values down to 0.23 ppm. The highest value for Ni + Cu (2.95%) was found near the point of reversal in a sample with a mixture of b- and d-growth. A b-growth sample had the lowest Mn/Fe ratio, 1.7.

If we plot the distance of samples from the nodule center vs. Mn/Fe ratio, we clearly recognize that hydrogenetic growth structures of b- and b-d-type with very low Mn/Fe ratios occur mainly near the nodule center (Figure 8.19). Diagenetic growth structures of h-type with very high Mn/Fe ratios were found only at a distance of >40 mm from the center.

With increasing size, nodules tend to grow more asymmetrically; the S-type shape changes to the D-type (see Section 8.4.2.4.). Nodule 165 KG/3, with a maximum diameter of 17 cm, reveals a distinct asymmetric distribution of chemical composition (Figure 8.20). The center with symmetric growth is characterized by low to medium Mn/Fe values, while the asymmetrically grown outer part shows very low Mn/Fe values at the top and very high Mn/Fe values at the bottom.

The REE distribution in hydrogenetic Mn-oxide material in the Peru Basin is distinctly different from that of diagenetic material.[19,53] Samples from diagenetic nodules reveal NASC-normalized values between 0.5 and 1.5, while hydrogenetic crusts have values ranging between 2 and 9. A plot of Er (erbium) vs. Mn/Fe also demonstrates the different chemical characteristic of each type of Mn oxide (Figure 8.21).

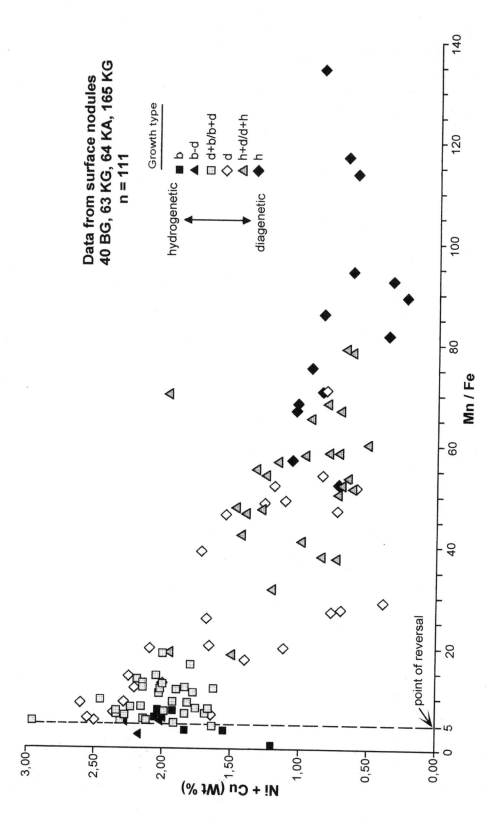

FIGURE 8.18 Distribution of different types of growth layers within the diagram Mn/Fe vs. Ni + Cu. b = laminated to columnar, d = dendritic, h = dense, laminated.

TABLE 8.4
Chemical Composition of Different Growth Structures of Manganese Nodules

Nodule	Sample	Burial Depth (cm)	Growth Structure	Distance from Surface mm	Center mm	Mn/Fe %	Ni+Cu %	Fe %	Mn %	Cu %	Ni %	Co %
40 KG/3	1		b		14	1.70	1.34	11.70	19.50	0.43	0.91	0.120
63 KG/2	21a		b+d		0	6.60	2.27	4.55	30.00	0.69	1.58	0.077
63 KG/2	22a		d		20	14.20	2.24	2.40	34.00	0.54	1.70	0.068
63 KG/2	23a		h		40	81.80	0.36	0.55	45.00	0.14	0.22	0.082
63 KG/2	24a		d+b		35	9.20	1.92	3.55	32.50	0.49	1.43	0.069
63 KG/3	1		d+b	2	86	8.00	1.75	4.00	32.00	0.50	1.25	0.085
63 KG/3	2		d	6	82	27.40	0.70	1.55	42.50	0.24	0.46	0.079
63 KG/3	3		d	10	78	27.00	0.77	1.50	40.50	0.27	0.50	0.086
63 KG/3	4		d+h	12	76	37.40	0.75	1.15	43.00	0.26	0.49	0.085
63 KG/3	5		d+h	13	75	38.00	0.85	1.10	41.75	0.31	0.54	0.072
63 KG/3	6		d	16	72	17.50	1.40	2.25	39.50	0.42	0.98	0.059
63 KG/3	7		d	18	70	19.70	2.09	1.85	36.50	0.54	1.55	0.069
63 KG/3	8		h	22	66	67.00	1.03	0.65	43.50	0.32	0.71	0.053
63 KG/3	9		h+d	27	61	41.10	1.00	0.98	40.25	0.33	0.67	0.060
63 KG/3	10		h	31	57	57.00	1.06	0.73	41.50	0.28	0.78	0.069
63 KG/3	11		d	35	53	19.90	1.12	2.00	39.75	0.36	0.76	0.098
63 KG/3	12		d+b	50	38	9.10	1.81	3.70	33.75	0.58	1.23	0.073
63 KG/3	13		d+b	62	26	7.00	1.68	4.40	30.75	0.58	1.10	0.074
63 KG/3	14		d+b	73	15	7.00	1.83	4.15	29.00	0.62	1.21	0.110
63 KG/3	15		h	3	60	52.00	0.71	0.75	39.00	0.23	0.48	0.081
63 KG/3	16		h	8	55	62.50	0.78	0.68	42.50	0.25	0.53	0.054
63 KG/3	17		h+d	13	48	47.60	1.48	0.83	39.50	0.37	1.11	0.078
63 KG/3	18		d+b	18	43	16.70	1.79	2.25	37.50	0.46	1.33	0.085
63 KG/3	19		d+b	20	41	12.10	1.83	3.05	37.00	0.53	1.30	0.075
63 KG/3	20		d+b	23	38	11.10	1.77	3.30	36.75	0.57	1.20	0.067
63 KG/3	21		d+b	26	35	18.80	1.99	2.10	39.50	0.69	1.30	0.054
63 KG/3	22		d+b	30	31	12.40	2.14	2.90	36.00	0.71	1.43	0.073
63 KG/3	23		d+b	33	28	11.00	2.02	3.10	34.25	0.65	1.37	0.063
63 KG/3	24		d+b	38	23	13.00	2.14	2.70	35.25	0.68	1.46	0.061
63 KG/3	25		d+b	42	19	12.60	2.00	2.85	36.00	0.67	1.33	0.084
63 KG/3	26		d+b	46	15	11.70	2.01	3.00	35.00	0.69	1.32	0.100
63 KG/3	27		d	26	77	51.60	0.59	0.80	41.25	0.16	0.43	0.057
63 KG/3	28		h	23	82	86.00	0.83	0.50	43.00	0.17	0.66	0.066
63 KG/3	29		d+h	18	87	50.40	0.73	0.81	40.85	0.20	0.53	0.070
63 KG/3	30		h+d	12	92	78.60	0.64	0.55	43.25	0.20	0.44	0.079
63 KG/3	31		d	6	97	54.00	0.84	0.75	40.50	0.21	0.63	0.074
63 KG/3	32		h	2	104	89.40	0.23	0.48	42.90	0.11	0.12	0.058
63 KG/4	41		d		0	9.00	2.28	3.80	34.00	0.90	1.38	0.085
63 KG/4	42		d		25	5.30	2.02	4.85	25.75	0.57	1.45	0.093
63 KG/4	43		h+d		40	60.20	0.51	0.66	39.75	0.20	0.31	0.094
63 KG/4	44		b		10	5.50	2.01	5.00	27.25	0.71	1.30	0.099
63 KG/5	51		b+d	116	5	11.90	1.62	3.00	35.75	0.85	0.77	0.093
63 KG/5	52		b	94	30	6.20	2.05	4.70	29.25	0.80	1.25	0.088
63 KG/5	53		d+b	89	35	7.60	2.33	4.05	30.75	0.75	1.58	0.069
63 KG/5	54		d+h	65	60	42.20	1.44	0.97	41.00	0.52	0.92	0.051
63 KG/5	55		h+d	60	65	55.20	1.33	0.77	42.50	0.46	0.87	0.073
63 KG/5	56		h+d	55	70	31.80	1.16	1.30	41.25	0.35	0.81	0.052

TABLE 8.4 (continued)
Chemical Composition of Different Growth Structures of Manganese Nodules

Nodule	Sample	Burial Depth (cm)	Growth Structure	Distance from Surface mm	Center mm	Mn/Fe %	Ni+Cu %	Fe %	Mn %	Cu %	Ni %	Co %
63 KG/5	57		h+d	51	75	65.30	0.93	0.67	43.75	0.31	0.62	0.050
63 KG/5	58		d	48	77	47.00	0.73	0.92	43.25	0.20	0.53	0.061
63 KG/5	59		h	45	80	94.50	0.62	0.47	44.40	0.18	0.44	0.057
63 KG/5	60		d+h	42	82	66.90	0.71	0.62	41.50	0.16	0.55	0.050
63 KG/5	61		h	40	85	117.10	0.66	0.38	44.50	0.16	0.50	0.048
63 KG/5	62		d	28	95	70.70	0.81	0.58	41.00	0.21	0.60	0.045
63 KG/5	63		h	22	103	113.80	0.59	0.38	43.25	0.16	0.43	0.050
63 KG/5	64		h	19	105	134.40	0.84	0.32	43.00	0.16	0.68	0.052
63 KG/5	65		d	15	110	46.20	1.54	0.78	36.00	0.29	1.25	0.057
63 KG/5	66		d	9	115	52.00	1.19	0.75	39.00	0.31	0.88	0.057
63 KG/5	67		d+h	3	120	68.30	0.80	0.60	41.00	0.23	0.57	0.064
63 KG/5	68		b	58	13	7.70	2.03	4.35	33.50	0.73	1.30	0.078
63 KG/5	69		b+d	45	28	8.30	2.15	3.80	31.50	0.60	1.55	0.065
63 KG/5	70		d	24	48	25.60	1.68	1.50	38.40	0.53	1.15	0.062
63 KG/5	71		h	13	62	75.40	0.92	0.58	43.75	0.22	0.70	0.059
63 KG/5	72		d	9	68	49.00	1.11	0.75	36.75	0.19	0.92	0.048
63 KG/5	73		h+d	6	70	79.10	0.67	0.55	43.50	0.19	0.48	0.051
63 KG/5	74		d+h	2	73	70.00	1.97	0.60	42.00	0.17	1.80	0.057
64 KA/1	11a	10	d		0	7.00	2.35	4.65	32.75	0.85	1.50	0.077
64 KA/1	12a	10	b-d		5	6.10	2.00	4.80	29.50	0.70	1.30	0.074
64 KA/1	13a	10	d		25	38.90	1.72	0.95	37.00	0.32	1.40	0.076
64 KA/1	14a	10	h		50	92.60	0.33	0.47	43.50	0.11	0.22	0.088
64 KA/1	15a	10	d		30	6.70	1.65	4.70	31.50	0.45	1.20	0.087
64 KA/11	111	36	d		0	7.20	1.97	4.30	30.75	0.95	1.02	0.108
64 KA/11	112	36	d		15	4.80	1.81	5.10	24.25	0.88	0.93	0.184
64 KA/4	41	290	d		0	8.90	1.48	3.45	30.75	0.72	0.76	0.137
64 KA/4	42	290	d		25	7.40	1.48	4.00	29.50	0.33	1.15	0.168
64 KA/4	43	290	d		38	8.40	1.93	3.60	30.25	0.95	0.98	0.210
64 KA/4	44	290	h		42	54.50	0.70	0.73	39.75	0.19	0.51	0.089
64 KA/4	45	290	h+d		52	58.90	1.19	0.69	40.65	0.65	0.54	0.117
64 KA/5	51	347	d		0	10.10	1.56	3.05	30.75	0.74	0.82	0.108
64 KA/5	52	347	h+d		20	34.90	1.78	1.16	40.50	0.93	0.85	0.122
64 KA/5	53	347	d		32	10.10	2.55	3.45	35.00	1.00	1.55	0.201
64 KA/5	54	347	h+d		36	30.80	1.47	1.25	38.50	0.80	0.67	0.144
64 KA/5	55	347	d+h		44	19.00	1.53	1.60	30.40	0.80	0.73	0.104
64 KA/10	101	445	d		0	8.90	1.14	3.70	32.75	0.64	0.50	0.114
64 KA/10	102	445	d+h		30	19.90	2.20	1.80	35.80	1.03	1.17	0.142
64 KA/10	103	445	h		35	26.30	1.80	1.50	39.50	1.00	0.80	0.122
64 KA/10	104	445	d		30	9.80	1.78	3.00	29.50	0.93	0.85	0.168
64 KA/10	105	445	d+h		55	11.20	2.48	2.45	27.50	1.18	1.30	0.154
64 KA/2	1	480	d+h	2		22.60	1.27	1.60	36.25	0.49	0.78	0.110
64 KA/2	2	480	d	5		12.00	1.39	2.95	35.50	0.50	0.89	0.130
64 KA/2	3	480	h	8		35.00	1.01	1.25	43.75	0.44	0.57	0.080
64 KA/2	4	480	d	12		22.60	1.01	1.70	38.50	0.41	0.60	0.930
64 KA/2	5	480	d	15		12.20	0.98	3.15	38.50	0.41	0.57	0.120
64 KA/2	6	480	d	20		14.90	1.44	2.45	36.50	0.52	0.92	0.110
64 KA/2	7	480	h	26		33.30	1.49	1.23	41.00	0.53	0.96	0.082

TABLE 8.4 (continued)
Chemical Composition of Different Growth Structures of Manganese Nodules

Nodule	Sample	Burial Depth (cm)	Growth Structure	Distance from Surface mm	Center mm	Mn/Fe %	Ni+Cu %	Fe %	Mn %	Cu %	Ni %	Co %
64 KA/2	8	480	h	31		35.20	1.18	1.20	42.25	0.45	0.73	0.092
64 KA/2	9	480	h+d	33		26.90	1.21	1.55	41.75	0.46	0.75	0.093
64 KA/2	10	480	h	36		43.70	1.02	1.03	45.00	0.44	0.58	0.067
64 KA/2	11	480	d	40		14.90	1.11	2.65	39.50	0.44	0.67	0.081
64 KA/2	12	480	d+h	44		29.50	1.14	1.40	41.25	0.42	0.72	0.092
64 KA/2	13	480	d+h	47		25.30	1.08	1.60	40.50	0.42	0.66	0.099
64 KA/2	14	480	h	48		54.20	1.04	0.84	45.50	0.42	0.62	0.095
64 KA/2	15	480	d	50		24.00	0.98	1.70	40.75	0.38	0.60	0.068
64 KA/2	16	480	h	52		58.30	0.90	0.78	45.50	0.39	0.51	0.070
64 KA/2	17	480	h	54		61.30	0.80	0.73	44.75	0.37	0.43	0.078
64 KA/2	18	480	h	59		75.40	0.70	0.62	46.75	0.36	0.34	0.065
64 KA/2	19	480	d+h	65		45.90	0.87	0.97	44.50	0.46	0.41	0.082
64 KA/2	20	480	d	72		19.10	1.00	2.00	38.25	0.52	0.48	0.068
64 KA/3	31	540	d	0		5.30	1.54	5.10	27.00	0.66	0.88	0.121
64 KA/3	32	540	h+d	25		49.70	0.79	0.88	43.75	0.45	0.34	0.102
64 KA/3	33	540	h	47		80.90	0.82	0.55	44.50	0.41	0.41	0.087
64 KA/3	34	540	d+h	60		26.00	0.84	1.50	39.00	0.39	0.45	0.113
64 KA/3	35	540	d	20		6.20	1.65	4.50	27.75	0.69	0.96	0.150
64 KA/7	71	590	d		0	4.50	1.40	5.70	25.90	0.62	0.78	0.114
64 KA/7	72	590	h+d		20	47.30	1.03	0.94	44.50	0.39	0.64	0.078
64 KA/7	73	590	h		28	71.10	0.73	0.64	45.50	0.36	0.37	0.070
64 KA/7	74	590	h+d		40	49.20	1.21	0.90	44.25	0.49	0.72	0.092
64 KA/7	75	590	d		65	18.60	1.57	1.95	36.25	0.62	0.95	0.076
64 KA/7	76	590	d+h		85	14.80	1.34	2.55	37.75	0.54	0.80	0.115
64 KA/7	77	590	d+h		90	14.10	0.90	2.75	38.75	0.41	0.49	0.126
165 KG/1	11a		d		10	8.80	2.60	3.90	34.25	1.15	1.45	0.085
165 KG/1	12a		d		25	5.30	2.50	5.45	28.75	0.85	1.65	0.092
165 KG/1	13a		h+d		30	18.60	1.50	2.00	37.15	0.43	1.07	0.068
165 KG/3	1		d	2	75	28.90	0.39	1.10	31.75	0.12	0.27	0.046
165 KG/3	2		d+h	4	73	51.30	0.62	0.76	39.00	0.16	0.46	0.070
165 KG/3	3		d+h	6	71	53.50	0.66	0.71	38.00	0.15	0.51	0.044
165 KG/3	4		h	7	70	70.60	0.86	0.63	44.50	0.17	0.69	0.053
165 KG/3	5		h+d	8	69	58.60	0.73	0.67	39.25	0.16	0.57	0.063
165 KG/3	6		d+h	10	67	52.00	0.70	0.77	40.00	0.18	0.52	0.050
165 KG/3	7		h+d	12	65	58.60	0.80	0.67	39.25	0.21	0.59	0.058
165 KG/3	8		d	17	60	48.10	1.27	0.78	37.50	0.27	1.00	0.049
165 KG/3	9		h+d	22	55	56.80	1.17	0.70	39.75	0.28	0.89	0.055
165 KG/3	10		h+d	23	54	58.00	0.97	0.69	40.00	0.25	0.72	0.055
165 KG/3	11		h	25	52	67.70	1.02	0.62	42.00	0.29	0.73	0.079
165 KG/3	12		d+h	27	50	47.50	1.28	0.81	38.50	0.35	0.93	0.075
165 KG/3	13		d+h	29	48	54.20	1.26	0.72	39.00	0.36	0.90	0.068
165 KG/3	14		d+h	31	46	46.60	1.41	0.81	37.75	0.39	1.02	0.051
165 KG/3	15		d	34	43	20.30	1.66	1.75	35.50	0.36	1.30	0.055
165 KG/3	16		b+d	37	40	7.80	1.66	4.40	34.25	0.41	1.25	0.095
165 KG/3	17		d+h	39	38	19.00	1.95	2.00	38.00	0.60	1.35	0.041
165 KG/3	18		d+b	41	36	14.50	2.04	2.45	35.50	0.59	1.45	0.059
165 KG/3	19		d+b	43	34	4.50	1.63	6.25	28.00	0.40	1.23	0.097

TABLE 8.4 (continued)
Chemical Composition of Different Growth Structures of Manganese Nodules

Nodule	Sample	Burial Depth (cm)	Growth Structure	Surface mm	Center mm	Mn/Fe %	Ni+Cu %	Fe %	Mn %	Cu %	Ni %	Co %
				Distance from								
165 KG/3	20		d+b	47	30	7.30	1.98	4.50	33.00	0.60	1.38	0.058
165 KG/3	21		d+b	51	26	13.70	2.18	2.45	33.50	0.73	1.45	0.059
165 KG/3	22		d+b	58	19	11.70	1.89	2.70	31.50	0.66	1.23	0.085
165 KG/3	23		d	77	0	12.80	2.00	2.50	32.00	0.75	1.25	0.051
165 KG/3	24		b	2	49	3.60	1.55	6.90	24.50	0.45	1.10	0.088
165 KG/3	25		d+b	5	46	5.10	1.91	5.75	29.00	0.66	1.25	0.088
165 KG/3	26		d+b	9	42	5.90	2.13	4.70	27.50	0.68	1.45	0.048
165 KG/3	27		d+b	15	36	8.20	2.23	3.90	32.00	0.73	1.50	0.064
165 KG/3	28		d+b	19	32	5.70	2.11	5.00	28.50	0.73	1.38	0.057
165 KG/3	29		d+b	23	28	9.60	2.45	3.40	32.50	0.80	1.65	0.065
165 KG/3	30		d+b	26	25	5.60	2.30	4.90	27.50	0.70	1.60	0.055
165 KG/3	31		d+b	34	17	7.20	2.33	4.15	30.00	0.83	1.50	0.071
165 KG/3	32		d	51	0	11.90	2.20	2.65	31.50	0.75	1.45	0.065
165 KG/4	41		b		10	7.50	1.92	4.00	30.00	0.80	1.12	0.102
165 KG/4	42		b		10	3.60	1.83	7.10	25.25	0.68	1.15	0.117
165 KG/4	43		b-d		7	2.70	2.18	8.60	23.25	0.83	1.35	0.154
165 KG/4	44		d+b		0	5.30	2.95	5.45	28.75	1.25	1.70	0.102
165 KG/5	51		d		0	5.80	2.55	5.70	32.15	1.10	1.45	0.105
165 KG/5	52		b-d		10	5.30	2.28	5.70	30.25	0.83	1.45	0.038

Maximum diameters of nodules: 40 KG/3: 5.3cm; 63 KG/2: 8.2cm; 63 KG/3: 16.0cm; 63KG/4: 6.4cm;
63 KG/5: 20.0cm; 64 KA/1: 9.0cm; 64 KA/2: 11.0cm; 64 KA/3: 9.5cm;
64 KA/4: 10.3cm; 64 KA/5: 9.7cm; 64KA/7: 13.0cm;64 KA/10: 10.1cm;
64 KA/11: 5.0cm
165 KG/1: 8.0cm; 165 KG/3: 15.0cm; 165 KG/4: 2.2cm; 165 KG/5: 3.0cm

TABLE 8.5
Mean Values of Mn/Fe Ratio and Ni+Cu Concentration
of Different Growth Structures of Surface and
Buried Nodules

Growth Structures	n	MV	Mn/Fe Max	Min	MV	Ni+Cu% Max	Min
			Surface nodules				
b	7	5.11	7.70	1.70	1.82	2.05	1.34
b+d/d+b/b–d	35	9.15	18.80	2.70	2.04	2.95	1.62
d	27	27.06	70.70	5.30	1.56	2.55	0.39
d+h/h+d	27	52.10	79.10	11.20	1.04	2.20	0.51
h	15	84.12	134.40	26.30	0.72	1.80	0.23
			Buried nodules				
d	20	11.50	24.00	4.50	1.49	2.55	0.62
d+h/h+d	17	30.94	58.90	14.10	1.31	2.48	0.79
h	12	52.43	80.90	33.30	1.02	1.49	0.70

MV = mean value

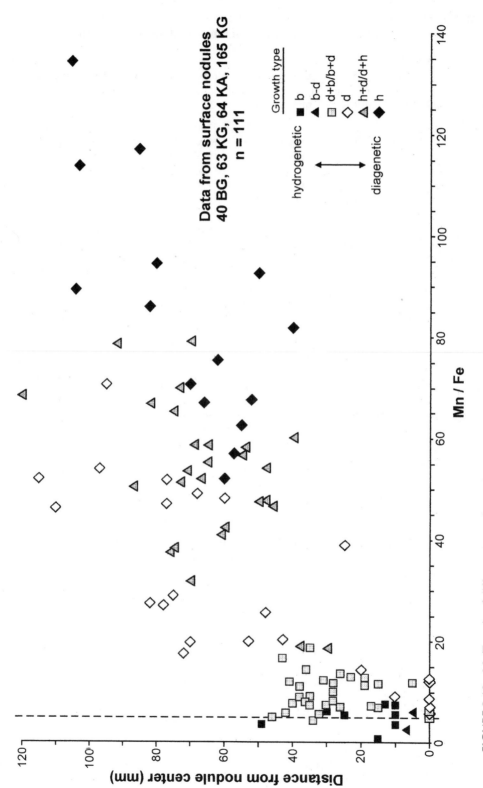

FIGURE 8.19   Mn/Fe ratio of different types of growth structure *vs.* distance from nodule center. For explanation of symbols see Figure 8.18.

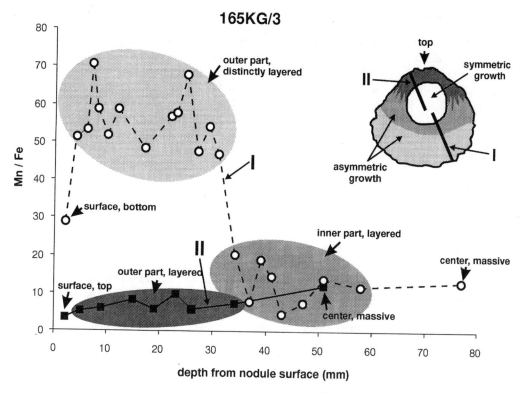

FIGURE 8.20 Large nodule 165 KG/3 (17 cm maximum diameter) with asymmetric growth indicated by asymmetric distribution of Mn/Fe ratios.

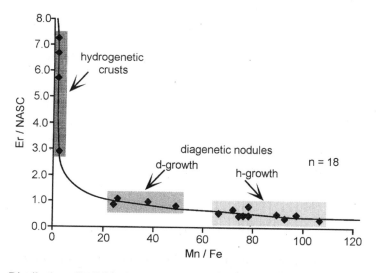

FIGURE 8.21 Distribution of individual growth layers of diagenetic nodules and hydrogenetic crusts within the diagram Mn/Fe vs. rare earth element Er.

8.4.2.8   Growth Rates

Heye and Marchig found a close correlation between nodule composition and growth rates.[53] Halbach et al. assume accretion rates of >100 mm/Ma for large, cauliflower-shaped nodules from the Peru Basin due to the high Mn/Fe ratio.[50,54] Reyss, Marchig, and Ku radiometrically determined the accretion rate of the outermost 25 mm of a spheroidal nodule 15 cm in diameter from location 32 in the SO-11/1 area with high Mn/Fe ratios up to 99.[55] The accretion rate based on U and Th series isotopes was as high as 168 ± 24 mm/Ma. Additional radiometric dating of a nodule from the southern part of the SO-11/1 area with a Mn/Fe ratio of 3 indicated a slow growth rate of 4.5 mm/Ma.

Lemaitre et al. and Lemaitre discuss the differences in chemical composition between the two opposite sides of nodule 262 from location 31.[56,57] The upper side of this nodule, which has a maximum diameter of 12 cm, contains hydrogenous components (Mn/Fe ratio = 3.1), whereas the chemical composition of the bottom side is strongly influenced by a diagenetic supply of elements from the underlying sediment (Mn/Fe ratio = 78.3). This asymmetrically grown nodule with extremely different top and bottom characteristics was radiochemically analyzed by Reyss et al.[58] The bottom side showed fast growth rates of 190 ± 36 mm/Ma, while the top side accreted with low rates of 6.7 ± 0.8 mm/Ma. The symmetrically grown center showed an average accretion rate of 30 mm/Ma, which started 1.5 Ma ago. About 180 ka ago the growth became asymmetric, with slow hydrogenetic growth rates on top and fast diagenetic ones on the bottom. During the last 180 ka, the nodule had not been rotated.

Bollhöfer and Bollhöfer et al. determined nodule growth rates for nodules from the SO-79 survey area SPN.[59,60] Thorium and uranium isotopes were measured by alpha spectrometry and thermal ionization mass spectrometry (TIMS). Nodules 62 KD and 63 KG, with maximum diameters of 10 cm and 13 cm, respectively, showed the same mean growth rate of 110 mm/Ma for the outer 27 mm and 25 mm, respectively, which are composed of a sequence of alternating dendritic layers (d-growth) and dense, laminated layers (h-growth). Nodule 62 KD especially has high accumulation rates up to about 250 mm/Ma for h-growth layers with high Mn/Fe ratios (up to 150) and high specific gravity (up to 2.5 g/cm³). The authors assume that these h-growth layers formed during interglacial stages, and the d-growth layers during glacials. The inner part of nodule 63 KG, representing a zone between 33 mm to 50 mm below the surface and with a mixture of d and b-growth, had accumulation rates of about 40 mm/Ma. Radiometric dating of nodule 106 KG from a ridge position in the SPN area (maximum diameter 7.5 cm; water depth 4,025 m) revealed a growth rate of about 5 mm/Ma for the outer 4 mm, which had laminated to columnar b-growth.

The frequent inclusion of siliceous microfossils in the nodules, mentioned in Section 8.4.2.4., further indicates high growth rates of h-growth layers. Imprints of 19 different species of diatoms were found in nodule 62 KD.[61] From the stratigraphic range of the diatoms a maximum age of 1.8 Ma is deduced for that nodule, which corresponds well with the age that can be estimated from radiometrically determined growth rates.

Even large nodules with a long period of asymmetric growth may be turned over. For example, nodules 62 KD and 63 KG showed a thin layer (0.5 to 2.0 mm) of laminated to columnar hydrogenetic b-growth covering the diagenetically grown sequence of a few cm. Bollhöfer determined growth rates of about 40 mm/Ma for this hydrogenetic layer.[59]

8.4.2.9   Buried Nodules

Stoffers et al., Glasby et al., and Thijssen et al. described 27 manganese nodules buried between depths of 0.73 and 2.50 m in four long box cores from the SO-11 area.[7,34,47] The buried nodules generally occur in yellowish-brown calcareous to siliceous mud. They are not associated with any obvious sedimentation hiatus. Fifty percent of the buried nodules are in the size class >6 cm, and 21% are in the size class >10 cm. This indicates that in the Peru Basin, on average, buried nodules are larger than surface nodules.

The morphology and internal structure of buried nodules is similar to that of surface nodules. The surface texture, however, is quite different, being more porous.

Buried nodules do not show marked differences in chemical and mineralogical composition compared to surface nodules.[34] Nedjatpoor et al., however, described a todorokite/birnessite ratio of 82/18 for 12 surface nodules, while in nine buried nodules the ratio was 100/0.[62] They ascribed this mineralogical difference to different thermal stability characteristics of Mn oxide in buried and surface nodules, respectively, found during heating experiments. It is assumed that aging of nodules during burial leads to expulsion of some of the water from the todorokite and that this in turn enhances the stability of the todorokite.

One hundred and seventy-seven nodules from SO-79 and SO-106 were found buried in the sediment deeper than 20 cm. The deepest burial of a nodule was observed in 217 KL at 16.23 m. Most of the buried nodules occur in Quaternary sediments, only three of them in Tertiary sediments. Buried nodules occur more frequently in sediment cores from basins than from slopes.

As soon as nodules are buried a few cm below the redox boundary, a change in their color from dark-brown to greyish-brown is observed which, however, is not accompanied by a change in mineral and chemical composition. Only nodules buried deeper in suboxic sediments show a distinct change in surface texture and composition. Buried nodules, in general, have a porous and brittle surface. A surface texture of r and r.s-type predominates, while s-type is very rarely found.

Compared to surface nodules, SO-79 and SO-106 buried nodules are poorly sorted, showing a size distribution over a wide range with a predominance of large nodules with maximum diameter >6 cm (compare Figures 8.13 and 8.22). This observation had already been made by Stoffers et al. and Thijssen et al.[7,34] Furthermore, the histogram of buried nodules shows a weak indication of bimodality which had also been observed by Finney et al. in the Guatemala Basin.[46] Similar to surface nodules, S and D nodules prevail in the size class >6 cm of buried nodules. However, this predominance is about 100% higher in buried nodules. Out of 19 large buried nodules (>9 cm), 13 showed asymmetric and only 6 of them symmetric growth. This indicates a concentration of asymmetric growth among buried nodules. Ten of 11 samples from buried nodules showed a mineral composition predominantly of todorokite and only one of them of birnessite, while from 24 surface nodules the ratio of todorokite/birnessite was 1:1 (Table 8.6).

Strongly altered nodules were found buried in the following sediment cores: 54 KG/46 cm, 80 KG/40 cm, 184 KL/915 cm, 217 KL/805 cm, 229 KL/942 cm, and 278 KA/322 + 390 + 412 cm. Within these nodules, and at their surface, we observe more or less extended zones where the Mn oxide is replaced by a whitish to yellowish material. This whitish material in samples 54-1, 80-1, 217-1+2, and 229-1 is composed of amorphous material (a) and smectite (s) (Table 8.6). Its chemical composition is characterized by high concentration of SiO_2 (37.8 to 55.4%), medium high concentration of Fe_2O_3 (5.6 to 17.0%), very low concentration of MnO (0.3 to 9.7%), and a strong enrichment of Cu (up to 18,715 ppm) and Ni (up to 4,442 ppm) (Table 8.7). In general, the same tendency of enrichment and depletion of elements as observed in the whitish material was found in most of the buried nodules (Table 8.5). Sample 261-1+2 showed extremely high values of Cu and Ni (up to 38,086 and 9,531 ppm, respectively). Apparently, therefore, the alteration of buried nodules results in a distinct dissolution and expulsion of manganese, an enrichment of Fe and Si, and a strong concentration of Cu and Ni. The dissolution and alteration of Mn oxide increasingly destroys the internal growth structures, which may finally result in a soft black powder replacing the nodule.

Three nodules occur very deeply buried: 184 KL/1,255 cm, 217 KL/1,623 cm, and 229KL/1,665 cm. These nodules are distinctly less altered than the buried nodules mentioned above. They show only a few inclusions of whitish material, and the chemical composition is similar to that of surface nodules (samples 217-3 and 278-1 in Table 8.7). The mineralogical composition of sample 217-3, however, is todorokite (Table 8.6). These three deeply buried nodules occur in dark-brown sediments of Tertiary age, while the younger buried nodules were found in yellowish-brown

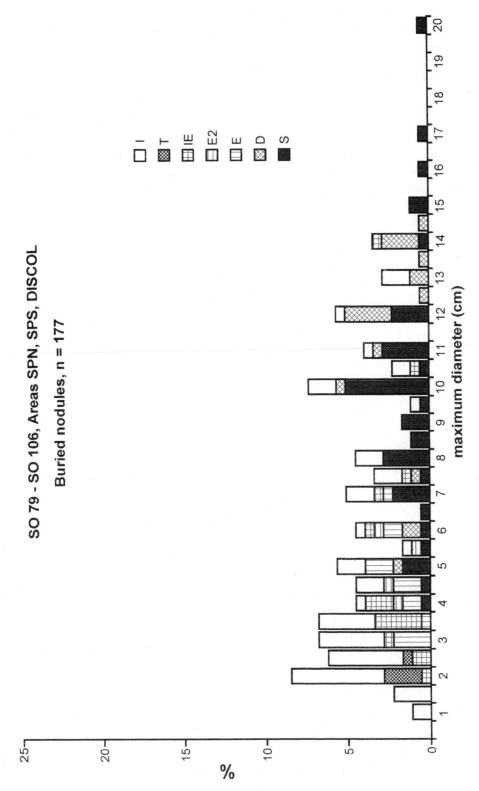

FIGURE 8.22   Size distribution of buried nodules with indication of their shape in areas SPN, SPS, and DISCOL. For explanation of symbols of nodule shape see Figure 8.12.

TABLE 8.6
Mineral Composition of Different Growth Layers of Surface and Buried Nodules

Sample	Growth Structure	XRD	Burial Depth	Sample	Growth Structure	XRD	Burial Depth	Sample	Growth Structure	XRD	Burial Depth
01-1	h	t	0	62-2	d	t,b	0	156-3	d	t	20
01-2	h	b	0	62-3	h	b	0	156-4	d	b	20
10-1	d+b	b,t	0	62-4	d	t	0	217-1	—	s	805
10-3	d	t	0	62-5	d+h	b	0	217-2	—	a	805
33-1	d+h	t	0	63-1	d+b	b	0	217-3	d+h	t	1623
33-2	d+h	t	118	63-2	h	b	0	229-1	—	s	942
33-3	d+h	b,t	424	64-1	d+h	b	0	261-1	d+h	t	200
52-1	d+h	t	0	64-2	d+h	t	170	261-2	d	t	422
53-1	d+h	t	616	64-3	d+h	t	540	278-1	h+d	b,t	0
54-1	—	a,s	46	80-1	—	a,s	40	278-2	d+h	t	106
59-1	h	t	0	106-1	d	b	0	278-3	d+h	t	390
59-2	d	t	0	106-2	b	t	0	278-4	d	t	417
59-3	h	b	0	156-1	d+b	t	4				
59-4	d	b,t	0	156-2	b+d	t	4				

Growth structure: b = laminated to columnar, d = dendritic, h = dense, laminated.

XRD: t = todorokite, b = birnessite, a = amorphous, s = smectite.

Quaternary sediments. Clearly, the more oxic conditions in the older sediments (Figure 8.5) has resulted in less alteration of the buried nodules than in the overlying suboxic sediments.

This conclusion is supported by data on nondestructive color measurements with a Minolta Chromatometer CR-300. A correlation between the red-green color values of sediment cores and the intensity of alteration of buried nodules was found. Nodules from the Tertiary with highest red values show only few compositional changes compared to surface nodules, while buried nodules from Quaternary sediment sections with low red values are strongly altered and nodules from Quaternary sections with higher red values are less altered again.

8.5 DISCUSSION

8.5.1 FACTORS CONTROLLING SURFACE DISTRIBUTION OF NODULES

Nodule abundance (kg/m^2) is mainly determined by two processes, the supply of nuclei and the growth rates of nodules.[19] Nuclei are predominantly supplied from elevated outcrops. In area D, the maximum nodule number (n/m^2) occurs on slopes in water depths of about 4,000 m (Figure 8.9). In shallower depths, the median accretion rate of nodules is much lower than in basins. Therefore, these nodules are small and are not buried as frequently as in basins.

The maximum nodule number coincides with the upper lobe of the abundance maximum. The lower lobe, at about 4,150 m, coincides with the maximum abundance/number ratio (Figure 8.10). This ratio, reflecting the median weight of nodules in a nodule assemblage, is a good indicator of median nodule growth rates. Apparently, the main maximum in nodule abundance in basins is mainly due to maximum growth rates immediately above the CCD. Furthermore, the main maximum of nodule abundance more or less coincides with the maximum Corg concentration in surface sediments (Figure 8.7). Generally, increased Corg concentrations result in increased decay of Corg within the sedimentary column followed by dissolution of Mn oxide, upward Mn flux in the interstitial pore water, and finally reprecipitation at the redox boundary (Figure 8.5). This process leads to rapidly growing diagenetic nodules. Therefore, the observed growth rate maximum near

TABLE 8.7
Chemical Composition of Different Growth Layers of Buried and Surface Nodules

Sample	Burial Depth cm	Growth Structure	SiO_2 (%)	TiO_2 (%)	Al_2O_3 (%)	Fe_2O_3 (%)	MnO (%)	MgO (%)	CaO (%)	Na_2O (%)	K_2O (%)	P_2O_5 (%)	LOI (%)	As ppm	Ba ppm	Co ppm	Cu ppm	Mo ppm	Ni ppm	Sr ppm	V ppm	W ppm	Zn ppm
54-1	46		55.12	0.34	9.13	5.62	1.99	2.47	1.10	4.45	1.83	0.24	15.55	7	9277	46	690	4	773	365	116	10	225
80-1	40		55.43	0.34	8.98	7.17	0.61	2.68	1.22	4.38	1.78	0.24	15.38	4	7685	27	410	<2	332	314	100	8	212
217-1	805		50.80	0.18	4.27	17.05	0.48	4.81	1.24	2.87	1.55	0.13	15.43	7	3101	214	1733	14	883	163	53	4	272
217-2	805		37.84	0.43	6.52	13.75	9.75	3.41	1.11	3.04	1.28	0.23	18.62	74	3309	961	18715	48	4442	463	100	333	678
217-3	1623	d + h	9.05	0.11	1.89	2.67	46.51	3.60	0.96	1.81	0.74	0.14	22.62	57	5335	914	8749	162	4693	492	607	269	729
229-1	942		51.30	0.21	5.25	15.90	0.34	5.09	0.65	3.36	1.66	0.14	15.07	6	3575	121	723	15	632	156	60	4	256
261-1	200	d + h	12.63	0.15	2.91	1.71	39.48	3.23	0.95	1.72	0.86	0.17	24.66	27	3044	553	38086	107	9359	285	363	252	993
261-2	422	d	14.87	0.20	3.61	3.57	35.79	3.59	1.21	1.72	1.01	0.19	25.36	46	3307	773	22251	92	9531	384	227	142	1107
278-1	0	h + d	7.51	0.09	1.98	1.28	45.10	2.10	1.39	2.93	0.60	0.09	23.18	21	2340	38	1366	380	1983	325	350	40	404
278-2	106	d + h	13.47	0.19	3.51	3.71	38.03	3.22	1.13	2.09	0.98	0.18	25.16	29	3975	552	8187	113	5013	503	547	141	678
278-3	390	d + h	16.09	0.20	3.92	4.62	37.59	3.48	1.06	2.15	1.26	0.16	23.80	33	5667	529	6930	123	5952	521	627	116	834
278-4	417	d	22.88	0.33	5.13	8.19	25.39	3.52	1.23	2.42	1.21	0.23	23.90	46	3342	991	9850	58	7309	496	499	51	857

Growth structure: d = dendritic, h = dense, laminated

the CCD is primarily due to the occurrence of a Corg maximum at that depth. However, if we compare Corg content of surface sediments with average values of nodule growth rates, we must keep in mind that during the long growth history of a nodule the chemical environment of the surface sediments has changed repeatedly. During glacials, in the Peru Basin, the accumulation rate of Corg was higher and the position of the CCD presumably was up to 200 m deeper than today.[16]

It is surprising that the maximum nodule abundance occurs in basins where the assemblage of surface nodules loses a great number of large nodules because they become buried. Apparently, the high accretion rate more than compensates for the loss of nodules by burial. Nodule assemblages in the upper lobe of nodule abundance are characterized by B-type size distribution representing mainly hydrogenetic growth, while the diagenetic A-type is typical of the lower lobe near the CCD (Figure 8.9). Cronan and Hodkinson observed a similar maximum of diagenetic nodule growth near the CCD on the Aitutaki–Jarvis transect in the central South Pacific.[41]

8.5.2 Factors Controlling Morphology, Size Distribution, Growth Sequence, Growth Rate, and Nucleus Supply

Due to increased accretion rates (up to 250 mm/Ma) in deeper waters, basin nodules develop a more or less regular shape and D- and S-types predominate (Figure 8.12). Irregularly shaped nodules prevail in shallower waters where, due to up to 50 times lower accretion rates, the irregular shape of nuclei strongly determines the nodule shape. D-type nodules, which show an asymmetric internal growth, develop from S-type nodules. This asymmetric growth is due to increasing immobility as nodule size increases.[41] Polylobate P-nodules occur mainly in shallower waters, and their smooth surface texture enhances a rapid agglutination due to the attachment of sessile foraminifera in the area of contact between two subnodules.[63]

There are nodule assemblages in the Peru Basin with a great number of E2 nodules of similar maximum diameter. At some time in the past, ellipsoidal E nodules must have been fragmented, generating E2 nodules. This process of fragmentation may have been due to shrinkage by aging or to increased movement by bioturbation or bottom currents. Nodules with a smoother surface texture on the bottom side than on the top side are typical of the Peru Basin and indicate extreme diagenetic growth conditions.They are mainly restricted to the northern areas D and SPN and only rarely occur in area SPS (Table 8.1).

The distinct difference in size distribution between A- and B-type nodules is explained mainly by the difference in nodule accretion rates at deeper and shallower water depths, respectively (Figure 8.16). The well-sorted size distribution in B-type nodule histograms reflects the low hydrogenetic accretion rates of irregularly shaped nodule types, e.g., E and I, while the poorly sorted size distribution in A-type nodule histograms with a predominance of regular shaped S- and D-type nodules is due to the generally increased diagenetic accretion rates and especially due to the great difference in accretion rate near the redox boundary and at the sediment surface, respectively. Actually, as soon as a nodule grows up to a size >7 cm, its accretion rate is much higher than that of the small nodules in the same assemblage. This results in a flattening of the size maximum and a shift to greater diameters. The same explanation is true for the difference of size distribution between E- and I-type and S- and D-type nodules, respectively (Figure 8.14).

The well-sorted size distribution in B-type nodule histograms indicates a single event of nodule formation and a uniform process of growth, while the poorly sorted A-type nodule histograms are the result of a great variety of processes of nodule formation and growth. Thus, size distribution is a good indicator of the growth history of the nodules.

If we plot histograms of individual nodule types, we recognize that some of them reveal similar size distributions, indicating similar growth histories, e.g., S and D, or E- and I-types, while that of D- and I-types is very different (Figure 8.14). The similarity of size distribution of a specific nodule type, e.g., IE, within different areas, indicates similar growth history of that type (Figure 8.15).

The concentration of hydrogenetic Mn/Fe-poor b and b-d growth layers near the nodule center and the restricted occurrence of diagenetic Mn/Fe-rich h-layers at a great distance from the nodule center in large nodules clearly indicates that growth type strongly depends on the size of the nodule (Figure 8.19). Very often we find large as well as small nodules within the same grab sample. Both types of nodules show a different type of growth sequence depending on their size. The size determines the degree of immersion and hence the influence of the special environment at the redox boundary upon the diagenetic growth on the bottom surface of the nodule, especially well demonstrated in large asymmetric nodules (Figure 8.20). Biogenic rotation of large nodules is much more difficult and rare than of small nodules, which enhances asymmetric growth in them.

Asymmetric distribution of nodule composition corresponds to asymmetric distribution of accretion rate (see Section 8.4.2.8.).[58] From the radiometric dating by Bollhöfer, two large nodules (62 KD and 63 KG) with distinctly asymmetric growth were selected because a comparability of growth sequence can be expected only for such types of nodules, while the repeated rotation of smaller symmetrically grown nodules will have disturbed the growth sequence.[59] Although these nodules have not been rotated for a long time, they must have been repeatedly lifted by jerky movements to avoid burial. As the geochemical environment within the semiliquid surface layer changes very distinctly from top to bottom, the lower surface of the nodule will be exposed to a quite different environment after a minor push upward by benthic organisms, resulting in the accretion of a growth layer of a different type. Therefore, it is questionable to relate the different growth layers of the nodules to different climatic conditions (Section 8.4.2.8.).

At the redox boundary, where O_2 diffusing from the surface largely disappears and Mn^{2+} from rising pore water is oxidized, MnO_2 precipitates mainly in the zone of overlap (Figure 8.5). Here, optimum diagenetic conditions for nodule growth exist. This is the reason for the rapid growth, especially of the h-growth layers. Only large nodules reach this depth in the sediment and, therefore, the growth rate increases in an exponential way as the size of the nodules increases. Due to the enormous difference in growth rate, up to 250 mm/Ma for h-growth layers and about 5mm/Ma for b-growth layers, diagenetic growth overwhelms hydrogenetic growth.[55,59] Therefore, in large nodules it is difficult to identify hydrogenetic b-growth under the microscope and still more difficult to separate pure b-growth material for analysis.

If we compare the nodule nucleus types in the Peru Basin (SO-79 + SO-106) and the Clarion–Clipperton Zone (SO-25), we recognize a great difference in frequency of occurrence (Table 8.2). The predominance of S-type nuclei (52.9%) in area SO-25 is due to an extended supply of fragments from indurated sediment horizons at hiatuses.[33,64] Most of the N-nuclei (26.6%) of this area are fragments of more or less hydrogenetically densely grown nodules, which do not fragment easily. In the Peru Basin, however, most of the nodules grow diagenetically dendritically, which enhances fragmentation and, therefore, N-nuclei (51.1%) predominate. P-nuclei observed in SO-25 areas were not found in the Peru Basin. Apparently, in the Peru Basin there is no supply of pumice by surface currents from the South American continent, although this basin is much closer to South America than the SO-25 area.

8.5.3 FACTORS CONTROLLING NODULE COMPOSITION

The high bioproduction at the equator is responsible for the transfer of organic matter to the seafloor. Therefore, in the Peru Basin organic matter in surface sediments increases toward the north. Decomposition of organic matter results in a geochemical gradient from south to north.[6] The generally lower enrichment of Cu in nodules from the Peru Basin compared to those from the CCZ has been suggested as being caused by a lower ratio of siliceous to calcareous skeletons due to the position of the seafloor near the CCD.[50] Cronan described nodules rich in Ni and Cu from the western equatorial Pacific Ocean mainly in a transitional area below weak to moderate bioproduction (50 to 100 g C/m^2/a).[65]

For bulk samples of individual nodules the increasing concentration of Ni + Cu with increasing Mn/Fe ratio up to a ratio of 5 is explained by a varying mixture of hydrogenetic and early diagenetic Mn oxide growth.[6,54] To the right of this so-called point of reversal (Figure 8.18), the decreasing Ni + Cu content with increasing Mn/Fe ratios is explained by an enhanced incorporation of Mn^{2+} ions instead of Ni^{2+} and Cu^{2+} in the todorokite lattice. An increased drop of Eh in the peneliquid surface sediments is assumed to be responsible for an intensive remobilization and supply of Mn^{2+}, which eventually results in extreme diagenetic nodule growth.

By separating and analyzing individual growth layers instead of bulk samples, it was intended to better recognize the history of growth conditions of nodules. The distinct clustering of different growth layers in Figure 8.18 supports this assumption. Each type of growth structure represents a special type of growth environment. However, we must realize that even such individual growth layers are always composed of a great number of different microlayers and only reflect an average environment, while the microlayers actually represent the pure microenvironment.[44]

The distinct negative correlation between Mn/Fe values and Er content clearly demonstrates the different growth conditions of nodules and crusts (Figure 8.21). The high concentration of REE in hydrogenetic crusts is due to the long duration of exposure to the sea water of slowly growing Mn oxide, while diagenetic nodules mainly grow within the sediment at a much higher rate.[66,67]

8.5.4 FATE OF BURIED NODULES

Stoffers et al. discuss whether the smooth to microbotryoidal surface texture of buried nodules results from overgrowth or removal of ferromanganese oxide after burial.[34] Overgrowth, however, as was observed by von Stackelberg on buried nodules from the CCZ, was not seen on buried Peru Basin nodules, while a wide range of dissolution was found.[64] The suboxic environment of the Quaternary sediments below the oxic surface layer supports dissolution and alteration of buried nodules, which distinctly increases the Mn flux to the sediment surface, while the oxic environment of the Tertiary sediments prevents compositional changes in the more deeply buried nodules (Tables 8.6 and 8.7). Apparently, intensity of dissolution and alteration of buried nodules is not a function of burial depth or age but of the chemical environment in the host sediment. Suboxic environmental conditions may also be responsible for the disappearance of birnessite in buried nodules because birnessite is more unstable in a suboxic environment than todorokite.[68,69] Therefore, the assumption of Nedjatpoor et al. that aging of buried nodules leads to expulsion of crystal water and enhanced stability of todorokite is questionable, because it requires that buried nodules are older than surface nodules.[62] A surface nodule of 10-cm diameter, however, is about 700 ka old, if we assume a mean accumulation rate of 70 mm/Ma for a mainly diagenetically grown nodule. This nodule was always kept at the surface by biogenic lifting and started to grow about 350 cm below the recent sediment surface, if we calculate with a mean sedimentation rate of 5 mm/ka for the Quaternary sediments of the Peru Basin.

Thus, buried nodules must not be older than surface nodules; they normally represent an early growth stage of surface nodules.[64] Burial may be due to a statistical selection where a few nodules, especially from the prevailing size class, failed to be lifted. This model of burial was observed in the CCZ where low accretion rates prevail and may also be true for slope and ridge positions in the Peru Basin. In deeper waters of the Peru Basin where high accretion rates, especially of deeply immersed large nodules, occur, the larger very often asymmetrically grown nodules always become buried as soon as they are stuck in the stiff suboxic sediment. This is the reason for the increased percentage of large, asymmetrically grown buried nodules compared to surface nodules. A similar observation was made by Huh and Ku for nodules at Site H in the Guatemala Basin.[43] Apparently, therefore, in the Peru Basin two types of burial exist: one on the basis of statistical selection and one by immersion, which may explain the bimodality of the composite histogram of buried nodules (Figure 8.22). The model of burial by immersion does not accord with the assumption of Heath that growth rates of nodules and the probability of their burial are independent of size.[70]

The ratio of buried nodules to surface nodules, in numbers as well as in weight, is much higher in the Peru Basin than in the CCZ. This is true, although in the Peru Basin we observe a more or less strong dissolution of buried nodules within the suboxic Quaternary.

8.5.5   BENTHIC LIFTING VERSUS CLIMATIC OSCILLATION

It is the general opinion that concentric layering of nodules is the result of climatic, oceanographic, or sedimentological variations changing the growth environment. For example, Glasby et al. explained the complex internal structure and banding of nodules by changes in bottom-current erosion during the Pleistocene.[47] Reyss et al. assume that the shift from symmetrical growth to asymmetrical growth in a nodule from the northern Peru Basin 180 Ka ago is due to sedimento-logical and oceanographic changes at that time.[58] However, during detailed studies of sediment cores from the Peru Basin, Weber et al. did not observe any such changes at that time.[16] Detailed study of SO-79 and SO-106 nodules indicates that asymmetric growth mainly starts when nodule size exceeds 7 cm and not so much when oceanographic conditions change. Halbach et al. explained the close interbedding of early diagenetic A1 and A2 substances in nodule microlayers by variation in pH controlled by the release of H^{2+} ions during precipitation reactions.[28,52] This would require a chemical reaction independent of changes in the sedimentary environment. Dymond et al. assume that major pulses of bioproductivity followed by decomposition of organic matter and dissolution of manganese micronodules are responsible for episodic suboxic accretion of nodules from site H of the Guatemala Basin.[44] Therefore, suboxic accretion should not occur today. Finney et al. assumed a nonsteady-state flux of biogenic matter in the Guatemala Basin with an oscillation in the depth of the former redox boundary between 5 and 25 cm with a periodicity of about 100 ka over the past 400 ka.[71] These oscillations must have determined the nodule growth. They certainly also existed in the Peru Basin where a 100 ka cyclicity in biogenic flux indicated by carbonate variations was described by Weber et al.[16] Finally, Bollhöfer published the results of radiometric dating and geochemical analyses of two nodules (62KD and 63KG) from the SPN area.[59] He assumed that variability in nodule growth is determined by oscillations in bioproductivity in the surface water and the O_2 content of the bottom water. Both nodules, located at a distance of 10 km apart in area SPN, show a similar trend of growth characteristics vs. age.

We must now carefully consider whether climatic oscillation or stepwise lifting by benthic organisms is more important in the variability of nodule growth. In Sections 8.5.2. and 8.5.3. it was demonstrated that nodule size strongly determines the accretion rate, the type of internal growth structures, and nodule composition by the different depth of immersion of large and small nodules. Suboxic accretion (h-growth) is not restricted to periods of high bioproductivity as was suggested by Dymond et al.[44]; it also occurs today. During a 100 ka climatic oscillation such as that mentioned above, about 0.5 m sediment would have accumulated, and a number of pushes by benthic organisms would have been needed to keep a nodule at the sediment surface. Apparently, the pulses of benthic lifting are much more frequent than the climate-induced oscillations. Due to the steep geochemical gradient between the redox boundary and the sediment surface, each push upward must have preferentially exposed the lower nodule surface to a different environment, which must have been documented by a change in growth type producing a fine layering. The sawtooth-shaped development of growth layers (see Section 8.4.2.4.) supports this assumption.

There is no question that climate-induced oscillations can also influence the nodule growth, but they have been superimposed on the bioturbation-induced microlayered growth pattern with a much longer cyclicity.

8.5.6   OPEN QUESTIONS

The areas investigated show great differences in depth of the CCD and in the maximum abundance of nodules, as well as in values of mass accumulation rates. This depends mainly on bioproductivity of surface waters, water depth, current direction and intensity, and O_2 concentration of bottom waters. In area D the bottom current is directed to the NW. In area SPN there are weak indications of a southward-directed current.[23,24] This may explain the differences between areas SPN and D and the occurrence of an asymmetric distribution of sedimentation around a seamount in SPN. To better understand the sedimentary regime in the Peru Basin, further bottom-current measurements are needed.

It is believed that stepwise movements of nodules by bioturbation are responsible for the microlayered internal structure of nodules, and climate-induced oscillation for superimposed growth variations. However, we are far from being able to quantify both processes. Further detailed studies are needed.

Dense, laminated h-growth layers indicate special diagenetic growth conditions. We need, however, an explanation of why these h-growth layers of very pure Mn oxide up to 10 mm thick with a smooth surface texture are formed instead of normal dendritic d-growth. Von Stackelberg assumed an open space forming immediately after the nodule is pushed upward by burrowing organisms, which enhances a rapid and undisturbed Mn oxide growth.[64] Similar growth conditions have been described for hydrothermal crusts of densely laminated Mn oxide material by Usui et al. and Hein et al.[72,73] We may consider further that radial crystallization pressure due to relatively high growth rates are responsible for these special growth types. Both assumptions, however, are very tentative and need further investigation.

Clearly, the degree of dissolution and alteration of buried nodules depends mainly on the chemical environment of the host sediments. We need more investigations of both nodules in relation to their host sediment in order to assess the Mn flux which determines nodule growth.

ACKNOWLEDGMENTS

The contributions of the following staff members of BGR are greatly appreciated: H. Schwetje for AAS analyses, H. Rösch for X-ray diffraction analyses, U. Siewers for XRF and ICP-MS analyses, H. Rask and H. Karmann for computerized data processing, line drawing, and photographing nodules, and E. Müller for typing and laying out the manuscript. The financial support of the Bundesministerium für Bildung, Forschung und Technologie is acknowledged.

REFERENCES

1. Agassis, A., Report on the scientific results of the expedition to the tropical Pacific of the steamer ALBATROSS, *Memoir of the Museum of Comparative Zoology*, 26, 1, 1902.
2. Glasby, G. P., Manganese nodules in the South Pacific: a review, *N. Z. Journal of Geology and Geophysics*, 19, 707, 1976.
3. Piper, D. Z. and Williamson, M. E., Composition of Pacific Ocean ferromanganese nodules, *Marine Geology*, 23, 285, 1977.
4. Skornyakova, N. S., Zonal regularities in occurrence, morphology and chemistry of manganese nodules of the Pacific Ocean, in *Marine Geology and Oceanography of the Pacific Manganese Nodule Province*, Bischoff, J. L. and Piper, D. Z. (Eds.), 9, Plenum Press, 699, 1979.
5. Thijssen, T., Glasby, G. P., Schmitz-Wiechowski, A., Friedrich, G., Kunzendorf, H., Mueller, D., and Richter, H., Reconnaissance survey of manganese nodules from the northern sector of the Peru Basin, *Marine Mining*, 2, 382, 1981.
6. Halbach, P., Marchig, V., and Scherhag, C., Regional variations in Mn, Ni, Cu, and Co of ferromanganese nodules from a basin in the Southeast Pacific, *Marine Geology*, 38, M1-M9, 1980.
7. Thijssen, T., Glasby, G. P., Friedrich, G., Stoffers, P., and Sioulas, A., Manganese nodules in the central Peru Basin, *Chemie der Erde*, 44, 1, 1985.
8. Steinkamp, K., Manganknollenforschung im Pazifischen Ozean, Industrielle Explorationsfahrten, *Geologisches Jahrbuch*, D93, 43, 1991.

9. Dubinin, A. V. and Strekopytor, S.V., Geochemistry of rare earth elements in the formation of ferro-manganese nodules in the Peruvian trough of the Pacific Ocean, *Lithology and Mineral Resources*, 29, 4, 322, 1994.

10. von Stackelberg, U., Manganknollenforschung im Pazifischen Ozean, *Geologisches Jahrbuch*, D93, 41, 1991.

11. Thiel, H. and Schriever, G., Deep sea mining, environmental impact and the DISCOL project, *AMBIO*, 19 (5), 245, 1990.

12. Foell, E. J., Thiel, H., and Schriever, G., Discol: A long-term, large-scale, disturbance-recolonization experiment in the abyssal eastern tropical South Pacific Ocean, *Proceedings of the 2nd Annual Offshore Technology Conference*, Houston, Texas, 7-10 May 1990. OTC Paper 6338 (2), 497, 1990.

13. Bluhm, H., Monitoring megabenthic communities in the abyssal manganese nodule sites of the East Pacific Ocean in association with commercial deep-sea mining, *Aquatic Conservation: Marine and Freshwater Ecosystems*, 4, 187, 1994.

14. Thiel, H. and Schriever, G., Environmental consequences of deep-sea mining, *International Challenges*, 13, (1), 54, 1993.

15. Schriever, G., DISCOL-Disturbance and recolonization experiment of a manganese nodule area of the southeastern Pacific, *Proceedings of the ISOPE-Ocean Mining Symposium* (1995), Tsukuba, Japan, 163, 1995.

16. Weber, M. E., Wiedicke, M., Riech, V., and Erlenkeuser, H., Carbonate preservation history in the Peru Basin: paleoceanographic implications, *Paleoceanography*, 10(4), 775, 1995.

17. Wiedicke, M. and Weber, M. E., Small-scale variability of seafloor features in the northern Peru Basin: Results from acoustic survey methods, *Marine Geophysical Research*, 18, 507, 1996.

18. Thiel, H. and Forschungsverbund Tiefsee-Umweltschutz, The German environmental impact research for manganese nodule mining in the SE Pacific Ocean, *Proceedings of the ISOPE-Ocean Mining Symposium* (1995), Tsukuba, Japan, 39, 1995.

19. von Stackelberg, U., Growth history of manganese nodules and crusts of the Peru Basin, in *Manganese Mineralization: Geochemistry and Mineralogy of Terrestrial and Marine Deposits,* Nicholson, K., Hein, J. R., Bühn, B., and Dasgupta, S. (Eds.), *Geological Society Special Publication*, 119, 153, 1997.

20. Schriever, G., Koschinsky, A., and Bluhm, H., Cruise report ATESEP, SONNE cruise 106, Institut für Hydrobiologie und Fischereiwissenschaften, Hamburg, 1996.

21. Wyrtki, K., Equatorial currents in the Pacific 1950–1970 and their relations to the trade winds, *Journal of Physical Oceanography*, 4, 372, 1974.

22. Lonsdale, P., Abyssal circulation of the southeastern Pacific and some geological implications, *Journal of Geophysical Research*, 81, (6), 1163, 1976.

23. Klein, H., Near-bottom currents in the deep Peru Basin, DISCOL Experimental Area, *Deutsche Hydrographische Zeitschrift*, 45, 31, 1993.

24. Klein, H., Current measurements in the Peru Basin, Data Report, Bundesamt für Seeschiffahrt und Hydrographie, Hamburg, 1996.

25. Toggweiler, J. R., Dixon, K., and Broecker, W. S., The Peru upwelling and the ventilation of the south Pacific thermocline, *Journal of Geophysical Research*, 96, C11, 20, 467, 1991.

26. Rea, D. K., Evolution of the East Pacific Rise between 3°S and 13°S since the Middle Miocene, *Geophysical Research Letters*, 5, (7), 561, 1978.

27. Mammerickx, J., Anderson, R. N., Menard, H. W., and Smith, S. M., Morphology and tectonic evolution of the East-Central Pacific, *Geol. Soc. Am. Bull.*, 86, 111, 1980.

28. Halbach, P., Scherhag, C., Hebisch, U., and Marchig, V., Geochemical and mineralogical control of different genetic types of deep-sea nodules from the Pacific Ocean, *Mineral Deposita*, 16, 59, 1981.

29. Erlandson, D., Hussong, D., and Campell, J., Sediments and associated structure of the northern Nazca Plate, in *Nazca Plate: Crustal Formation and Andean Convergence,* Kulm, L., Dymond, J., Dasch, E. J., Hussong, D., and Roderick, R. (Eds.), *Geol. Soc. Am. Mem.*, 154, 295, 1981.

30. Meylan, M., Field description and classification of manganese nodules, in *Ferromanganese Deposits on the Ocean Floor*, Hawaii Institute of Geophysics, Techn. Report, 74-9, 158, 1974.

31. Moritani, T., Maruyama, S., Nohara, M., Matsumoto, K., Ogitsu, T., and Moriwaki, H., Cruise report, *Description, Classification and Distribution of Manganese Nodules,* Geol. Surv. Japan, 8, 136, 1977.

32. Usui, A., Variability of manganese nodule deposits: the Wake to Tahiti transect, *Cruise Report, Geol. Surv. Japan*, 18, 138, 1982.

33. Von Stackelberg, U. and Marchig, V., Manganese nodules from the equatorial North Pacific Ocean, *Geologisches Jahrbuch*, D87, 123, 1987.

34. Stoffers, P., Sioulas, A., Glasby, G. P., and Thijssen, T., Geochemical and sedimentological studies of a box core from the western sector of the Peru Basin, *Marine Geology*, 49, 225, 1982.

35. Stoffers, P., Sioulas, A., Glasby, G. P., Schmitz, W., and Mangini, A., Sediments and micronodules in the northern and central Peru Basin, *Geologische Rundschau*, 73, 1055, 1984.

36. Stoffers, P., Glasby, G. P., and Frenzel, G., Comparison of the characteristics of manganese micron-odules from the equatorial and south-west Pacific, *TMPM Tschermaks Min. Petr. Mitt.*, 33, 1, 1984.

37. Riech, V., Marchig, V., Weber, M. E., Wiedicke, M., and Cepek, P., A six Ma sedimentary record in the Peru Basin: geochemical indications for a significant redox shift at the end of the Pliocene, *M Geolog. Rundschau*, in press.

38. Marchig, V. and Reyss, J. L., Diagenetic mobilisation of manganese in Peru Basin sediments, *Geochimica et Cosmochimica Acta*, 48, 1349, 1984.

39. Mangini, A., Stoffers, P., and Botz, R., Periodic events of bottom transport of Peru Basin sediment during the late Quaternary, *Marine Geology*, 76, 325, 1987.

40. Weber, M., von Stackelberg, U., Marchig, V., Wiedicke, M., and Grupe B., Variability of surface sediments in the Peru Basin: Dependence on water, productivity, bottom water flow, and seafloor topography, *Deep Sea Research*, in press.

41. Cronan, D. S. and Hodkinson, R. A., Element supply to surface manganese nodules along the Aitutaki-Jarvis Transect, South Pacific, *Journal of the Geological Society*, London, 151, 39, 1994.

42. König, I., unpublished data, 1997.

43. Huh, Ch. A. and Ku, T. L., Radiochemical observations on manganese nodules from three sedimentary environments in the north Pacific, *Geochimica et Cosmochimica Acta*, 48, 951, 1984.

44. Dymond, J., Lyle, M., Finney, B., Piper, D. Z., Murphy, K., Conrad, R., and Pisias, N., Ferromanganese nodules from the MANOP Sites H, S, and R-Control of mineralogical and chemical composition by multiple accretionary processes, *Geochimica et Cosmochimica Acta*, 48, 931, 1984.

45. Sorem, R. K. and Fewkes, R. H., *Manganese Nodules*, Plenum, New York, 1979, 1.

46. Finney, B., Heath, G. R., and Lyle, M., Growth rates of manganese-rich nodules at MANOP Site H (Eastern North Pacific), *Geochimica et Cosmochimica Acta*, 48, 911, 1984.

47. Glasby, G. P., Stoffers, P., Sioulas, A., Thijssen, T., and Friedrich, G., Manganese nodule formation in the Pacific Ocean: a general theory, *Geo-Marine Letters*, 2, 47, 1982.

48. Marchig, V., von Stackelberg, U., Wiedicke, M., Durn, G., and Milovanovic, D., Hydrothermal activity associated with off-axis volcanism in the Peru Basin, *Marine Geology*, 159, 179, 1999.

49. Thiel, H., Schriever, G., Bussau, C., and Borowski, C., Manganese nodule crevice fauna, *Deep Sea Research*, 40, No. 2, 419, 1993.

50. Halbach, P., Hebisch, U., and Scherag, Ch., Geochemical variations of ferromanganese nodules and crusts from different provinces of the Pacific Ocean and their genetic control, *Chemical Geology*, 34, 3, 1981.

51. Glasby, G. P. and Thijssen, T., The nature and composition of the acid-insoluble residue and hydroly-sate fraction of manganese nodules from selected areas in the Equatorial and S. W. Pacific, *TMPM Tschermaks, Min. Petr. Mitt.*, 30, 205, 1982.

52. Halbach, H., Giovanoli, R., and von Borstel, D., Geochemical processes controlling the relationship between Co, Mn, and Fe early diagenetic deep-sea nodules, *Earth and Planetary Science Lett.*, 60, 226, 1982.

53. Heye, D. and Marchig, V., Relationship between the growth rate of manganese nodules from the Central Pacific and their chemical constitution, *Marine Geology*, 23, M19, 1977.

54. Halbach, P. and Puteanus, D., Geochemical trends of different genetic types of nodules and crusts, in *The Manganese Nodule Belt of the Pacific Ocean*, Halbach, P., Friedrich, G., and von Stackelberg, U. (Eds.), Ferdinand Enke, 1988, 61.

55. Reyss, J. L., Marchig, V., and Ku, T., Rapid growth of a deep-sea manganese nodule: effect of diagenetic source of manganese, *Nature*, 295, 101, 1982.

56. Lemaitre, N., Reyss, J. L., and Marchig, V., Géologie marine — Différences de composition chimique entre les faces opposées d'un nodule de manganèse orienté du Bassin du Pérou, *Comptes-Rendus*, *Acad. Sc. Paris*, 298, Serie II, no. 9, 407, 1984.

57. Lemaitre, N., *Les nodules polymetalliques du basin du Peru*, These, Université de Paris-Sud, Centre d'Orsay, 1987.
58. Reyss, J. L., Lemaitre, N., Ku, T. L., Marchig, V., Southon, J. R., Nelson, D. E., and Vogel, J. S., Growth of a manganese nodule from Peru Basin: a radiochemical anatomy, *Geochimica et Cosmochimica Acta*, 49, 2401, 1985.
59. Bollhöfer, A., Uranreihen-Datierung diagenetischer Manganknollen mittels Thermionen-Massenspektrometrie (TIMS): klimainduzierte Wachstumsschwankungen im Spätquartär, Dissertation, Universität Heidelberg, 133 pp., 1996.
60. Bollhöfer, A., Eisenhauer, A., Frank, N., Pech, D., and Mangini, A., Thorium and uranium isotopes in a manganese nodule from the Peru basin determined by alpha spectrometry and thermal ionization mass spectrometry (TIMS): Are manganese supply and growth related to climate? *Geologische Rundschau*, 85, 577, 1996.
61. Fenner, J., unpublished data, 1995.
62. Nedjatpoor, M., Stoffers, P., and Glasby, G. P., Influence of ageing effects on manganese nodule mineralogy, *Neues Jahrbuch für Mineralogie*, Mh., 5, 204, 1985.
63. Von Stackelberg, U., Significance of benthic organisms for the growth and movement of manganese nodules, Equatorial North Pacific, *Geo-Marine Letters*, 4, 37, 1984.
64. Von Stackelberg, U., Pumice and buried manganese nodules from the equatorial North Pacific Ocean, *Geologisches Jahrbuch*, D 87, 229, 1987.
65. Cronan, D. S., Controls on the nature and distribution of manganese nodules in the western equatorial Pacific Ocean, in Marine Minerals, Teleki, P. G., Dobson, M. R., Moore, J. R., and von Stackelberg, U. (Eds.), *Nato ASI Series*, 177, 1987.
66. Glasby, G. P., Mechanisms of enrichment of the rarer elements in marine manganese nodules, *Marine Chemistry*, 1, 105, 1973.
67. Murphy, K. and Dymond, J., Rare earth element fluxes and geochemical budget in the eastern equatorial Pacific, *Nature*, 307, 444, 1984.
68. Glasby, G. P., The mineralogy of manganese nodules from a range of marine environments, *Marine Geology*, 13, 57, 1972.
69. Cronan, D. S., Manganese nodules and other ferromanganese oxide deposits from the Atlantic Ocean, *Journal of Geophysical Research*, 80, 3831, 1975.
70. Heath, G. R., Burial rates, growth rates, and size distribution of deep-sea manganese nodules, *Science*, 205, 903, 1979.
71. Finney, B., Lyle, M. W., and Heath, G. R., Sedimentation at MANOP Site H (Eastern Equatorial Pacific) over the past 400,000 years: climatically induced redox variations and their effects on transition metal cycling, *Paleoceanography*, 3, 2, 169, 1988.
72. Usui, A., Yuasa, M., Yokota, S., Nohara, M., Nishimura, A., and Murakami, F., Submarine hydrothermal manganese deposits from the Ogasawara (Bonin) arc, off the Japan Island, *Marine Geology*, 73, 311, 1986.
73. Hein, J. R., Schulz, M. S., and Kang, J.-K., Insular and submarine ferromanganese mineralization of the Tonga-Lau region, *Marine Mining*, 9, 305, 1990.
74. Lyle, M., Composition maps of surface sediments of the eastern tropical Pacific Ocean, *Proceedings of the Ocean Drilling Program, Initial Reports*, 138, 101, 1992.

9 Cobalt-Rich Ferromanganese Crusts in the Pacific

James R. Hein, Andrea Koschinsky, Michael Bau, Frank T. Manheim, Jung-Keuk Kang, and Leanne Roberts

ABSTRACT

Co-rich Fe–Mn crusts occur throughout the Pacific on seamounts, ridges, and plateaus where currents have kept the rocks swept clean of sediments at least intermittently for millions of years. Crusts precipitate out of cold ambient sea water onto hard-rock substrates forming pavements up to 250 mm thick. Crusts are important as a potential resource for Co, Ni, Pt, Mn, Tl, Te, and other metals, as well as for the paleoclimate signals stored in their stratigraphic layers. Crusts form at water depths of about 400 to 4000 m, with the thickest and most Co-rich crusts occurring at depths of about 800 to 2500 m, which may vary on a regional scale. Gravity processes, sediment cover, submerged and emergent reefs, and currents control the distribution and thickness of crusts on seamounts. Crusts occur on a variety of substrate rocks that generally decrease in the order, breccia, basalt, phosphorite, limestone, hyaloclastite, and mudstone. Because of this wide variety of substrate types, crusts are difficult to distinguish from the substrate using remotely sensed data, such as geophysical measurements, but are generally weaker and lighter-weight than the substrate. Crusts can be distinguished from the substrates, however, by their much higher gamma radiation levels. The mean dry bulk density of crusts is 1.3 g/cm^3, the mean porosity is 60%, and the mean surface area is extremely high, 300 m^2/g. Crusts generally grow at rates of 1 to 10 mm/Ma. Crust surfaces are botryoidal, which may be modified to a variety of forms by current erosion. In cross-section, crusts are generally layered, with individual layers displaying massive, botryoidal, laminated, columnar, or mottled textures. Characteristic layering is persistent regionally in the Pacific. Crusts are composed of ferruginous vernadite (δ-MnO$_2$) and X-ray amorphous Fe oxyhydroxide, with moderate amounts of carbonate fluorapatite (CFA) in thick crusts and minor amounts of quartz and feldspar in most crusts. Elements most commonly associated with the vernadite phase include Mn, Co, Ni, Cd, and Mo, whereas those most commonly associated with Fe oxyhydroxide are Fe and As. Detrital phases are represented by Si, Al, K, Ti, Cr, Mg, Fe, and Na; the CFA phase by Ca, P, Sr, Y, and CO$_2$; and a residual biogenic phase by Ba, Sr, Ce, Cu, V, Ca, and Mg. Crusts contain Co contents up to about 2.3%, Ni to 1%, and Pt to 3 ppm, with mean Fe/Mn ratios of 0.6 to 1.3. Fe/Mn decreases, whereas Co, Ni, Ti, and Pt increase in central Pacific crusts and Fe/Mn, Si, and Al increase in continental margin crusts and in crusts with proximity to west Pacific volcanic arcs. Vernadite and CFA-related elements decrease, whereas Fe, Cu, and detrital-related elements increase with increasing water depth of crust occurrence. Cobalt, Ce, Tl, and maybe also Ti, Pb, and Pt are strongly concentrated in crusts over other metals because of oxidation reactions. Total rare earth elements (REEs) commonly vary between 0.1% and 0.3% and are derived from sea water along with other hydrogenetic elements, Co, Mn, Ni, etc. Platinum, Rh, Ir, and some Ru in crusts are also derived from sea water, whereas Pd and the remainder of the Ru derive from detrital minerals. The older parts of thick crusts were phosphatized during at least two global phosphogenic events during the Tertiary, which mobilized and redistributed elements in those parts of the crusts.

0-8493-8429-X/00/$0.00+$.50
© 2000 by CRC Press LLC

Silicon, Fe, Al, Th, Ti, Co, Mn, Pb, and U are commonly depleted, whereas Ni, Cu, Zn, Y, REEs, Sr, and Pt are commonly enriched in phosphatized layers compared to younger nonphosphatized layers. The dominant controls on the concentration of elements in crusts include the concentration of metals in sea water and their ratios, colloid surface charge, types of complexing agents, surface area, and growth rates. Crusts act as closed systems with regard to the isotopic ratios of Be, Nd, Pb, Hf, Os, and U-series, which in part have been used to date crusts and in part used as isotopic tracers of paleoceanographic and paleoclimatic conditions. Those tracers are especially useful in delineating temporal changes in deep-ocean circulation. Research and development on the technology of mining crusts are only in their infancy. Detailed maps of crust deposits and a better understanding of small-scale seamount topography are required to design the most appropriate mining equipment.

9.1 INTRODUCTION

Iron–manganese oxyhydroxide crusts (Figure 9.1), hereafter called Fe–Mn crusts, are ubiquitous on hard-rock substrates throughout the ocean basins. They form at the seafloor on the flanks and summits of seamounts, ridges, plateaus, and abyssal hills where the rocks have been swept clean of sediments at least intermittently for millions of years. Crusts form pavements up to 250 mm thick on rock outcrops, or coat talus debris. Fe–Mn crusts form by hydrogenetic precipitation from cold ambient bottom waters (Figure 9.1A, C, D), or by a combination of hydrogenetic and hydrothermal precipitation in areas of hydrothermal venting (Figure 9.1B), such as near oceanic spreading axes, volcanic arcs, and hotspot volcanoes. Fe–Mn crusts contain subequal amounts of Fe and Mn and are especially enriched in Mn, Co, Pb, Te, Bi, and Pt relative to their lithospheric and sea water concentrations (Table 9.1). There are two primary practical interests in Fe–Mn crusts, the first being their economic potential for Co, but also for Mn, Ni, and Pt, and possibly also Ti, rare earth elements (REEs), Te, Tl, P, and others. The second interest is the use of crusts as recorders of the past 60 Ma of oceanic and climatic history.

9.1.1 Classification

Up until the late 1970s Fe–Mn crusts were usually not distinguished from Fe–Mn nodules. If a distinction was made, crusts were called seamount nodules. However, there are distinct differences between Fe–Mn nodules and crusts, other than just form (Table 9.2). Nodules nucleate on small bits of rock, bone, or old nodule fragments on the surface of sediments. Nodules commonly form by both diagenetic and hydrogenetic processes and thus their composition reflects input from both sea water and sediment pore water sources, the latter being a substrate contribution, which is not found in crusts.[1] However, some nodules form predominantly by diagenetic or hydrogenetic processes. Nodules have sometimes been referred to as hydrogenous, regardless of their origin; consequently, we use the term hydrogenetic to avoid any confusion about a substrate contribution to crusts. Generally, crusts and nodules have different mineralogical (vernadite vs. todorokite and vernadite) and chemical compositions (for example, high Co vs. high Cu) because of their genetic differences as well as differences in water depths of formation, although there is much overlap.

Iron and manganese hydroxides and oxyhydroxides may also form by hydrothermal processes below the seafloor (Table 9.2). These deposits usually consist of stratabound layers or manganese-cemented volcaniclastic sediments and are distinctly different in texture and composition from hydrogenetic Fe–Mn crusts, especially with regard to the extreme fractionation of Mn and Fe (Table 9.1) and low multiple-trace metal contents in the hydrothermal deposits.[2] In the hydrothermal deposits, Fe/Mn varies from about 24,000 (up to 58% Fe) for hydrothermal ironstones to 0.001 (up to 52% Mn) for hydrothermal stratabound Mn oxides (means in Table 9.1[2]). Many other minor types of iron and manganese mineralization occur in the ocean basins (Table 9.2), but the present work is confined to a discussion of Co-rich hydrogenetic Fe–Mn crusts (Figure 9.1A, C, D).

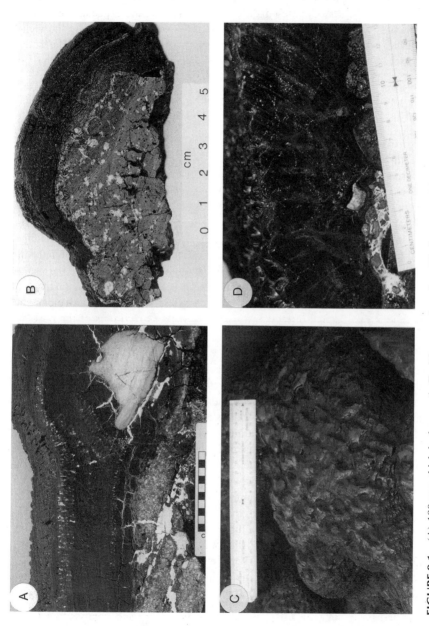

FIGURE 9.1 (A) 180-mm-thick hydrogenetic Fe–Mn crust from Lomilik Seamount, Marshall Islands, which grew on a phosphatized hyaloclastite substrate and contains fractures filled with CFA; distinct growth layers can be seen, some of which are very porous, with pores and vugs filled with either carbonate sediment or CFA. (B) Mixed hydrothermal–hydrogenetic Fe–Mn crust from Gorda Ridge, northeast Pacific; the crust shows distinct growth layers and Fe-rich laminae near the base; the substrate is amygdaloidal basalt. (C) Current polished and fluted surface of a botryoidal Fe–Mn crust from the Marshall Islands, water depth 2,090 m. (D) Same crust as in (C) showing 30- to 50-mm-tall columns in a 40- to 50-mm-thick crust; note that the columns are separated by detrital-rich material (pale gray). Scales are in centimeters.

TABLE 9.1
Contents of Mn, Fe, Co, Ni, and Pt (wt. %) in Marine Fe, Mn, and Fe–Mn Deposits Compared to Contents in and Enrichments Over Sea Water and the Earth's Crust

	Sea Water	Lithosphere	Fe–Mn Crusts	Stratabound Mn	Seamount Ironstones
Fe/Mn	1.2	57	0.7	0.04	5100
Mn	5.0×10^{-9}	0.095	26	47	0.01
Mn/sea water	—	1.9×10^{7}	5.2×10^{9}	9.4×10^{9}	2.0×10^{6}
Mn/lithosphere	—	—	274	495	—
Fe	6.0×10^{-9}	5.4	19	1.8	51
Fe/sea water	—	9.0×10^{8}	3.2×10^{9}	3.0×10^{8}	8.5×10^{9}
Fe/lithosphere	—	—	3.5	—	9.4
Co	1.0×10^{-10}	2.5×10^{-3}	0.70	0.017	$<2.3 \times 10^{-4}$
Co/sea water	—	2.5×10^{6}	7.0×10^{8}	1.7×10^{7}	$<2.3 \times 10^{5}$
Co/lithosphere	—	—	280	6.8	—
Ni	5.0×10^{-8}	8.0×10^{-3}	0.48	0.10	0.024
Ni/sea water	—	1.6×10^{5}	9.6×10^{6}	2.0×10^{6}	4.8×10^{5}
Ni/lithosphere	—	—	60	12.5	3.0
Pt	2.4×10^{-11}	4.0×10^{-7}	5×10^{-5}	3.0×10^{-6}	1.1×10^{-6}
Pt/sea water	—	1.7×10^{4}	2.1×10^{6}	1.3×10^{5}	4.6×10^{4}
Pt/lithosphere	—	—	125	7.5	2.8

9.1.2 DISTRIBUTION

Fe–Mn crusts have been recovered from seamounts in the Pacific as far north as the Aleutian Trench and as far south as the Pacific sector of the Circum–Antarctic Ridge. However, the most detailed studies have been on crusts from the equatorial Pacific, mostly from the Exclusive Economic Zones (EEZ) of the Federated States of Micronesia, Marshall Islands, Kiribati, and the U.S. (Hawaii, Johnston Island, and California), and from international waters in the Mid-Pacific Mountains. Compared to the estimated 50,000 or so seamounts that occur in the Pacific, the Atlantic and Indian oceans contain fewer seamounts, and most crusts there are associated with the spreading ridges. Crusts associated with those spreading ridges usually have a hydrothermal component that may be large near active venting, but which regionally is generally a small (<30%) component of the crusts formed along most of the ridges.[3] Those types of hydrogenetic–hydrothermal crusts are also common along the active volcanic arcs in the west Pacific,[4,5] the spreading ridges in back-arc basins of the west and southwest Pacific, spreading centers in the south and east Pacific, and active hotspots in the central (Hawaii) and south (Pitcairn) Pacific. Very few (<15) of the approximately 50,000 seamounts in the Pacific have been mapped and sampled in detail, and none of the larger ones, which are comparable in size to continental mountain ranges (Figure 9.2).

Fe–Mn crusts occur at water depths of about 400 to 4000 m, but most commonly occur at depths from about 1000 to 3000 m. The most Co-rich crusts occur at water depths from 800 to 2200 m, which mostly encompasses the oxygen minimum zone (OMZ). The thickest crusts occur at water depths of 1500 to 2500 m, which corresponds to the depths of the outer summit area and upper flanks of most Pacific Cretaceous seamounts. The water depths of thick high-Co-content crusts vary regionally and are generally shallower in the South Pacific where the OMZ is less well developed; there, the maximum Co contents and thickest crusts occur at about 1000 to 1500 m.[6]

TABLE 9.2
Classification of Marine Ferromanganese Oxide Deposits by Form, Processes of Formation, and Environment of Formation; Most Common Types in Bold

	Hydrogenetic	Hydrothermal	Diagenetic	Hydrogenetic and Hydrothermal	Hydrogenetic and Diagenetic	Replacement
Nodules	Abyssal plains, oceanic plateaus, seamounts[1]	Submerged calderas and fracture zones	Abyssal plains, oceanic plateaus	Submerged calderas	**Abyssal plains, oceanic plateaus**[1]	All areas (nodule nuclei)
Crusts	**Midplate volcanic edifices**[2]	**Active spreading axes, volcanic arcs, fracture zones, midplate edifices**	—	**Active volcanic arcs, spreading axes, off axis seamounts, fracture zones**	Abyssal hills	Midplate edifices (crust substrate rock)
Sediment-hosted stratabound layers and lenses	—	**Active volcanic arcs, large midplate volcanic edifices, sediment-covered spreading axes**	Continental margins[3]	—	—	Continental margins, volcanic arcs, midplate edifices
Cements	Midplate volcanic edifices[4]	**Active volcanic arcs, large midplate volcanic edifices**[5]	Midplate volcanic edifices[4]	—	Midplate volcanic edifices[4]	Volcanic arcs, midplate edifices
Mounds and chimneys	—	Back-arc basins, spreading centers, volcanic arcs	—	—	—	—

[1]Less common on ridges, continental slope and shelf; [2]Includes seamounts, guyots, ridges, plateaus; [3]Fe and Mn carbonate lenses and concretions; [4]Mostly fracture and vein fill, cement for volcanic breccia; [5]Mostly cement for breccia, sandstone, and siltstone.

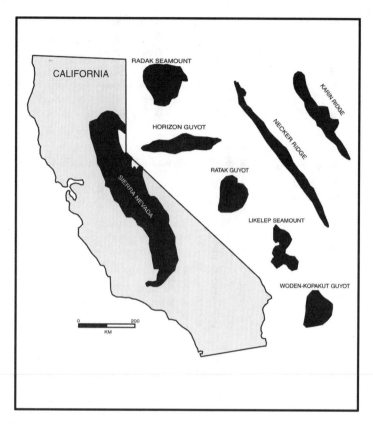

FIGURE 9.2   Comparison of the size of typical central Pacific seamounts and ridges sampled for this study compared to the size of the Sierra Nevada of California. Necker Ridge is in the Hawaii EEZ; Karin Ridge and Horizon Guyot are in the Johnston Island EEZ; and the others are in the Marshall Islands EEZ.

Crusts become thinner with increasing water depth because of mass movements and reworking of the deposits on the seamount flanks. Most crusts on the middle and lower seamount flanks consist of encrusted talus rather than encrusted rock outcrop, the latter however typically has thicker crusts.[7] Many seamounts and ridges are capped by pelagic sediments and therefore do not support the growth of crusts on the summit. Other volcanic edifices are capped by limestone (drowned reefs), which commonly supports thinner crusts than nearby volcanic and volcaniclastic rocks[7-10] because of the younger age of the limestones and therefore shorter time for crust growth coupled with the instability and mass wasting of the limestone. Crusts are usually thin down to as much as 3000 m water depth on the submarine flanks of islands and atolls because of the large amounts of debris that are shed down the flanks by gravity processes.[11] Reworked crust fragments occur as clasts in breccia, which is one of the most common rock types on seamount flanks.[7] Regional mean crust thicknesses for most of the Pacific fall between 20 and 40 mm. Only rarely are very thick crusts (>80 mm) found, for which initial growth may approach within 10 to 30 Ma the age of the Cretaceous substrate rock. Clearly, while most Pacific seamounts are 65 to 95 Ma old, most crusts collected on those seamounts represent less than 25 Ma of growth because of reworking and episodic sediment cover.

The distribution of crusts on individual seamounts is poorly known. Seamounts generally have either a rugged summit with moderately thick to no sediment cover (0 to 150 m) or a flat summit (guyot) with thick to no sediment cover (0 to 500 m). The outer summit margin and the flanks may be terraced with shallowly dipping terraces headed by steep slopes meters to tens of meters high. Talus piles commonly accumulate at the base of the steep slopes and at the foot of the seamounts;

thin sediment layers may blanket the terraces, alternately covering and exhuming Fe–Mn crusts. Other seamount flanks may be uniformly steep up to 20°, but most seamount flanks average about 14°.[7,12] The thickest crusts occur on summit outer-rim terraces and on broad saddles on the summits. Estimates of sediment cover on various seamounts range from 15% to 75%, but sediment cover probably averages about 50%. Crusts are commonly covered by a thin blanket of sediments in the summit region and on flank terraces. It is not known how much sediment can accumulate before crusts stop growing. Crusts have been recovered from under up to 2 m of sediment without apparent dissolution.[13-16] Based on coring results, Yamazaki[17] estimated that there are 2 to 5 times more crust deposits on seamounts than estimates based on exposed crust outcrops because of their coverage by a thin blanket of sediment. Those thinly veiled crusts would be within reach of mining operations.

9.1.3 HISTORICAL PERSPECTIVE

Ferromanganese crusts are members of a family of Fe–Mn precipitates and accretionary deposits that form to greater or lesser extents wherever oxygen-containing aqueous solutions occur. Emmanuel Swedenborg's[18] great monograph on iron (*De Ferro*) in 1734 was the first, and in some respects a surprisingly modern, summary of the origin of Fe–Mn concretions. For example, his account of lake and bog ores, then important raw materials for iron production, correctly attributed their formation to precipitation from solutions enriched in Fe and Mn by leaching of soils and organic matter. The next major advance occurred through the discovery of deep-ocean nodules and crusts by the Challenger Expedition of 1873–6.[19,20] The Challenger's dredge hauls yielded not only classical black Fe–Mn nodules from abyssal depths (4500 to 6000 m), but also recovered a variety of coatings, layers, and crusts of Fe–Mn oxide material from lesser depths, some as shallow as 370 m. The chemist on board, J.Y. Buchanan, and co-workers demonstrated that those samples contained significant minor amounts of Cu, Ni, and Co,[21] as well as the then newly discovered elements such as Th and Tl, which would not be further investigated for another 60 years.

After World War II, E.D. Goldberg, G. Arrhenius, and co-workers at Scripps Institution of Oceanography studied Fe–Mn nodules and established the concept of metal scavenging on active catalytic surfaces of Mn oxides on the seafloor.[22,23] Most nodules studied were from abyssal depths and had 0.30% Co or less. A few samples, from ocean ridge areas, had as little as 0.03% Co, whereas samples from Cape Johnson and Sylvania seamounts had Co contents as high as 0.70%.

Interest in the economic potential of abyssal Fe–Mn nodules, ignited by John Mero in the early 1960s,[24,25] developed into massive international activities by many consortia that peaked in the middle 1970s. Mero[26] and Russian investigations led by P.L. Bezrukov[27,28] noted that topographic highs in the central Pacific had deposits with the highest mean Co contents (1.2%) of any region in the world's oceans. Based on existing data and new analyses performed as part of a Ph.D. dissertation at Imperial College, University of London, Cronan[29,30] drew attention to an inverse correlation between water depth and Co content of nodules recovered from seamounts. The depth relationship for Co in nodules is apparently not universal, however, as can be seen in statistical compilations of Fe–Mn nodule data in the Scripps Institution of Oceanography Nodule Data Bank.[31] However, some apparent exceptions to that relationship, for example those noted around the Hawaiian Islands by Frank et al.[8] and Craig et al.,[32] were later shown with additional data to confirm the Co-water depth relationship.[33]

Fe–Mn crusts had begun to be distinguished from nodules in the 1970s,[8,13,34,35] and their economic potential was recognized.[32,36] The first systematic investigations of Co-rich crust areas were carried out in the Line Islands south of Hawaii by the 1981 German Midpac I cruise on the *R.V. Sonne*, headed by Peter Halbach of the Technical University, Clausthal-Zellerfeld. That cruise made breakthrough discoveries using large dredges coupled with seismic profiling and bottom photography.[12] Subsequent German, U.S. Geological Survey, and University of Hawaii cruises to submarine edifices in the central Pacific refined our knowledge of crust distributions and chemical relationships. They showed that crusts were especially enriched in Co, Fe, Ce, Ti, P, Pb, As, and especially Pt, but relatively lower in Mn, Ni, Cu, and Zn than nodules.[33,37-40]

International complications partly linked to the location of abyssal nodules in international waters contributed to the withdrawal of commercial consortia from active preparations for nodule mining in the late 1970s. The extensive occurrences of Fe–Mn crusts within the EEZ of coastal nations (200 nautical miles) provided an incentive for mineral interests within areas of national jurisdiction.[41] This applied not only to Pacific seamount deposits, but also to Fe–Mn crusts on continental margin terraces that have been kept free of detrital sedimentation by currents.[42,43]

Other systematic Fe–Mn minerals investigations during the 1970s and 1980s included a series of cruises by the Japanese Geological Survey, directed mainly toward abyssal nodule deposits [e.g., Reference 44], crust studies centered in the Japanese EEZ (Izu-Bonin arc; see Usui and Someya[5] for compilation of data and references), and in international waters of the Mid-Pacific Mountains. Other Asian nations like Korea and China initiated Fe–Mn crust research with the U.S. and other cooperators. [e.g., References 45, 46] A detailed database and review of the chemical composition of crusts based on data up to 1987 was published in 1989,[47] and is also available on NOAA and MMS Marine Minerals CD-ROM, NGDC Data Set #0827, and web site http://www.ngdc.noaa.gov/mgg/geology/mmdb/.

9.2 FE–MN CRUST CHARACTERISTICS

9.2.1 Textures and Physical Properties

Crust surfaces exposed directly to the seafloor are botryoidal, with botryoids varying in size from microbotryoidal (millimeter size) to botryoidal (centimeter size). Fresh growth surfaces are characterized by a fractal distribution of botryoids with extreme surface area. Under conditions of high current flow, the botryoids are modified, either by smoothing or by accentuation of the relief by erosion around the margins of the botryoids, in places producing mushroom-shaped forms. With strong unidirectional flow, the botryoids become polished and fluted (Figure 9.1C). Crusts on the sides of rocks are commonly more protected from current activity and grow at a slower rate, thereby acquiring very high Co contents; textures of those side crusts are very porous and granular.

Crust profiles vary according to thickness and regional oceanographic conditions. Thin crusts (<40 mm) are usually black and massive, botryoidal, or laminated. Thicker crusts (40 to 80 mm) commonly have at least two distinct layers, a lower black, massive, dense layer that is phosphatized and an upper black-to-brown layer that is more porous, with laminated, mottled, botryoidal, and/or columnar textures. The thickest crusts (>80 mm) may have up to eight distinct macroscopic layers, the lower several layers of which may be phosphatized. Thick crusts in the equatorial Pacific typically have four to five macroscopic layers that are consistent over large regions: (1) the uppermost layer is black, dense, and has a botryoidal, columnar, or laminated texture; (2) the next layer is Fe-stained brown, very porous to vuggy with vugs elongate perpendicular to layers; vugs are filled with infiltered carbonate sediment and the texture is mottled or columnar; (3) black, dense to sparsely vuggy, with columnar, mottled, laminated, and/or botryoidal textures; (4) Fe-stained brown along extensive network of microfractures, with columnar, botryoidal, or laminated texture and impregnated with CFA; and (5) black, dense, laminated, mottled, botryoidal, and columnar textures; impregnated by CFA and cut by CFA-filled veins.

In polished thin sections, textures consist of alternating laminated, columnar, botryoidal, and mottled textures. Columns range in height from millimeters to 50 mm (Figure 9.1D[48]), with detrital grains separating columns and actually promoting columnar growth.[49] Mottled layers are the most porous and detritus-rich. The various textures probably reflect bottom water conditions at the time of precipitation of the oxides. Mottled, columnar, botryoidal, and laminated textures probably represent progressively decreasing energy in the depositional environment.[50]

An important consideration in the exploration and exploitation of potential crust resources is the contrast in physical properties between crusts and substrate rocks. Those comparisons are complicated by the fact that crusts grow on a wide variety of substrates, including breccia, basalt, phosphorite, limestone, hyaloclastite, and mudstone, in that order of abundance. Phosphorite and fresh basalt are strong, competent rocks and contrast significantly with crusts, which are weak, light-weight, and porous (Table 9.3); the other rock types, including altered basalt, may not contrast much in physical properties with Fe–Mn crusts. In general, crusts are much more porous (mean 60%) than most substrate rocks and have an extreme amount of specific surface area, which averages about 300 m^2/g (Table 9.3), similar to the surface area of silica gel. Interestingly, the surface area decreases by up to 20% when measured one month after collection of the crust and up to 40% after two months.[51] This clearly shows that many physical properties that are measured a long time after collection of the crusts may not closely approximate *in situ* crust properties. The mean wet bulk density of crusts is 1.90 g/cm^3, and the mean dry bulk density is 1.30 g/cm^3. The P-wave velocity of crusts may be less or more than that of sedimentary substrate rocks, but is generally less than that of basalt. This variable contrast will make it difficult to develop sonic devices for measuring *in situ* crust thicknesses. The most distinctive property of Fe–Mn crusts is their gamma radiation level, which averages 475 net counts/min in contrast to sedimentary rock substrates[101] and basalt substrates (Reference 146; Table 9.3). Gamma radiation may be a useful tool for crust exploration under thin sediment cover and for measuring crust thicknesses *in situ*.

9.2.2 MINERALOGY

The mineralogy of bulk crusts is relatively simple compared to hydrothermal and diagenetic Fe and Mn deposits. The dominant crystalline phase is Fe-rich δ-MnO$_2$ (ferruginous vernadite[14,52,53]) with generally two X-ray reflections at about 1.4 Å and 2.4 Å that vary widely in sharpness as the result of crystallite size and Mn content. δ-MnO$_2$ generally makes up more than 95% of the X-ray crystalline phases, the remainder being detrital minerals such as quartz, plagioclase, K-feldspar, pyroxene, phillipsite, and authigenic carbonate fluorapatite (CFA; Table 9.4). The older parts of thick crusts are phosphatized and may contain up to 30% CFA in that part of the crust, but CFA is generally less than 10% of thick bulk crusts. Another major phase in crusts is X-ray amorphous Fe oxyhydroxide (δ-FeOOH, feroxyhyte[52]), which is commonly epitaxially intergrown with δ-MnO$_2$.[54] It is not clear whether or not small amounts of Mn may substitute for Fe in the FeOOH. In about 6% of 640 samples analyzed, the feroxyhyte crystallizes as goethite in the older parts of thick crusts. In Pacific crusts, the quartz and part of the plagioclase are eolian, whereas the remainder of the plagioclase and the other volcanogenic minerals derive from local outcrops.

Todorokite, which is common in diagenetic Fe–Mn nodules and hydrothermal Mn deposits, is rare in hydrogenetic crusts. Of 640 XRD analyses done by the USGS on Pacific crusts, only 5% (2% if offshore California samples are excluded) contain todorokite; 30% of crust samples from offshore California contain todorokite, which, because of very high biological productivity, may reflect the lower oxidation potential of sea water there compared to Pacific sites farther to the west. CFA occurs in 28% of the crust samples analyzed, but not in any crusts from the east (offshore California) or far north Pacific; if those crusts are excluded, 34% of central and west Pacific samples contain CFA.

9.2.3 AGES AND GROWTH RATES OF FE–MN CRUSTS

Hydrogenetic Fe–Mn crusts grow at incredibly slow rates of <1 to about 10 mm/Ma, with the most common rates being from 1 to 6 mm/Ma (Table 9.5; Figure 9.3). These slow growth rates allow for the adsorption of large quantities of trace metals by the oxyhydroxides at the crust surface. Growth rates were first measured using either U-series or Be radioisotopes (which give reliable ages for the outermost 2 and 20 mm of crusts, respectively) or by radiometric or paleontological

TABLE 9.3
Physical Properties of Fe–Mn Crusts and Substrate Rocks

Physical Properties	n	Fe–Mn Crusts	n	Sed. Rock Substrate	n	Basalt Substrate	Ref.
Porosity (volume%)							
Range	13	52–66	8	18–47	2	15–37	7
Mean	13	61	8	39	2	26	7
	—	41–74	—	7–69	—	7–67	118,119
	—	55	—	35	—	—	120
Wet bulk density (g/cm³)							
Range	13	1.83–2.04	8	2.04–2.57	2	2.22–2.62	7
Mean	13	1.92	8	2.18	2	2.42	7
Range	18	1.90–2.44	7	1.59–2.68	23	2.06–2.66	121
Mean	18	2.00	7	1.90	23	2.34	121
Dry bulk density (g/cm³)							
Range	13	1.18–1.48	8	1.56–2.38	2	1.84–2.46	7
Mean	13	1.29	8	1.78	2	2.15	7
	—	1.31	—	—	—	—	120
	—	1.04–2.17	—	1.44–2.92	—	0.78–2.74	118,119
Grain density (g/cm³)							
Range	13	3.03–3.87	—	—	—	—	7
Mean	13	3.48	—	—	—	—	7
	—	2.70–3.44	—	2.54–2.81	—	2.32–2.95	119
Specific surface area (m²/g)							
Range	15	250–381	—	—	—	—	51
Mean	15	323	—	—	—	—	51
Compressive strength (MPa)	—	8.36	—	3.66–32.6	—	—	120
	—	0.5–25.0	36	0.1–52.3	—	0.37–71.0	118,119
Sea water saturated	—	5.39–8.92	—	1.71–12.4	—	165–219	120
Tensile strength (MPa)	—	1.75	—	4.5	—	—	120
	—	0.1–2.3	—	0.1–4.5	—	0.1–18.9	118,119
Sea water saturated	—	0.45	—	0.23–0.48	—	11.9	120
Cohesive strength (MPa)	—	2.9	—	7.8	—	—	120
Sea water saturated	—	1.5	—	0.4–2.3	—	26.4	120
Shear strength (MPa)	—	1.26–2.5	—	—	—	—	119,122
Angle of internal friction	—	42°	—	52°	—	—	120
Sea water saturated	—	76°	—	61°–77°	—	76°	120
Young's modulus (GPa)	—	2.15	—	0.31–10.1	—	—	120
Sea water saturated	—	3.11–4.25	—	0.62–4.76	—	51.3–63.7	120
P-wave velocity (km/s)	—	2.09–3.39	—	1.76–5.86	—	—	118,119
Range, parallel to bedding	18	2.46–4.19	5	1.80–4.35	23	3.14–5.14	121
Mean, parallel to bedding	18	3.36	5	2.57	23	4.13	121
Range, perpendicular to layers	17	2.07–3.86	4	1.78–4.34	23	3.02–4.93	121
Mean, perpendicular to layers	17	3.16	4	2.56	23	3.99	121
	—	2.26	—	1.01–3.45	—	—	120
Sea water saturated	—	2.72–2.78	—	2.07–2.87	—	5.76–5.80	120
S-wave velocity-saturated (km/s)	—	1.35–1.83	—	1.15–1.67	—	3.46–3.57	120
Gamma radiation (net counts/min)							
Range	18	271–800	6	52–137	23	21–366	121
Mean	18	475	6	101	23	146	

dash means not reported

TABLE 9.4
Mineralogy of Pacific Fe–Mn Crusts

Dominant	Common	Less Common	Uncommon or Uncertain
δ-MnO_2 (Fe-vernadite), Fe oxyhydroxide (feroxyhyte)	CFA, quartz, plagioclase, smectite	Phillipsite, goethite, todorokite, calcite, K-feldspar, pyroxene, opal-A, barite, amphibole, magnetite, amorphous aluminosilicates	Halite, illite, clinoptilolite, lepidocrocite, manjiroite, manganite, palygorskite, chlorite, dolomite, stevensite, kutnohorite, mordenite, natrojarosite, hematite, manganosite, maghemite, lithiophorite, analcite

dating of the substrate rock (and assuming that the substrate age is equivalent to the age of the base of the crust). With both methods, the growth rates and ages of the crusts are extrapolated and do not take into account changes in growth rates, growth hiatuses, or in the later method, the time between formation of the substrate rock and the beginning of growth of the crusts. For equatorial Pacific Cretaceous seamounts, the age of substrate rocks and crusts can vary by as much as 60 Ma, although in rare circumstances the ages of very thick crusts may approach those of the Cretaceous substrates. Subsequent to those initial dating techniques, $^{10}Be/^{9}Be$ ratios were used to date crusts, and that ratio along with ^{10}Be decay are the most reliable and widely used techniques today. However, using Be isotope techniques requires that the age of the base of crusts thicker than about 20 mm be determined by extrapolation using the growth rate(s) determined for the outer 20 mm.

Ratios of $^{87}Sr/^{86}Sr$ were also used to date crusts by comparing the ratios in the various crust layers with the ratios that define the Cenozoic sea water curve.[55] That technique (and the other isotopic techniques) assumes that there has been no post-depositional exchange of the isotopes between the crust and sea water. Ingram et al.[56] showed that in fact there is exchange of Sr with sea water in some of the most porous, detritus-rich, and CFA-impregnated crust layers. They provided criteria to choose the most appropriate layers for dating, but, for unknown reasons, using those criteria still did not produce reliable ages for all the selected crust layers. Consequently, that technique has been mostly abandoned because Sr is apparently far more mobile in crusts than are Be, U, and Th. Ratios of $^{187}Os/^{186}Os$ may provide a reliable dating tool for crusts as old as 65 Ma by comparing the ratios in various crust layers with ratios that define the Cenozoic sea water curve.[57] However, additional data are required on Os isotopes in the oceans before that technique can be applied to age-date crusts. Nannofossil biostratigraphy has been used to date crusts from impressions and molds of nannofossils left in crust layers after replacement of the carbonate by Fe–Mn oxyhydroxides.[58,59] That technique, although reliable, is time consuming to perform and consequently has not been widely used. Finally, empirical equations have been developed to date marine Fe–Mn oxide deposits by Scott et al.,[60] Lyle,[61] Manheim and Lane-Bostwick,[62] and Puteanus and Halbach,[63] the latter two of which are most commonly used to date hydrogenetic crusts. The Co-chronometer of Manheim and Lane-Bostwick[62] was derived from and applied to both Fe–Mn crusts and pelagic sediments, whereas the method of Puteanus and Halbach[63] was derived from central Pacific crusts having Co contents >0.24%. Those equations usually give minimum ages for the base of crusts and produce growth rates that are generally faster than those determined from isotopic techniques (compare Tables 9.5 and 9.6), although the Manheim and Lane-Bostwick[62] equation does generally produce rates more in line with those determined by isotopic methods. It is clear that additional techniques are needed to accurately date thick crusts, and the best opportunity may be development of Os isotope stratigraphy, although $^{40}Ar/^{39}Ar$, K–Ar, and paleomagnetic reversal stratigraphy should also be looked into. A significant problem with thick crusts is that the inner layers were phosphatized by a diagenetic process that promoted the mobilization of many elements.[53,64] However, the remobilization of elements apparently did not affect Nd and Pb isotopic ratios[65] and also may not have affected Os isotopic ratios.

TABLE 9.5

Isotope-Determined Growth Rates of Hydrogenetic Fe–Mn Crusts

Location	Water Depth (m)	Crust Thickness (mm)	Dated Interval (mm)	Technique	Growth Rates (mm/Ma)	Extrap. Age (Ma)	Ref.
Horizon Guyot	1790	55	3–51	$^{87}Sr/^{86}Sr$	3.4–1.6–5.2–1.8–8.8	18	50,55
Horizon Guyot	2400	3	0–2.2	U-series	2.2, 2.5	1	123
Horizon Guyot	1800	10	0–1.9	U-series	1.1, 2.5, 4.6	4	123
Necker Ridge	2100	51	12–49	$^{87}Sr/^{86}Sr$	5.4–0.7–2.5	23	56
Necker Ridge	2100	25	0–1.4	U-series	2.7, 2.0, 3.1	13	123
Necker Ridge	2350	25	0–0.9	U-series	0.8, 2.0, 3.1	13	123
Marshall Is.	2900	57	1–57	$^{87}Sr/^{86}Sr$	1.5–5.0	22	56
Marshall Is.	1863	15	0–1.2	U-series	6.6:6.8, 7.4:7.8	2	66
Marshall Is.	1800	160	0–21	$^{10}Be/^{9}Be$	1.4–2.7	62	80
Central Pacific	2300	115	0–21	$^{10}Be/^{9}Be$	2.1	55	80
Central Pacific	1278	30	0–27	^{10}Be; $^{10}Be/^{9}Be$	3.0; 2.7	10	124
Central Pacific	1809	25	0–18	^{10}Be; $^{10}Be/^{9}Be$	3.2; 3.8	7–8	124
Central Pacific	3196	30	0–28	^{10}Be; $^{10}Be/^{9}Be$	4.0; 3.0	8–10	124
Central Pacific	3280	—	0–1	U-series	2.7	—	70
Central Pacific	1240	—	0–1	U-series	1.2	—	70
Central Pacific	1190	—	0–1	U-series	2.7	—	70
Central Pacific	2860	—	0–1	U-series	2.6	—	70
Central Pacific	1120	—	0–1	U-series	0.8	—	70
Central Pacific	2100	—	0–1	U-series	2.7	—	70
Central Pacific	1510	27	0–15	^{10}Be	2.1	13	125
Central Pacific	1240	12	0–12	^{10}Be	2.5–3.7	4	125
Central Pacific	2100	70	0–8	^{10}Be	1.1	64	125
Central Pacific	1550	50	0–45	^{10}Be	4.3–1.4–4.4	18	125
Same crust	1550	50	0–1	U-series	3.4	15	126
Central Pacific	5211	—	0–10	^{10}Be	1.3	—	125
Central Pacific	2425	95	0–11	^{10}Be	1.1	86	89
Central Pacific	1550	55	0–45	^{10}Be	4.3–1.4–4.4	19	63
NW Pacific	1634	40	0–12	^{10}Be	2.0	20	124
NW Pacific	2345	11	0–11	^{10}Be; $^{10}Be/^{9}Be$	8.0; 8.0	1	124
N Equat. Pacific	≈4000	45	0–45	^{10}Be	2–19	8	94
NE Pacific	3840	50	0–17	^{10}Be; $^{10}Be/^{9}Be$	1.6; 1.5	31–33	127
NE Pacific	4830	250	0–1.5; 0–35; 0–35	^{230}Th; ^{10}Be; $^{10}Be/^{9}Be$	6.4–1.8; 2.7–4.8; 2.3–3.8	55–65	111
Same crust	4830	250	0–1.4	U-series	6.6–6.1–5.8	40	126
E Equat. Pacific	3500	30	0–1.9	U-series	0.8, 4.6, 2.7	11	123
E Equat. Pacific	3710	—	0–1	U-series	4.3, 9.6	—	67
Far S Pacific	3700	20	0–2.7	U-series	16	1	34
SE Pacific	3800	11	0–2.9	U-series	9	1	34
Cen. S Pacific	4700	3	0–3.0	U-series	5	<1	34
Cen. S Pacific	4020	25	0–24; 0–13	^{10}Be; ^{26}Al	2.8; 2.3	9–11	128
SW Pacific	1120	20	0–16	^{10}Be; $^{10}Be/^{9}Be$	1.0; 1.0	20	124
SW Pacific	5150	—	0–10	^{10}Be	1.9	—	125
South China Sea	1000	3.5	0–3.5	U-series; ^{10}Be	1.5; 1.6	2.3	129
NE Atlantic	—	38	0–38	^{10}Be	3.0	12	130
NE Atlantic	1500	38	0–38	^{10}Be	4.5	9	130
Equat. E Atlantic	1800	50	0–23	^{10}Be; $^{10}Be/^{9}Be$	1.4; 1.3	36–38	127
Equat. E Atlantic	990	1	0–0.4	U-series	1.0	1	131
Equat. E Atlantic	2800	15	0–0.2	U-series	2.0	8	131
Equat. E Atlantic	1800	20	0–0.1	U-series	1.4	14	131

TABLE 9.5
Isotope-Determined Growth Rates of Hydrogenetic Fe–Mn Crusts

Location	Water Depth (m)	Crust Thickness (mm)	Dated Interval (mm)	Technique	Growth Rates (mm/Ma)	Extrap. Age (Ma)	Ref.
NW Atlantic	1850	120	0–12	$^{10}Be/^9Be$	1.6	75	132
NW Atlantic	2665	91	0–22	$^{10}Be/^9Be$	2.4	38	132
SW Atlantic	2394	50	0–41	^{10}Be	1.4	36	124
SW Atlantic	4388	20	0–15	^{10}Be; $^{10}Be/^9Be$	1.8; 1.3	11–15	124
SE Atlantic	3040	4	0–0.6	U-series	1.2	3	131
Central Indian	5385	30	0–1.3	U-series	2.3	13	133
Central Indian	5250	72	0–1.7	U-series	2.8	26	133
Same crust	5250	72	0–33	$^{10}Be/^9Be$	2.8	26	132
W Central Indian	5438	35	0–1.2	U-series	2.9:3.8, 7.2:8.4	4–12	67
Same crust	5438	35	0–17	$^{10}Be/^9Be$	1.6	22	132
SW Indian	4325	8	0–0.3	U-Series	1.4	6	131
SW Indian	4052	18	0–18	^{10}Be; $^{10}Be/^9Be$	1.9; 1.4	9–13	124
Underside crust	4052	32	0–32	^{10}Be; $^{10}Be/^9Be$	5.2; 4.0	6–8	124

A dash separating growth rates indicates changes with depth in the crust starting from the surface layer; a semicolon separating growth rates indicates rates determined by the different techniques listed; growth rates separated by a comma indicate rates determined by different U-series isotopes; growth rates separated by a colon indicate rates from different profiles in the same crust; underlined values are preferred by indicated reference; dash means data not available; Extrap. means extrapolated age of the beginning of crust growth

Various U-series- and Be isotope-generated growth rates of the same crusts may or may not produce consistent results (Table 9.5). For example, in Table 9.5 the U-series- and Be isotope ratio-produced growth rates are identical for the same central Indian and South China Sea crusts, but quite different for the west-central Indian and the northeast Pacific (4830 m) crusts. Those differences may reflect real differences in growth rates between the outer millimeter and outer centimeters of the crusts. Different isotopes within the U-series commonly do not produce identical growth rates, for example, note the east equatorial Pacific sample (3500 m), and may even produce different rates for different profiles within a single crust (see west-central Indian and Marshall Islands, 1863 m, samples in Table 9.5). Chabaux et al.[66,67] discussed the efficacy of the various U-series isotopes and provided criteria to establish whether the U-series isotopes display open- or closed-system behavior. Beryllium isotope ratios and ^{10}Be decay usually produce similar, but not identical, growth rates (Table 9.5), and the isotopic ratio may be preferable (see von Blanckenburg et al.[68] for discussion). The distribution of isotope-determined growth rates does not change with isotopic technique used or with geographic location (Figure 9.3). No correlation exists in this compiled dataset between growth rates of the outermost layer of the crusts and water depth of occurrence.

9.2.4 CHEMICAL COMPOSITION

All USGS chemical data in this chapter (Table 9.7) are normalized to 0% H_2O^- because hygroscopic water varies markedly depending on analytical conditions. Hygroscopic water can vary up to 30 wt.% and thereby affects the contents of all other elements. Compositions normalized to 0% H_2O^- can be more meaningfully compared and also more closely represent the grade of the potential ore. Unfortunately, water contents are not provided in many published reports, so we were unable to correct compiled data listed in Table 9.7. Mean chemical compositions are provided for crusts that occur in the areas marked on Figure 9.4, which correspond to the different columns in Table 9.7.

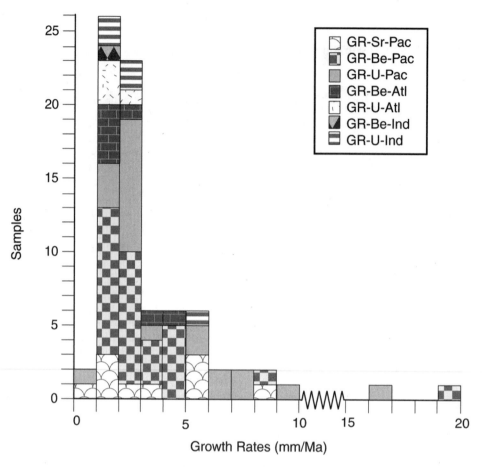

FIGURE 9.3 Range of isotopically determined growth rates of hydrogenetic Fe–Mn crusts. A hydrothermal component may exist in the two crusts with the fastest rates; GR = growth rates as determined by Sr, Be, and U isotopes for the Pacific (Pac), Atlantic (Atl), and Indian (Ind) oceans; sources of data are in Table 9.5

TABLE 9.6
Growth Rates (mm/Ma) Determined from
Empirical Equations

Location	n	Equation	Range	Mean	Ref
Micronesia	24	MLB	0.9–5.8	2.7	87
Micronesia	24	PH	3.7–130	21.7	87
Micronesia	42	PH	2.1–9.8	5.0	83
Marshall Islands	61	PH	1.5–16.0	4.7	46
Surface scrapes	17	PH	—	1.7	46
Johnston I. EEZ	97	PH	1.6–10.0	3.8	81
Johnston I. EEZ	12	PH	3.0–32.0	12.0	82
Surface scrapes	14	PH	3.0–13.0	8.0	82
S of Australia	12	MLB	0.5–23.6	8.0	134

MLB = equation of Manheim and Lane-Bostwick[63]; PH = equation of Puteanus and Halbach[62]

TABLE 9.7

Mean Chemical Composition of Fe–Mn Crusts from Various Parts of the Pacific Compared to Indian and Atlantic Ocean Crusts and C–C and Other Abyssal Fe–Mn Nodules

	FSM-Palau n = 35	Marshall Is. n = 77	NW of Marshall Is. n = 46	Johnston I. n = 99	Calif. Borderland n = 60	NW Pacific n = 1478	Hawaii n = 203	Far North Pacific n = 5	South Pacific 0–25° Lat n = 330	Far South Pacific >25° Lat n = 26	Atlantic n = 21	Indian n = 9	C–C Zone Nodules n = x	Other Abyssal Nodules n = x
Fe/Mn	1.02	0.64	0.76	0.72	1.27	0.68	0.77	1.10	0.69	1.00	1.80	1.50	0.27	0.69
Fe wt. %	21.7	16.9	16.7	19.3	21.4	15.1	16.9	19.6	15.8(249)	17.6	22.9	20.2	6.9	12.7
Mn	21.3	26.3	22.1	26.8	16.8	22.1	22.0	17.8	22.8	17.6	12.7	13.5	25.4	18.5
Si	5.7	3.0	5.4	4.6	11.4	3.7(273)	5.24(79)	5.0	3.5(230)	7.3(7)	5.2	6.8	7.6	8.8
Na	1.9	1.9	1.7	1.9	2.0	1.6(175)	1.72(79)	2.0	0.97(18)	0.85(17)	1.3	1.8	2.8	2.1
Al	1.4	0.70	1.2	0.88	2.1	1.0(328)	1.31(150)	0.7	1.0(247)	3.3(7)	2.2	2.7	2.9	3.0
K	0.57	0.53	0.65	0.61	0.14	0.56(177)	0.65(79)	0.5	0.59(99)	0.32(17)	0.5	0.9	1.0	0.93
Mg	1.2	1.2	1.0	1.2	1.2	1.3(342)	1.21(104)	1.1	1.2(199)	1.3(7)	1.6	1.2	1.7	1.4
Ca	2.9	6.6	3.2	3.4	1.9	4.1(374)	2.5(104)	2.5	3.5(247)	2.3(15)	4.0	2.3	1.7	1.8
Ti	1.1	1.1	1.1	1.3	0.59	0.77(370)	1.2(168)	0.8	1.0(240)	0.79(7)	1.0	0.9	0.53	0.78
P	0.51	1.8	0.63	0.66	0.42	1.2(332)	0.44(76)	0.5	0.83(226)	0.49(7)	0.5	0.3	0.10A	0.10A
S	0.1(2)	0.24(4)	—	0.2(2)	0.1(4)	0.3	0.144(1)	0.2(2)	0.2(285)	—	0.2(10)	0.1(2)	—	—
Cl	0.9(2)	0.94(4)	—	1.1(2)	1.0(4)	—	0.96(1)	1.0(2)	1.2(285)	—	1.1(10)	1.3(3)	—	—
H_2O^+	10.3(32)	8.2(73)	8.6	9.3(97)	10.4(56)	—	6.0(40)	—	—	—	—	—	7.5A	7.5A
H_2O^-	0.0	0.0	0.0	0.0	0.0	—	14.6(1)	0.0	19.8(3)	—	0.0	0.0	—	—
CO_2	0.75(32)	1.2(73)	0.59	0.69(97)	0.42(56)	—	0.5(50)	—	—	—	—	—	0.16A	0.16A
LOI	36.1(3)	41.2(4)	—	38.0(2)	30.4(4)	—	16.4(29)	33.6	36.8(3)	14.9(7)	30.4	30.1	—	—
Ni ppm	3884	5716	3992	4896	3255	5403(1427)	3960	3085	5368	3585	2599	3025	12800	6300
Cu	1023	1057	1369	1185	747	1075(1418)	677	292	1072	713	774	1518	10200	4200
Co	4451	7888	5177	8054	3006	6372(1440)	7774	4323	7534	4485	3515	3346	2400	2400
Zn	751	864	652	771	686	680(335)	684(108)	528	668(57)	777	587	512	1400	900
Ba	1747	2121	1978	2049	4085	1695(266)	1785(104)	1819	1670(130)	1190(10)	1656	1494	2800	2000
Mo	451	569	422	532	379	455(172)	386(76)	737	374(18)	—	435	334	520	360
Sr	1691	1848	1538	1796	1372	1212(166)	945(51)	1646	995(18)	—	1371	1066	450	700
Ce	871(32)	1137(73)	1260	1306(97)	1167(56)	1105(48)	1684(50)	—	1079(35)	696(10)	—	—	530A	530A

TABLE 9.7 (continued)
Mean Chemical Composition of Fe–Mn Crusts from Various Parts of the Pacific Compared to Indian and Atlantic Ocean Crusts and C–C and Other Abyssal Fe–Mn Nodules

	FSM-Palau n = 35	Marshall Is. n = 77	NW of Marshall Is. n = 46	Johnston I. n = 99	Calif. Borderland n = 60	NW Pacific n = 1478	Hawaii n = 203	Far North Pacific n = 5	South Pacific 0–25° Lat n = 330	Far South Pacific >25° Lat n = 26	Atlantic n = 21	Indian n = 9	C–C Zone Nodules n = x	Other Abyssal Nodules n = x
Y ppm	208	267	215	228	175	166(104)	185(51)	204	131(1)	—	177	154	133A	133A
V	732	724	645	712	616	515(163)	648(104)	748	687(182)	—	855	573	470	480
Pb	1474	1799	1689	1871	1322	1777(307)	1957(109)	1659	741(166)	1654	1133	1027	450	820
Cr	33	15(73)	15.3	9.2(98)	33	22.4(67)	180(16)	17.1(2)	44.2(18)	24.35(6)	26.3(19)	33.9(7)	27	25
Cd	2.8	4.1	3.1	3.5	3.4	4.6(63)	5.8(79)	4.1(2)	3.9(18)	—	3.6(13)	4.9(4)	12	11
As	279	244	237	289	273	165(106)	1141(76)	245	179(313)	889(10)	312	158	159A	159A
Li	4.0(3)	3.6(4)	—	3.5(2)	15.8(4)	63(2)	—	5.2(4)	2.3(285)	—	27.5	16.0(7)	160A	160A
Be	6.3(3)	5.8(4)	—	7.0(2)	3.7(4)	—	3(1)	4.8	3.8(18)	—	8.3	5.7	4A	4A
B	234(2)	129(4)	—	208(2)	202(4)	115(37)	141(1)	194(2)	148(1)	—	243(10)	221(3)	273A	273A
Sc	10.1(3)	9.2(4)	—	9.4(2)	9.6(4)	6.4(64)	5(1)	10.3	13.4(29)	7.91(10)	19.0	14.4	10A	10A
Ga	10.8(2)	11.2(4)	—	12.4(2)	16.9(4)	—	9(1)	13.1(2)	9.0(1)	—	12.7(11)	14.5(5)	11A	11A
Se	—	—	—	—	—	0.40(23)	—	—	—	—	—	—	52A	52A
Br	32.0(2)	32.0(4)	—	35.9(2)	38.3(4)	—	43(1)	30.5(2)	10.5(11)	31.75(10)	36.5(10)	52.4(3)	0.05A	0.05A
Ag	<4	<0.2	—	<0.2	<0.2	0.717(32)	<0.2	<4	<0.2	<0.2	<4	<4	0.10A	0.10A
In	<0.5	<0.5	—	<0.5	<0.5	—	<0.5	<0.5	<0.5	<0.5	0.6(1)	0.7(1)	—	—
Sb	43.3(2)	45.5(4)	—	45.3(2)	34.7(4)	24.4(47)	39(1)	50.5(2)	32.2(28)	31.93(10)	62.7(10)	39.5(3)	37A	37A
Hf	7.3(2)	8.4(4)	—	10.5(2)	9.5(4)	7.4(6)	4.9(1)	6.6(2)	10.1(29)	8.55(7)	14.8(10)	53.8(3)	6A	6A
W	102(2)	101(4)	—	82.8(2)	58.1(4)	93.3(44)	150(1)	136(2)	107(1)	56.44(10)	81.7(10)	82.0(3)	76A	76A
Te	20.1(2)	43.6(4)	—	20.8(2)	6.6(4)	—	36(1)	11.2(2)	40(1)	—	27.4(9)	22.2(2)	216A	216A
Tl	125(2)	144(4)	—	130(2)	57.6(4)	—	186(1)	119(2)	189(1)	—	81.2(10)	107(3)	169A	169A
Bi	23.7(2)	53.2(4)	—	41.4(2)	6.2(4)	—	46.3(1)	12.0(2)	31.8(1)	—	14.6(10)	23.9(3)	21A	21A
Th	11.4(2)	10.2(4)	—	17.7(2)	27.9(4)	33.0(12)	9.7(1)	37.0	2.1(1)	21.2(10)	54.2	60.2(6)	28A	28A
U	12.5(2)	13.3(4)	—	12.7(2)	8.2(4)	9.6(2)	9.7(1)	16.4(2)	10.2(1)	9.53(8)	11.0(10)	10.3(3)	6.8A	6.8A
Rb	14.0(1)	—	—	16.5(2)	20.3(3)	—	14(1)	—	—	—	14.0(3)	—	15A	15A
Zr	649(2)	597(4)	—	802(2)	662(4)	172(76)	369(1)	611(2)	475(1)	—	701(10)	722(3)	350	620
Nb	49.1(3)	62.8(4)	—	61.2(2)	41.5(4)	—	47(1)	33.7(4)	52(1)	—	54.4	48.6	74A	74A

														0.013A	0.013A
F	—	—	—	—	—	0.34(8)	—	—	—	—	—	—	—	0.013A	0.013A
Cs	<2	<2	—	<2	<2	—	<2	<2	<2	1.8(3)	<2	<2		157A	157A
La	117(17)	237(25)	343(12)	275(14)	251(16)	202(66)	—	—	200(35)	185(9)	—	—		36A	36A
Pr	24.9(17)	37.9(25)	60.6(12)	52.5(14)	56.4(15)	106(11)	—	—	—	—	—	—		158A	158A
Nd	108(17)	170(25)	238(12)	210(14)	289(16)	162(25)	—	—	226(27)	148(10)	—	—		35A	35A
Sm	23.3(17)	30.6(25)	47.0(12)	43.6(14)	50.0(15)	41.6(26)	—	—	53.5(28)	23.1(10)	—	—		9A	9A
Eu	6(17)	7.6(25)	10.7(12)	11.3(14)	12.3(16)	9.90(26)	—	—	9.3(35)	72.2(10)	—	—		32A	32A
Gd	26(17)	34.4(25)	49.2(12)	49.3(14)	48.4(15)	26(13)	—	—	—	—	—	—		5.4A	5.4A
Tb	4.2(17)	5.3(25)	7.9(12)	7.8(14)	7.8(15)	7.53(26)	—	—	6.2(28)	5.99(10)	—	—		31A	31A
Dy	20.1(17)	34.1(21)	47.2(12)	47.8(14)	45.0(15)	57.8(11)	—	—	—	—	—	—		4A	4A
Ho	4.7(17)	7.4(25)	9.7(12)	10.0(14)	8.8(16)	6.6(19)	—	—	—	—	—	—		18A	18A
Er	13.1(17)	21.4(25)	27.0(12)	27.8(14)	23.1(15)	31.9(11)	—	—	—	—	—	—		2.3A	2.3A
Tm	1.9(17)	3.1(25)	3.9(12)	4.5(14)	3.5(15)	4.3(19)	—	—	—	—	—	—		20A	20A
Yb	12.3(17)	20.1(25)	23.7(12)	27.0(14)	22.7(16)	17.7(27)	—	—	19.8(35)	19.3(10)	—	—		1.8A	1.8A
Lu	0.22(17)	3.8(4)	—	3.8(14)	3.0(15)	3.3(26)	—	—	3.3(35)	2.9(10)	—	—		0.75A	0.75A
Hg ppb	10.3(2)	11.0(1)	—	59.4(1)	360(4)	—	87(1)	56.6(2)	30.3(1)	—	154(10)	53.8(3)		0.15A	0.15A
Pt	239(32)	634	501(19)	244	82	777(113)	174(25)	—	—	—	—	—		97A	97A
Pd	1.7(9)	2.6(13)	<5.4(19)	—	—	—	2.5(23)	—	—	—	—	—		6.2A	6.2A
Rh	11.6(10)	21.6(20)	23.2(19)	—	—	—	14.9(25)	—	—	—	—	—		—	—
Ru	15.9(10)	14.0(18)	21.3(19)	—	—	—	—	—	—	—	—	—		—	—
Ir	4.1(10)	5.9(18)	7.8(19)	—	—	—	—	—	—	—	—	—		9.1A	9.1A

Pacific areas outlined in Figure 9.4; dash means not analyzed; NW Pacific data from Usui and Someya[5]; Hawaii data from De Carlo et al.,[33] Chave et al.,[100] Hein et al.,[38] Craig et al.,[32] Frank et al.[8]; South Pacific data from Walter et al.,[135] Grau and Kudrass,[91] De Carlo and Fraley,[90] Meylan et al.,[136] Cronan and Hodkinson,[137] Le Suave et al.,[85] Puteanus et al.,[93] Cronan and Hodkinson,[138] Cronan et al.,[139] De Carlo et al.,[40] Aplin and Cronan,[97] Lyle et al.[35]; Far South Pacific data from Exon,[134] Glasby et al.,[140] Glasby and Wright,[141] Bolton et al.[14,15]; C-C Zone (Clarion–Clipperton) and other abyssal nodule data from Haynes et al.[142]: n = 100–1000 for Si, Na, Al, K, Mg, Ca, Ti, Ba, Mo, Sr, V, Cr, Cd, As, P, Ce, Y; n = 1000–2000 Zn, Pb; n = 2000–4000 for Fe, Mn, Ni, Cu, Co; A = means for the entire Pacific[142]

USGS data (columns 1–5,8,11,12) normalized to 0% H_2O; n in parentheses differ from those in heading

Hydrogenetic Fe–Mn crusts generally have Fe/Mn ratios between 0.4 and 1.2, most commonly 0.7 ± 0.2, whereas mixed hydrogenetic and hydrothermal crusts and continental margin hydrogenetic crusts have ratios between 1 and 3, mostly 1.3 to 1.8 (Table 9.7). Cobalt is the metal with the greatest economic potential in crusts and ranges from about 0.2 to 2.3% in bulk crusts and averages between 0.3% and 0.8% for various parts of the Pacific (Table 9.7). Cobalt is also considered the element most characteristic of hydrogenetic precipitation in crusts[12,69] and is considered to maintain a constant flux in the oceans,[70] regardless of water depth. Based on sequential leaching experiments, Koschinsky and Halbach[53] provided the following list of elements with decreasing degrees of hydrogenesis: Co = Mn>Ni>Zn = Pb = Cu>Fe>Ti. Nickel and Pt are also considered of economic importance and range up to 1.0% and 1 ppm, respectively, for bulk crusts. Platinum ranges up to 3 ppm for individual crust layers.[4,5] Elements most strongly enriched over abyssal Fe–Mn nodules include Co, Pt, Pb, V, P, Ca, Ti, Sr, Te, and rare earth elements (REEs), whereas nodules are more enriched in Cu, Ni, Zn, Al, K, and Cd. Crusts are enriched over sea water in all elements except Br, Cl, and Na; enrichments over sea water between 10^8 and 10^{10} times include Bi, Co, Mn, Ti, Fe, Te, Pb, and Th, and between 10^6 and 10^8 times include Sn, Hf, Zr, Al, Y, Sc, Tl, Ni, Ca, Nb, In, Cu, Ge, Zn, W, and Ta. Crusts are enriched over lithospheric concentrations about 30,000 times for Te and 100 to 500 times for Mo, Tl, Sb, Co, Mn, Bi, As, Se, and Pb. Crusts may not only have an economic potential for Co, Ni, Mn, and Pt, but also for Te, Tl, Zr, P, and Ti.

Elements in crusts have different origins and are associated with different crust mineral phases.[39,33,40,71] Generally, elements are associated with five phases in crusts, δ-MnO_2, Fe oxyhydroxide, detrital or aluminosilicate, CFA, and residual biogenic phases. Manganese, Co, Ni, Cd, and Mo are invariably associated with the δ-MnO_2 phase. In addition, in more than 40% of the regions studied, Pb, V, Zn, Na, Ca, Sr, Mg, and Ti are also associated with that phase. Iron and As are most commonly the only elements associated with the Fe oxyhydroxide phase, although less commonly V, Cu, Pb, Y, P, Cr, Be, Sr, Ti, and Ce have also been reported to be associated with that phase. The detrital phase always includes Si, Al, and K, and commonly also Ti, Cr, Mg, Fe, Na, and Cu. The CFA phase invariably includes Ca, P, and CO_2, and also commonly Sr and Y; Mo, Ba, Ce, and Zn may also be associated with the CFA phase in some regions. The residual biogenic phase includes Ba, Sr, Ce, Cu, V, Ca, and Mg, and in some regions also Fe, As, Na, Mo, Y, P, CO_2, Pb, Ti, and Ni. Other phases that have been reported to occur in hydrogenetic crusts from various regions include Ti hydroxide (Ti, Fe); calcite (Ca, CO_2, Sr), chrome spinel (Cr, Mg, As), and todorokite (Mn, Ni, Cu, Zn). Iron is the most widely distributed element and occurs intermixed in the δ-MnO_2 phase; is the main constituent in the Fe oxyhydroxide phase; occurs in the detrital phase in minerals such as pyroxene, amphibole, smectite, magnetite, and spinels; and is in the residual biogenic phase. The strength of correlations between Fe and other elements depends on the relative abundance of Fe in the various phases. The CFA phase only occurs in thick crusts because the inner layers of those crusts have been phosphatized. In thin crusts or the surface scrapes of thick crusts, Ca, P, and CO_2 are associated with the δ-MnO_2 and/or residual biogenic phases. CFA-associated elements as well as Pt, Rh, and Ir, generally increase with increasing crust thickness. In contrast, Co and elements associated with the detrital phase usually decrease with increasing crust thickness.[12,71]

9.2.4.1   Rare Earth Elements and Yttrium

Hydrogenetic Fe–Mn crusts are characterized by high Y (10^2 ppm range) and REE (10^3 ppm range) contents. Concentrations of the trivalent REEs appear to increase in crusts with increasing water depth of occurrence.[72,73] Irrespective of their specific compositions, hydrogenetic Fe–Mn crusts from the major oceans show remarkably similar REE patterns. Shale-normalized REE patterns increase from La to Gd, decrease slightly from Gd to Lu, and commonly show positive anomalies of La, Ce, Eu, and Gd (Figure 9.5); moreover, Fe–Mn crusts typically display subchondritic Y/Ho ratios of about 20.[74] While Ce isotopic data ($^{138}Ce/^{142}Ce$) are not yet available for crusts (see

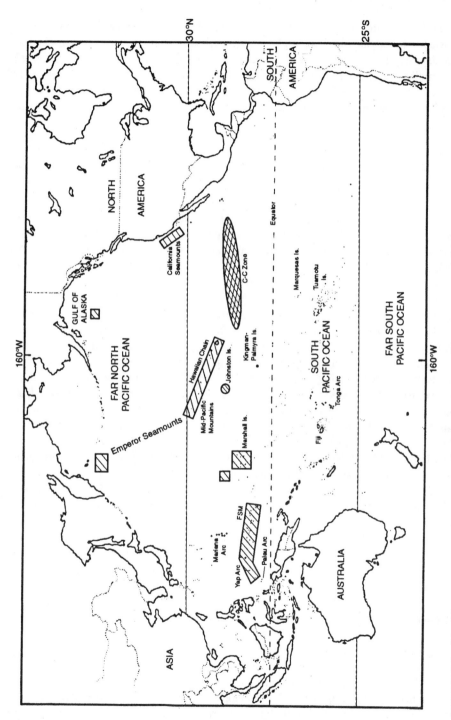

FIGURE 9.4 Lined boxes are areas with mean Fe–Mn crust compositional data listed in Table 9.7. The far North Pacific data are from the two indicated boxes, whereas the far South Pacific data cover areas mostly offshore New Zealand and southern Australia; the South Pacific data span most of the area between South America and the West Pacific arcs and between the equator and 25° South latitude; The crossed-hatched oval area is the Clarion–Clipperton prime Fe–Mn nodule zone.

Amakawa et al.,[75] for Fe–Mn nodule data), crust Nd isotopic ratios (^{143}Nd/^{144}Nd) cover a range roughly that of ambient sea water (e.g., Reference 76), indicating that Nd in crusts was derived from sea water, which in turn was supplied with Nd from fluvial and eolian input from continental sources.

The apparent bulk coefficients for partitioning of Y and REEs between Fe–Mn crusts and sea water (KD crust/sea water) are similarly high for La to Sm, decrease from Sm to Lu, display a strong positive anomaly for Ce, and show negative anomalies for La and Gd (Figure 9.6); KD(Y) is significantly smaller than the KD of any REE.[73,74] This indicates that despite the coherent overall geochemical behavior of Y and REEs, scavenging of those elements from sea water by Fe–Mn oxyhydroxides is accompanied by pronounced fractionation within the REE series and between Y and the REE.

Except for redox-sensitive Ce, Y and REE distributions in Fe–Mn crusts represent the exchange equilibrium between Y and REEs dissolved in sea water and Y and REEs sorbed on the surface of Fe–Mn oxyhydroxide particles. Crust Y and REE distributions, therefore, may be described as the net effect of the two competing processes of solution complexation (predominantly with CO_3^{2-}) and surface complexation (roughly similar to hydroxide complexation). Although the stabilities of both types of complexes increase with increasing REE atomic number, the much stronger increase of the carbonate complexes results in an overall marine particle reactivity that decreases from La to Lu.[77] Bau et al.[74] suggested that the negative anomalies for KD(La) and KD(Gd), and the low value for KD(Y) are due to the anomalously low stability of their respective hydroxide complexes.

Cerium is the only REE that in near-surface environments may occur as a tetravalent ion, and the contrasting charge and size of Ce^{4+} compared to its trivalent REE neighbors leads to decoupling of Ce(IV) from the other REEs. In the marine environment, Ce uptake by Mn and Fe oxyhydroxides is accompanied by Ce oxidation, which is mediated by microbial processes and/or a surface catalysis, a reaction that would otherwise be very sluggish. [e.g., Reference 78] Oxidative scavenging of Ce results in its preferential uptake and development of positive Ce anomalies on shale-normalized REE and KD(REE) plots (Figures 9.5 and 9.6). Hence, while uptake of Y(III) and REEs(III) is governed by the fast kinetics of adsorption–desorption reactions, Ce uptake is additionally affected by the kinetics of Ce oxidation. The impact of this difference on the Y and REE distributions in Fe–Mn–oxyhydroxide precipitates is most apparent in mixed hydrogenetic–hydrothermal Fe–Mn crusts formed from the fallout from hydrothermal plumes. Such mixed-origin crusts often show distributions of Y(III) and REE(III) that are indistinguishable from those of hydrogenetic crusts *sensu strictu*, except they show no or even a negative Ce anomaly and may have less ΣREEs.[79] Although part of the Fe and Mn in those crusts is of hydrothermal origin, their Y and REE contents are derived via scavenging from ambient sea water. Thus, with respect to Fe, Mn, and some trace metals, those crusts may be of mixed origin, but with respect to Y and REEs they are strictly hydrogenetic; this clearly indicates that a continuum exists from hydrothermal to hydrogenetic crusts. Although both types of crusts grow slowly enough to reach exchange equilibrium for Y(III) and REEs(III), the growth rates of the mixed crusts are too high to allow for a preferential uptake of Ce that is pronounced enough to compensate for the strong negative Ce anomaly of ambient sea water.[79] Consequently, despite oxidative scavenging of Ce in an oxidizing environment as indicated by positive anomalies for KD(Ce) (Figure 9.6), such crusts show no or negative Ce anomalies in REE patterns (Figure 9.5).

The short residence time of Nd in the oceans allows for the regional temporal variations of Nd isotopic ratios recorded during growth of hydrogenetic Fe–Mn crusts to be used as a paleoceanographic proxy. Although such paleoceanographic studies have gained broad attention over recent years [e.g., References 55 and 80], it should be emphasized that the original Y, REE, and Nd-isotopic compositions may be obliterated by diagenetic events such as phosphatization. However, those changes may easily be recognized by a change from subchondritic Y/Ho ratios typical of hydrogenetic Fe–Mn crusts to superchondritic ratios of up to 50 shown by phosphatized crusts.[64,74]

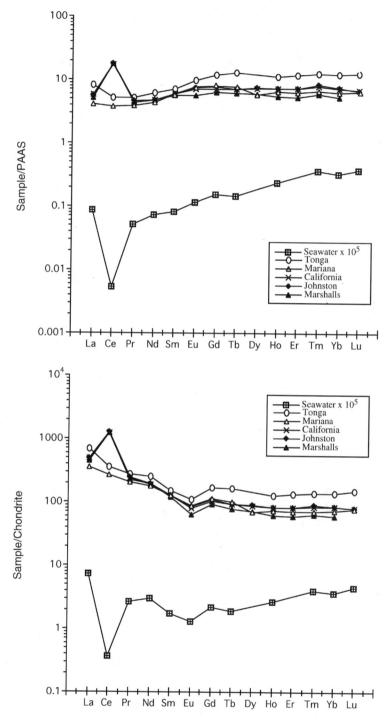

FIGURE 9.5 Shale (PAAS, Post-Archean Australian shales;[143])- and chondrite[144]-normalized REE patterns of typical hydrogenetic crusts (Johnston I., California borderland, and Marshall Is.) and mixed hydrogenetic–hydrothermal crusts (Tonga Arc and Mariana Arc) from the Pacific, compared to the sea water pattern at 1,250 m water depth.[145]

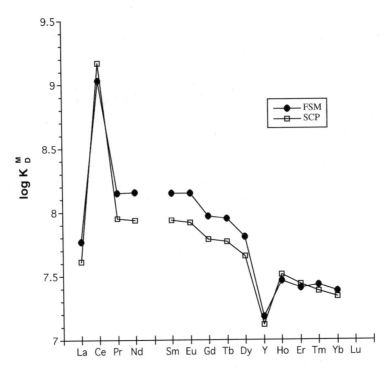

FIGURE 9.6 Partitioning between nonphosphatized crusts from FSM (Table 9.7) and SCP (South-Central Pacific data from Bau et al.[74]) and sea water REEs contents averaged from 1000, 1200, 1500, 1600, 1800, and 2000 m water depths.[74]

9.2.4.2 Platinum Group Elements

Platinum group elements (PGEs, Pt, Pd, Rh, Ru, Ir) are highly enriched in Fe–Mn crusts over lithospheric abundances (with the exception of Pd) and over concentrations in Fe–Mn nodules (Table 9.7). Platinum has been considered a potential by-product of Co mining of crusts and has mean values for different areas of up to 0.8 ppm and individual sample values of up to 3 ppm.[5,9,46,81-85] Rh, Ru, and Ir contents are up to 100, 32, and 22 ppb, respectively, whereas Pd values (1 to 5 ppb) are usually at or just above lithospheric contents. Most PGEs correlate with crust thickness and are concentrated in the inner part of thick crusts. Most PGEs commonly correlate inversely with water depth of occurrence of the crusts.[86]

The Marshall Islands EEZ and areas to the north and northwest (Figure 9.4) contain crusts with the highest PGE contents; crusts from French Polynesia also have high PGE contents. Platinum contents increase markedly in crusts from the west Pacific compared to the central and east Pacific: east Pacific (82 ppb), central Pacific (≈170 to 250 ppb), and west Pacific (500 to 600 ppb); then Pt contents decrease again adjacent to the west Pacific arcs (Table 9.7[2]). Platinum, Rh, Ir, and in some regions, Ru comprise part of the δ-MnO_2 phase, whereas Pd is commonly part of the detrital phase.[9,46,85,86] Part of the Ru may also be part of the detrital phase, and both Ir and Ru occur with the residual biogenic phase in some regions. Platinum, Ir, and Rh are derived predominantly from sea water, whereas Pd and much of the Ru are derived from clastic debris, the remainder of the Ru being derived from sea water.[9,46,81,82,87] The extraterrestrial component (meteorite debris) in bulk crusts is small. However, meteorite debris may be concentrated locally in the crusts by formation of dissolution unconformities or by proximity of the crust to meteorite fallout during formation of the layer. Localized extraterrestrial debris-rich horizons, however, do not alter the overall hydrogenetic signature of PGEs in crusts. The high Pt and Ir contents in some crust layers occur over many stratigraphic millimeters of the inner parts of the crusts, which represent millions of years of accretion of Fe–Mn oxyhydroxides and therefore cannot be explained as the result of meteorite impacts, as those are essentially instantaneous events and would form only a very thin parting in crusts. In addition, the PGE ratios are nonchondritic, with Fe–Mn crust compositions showing more than an order of magnitude more Pt relative to Ir and Rh relative to Ir. The extraterrestrial component in bulk crusts cannot be more than about 15 to 20% of the total PGEs present. Platinum is a redox-sensitive element, and its changing concentration probably reflects changing sea water redox conditions, and diagenesis within the older generation of crusts, including maturation of oxide phases and phosphatization.[2]

The mechanisms of incorporation of hydrogenetic PGEs into crusts are unknown. Both oxidation ($Pt^{2+} \Leftrightarrow Pt^{4+}$ [2,86]) and reduction ($Pt^{2+} \Leftrightarrow Pt^0$ [84]) reactions have been proposed as mechanisms to fix and enrich Pt in crusts. The reduction reaction would be coupled with the oxidation of Mn^{2+} to Mn^{4+} after the surface adsorption of Pt and destruction of its tetrachloro complex.[84] However, as pointed out by Halbach et al.,[84] that coupled reaction cannot proceed under present-day dissolved Mn^{2+} sea water concentrations, even in the OMZ where dissolved Mn contents are highest. Consequently, that mechanism does not explain the enrichment of Pt during much of the growth history of crusts, but may have promoted enrichment of Pt in some of the inner crust layers that formed when there may have been an expanded and intensified OMZ. Alternatively, Halbach et al.[84] suggested that surface adsorption of the tetrachloro Pt complexes onto positively charged iron hydroxide particles may occur. However, that explanation is difficult to reconcile with the common correlation of Pt with the δ-MnO_2 phase. Additional research is required to clarify the mechanisms responsible for the incorporation and enrichment of Pt and the other PGEs in crusts.

9.2.5 PHOSPHATIZATION OF FE–MN CRUSTS

Most thick hydrogenetic Fe–Mn crusts that formed in the open Pacific consist of two growth generations: a phosphatized older generation and a younger nonphosphatized generation. Phosphatization of older crust layers was widespread, and Halbach et al.[37] found crusts with an older phosphatized generation at many locations in the central Pacific. Phosphatized crusts have also been found on Schumann Seamount, located NE of Hawaii,[59,88] the Hawaiian Archipelago,[89] the Marshall Islands,[9] and the Johnston Island EEZ.[71] Phosphatized crusts from the South Pacific are also common in the EEZ of Kiribati, Tuvalu, Cook Islands, and Tuamotu Islands.[40,48,49,64,90-92] Crusts from relatively deep water (>2,800 m), such as from the Teahitia–Mehetia region[93] and from the Tasman Rise,[14] show only minor or no phosphatization, which is also demonstrated by a decrease in CFA-associated elements with increasing water depth of crust occurrence (see next section). Phosphatized crusts are also not found in most continental margin environments, such as in the EEZ of western North America.

Precipitation of the old crust generation began for various crusts between about 55 and 25 Ma ago [e.g., References 56, 80, and 94] and was interrupted by several Cenozoic phosphogenic events.[95,96] Phosphogenesis entailed CFA impregnation of the older crust and formation of phosphorite partings within and at the top of the older generation of some crusts. Phosphatization took place by CFA replacement of calcite infiltered into crust pore spaces, by direct precipitation of CFA in pore space, and by replacement of other crust phases, most commonly Fe oxyhydroxide. The thickness of the phosphatized layer can be as large as 12 cm, and Ca and P concentrations can increase up to 15 wt.% and 5 wt.%, respectively (Figure 9.7). Growth of the younger crust generation started during the Miocene and continued to the present without interruption by further phosphogenic events.

According to Hein et al.,[96] two major and possibly several minor Cenozoic episodes of phosphogenesis were responsible for formation of phosphorite on equatorial Pacific seamounts. Those same phosphogenic events are suspected to have phosphatized the older generation of Fe–Mn crusts. Crust ages would allow for the major phosphogenic events centered on the Eocene–Oligocene ($\approx$34 MA) and Oligocene–Miocene boundaries ($\approx$24 Ma), as well as the minor middle Miocene event

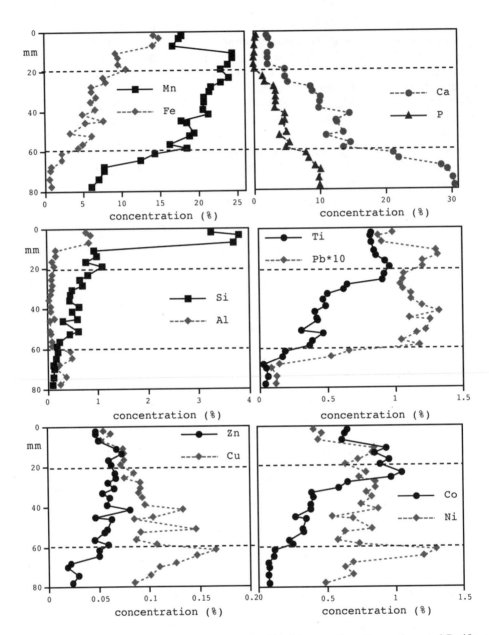

FIGURE 9.7 Element profiles of a strongly phosphatized Fe–Mn crust from the south central Pacific, consisting of an unphosphatized layer, a thick phosphatized layer, and a lower transition layer of phosphatized crust intergrown with phosphatized limestone substrate. Note in the phosphatized layers a strong increase of Ca and P contents, a decrease of Fe, Al, Si, and Ti, and a relative enrichment of Ni, Zn, and Cu, compared to Co.

at about 15 Ma, to have affected growth of Pacific Fe–Mn crusts. Extraction and dating of the CFA from older crust generations support the coeval phosphatization of substrate rocks and crusts (Chan, Hein, Koschinsky, unpublished data). Hein et al.[96] proposed that dissolved phosphorus derived from intense chemical weathering on continents accumulated in large quantities in the deep sea during stable, warm climatic conditions when sluggish oceanic circulation prevailed. As Antarctic glaciation expanded and oceanic circulation intensified, the phosphorus-rich deep waters were redistributed to intermediate water depths by upwelling at the seamounts and may have been temporarily stored in the OMZ. Koschinsky and Halbach[53] proposed that Fe–Mn crust precipitation took place below the OMZ where Mn^{2+}-rich and O_2-poor waters mixed with O_2-rich deep water. Halbach et al.[12,37] and Koschinsky et al.[64] linked the phosphogenesis of crusts with expansion of the OMZ, which was related to increased surface-water bioproductivity. The suboxic and phosphate-rich water layer reached down to the crust-covered slopes of the seamounts, inhibiting further crust precipitation and impregnating the crusts with CFA.

Phosphatization of the older Fe–Mn crust layers caused changes in their chemistry and mineralogy (Figure 9.7).[63,64] Besides dilution of the primary crust contents by CFA, certain elements were added and others removed. Depletion of elements in the phosphatized crust generation compared to nonphosphatized crust layers occurred in the approximate order $Si > Fe \geq Al \geq Th > Ti \geq Co > Mn \geq Pb \geq U$. In contrast, Ni, Cu, Zn, Y, REEs, Sr, Pt, and commonly Ba are enriched in phosphatized crusts. Manganese and Fe were partly fractionated during the dissolution of Fe–vernadite, and some of the Mn may have been reprecipitated as todorokite, which is more stable than vernadite under suboxic conditions. The secondary todorokite shows high Ni, Cu, and Zn contents, which are incorporated in the todorokite lattice in contrast to Co, possibly partly explaining the relative enrichment of Ni, Cu, and Zn compared to Co in phosphatized crusts (Figure 9.7), although many phosphatized crusts do not contain todorokite. Metals such as Pb, Sr, Y, Ba, and REEs may partly precipitate as stable, less soluble phosphate phases during phosphatization, whereas Ba forms barite. Iron released from Fe vernadite may partly reprecipitate as goethite or as Fe- and Ti-rich laminae. However, general depletions of all hydroxide-dominated elements (Fe, Al, Si, Ti, Th) are evident in phosphatized crusts (Figure 9.7).

Because primary crust precipitates were reorganized during phosphatization, the phosphatized layers may not reflect sea water composition at the time of initial crust formation. Consequently, isotopic and chemical (empirical equations) methods of dating crusts and the use of phosphatized Fe–Mn crusts in paleoceanographic studies should be done with caution.

9.2.6 LOCAL AND REGIONAL VARIATIONS IN COMPOSITION

The Fe/Mn ratios are lowest in crusts from the central and west-central Pacific and highest in crusts collected near continental margins and volcanic arcs in the Pacific and throughout the Atlantic and Indian Oceans. The detrital-associated elements (Si, Al) increase in crusts with proximity to continental margins (off western North America, far South Pacific) and volcanic arcs in the West Pacific, which have contents equivalent to those found in most crusts from the Atlantic and Indian Oceans (Table 9.7; Figure 9.8A). Within the central Pacific region, detrital-related elements are most abundant in the eastern part, along the Hawaiian and Line Islands. In contrast, Co, Ni, and Pt contents are highest in crusts from the central Pacific and lowest in crusts from along the spreading centers in the southeast Pacific, the continental margins, and along the volcanic arcs of the west Pacific (Table 9.7; Figure 9.8B, C).[39,47,62,71] Cobalt contents in Atlantic and Indian Ocean crusts are roughly equivalent to those in crusts from along the western margin of North America, but the Atlantic and Indian Ocean crusts generally have relatively more Fe (Figure 9.8B, C). Copper contents follow the trends for Co, Ni, and Pt, except for the Indian Ocean, where the highest mean value (1,518 ppm) is found. The reason for those high values is the much greater mean water depth for crusts collected from the Indian Ocean. Mean Ba content is about twice as high in northeast Pacific crusts as anywhere else in the global oceans. Those high Ba contents are the result of intense upwelling and high bioproductivity in that region.[2] (also Hein et al., unpublished data). Trends for mean Ti contents do not follow those of Al and Si, but rather follow those of Co, Ni, and Pt, which supports the idea that much of the Ti in crusts is a hydrogenetic phase.[9,53,97] Another interesting distribution is seen with P because it is not most enriched in areas where upwelling and bioproductivity are greatest (East Pacific, east equatorial Pacific), but rather is highest in crusts from the Marshall Islands and the northwest Pacific (Figure 9.8A; see also Hein et al.[71]). It is not clear why P does not reflect the high bioproductivity in the East Pacific as does Ba, or why CFA does not occur in crusts from that region. Cerium is generally lower in South Pacific crusts than it is in

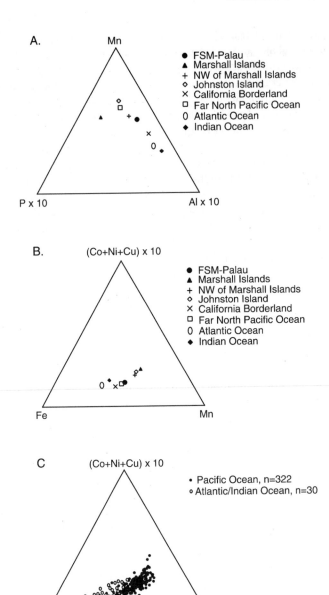

FIGURE 9.8 (A) Mn–Px10–Alx10 ternary diagram for mean compositions of USGS data listed in Table 9.7; note that crusts from all areas except the Marshall Islands have about the same relative amounts of P, but highly variable relative amounts of Al and Mn, depending on proximity to continental margins for Al and to the equatorial Pacific for Mn. (B) Mn–Fe–(Co+Ni+Cu)x10 ternary diagram after Bonatti et al.[146] for data as in (A); Note that crusts in proximity to continental margins and volcanic arcs are relatively enriched in Fe and depleted in trace metals, whereas central Pacific crusts are relatively enriched in Mn and trace metals, with the Marshall Islands crusts showing the greatest enrichments. (C) Complete dataset used in the means in B.

North Pacific crusts. Those elements concentrated in crusts by oxidation reactions (Co, Ce, Tl, and maybe Pb, Ti, and Pt) should be highest in the West Pacific where sea water oxygen contents are the highest. However, that relationship is only clearly defined for Pt and the Ce anomaly, probably because regional patterns are somewhat masked by local conditions where upwelling around seamounts mixes various water layers.[2]

Manganese, manganophile elements, and the Fe oxyhydroxide phase generally decrease, whereas detrital phase-related elements (Si, Al, Fe, Y) increase with increasing latitude.[38,71,98] The intensification of the OMZ in the equatorial zone of high bioproductivity allows for greater amounts of Mn remaining in solution in sea water and slower growth rates of crusts. The increase in Fe, Si, Al, and Y to the north is partly due to an increased supply of detritus by the trade winds. The same trends in Mn and manganophile elements should occur with longitude (increasing to the east) in the equatorial region because regional bioproductivity increases in that direction. However, a paucity of seamounts exists between Hawaii and North America, so that trend cannot be confirmed Pacific-wide; regardless, the trend is generally poorly developed regionally (e.g., Reference 87); contents of manganophile elements in California continental margin crusts are diluted by detrital and biogenic inputs, which ameliorates the increased δ-MnO_2-related elements expected to occur there. The residual biogenic phase elements generally increase with proximity to the equatorial zone of high bioproductivity and in the eastern Pacific, where productivity is yet higher.[71]

Manganese, manganophile elements, and CFA-related elements decrease, whereas Fe, Cu, and detrital-related elements increase with increasing water depth of occurrence of crusts.[12,33,37,38,98] Cobalt and other manganophile elements are enriched more than Mn is in shallow water.[97] Those relationships have been explained by an enhanced supply of dissolved Mn^{2+} in the OMZ (about 300 to 1500 m) and an increased supply of Fe in deeper waters (3,500 to 4,500) near the CCD as the result of dissolution of biogenic calcite.[95,99] However, the supply of significant amounts of Fe by biocalcite has been contested,[33,97] and much of the increased Fe, Al, and Si with increasing water depth may be the result of an increased supply of detrital phases. Enhanced Mn and Co supply to the OMZ may result from advection of metals released from anoxic continental margin sediments to the east.[97,100] However, even though that explanation may be valid for some east-central Pacific crusts, it apparently does not apply to crusts forming farther to the south and west.[62]

9.3 Fe–Mn CRUST FORMATION

Even though Fe–Mn crusts form by hydrogenetic precipitation, the exact mechanisms of metal enrichments in the water column and at the crust surface are poorly understood. The ultimate sources of metals to the oceans are river and eolian input, hydrothermal input, weathering of basalts, release of metals from sediments, and extraterrestrial input. Elements in sea water may occur in their elemental form or as inorganic and organic complexes. Those complexes may in turn form colloids that interact with each other and with other dissolved metals (e.g., References 53, 74, and 101). Thermodynamic, surface-chemical, and colloidal-chemical models show that most hydrogenetic elements in crusts occur as inorganic complexes in sea water (Figure 9.9).[53] Hydrated cations (Co, Ni, Zn, Pb, Cd, Tl, etc.) are attracted to the negatively charged surface of Mn oxyhydroxides, whereas anions and elements that form large complexes with low charge-density (V, As, P, Zr, Hf, etc.) are attracted to the slightly positive charge of Fe hydroxide surfaces.

Mixed Fe and Mn colloids with adsorbed metals precipitate onto hard-rock surfaces as poorly crystalline or amorphous oxyhydroxides, probably through bacterially mediated catalytic processes. Continued crust accretion after precipitation of that first molecular layer is autocatalytic but is probably enhanced to some degree by bacterial processes. Additional metals are incorporated into the deposits either by coprecipitation, or by diffusion of the adsorbed ions into the Mn and Fe oxyhydroxide crystal lattices. Cobalt is strongly enriched in hydrogenetic crusts because it is oxidized from Co^{2+} to the less soluble Co^{3+} at the crust surface, possibly through a disproportionation reaction.[102] Lead, Ti, and Tl, as well as Ce are also highly enriched in hydrogenetic deposits, probably by a similar oxidation mechanism;[53,72,97] however, sequential leaching studies suggested that much of the Pb may not be incorporated by that mechanism.[53]

Concentrations of elements in sea water are generally reflected in their concentrations in crusts, although there are many complicating factors. For example, Cu, Ni, and Zn occur in comparable concentrations in sea water,[103] yet Ni is much more enriched in crusts than either Cu or Zn. Copper

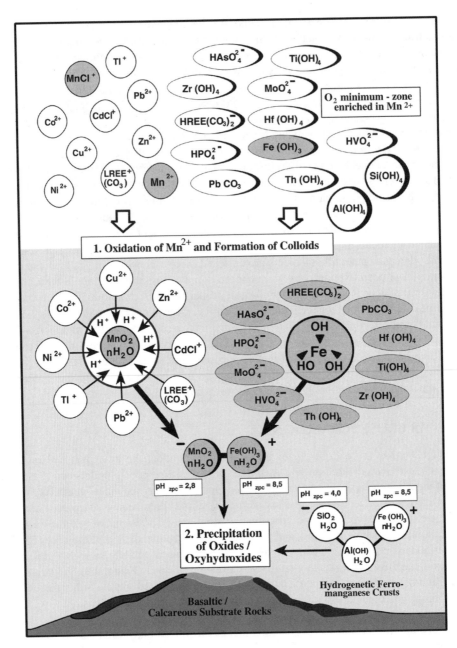

FIGURE 9.9 Colloidal–chemical model for formation of hydrogenetic crusts showing formation of complexes and colloidal phases, adsorption of metals, the pH of the zero point of charge (zpc), and precipitation of oxides on seamount rock substrates. (Modified from Koschinsky, A. and Halbach, P., Sequential leaching of ferromanganese precipitates: Genetic implications, *Geochimica et Cosmochimica Acta*, 59, 5113, 1995.)

contents may be relatively low in hydrogenetic crusts because it occurs mostly in an organically bound form in deep sea water, which is not readily incorporated into Fe and Mn metal oxyhydroxides.[104] Zinc contents may be relatively low in crusts compared to Ni because little Zn may be adsorbed onto crusts after precipitation of the oxyhydroxides, which follows the order of Ni>Co>Zn>Cu.[104] In contrast, comparable proportions of Mn:Fe:Co exist in deep sea water (0.5 to 1.0:1:0.02 to 0.05 [103,105]) as exist in Fe–Mn crusts (0.6 to 1.6:1:0.02 to 0.05; Tables 9.1 and 9.7).

The dominant controls on the concentration of elements in hydrogenetic crusts are the concentration of each element in sea water; element–particle reactivity; element residence times in sea water; the absolute and relative amounts of Fe and Mn in the crusts, which in turn are related to their abundance and ratio in colloidal flocs in sea water;[97] the colloid surface charge and types of complexing agents, which will determine the amount of scavenging within the water column;[53] the degree of oxidation of MnO_2 (O/Mn) — the greater the degree of oxidation the greater the adsorption capacity — which in turn depends on the oxygen content and pH of sea water;[106] the amount of surface area available for accretion, which at the surface of growing crusts is extremely large (mean 300 m^2/g), but which decreases with maturation of crusts;[51] the amount of dilution by detrital minerals and diagenetic phases; and growth rates. Elements that form carbonate complexes in sea water behave independently from those that form hydroxide complexes, which indicates their different modes of removal from sea water onto crust surfaces.[74] For example, Y and REEs are removed by formation of surface complexes, whereas Ti is probably removed by precipitation as a hydroxide. Very slow growth rates promote enrichment of minor elements by allowing time for extensive scavenging by the major oxyhydroxides. Accretion of oxyhydroxides will be slower where the OMZ intersects the seafloor than it will be above and below that zone, because Mn is more soluble in low-oxygen sea water. Crusts exposed at the seafloor may not necessarily be actively accreting oxyhydroxides[66] because of mechanical erosion or, less commonly in the contemporary oceans, because of sea water oxygen contents that are insufficient to permit oxidation of the major metals.

9.4 PALEOCEANOGRAPHIC AND PALEOCLIMATE STUDIES

The composition of Fe–Mn crusts reflects the composition of sea water at the time the metals precipitated. Therefore, Fe–Mn crusts are condensed stratigraphic sections in which several stratigraphic millimeters can represent a million years of oceanic history; the crusts reliably store the records of that history. Our ability to use crusts for paleoceanographic studies depends on the accuracy of dating techniques (see Section 9.2.3 on Ages and Growth Rates). Detailed sampling of crusts can possibly be done with a resolution as good as 3000 y,[107] but more commonly with resolutions of 10s to 100s of thousands of years.[65,80] Apparently, the content of most elements in crusts in addition to crust textures record local oceanographic conditions, whereas isotopic ratios of elements record regional or global conditions. Crusts should house a wide variety of records, including paleosea water chemistry and isotopes, paleocirculation patterns (isotopic tracers), paleowinds from incorporated eolian grains, paleoerosion rates of continents (isotopic tracers), history of hydrothermal input (isotopic and chemical tracers), paleo-pH and temperature (isotopic tracers), and history of extraterrestrial input (isotopic and chemical tracers). The isotopic tracers that have been used so far include Be, Nd, Pb, Hf, Os, Li, B, and U-series isotopes and chemical tracers such as Ba, Co, Pt, and Li.[67,68,76,108-110] For example, deep ocean circulation today is characterized by creation of deep water in the North Atlantic that has a characteristic isotopic composition (Figure 9.10). Then, that deep water travels south and then east into the Indian Ocean and finally farther east into the Pacific, where its fate in the far North Pacific is poorly understood. The isotopic compositions of Nd, Pb, and Be change along that route and temporal changes in the pattern of deep water flow can be mapped via isotopic signatures of those elements in crusts distributed globally.[68,76,108]

Many advantages exist in using crust stratigraphy for paleoceanographic studies rather than using carbonate sections in the deep sea, the heretofore common method of addressing paleoceanographic problems. Crusts are more widely distributed geographically and with water depth than are thick carbonate sections; crusts have higher concentrations of most elements and their isotopes; crusts are less prone to diagenetic isotopic exchange; and crusts have less detrital contamination. In contrast, carbonate sections are easier to date and generally have better temporal resolution.

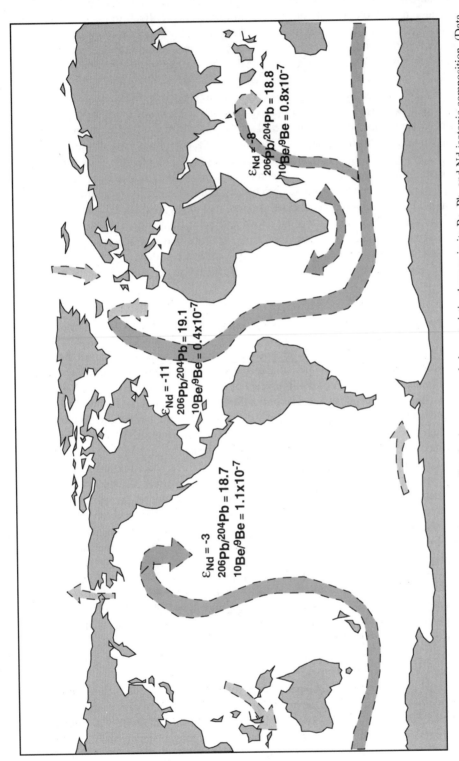

FIGURE 9.10 Schematic representation of deep water flow in the oceans and characteristic changes in its Be, Pb, and Nd isotopic composition. (Data from Albarède, F. and Goldstein, S. L., World map of Nd isotopes in seafloor ferromanganese deposits, *Geology*, 20, 761, 1992; von Blanckenburg, F., O'Nions, R. K., Belshaw, N. S., Gibb, A., and Hein, J. R., Global distribution of beryllium isotopes in deep ocean water as derived from Fe–Mn crusts, *Earth and Planetary Science Letters*, 141, 213, 1996; von Blanckenburg, F., O'Nions, R. K., and Hein, J. R., Distribution and sources of pre-anthropogenic lead isotopes in deep ocean water from Fe–Mn crusts, *Geochimica et Cosmochimica Acta*, 60, 4957, 1996.)

Crust element contents and crust textures have been used with limited success for the past 15 years in understanding paleoceanographic processes.[50,92,99,111] However, the isotopic tracer studies of crusts began only in the mid 1990s and offer a unique way to solve important problems in paleoceanography.

9.5 RESOURCE, TECHNOLOGY, AND ECONOMIC CONSIDERATIONS

During the early stages of exploration for Co-rich crusts, the main objectives are to find areally extensive, thick, and high-grade deposits. In later stages of exploration, the objective is to map the precise range of minable crusts. Consequently, there is a continuing refinement of detail secured on each seamount by successive iterations of sampling and surveying. The main field operations used for exploration include continuous mapping of seamounts using multibeam echosounder, side-scan sonar, and single- or multichannel seismic systems; systematic sampling using dredges and corers; bottom photo coverage; water column sampling; and analysis of crusts and substrates for composition and physical properties. Operations may also include collecting biological and ecological information that can be used in future environmental impact studies. These exploration activities require use of a large and sophisticated research vessel because of the large number of bottom acoustic beacons, large towed equipment, and volume of samples collected. During advanced stages of exploration, manned submersibles or ROVs may be used for closer observations and sampling of seamount deposits.

Based on data collected during the first six years of Fe–Mn crust studies, Hein et al.[9] listed 11 criteria for the exploration for and exploitation of Fe–Mn crusts: (1) large volcanic edifices shallower than 1,000 to 1,500 m; (2) substrates older than 20 Ma; (3) areas of strong and persistent current activity; (4) volcanic structures not capped by large atolls or reefs; (5) a shallow and well-developed OMZ; (6) slope stability; (7) absence of local volcanism; and (8) areas isolated from input of abundant fluvial and eolian debris. Exploitation criteria included (9) average Co contents $\geq 0.8\%$; (10) average crust thicknesses ≥ 40 mm; and (11) subdued small-scale topography. Depending on the mining systems used, crust thickness may turn out to be more important than grade; if true, then the crust thickness criterion would increase and grade criterion would decrease. That relationship is inevitable because Co grade decreases with increasing crust thickness.

9.5.1 MINING SYSTEMS

Crust mining is similar to Fe–Mn nodule mining in some ways and considerably different in others. Nodule mining concepts developed by mining consortia in the last decades consist of a hydraulic dredge and a slurry lift system.[112,113] Recovery of the nodules is relatively easy because they sit on a soft sediment substrate. In contrast, Fe–Mn crusts are weakly to strongly attached to substrate rock. The key to successful mining will be the ability to attain efficient crust recovery while avoiding collection of substrate rock, which would significantly dilute crust grade. Five crust mining functions include fragmentation, crushing, lifting, pickup, and separation. The proposed method for crust recovery[41] consists of a bottom-crawling vehicle attached to a surface mining vessel by means of a hydraulic pipe lift system and an electrical umbilical. The mining machine provides its own propulsion and travels at a speed of about 20 cm/s. The miner has articulating cutting devices that would allow crusts to be fragmented, while minimizing the amount of substrate collected (Figure 9.11). The hydraulic suction dredges are similar to trailing suction dredge heads commonly used with hopper dredges for sand and gravel mining. About 95% of the fragmented material would be picked up and processed through a gravity separator prior to lifting. Material throughput for the base-case mining scenario[41] is 1,000,000 t/y. That scenario allows 80% fragmentation efficiency and 25% dilution of crust with substrate during fragmentation as reasonable miner capabilities. The net recovery of crusts depends on fragmentation efficiency, pickup efficiency, and separation

MAJOR DIMENSIONS

Length : 13 m
Width   : 8 m
Height : 6 m
Weight : 100 t
Installed Power : 900 KW

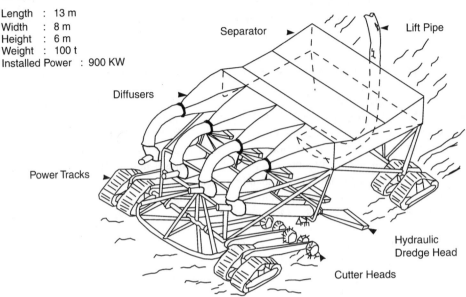

FIGURE 9.11   Schematic representation of a deep-sea mining vehicle for Fe–Mn crusts. (From DOI-MMS and DPED-State of Hawaii, Proposed marine mineral lease sale: Exclusive economic zone, adjacent to Hawaii and Johnston Island, Final Environmental Impact Statement, vols. I & II, 1990; designed by J. E. Halkyard, OTC Corporation).

losses. Fragmentation efficiencies depend on microtopography and depth of the cut. Pickup efficiencies also depend on seafloor roughness, but to a lesser extent than fragmentation efficiency, and on the size of fragmented particles and type of pickup device.

The Japan Resource Association[114] studied the applicability of a continuous line bucket (CLB) system as a method of crust mining. The CLB could be competitive in an area where crusts are easily separated from substrate rock, or where the substrate is soft enough to be removed by washing.[115] While there is some merit in the CLB's simplicity, the most likely commercial crust mining systems will probably use hydraulic lift together with a mechanical fragmentation system attached to a self-propelled collector. That type of system has a better likelihood of efficient crust recovery and substrate separation. The economics of ocean mining favors efficient mining techniques even at a cost penalty over simpler but less efficient systems. Some new and innovative systems that have been suggested for Fe–Mn crust mining include water jet stripping of crusts from the substrate, and *in situ* leaching techniques. Both suggestions offer promise and need to be further explored.

9.5.2   ECONOMICS

The importance of metals contained in Fe–Mn crusts to the world economy is reflected in their patterns of consumption.[116] The primary uses of Mn, Co, and Ni are in the manufacture of steel to which they provide unique characteristics. Cobalt is also used in the electrical, communications, aerospace, and engine and tool manufacturing industries, and its radioisotope is used in modern diagnostic and therapeutic medicine. Nickel is also used in chemical plants, petroleum refineries, electrical appliances, and motor vehicles. Supplies of these metals and other rare metals found in crusts are essential for maintaining the efficiency of modern industrial societies and in improving the standard of living in the 21st century.

TABLE 9.8
Fe–Mn Crust Mining and Processing,
Estimated Annual Revenues and Costs

Commodity	Output (t)	Prices (1985 dollars)	Amount ($ × 10⁶)
Pyrometallurgical Process Revenues			
Cobalt	5,710	11.70	133.6
Nickel	2,990	3.29	19.7
Copper	120	0.65	0.2
Fe–Mn	195,000	0.30	117.0
Total Revenues	—	—	207.5
Costs	—	—	−291.0
Net Revenues	—	—	−20.5
Hydrometallurgical Process Revenues			
Cobalt	5,365	11.70	125.5
Nickel	2,900	3.29	19.1
Copper	420	0.65	0.5
Total Revenues	—	—	145.1
Costs	—	—	−204.0
Net Revenues	—	—	−58.9

Source: DOI-MMS and DPED-State of Hawaii, Proposed marine mineral lease sale: Exclusive economic zone, adjacent to Hawaii and Johnston Island, Final Environmental Impact Statement, vols. I & II, 1990.

Most mineral industry analysts agree that the supply of Co is more uncertain than the supply of the other crust metals because most Co production has come heretofore from Zaire (57%) and Zambia (11%). This uncertainty in supply has caused industry to look for alternatives to Co, resulting in only a modest growth in the Co market over the past decade, and consequently, relatively low Co prices. If substantial alternative sources of Co supply are developed, there should be a greater incentive to reintroduce Co back into products and expand the market.[117]

A preliminary estimate of the economics of crust mining and processing operations for the State of Hawaii[41] indicated that crust mining and processing is not economical under present scenarios (Table 9.8). The minimum required return on the investment of US$750 million in mining and processing is on the order of US$100 million per year. Neither pyrometallurgical or hydrometallurgical processing options would be commercially viable under the current scenario of 700,000 dry t/y of crust production.[117] Outside of Japan, there has been limited research and development on mining technologies for crusts, and therefore economic analysis at this early stage is highly speculative. Despite these difficulties, China has held discussions with the State of Hawaii about the possibility of placing a plant on the islands for processing Fe–Mn oxide materials. Several evolving circumstances may change the economic environment and promote crust mining in the oceans, for example land-use priorities, freshwater issues, and environmental concerns in areas of mining on land.

There is a growing recognition that Co-rich Fe–Mn crusts are an important potential resource. Accordingly, it is necessary to fill the information gap concerning various aspects of crust mining through research, exploration, and technology development. The Marshall Islands, Johnston I. (U.S. EEZ), Kiribati, Federated States of Micronesia, and French Polynesia are the Pacific Island areas with the greatest known potential for crust resources. At present, the Marshall Islands EEZ has been the area most extensively studied.

9.6   RESEARCH FOR THE 21ST CENTURY

At the end of the 20th century, 20 years will have gone into studies of Co-rich Fe–Mn crusts. However, many questions remain to be answered and we recommend that research in the 21st century should include:

- Detailed mapping of selected seamounts, including analysis of microtopography
- Development of better dating techniques for crusts
- Determining the oceanographic and geologic conditions that produce very thick crusts
- Determining what controls the concentration of PGEs and other rare elements in crusts
- Determining how much burial by sediment is required to affect crust growth; and to what extent crusts occur on seamounts under a thin blanket of sediment
- Determining the role of microbiota in the formation and growth of crusts
- Determining the extent and significance of organic complexing of metals that comprise crusts
- Studying currents, internal tides, and upwelling around seamounts
- Complete environmental and ecological studies of seamount communities
- Determining to what extent the chemical and isotopic compositions of crusts reflect paleoceanographic and paleoclimatic conditions, and complete deciphering of those records
- Development of new mining technologies and processes of extractive metallurgy

REFERENCES

1. Hein, J. R. and Morgan, C. L., Influence of substrate rocks on Fe–Mn crust composition, *Deep-Sea Research I,* 46, 855, 1999.
2. Hein, J. R., Koschinsky, A., Halbach, P., Manheim, F. T., Bau, M., Kang, J.-K., and Lubick, N., Iron and manganese oxide mineralization in the Pacific, in *Manganese Mineralization: Geochemistry and Mineralogy of Terrestrial and Marine Deposits,* Nicholson, K., Hein, J. R., Bühn, B., and Dasgupta, S. (Eds.), Geological Society Special Publication No. 119, London, 1997, 123.
3. Bury, S. J., *The Geochemistry of North Atlantic Ferromanganese Encrustations,* Ph.D. thesis, Cambridge University, 265 pp., 1989.
4. Hein, J. R., Fleishman, C. L., Morgenson, L. A., Bloomer, S. H., and Stern, R. J., *Submarine Ferromanganese Deposits from the Mariana and Volcano Volcanic Arcs, West Pacific,* U.S. Geological Survey Open-File Report 87-281, 67 pp., 1987.
5. Usui, A. and Someya, M., Distribution and composition of marine hydrogenetic and hydrothermal manganese deposits in the northwest Pacific, in *Manganese Mineralization: Geochemistry and Mineralogy of Terrestrial and Marine Deposits,* Nicholson, K., Hein, J. R., Bühn, B., and Dasgupta, S. (Eds.), Geological Society Special Publication No. 119, London, 1997, 177.
6. Cronan, D. S., Criteria for the recognition of areas of potentially economic manganese nodules and encrustations in the CCOP/SOPAC region of the central and southwestern tropical Pacific, *South Pacific Marine Geological Notes,* CCOP/SOPAC, 3, 1, 1984.
7. Hein, J. R., Manheim, F. T., Schwab, W. C., and Davis, A. S., Ferromanganese crusts from Necker Ridge, Horizon Guyot, and S.P. Lee Guyot: geological considerations, *Marine Geology,* 69, 25, 1985.
8. Frank, D. J., Meylan, M. A., Craig, J. D., and Glasby, G. P., Ferromanganese deposits of the Hawaiian Archipelago, *Hawaii Institute of Geophysics Report,* HIG 76-14, 71 pp., 1976.
9. Hein, J. R., Schwab, W. C., and Davis, A. S., Cobalt- and platinum-rich ferromanganese crusts and associated substrate rocks from the Marshall Islands, *Marine Geology,* 78, 255, 1988.
10. Usui, A., Nishimura, A., and Iizasa, K., Submersible observations of manganese nodule and crust deposits on the Tenpo Seamount, northwestern Pacific, *Marine Georesources and Geotechnology,* 11, 263, 1993.
11. Moore, J., Normark, W. R., and Holcomb, R. T., Giant Hawaiian underwater landslides, *Science,* 264, 46, 1994.

12. Halbach, P., Manheim, F. T., and Otten, P., Co-rich ferromanganese deposits in the marginal seamount regions of the central Pacific basin — results of the Midpac '81, *Erzmetall,* 35, 447, 1982.

13. Morgenstein, M., Manganese accretion at the sediment-water interface at 400-2400 meters depth, Hawaiian Archipelago, in *Ferromanganese Deposits on the Ocean Floor,* Horn, D. R. (Ed.), Arden House, Harriman, New York, 1972, 131.

14. Bolton, B. R., Exon, N. F., Ostwald, H. R., and Kudrass, H. R., Geochemistry of ferromanganese crusts and nodules from the South Tasman Rise, southeast of Australia, *Marine Geology,* 84, 53, 1988.

15. Bolton, B. R., Exon, N. F., and Ostwald, J., Thick ferromanganese deposits from the Dampier Ridge and the Lord Howe Rise off eastern Australia, *BMR Journal of Australian Geology and Geophysics,* 11, 421, 1990.

16. Yamazaki, T., Igarashi, Y., and Maeda, K., Buried cobalt rich manganese deposits on seamounts, *Resource Geology Special Issue,* 17, 76, 1993.

17. Yamazaki, T., A re-evaluation of cobalt-rich crust abundance on the Pacific seamounts, *International Journal of Offshore and Polar Engineering,* 3, 258, 1993.

18. Swedenborg, E., *De Ferro (Opera Philosophica et Mineralia),* Friedrich Haeckel, Dresden and Leipzig, Quarto 455 pp., 1734.

19. Murray, J., Preliminary report on specimens of the sea bottom obtained in soundings, dredgings, and trawlings of H.M.S. Challenger in the years 1873-1875 between England and Valparaiso, *Proceedings of the Royal Society,* 24, 471, 1876.

20. Murray, J. and Renard, A. F., *Report on Deep Sea Deposits, Report on the Scientific Results of the Voyage of H.M.S. Challenger, Deep Sea Deposits,* 525 pp., 1891.

21. Buchanan, J. Y., On chemical and geological work done on board H.M.S. Challenger, *Proceedings of the Royal Society of London,* 24, 593, 1876.

22. Goldberg, E. D., Marine geochemistry, chemical scavengers of the sea, *Journal of Geology,* 62, 249, 1954.

23. Goldberg, E. D. and Arrhenius, G. O. S., Chemistry of Pacific pelagic sediments, *Geochimica et Cosmochimica Acta,* 26, 417, 1958.

24. Mero, J. L., *The Mining and Processing of Deep-Sea Manganese Nodules,* University of California, Institute of Marine Resources, 96 pp., 1959.

25. Mero, J. L., Ocean-floor manganese nodules, *Economic Geology,* 57, 747, 1962.

26. Mero, J. L., *The Mineral Resources of the Sea,* Elsevier, Amsterdam, 312 pp., 1965.

27. Skornyakova, N. S., Manganese concretions in sediments of the northeastern part of the Pacific Ocean, *Doklady Akademii Nauk,* SSSR, 130, 653, 1960 (in Russian).

28. Bezrukov, P. L. (Ed.), *Zhelezomargantsevye konkretsii Tikhogo okeans (Ferromanganese Nodules in the Pacific Ocean),* Trudy Instituta Okeanologii, Akademii Nauk, SSSR, Moscow, 301 pp., 1976.

29. Cronan, D. S., The Geochemistry of Some Manganese Nodules and Associated Pelagic Deposits, Ph.D. thesis, Imperial College, University of London, 1967.

30. Cronan, D. S., Deep-sea nodules: distribution and geochemistry, in *Marine Manganese Deposits,* Glasby, G. P. (Ed.), Elsevier, Amsterdam, 11-44, 1977.

31. Frazer, J. Z. and Fisk, M. B., Geological factors related to characteristics of seafloor manganese nodule deposits, Report for U.S. Department of Interior, Bureau of Mines, Scripps Institute of Oceanography Reference 79-19, 41 pp., 1980

32. Craig, J. D., Andrews, J. E., and Meylan, M. A., Ferromanganese deposits in the Hawaiian Archipelago, *Marine Geology,* 45, 127, 1982.

33. De Carlo, E. H., McMurtry, G. M., and Kim, K. H., Geochemistry of ferromanganese crusts from the Hawaiian Archipelago — I. Northern survey areas, *Deep-Sea Research,* 34, 441, 1987.

34. Kraemer, T. and Schornick, J. C., Comparison of elemental accumulation rates between ferromanganese deposits and sediments in the South Pacific Ocean, *Chemical Geology,* 13, 187, 1974.

35. Lyle, M., Dymond, J., and Heath, G. R., Copper-nickel-enriched ferromanganese nodules and associated crusts from the Bauer Basin, northwest Nazca Plate, *Earth and Planetary Science Letters,* 35, 55, 1977.

36. Glasby, G. P. and Andrews, J. E., Manganese crusts and nodules from the Hawaiian Ridge, *Pacific Science,* 31, 363, 1977.

37. Halbach, P., Sattler, C.-D., Teichmann, F., and Wahsner, M., Cobalt-rich and platinum-bearing manganese crust deposits on seamounts: nature, formation and metal potential, *Marine Mining,* 8, 23, 1989.

38. Hein, J. R., Manheim, F. T., Schwab, W. C., Davis, A. S., Daniel, C. L., Bouse, R. M., Morgenson, L. A., Sliney, R. E., Clague, D., Tate, G. B., and Cacchione, D. A., Geologic and geochemical data for seamounts and associated ferromanganese crusts in and near the Hawaiian, Johnston Island, and Palmyra Island Exclusive Economic Zones, U.S. Geological Survey Open File Report 85-292, 129 pp., 1985.

39. Hein, J. R., Morgenson, L. A., Clague, D. A., and Koski, R. A., Cobalt-rich ferromanganese crusts from the Exclusive Economic Zone of the United States and nodules from the oceanic Pacific, in *Geology and Resource Potential of the Continental Margin of Western North America and Adjacent Ocean Basins — Beaufort Sea to Baja California*, Scholl, D. W., Grantz, A., and Vedder, J. G. (Eds.), Circum-Pacific Council for Energy and Mineral Resources, Earth Science Series v. 6, Houston, Texas, 1987, 753.

40. De Carlo, E. H., Pennywell, P. A., and Fraley, C. M., Geochemistry of ferromanganese deposits from the Kiribati and Tuvalu region of the west central Pacific Ocean, *Marine Mining*, 6, 301, 1987.

41. DOI-MMS and DPED-State of Hawaii, Proposed marine mineral lease sale: Exclusive economic zone, adjacent to Hawaii and Johnston Island, Final Environmental Impact Statement, Vols. I & II, 1990.

42. Manheim, F. T., Pratt, R. M., and McFarlin, P. F., Composition and origin of phosphorite deposits of the Blake Plateau, in *Marine Phosphorites — Geochemistry, Occurrence, Genesis*, Bentor, Y. K. (Ed.), SEPM Special Publication 29, 1980, 117.

43. Manheim, F. T., Popenoe, P., Siapno, W., and Lane, C., Manganese-phosphorite deposits on the Blake Plateau, in *Marine Mineral Deposits — New Research Results and Economic Prospects*, Halbach, P. and Winter, P. (Eds.), Verlag Glückauf, Essen, 1982, 9.

44. Moritani, T. and Nakao, S. (Eds.), Deep sea mineral resources investigation in the western part of the central Pacific basin (GH78-1 cruise), Geological Survey of Japan Cruise Report 17, 281 pp., 1981.

45. Kang, J.-K., Mineralogy and internal structures of a ferromanganese crust from a seamount, central Pacific, *Journal Oceanographical Society of Korea*, 22, 168, 1987.

46. Hein, J. R., Kang, J-K., Schulz, M. S., Park, B-K., Kirschenbaum, H., Yoon, S-H., Olson, R. L., Smith, V. K., Park, D-W., Riddle, G. O., Quinterno, P. J., Lee, Y-O., Davis, A. S., Kim, S. R., Pringle, M. S., Choi, D-L., Pickthorn, L. B., Schlanger, S. O., Duennebier, F. K., Bergersen, D. D., and Lincoln, J. M., Geological, geochemical, geophysical, and oceanographic data and interpretations of seamounts and Co-rich ferromanganese crusts from the Marshall Islands, KORDI-USGS R.V. Farnella Cruise F10-89-CP, U.S. Geological Survey Open File Report 90-407, 246 pp., 1990.

47. Manheim, F. T. and Lane-Bostwick, C. M., Chemical composition of ferromanganese crusts in the world ocean: a review and comprehensive database, U.S. Geological Survey Open File Report 89-020, 200 pp., 1989.

48. Alvarez, R., De Carlo, E. H., Cowen, J., and Andermann, G., Micromorphological characteristics of a marine ferromanganese crust, *Marine Geology*, 94, 239, 1990.

49. Pichocki, C. and Hoffert, M., Characteristics of Co-rich ferromanganese nodules and crusts sampled in French Polynesia, *Marine Geology*, 77, 109, 1987.

50. Hein, J. R., Bohrson, W. A., Schulz, M. S., Noble, M., and Clague, D. A., Variations in the fine-scale composition of a central Pacific ferromanganese crust: paleoceanographic implications, *Paleoceanography*, 7, 63-77, 1992.

51. Stashchuk, M. F., Chervonetsky, D. V., Kaplun, E. V., Chichkin, R. V., Avramenko, V. A., Tischenko, P. Y., and Gramm-Osipov, L. M., Adsorption properties of ferromanganese crusts and nodules, in Data and Results from R.V. Aleksandr Vinogradov cruises 91-AV-19/1, north Pacific hydrochemistry transect; 91-AV-19/2, north equatorial Pacific Karin Ridge Fe–Mn crust studies; and 91-AV-19/4, northwest Pacific and Bering Sea sediment geochemistry and paleoceanographic studies, Hein, J. R., Bychkov, A. S., and Gibbs, A. E. (Eds.), U.S. Geological Survey Open File Report 94-230, 93, 1994.

52. Varentsov, I. M., Drits, V. A., Gorshkov, A. I., Sivtsov, A. V., and Sakharov, B. A., Mn-Fe oxyhydroxide crusts from Krylov Seamount (eastern Atlantic): mineralogy, geochemistry and genesis, *Marine Geology*, 96, 53, 1991.

53. Koschinsky, A. and Halbach, P., Sequential leaching of ferromanganese precipitates: Genetic implications, *Geochimica et Cosmochimica Acta*, 59, 5113, 1995.

54. Burns, R. G., and Burns, V. M., Mineralogy, in *Marine Manganese Deposits*, Glasby, G. P. (Ed.), Elsevier, Amsterdam, 1977, 185.

55. Futa, K., Peterman, Z. E., and Hein, J. R., Sr and Nd isotopic variations in ferromanganese crusts from the central Pacific: implications for age and source provenance, *Geochimica et Cosmochimica Acta,* 52, 2229-2233, 1988.

56. Ingram, B. L., Hein, J. R., and Farmer, G. L., Age determinations and growth rates of Pacific ferromanganese deposits using strontium isotopes, *Geochimica et Cosmochimica Acta,* 54, 1709, 1990.

57. Peucker-Ehrenbrink, B., Ravizza, G., and Hofmann, A. W., The marine $^{187}Os/^{186}Os$ record of the past 80 million years, *Earth and Planetary Science Letters,* 130, 155, 1995.

58. Janin, M. C., Biostratigraphie de concrétions polymétalliques de l'Archipel des Tuamotu, fondée sur les nannofossiles calcaires, *Bull. Soc. Géol. France,* 8, 79, 1985.

59. Cowen, J. P., De Carlo, E. H., and McGee, D. L., Calcareous nannofossil biostratigraphic dating of a ferromanganese crust from Schumann Seamount, *Marine Geology,* 115, 289, 1993.

60. Scott, R. B., Malpas, J., Rona, P. A., and Udintsev, G., Duration of hydrothermal activity at an oceanic spreading center, Mid-Atlantic Ridge (Lat. 26°N), *Geology,* 4, 233, 1976.

61. Lyle, M., Estimating growth rates of ferromanganese nodules from chemical compositions: Implications for nodule formation processes, *Geochimica et Cosmochimica Acta,* 46, 2301–2306, 1982.

62. Manheim, F. T. and Lane-Bostwick, C. M., Cobalt in ferromanganese crusts as a monitor of hydrothermal discharge on the Pacific seafloor, *Nature,* 335, 59, 1988.

63. Puteanus, D. and Halbach, P., Correlation of Co concentration and growth rate: a method for age determination of ferromanganese crusts, *Chemical Geology,* 69, 73-85, 1988.

64. Koschinsky, A., Stascheit, A., Bau, M., and Halbach, P., Effects of phosphatization on the geochemical and mineralogical composition of marine ferromanganese crusts, *Geochimica et Cosmochimica Acta,* 61, 4079, 1997.

65. Christensen, J. N., Halliday, A. N., Godfrey, L. V., Hein, J. R., and Rea, D. K., Climate and ocean dynamics and the lead isotopic records in Pacific ferromanganese crusts, *Science,* 277, 913, 1997.

66. Chabaux, F., Cohen, A. S., O'Nions, R. K., and Hein, J. R., $^{238}U-^{234}U-^{230}Th$ chronometry of Fe–Mn crusts: growth processes and recovery of thorium isotopic ratios of sea water, *Geochimica et Cosmochimica Acta,* 59, 633, 1995.

67. Chabaux, F., O'Nions, R. K., Cohen, A. S., and Hein, J. R., $^{238}U-^{234}U-^{230}Th$ disequilibrium in hydrogenous oceanic Fe–Mn crusts: Palaeoceanographic record or diagenetic alterations?, *Geochimica et Cosmochimica Acta,* 61, 3619, 1997.

68. von Blanckenburg, F., O'Nions, R. K., Belshaw, N. S., Gibb, A., and Hein, J. R., Global distribution of beryllium isotopes in deep ocean water as derived from Fe–Mn crusts, *Earth and Planetary Science Letters,* 141, 213, 1996.

69. Manheim, F. T., Marine cobalt resources, *Science,* 232, 601, 1986.

70. Halbach, P., Segl, M., Puteanus, D., and Mangini, A., Co-fluxes and growth rates in ferromanganese deposits from central Pacific seamount areas, *Nature,* 304, 716, 1983.

71. Hein, J. R., Schulz, M. S., and Gein, L. M., Central Pacific Cobalt-rich ferromanganese crusts: Historical Perspective and Regional Variability, in *Geology and Offshore Mineral Resources of the Central Pacific Basin,* Keating, B. H. and Bolton, B. R. (Eds.), Circum-Pacific Council for Energy and Mineral Resources, Earth Sciences Series, 14, New York, Springer Verlag, 1992, 261.

72. Aplin, A. C., Rare earth element geochemistry of Central Pacific ferromanganese encrustations, *Earth and Planetary Science Letters,* 71, 13, 1984.

73. De Carlo, E. H. and McMurtry, G. M., Rare-earth element geochemistry of ferromanganese crusts from the Hawaiian Archipelago, central Pacific, *Chemical Geology,* 95, 235, 1992.

74. Bau, M., Koschinsky, A., Dulski, P., and Hein, J. R., Comparison of the partitioning behaviours of yttrium, rare earth elements, and titanium between hydrogenetic marine ferromanganese crusts and sea water, *Geochimica et Cosmochimica Acta,* 60, 1709, 1996.

75. Amakawa, H., Ingri, J., Masuda, A., and Shimizu, H., Isotopic compositions of Ce, Nd, and Sr in ferromanganese nodules from the Pacific and Atlantic Oceans, the Baltic and Barents Seas and the Gulf of Bothnia, *Earth and Planetary Science Letters,* 105, 554, 1991.

76. Albarède, F. and Goldstein, S. L., World map of Nd isotopes in seafloor ferromanganese deposits, *Geology,* 20, 761, 1992.

77. Byrne, R. H. and Kim, K. H., Rare earth element scavenging in sea water, *Geochimica et Cosmochimica Acta,* 54, 2645, 1990.

78. Moffett, J. W., Microbially mediated cerium oxidation in sea water, *Nature,* 345, 421, 1990.

79. Kuhn, T., Bau, M., Blum, N., and Halbach, P., Origin of negative Ce anomalies in mixed hydrothermal-hydrogenetic Fe–Mn crusts from the Central Indian Ridge, *Earth and Planetary Science Letters,* 163, 207, 1998.

80. Ling, H. F., Burton, K. W., O'Nions, R. K., Kamber, B. S., von Blanckenburg, F., Gibb, A. J., and Hein, J. R., Evolution of Nd and Pb isotopes in central Pacific sea water from ferromanganese crusts, *Earth and Planetary Science Letters,* 146, 1, 1997.

81. Hein, J. R., Kirschenbaum, H., Schwab, W. C., Usui, A., Taggart, J. E., Stewart, K. C., Davis, A. S., Terashima, S., Quinterno, P. J., Olson, R. L., Pickthorn, L. G., Schulz, M. S., and Morgan, C. L., Mineralogy and geochemistry of Co-rich ferromanganese crusts and substrate rocks from Karin Ridge and Johnston Island, Farnella Cruise F7-86-HW, U.S. Geological Survey Open File Report 90-298, 80 pp., 1990.

82. Hein, J. R., Gramm-Osipov, L., Gibbs, A. E., Kalyagin, A. N., d'Angelo, W. M., Nachaev, V. P., Briggs, P. H., Bychkov, A. S., Davis, A. S., Gusev, V. V., Chezar, H., Gorbarenko, S. A., Bullock, J. H., Kraynikov, G. A., Siems, D. F., Mikhailik, E. V., Smith, H., Eyberman, M. F., Schutt, M. J., Belogla-zov, A. I., Mozherovsky, A. V., and Chichkin, R. V., Description and composition of Fe–Mn crusts, rocks, and sediments collected on Karin Ridge, R.V. Aleksandr Vinogradov cruise 91-AV-19/2, in Data and results from R.V. Aleksandr Vinogradov cruises 91-AV-19/1, north Pacific hydrochemistry transect; 91-AV-19/2, north equatorial Pacific Karin Ridge Fe–Mn crust studies; and 91-AV-19/4, northwest Pacific and Bering Sea sediment geochemistry and paleoceanographic studies, Hein, J. R., Bychkov, A. S., and Gibbs, A. E. (Eds.), U.S. Geological Survey Open-File Report 94-230, 1994, 39.

83. Hein, J. R., Zielinski, S. E., Staudigel, H., Chang, S.-W., Greene, M., and Pringle, M. S., Composition of Co-rich ferromanganese crusts and substrate rocks from the NW Marshall Islands and International waters to the north, Tunes 6 cruise, U.S. Geological Survey Open File Report 97-482, 65 pp., 1997.

84. Halbach, P., Kriete, C., Prause, B., and Puteanus, D., Mechanisms to explain the platinum concentration in ferromanganese seamount crusts, *Chemical Geology,* 76, 95, 1989.

85. Le Suave, R., Pichocki, C., Pautot, G., Hoffert, M., Morel, Y., Voisset, M., Monti, S., Amossé, J., and Kosakevitch, A., Geological and mineralogical study of Co-rich ferromanganese crusts from a submerged atoll in the Tuamotu Archipelago (French Polynesia), *Marine Geology,* 87, 227, 1989.

86. Halbach, P., Puteanus, D., and Manheim, F. T., 1984, Platinum concentrations in ferromanganese seamount crusts from the central Pacific, *Naturwissenshaften,* 71, 577, 1984.

87. Hein, J. R., Ahn, J-H., Wong, J. C., Kang, J.-K., Smith, V. K., Yoon, S.-H., d'Angelo, W. M., Yoo, S.-O., Gibbs, A. E., Kim, H.-J., Quinterno, P. J., Jung, M.-Y., Davis, A. S., Park, B.-K., Gillison, J. R., Marlow, M. S., Schulz, M. S., Siems, D. F., Taggart, J. E., Rait, N., Gray, L., Malcolm, M. J., Kavulak, M. G., Yeh, H.-W., Mann, D. M., Noble, M., Riddle, G. O., Roushey, B. H., and Smith, H., Geology, geophysics, geochemistry, and deep-sea mineral deposits, Federated States of Micronesia: KORDI-USGS R.V. *Farnella* Cruise F11-90-CP, U.S. Geological Survey Open File Report 92-218, 191 pp., 1992.

88. De Carlo, E. H., Paleoceanographic implications of rare earth element variability within a Fe–Mn crust from the central Pacific Ocean, *Marine Geology,* 98, 449, 1991.

89. McMurtry, G. M., Vonderhaar, D. L., Eisenhauer, A., and Mahoney, J. J., Cenozoic accumulation history of a Pacific ferromanganese crust, *Earth and Planetary Science Letters,* 125, 105, 1994.

90. De Carlo, E. H. and Fraley, C. M., Chemistry and mineralogy of ferromanganese deposits from the equatorial Pacific Ocean, in *Geology and Offshore Mineral Resources of the Central Pacific Basin,* Keating, B. H. and Bolton, B. R. (Eds.), Circum-Pacific Council for Energy and Mineral Resources, Houston, Texas, Earth Science Series, v. 15, 1990, 225.

91. Grau, R. and Kudrass, H. R., Pre-Eocene and younger manganese crusts from the Manihiki Plateau, southwest Pacific Ocean, *Marine Mining,* 10, 231, 1991.

92. Neumann, T. and Stüben, D., Detailed geochemical study and growth history of some ferromanganese crusts from the Tuamotu Archipelago, *Marine Mining,* 10, 29, 1991.

93. Puteanus, D., Glasby, G. P., Stoffers, P., Mangini, A., and Kunzendorf, H., Distribution, internal structure, and composition of manganese crusts from seamounts of the Teahitia-Mehetia hot spot, southwest Pacific, *Marine Mining,* 8, 245, 1989.

94. Segl, M., Mangini, A., Bonani, G., Hofmann, H. J., Morenzoni, E., Nessi, M., Suter, M., and Wölfli, W., ^{10}Be dating of the inner structure of Mn-encrustations applying the Zürich tandem accelerator, *Nuclear Instruments and Methods in Physics Research,* B5, 359, 1984.

95. Halbach, P. and Puteanus, D., The influence of the carbonate dissolution rate on the growth and composition of Co-rich ferromanganese crusts from central Pacific seamount areas, *Earth Planetary Science Letters,* 68, 73, 1984.

96. Hein, J. R., Yeh, H.-W., Gunn, S. H., Sliter, W. V., Benninger, L. M., and Wang, C.-H., Two major Cenozoic episodes of phosphogenesis recorded in equatorial Pacific seamount deposits, *Paleoceanography,* 8, 293, 1993.

97. Aplin, A. C. and Cronan, D. S., Ferromanganese oxide deposits from the central Pacific Ocean, I. Encrustations from the Line Islands Archipelago, *Geochimica et Cosmochimica Acta,* 49, 427, 1985.

98. Hodkinson, R. A. and Cronan, D. S., Regional and depth variability in the composition of cobalt-rich ferromanganese crusts from the SOPAC area and adjacent parts of the central equatorial Pacific, *Marine Geology,* 98, 437, 1991.

99. von Stackelberg, U., Kunzendorf, H., Marchig, V., and Gwozdz, R., Growth history of a large ferromanganese crust from the equatorial north Pacific nodule belt, *Geologisches Jahrbuch,* A75, 213, 1984.

100. Chave, K. E., Morgan, C. L., and Green, W. J., A geochemical comparison of manganese oxide deposits of the Hawaiian Archipelago and the deep sea, *Applied Geochemistry,* 1, 233, 1986.

101. Koeppenkastrop, D. and De Carlo, E. H., Sorption of rare earth elements from sea water onto synthetic mineral particles: an experimental approach, *Chemical Geology,* 95, 251, 1992.

102. Hem, J. D., Redox processes at surfaces of manganese oxide and their effects on aqueous metal ions, *Chemical Geology,* 21, 199, 1978.

103. Bruland, K. W., Trace elements in sea-water, in *Chemical Oceanography,* 8, Riley, J. P. and Chester, R. (Eds.), Academic Press, London, 1983, 157.

104. Takematsu, N., Sato, Y., and Okabe, S., Factors controlling the chemical composition of marine manganese nodules and crusts: a review and synthesis, *Marine Chemistry,* 26, 41, 1989.

105. Quinby-Hunt, M. S. and Turekian, K. K., Distribution of elements in sea water, EOS, *Transactions American Geophysical Union,* 64, 130, 1983.

106. Gramm-Osipov, L. M., Hein, J. R., and Chichkin, R. V., Manganese geochemistry in the Karin Ridge region: Preliminary physiochemical description, in Data and results from R.V. Aleksandr Vinogradov cruises 91-AV-19/1, north Pacific hydrochemistry transect; 91-AV-19/2, north equatorial Pacific Karin Ridge Fe–Mn crust studies; and 91-AV-19/4, northwest Pacific and Bering Sea sediment geochemistry and paleoceanographic studies, Hein, J. R., Bychkov, A. S., and Gibbs, A. E. (Eds.), U.S. Geological Survey Open File Report 94-230, 1994, 99.

107. Abouchami, W., Goldstein, S. L., Galer, S. J. G., Eisenhauer, A., and Mangini, A., Secular changes of lead and neodymium in central Pacific sea water recorded by a Fe–Mn crust, *Geochimica et Cosmochimica Acta,* 61, 3957, 1997.

108. von Blanckenburg, F., O'Nions, R. K., and Hein, J. R., Distribution and sources of pre-anthropogenic lead isotopes in deep ocean water from Fe–Mn crusts, *Geochimica et Cosmochimica Acta,* 60, 4957, 1996.

109. Godfrey, L. V., Lee, D.-C., Sangrey, W. F., Halliday, A. N., Salters, V. J. M., Hein, J. R., and White, W. M., The Hf isotopic composition of ferromanganese nodules and crusts and hydrothermal manganese deposits: implications for sea water Hf, *Earth and Planetary Science Letters,* 151, 91, 1997.

110. Burton, K. W., Birck, J.-L., Allègre, C. J., von Blanckenburg, F., O'Nions, R. K., and Hein, J. R., Osmium distribution and behaviour in the oceans recorded by Fe–Mn crusts, *Earth and Planetary Science Letters,* 171, 185, 1999.

111. Segl, M., Mangini, A., Bonani, G., Hofmann, H. J., Nessi, M., Suter, M., Wölfli, W., Friedrich, G., Plüger, W. L., Wiechowski, A., and Beer, J., [10]Be dating of a manganese crust from central North Pacific and implications for ocean palaeocirculation, *Nature,* 309, 540, 1984.

112. Welling, C. G., An advanced design deep sea mining system, Offshore Technology Conference, Paper 4094, Houston, TX, 1981.

113. Halkyard, J., Ocean engineering challenges in deep-sea mining, Society of Naval Architects and Marine Engineers, Proceedings, STAR Symposium, Honolulu, HI, 1982.

114. Japan Resources Association, Study report for exploitation of Co-rich manganese crust, unpublished report, 1985.

115. Masuda, Y., Crust mining plans of the Japan Resources Association, *Marine Mining,* 10, 95, 1991.

116. Bernhard, H. H. and Blissenbach, E., Economic importance, in *The Manganese Nodule Belt of the Pacific Ocean,* Halbach, P., Friedrich, G., and von Stackelberg, U. (Eds.), Ferdinand Enke Verlag Stuttgart, 1988, 4.

117. Callies, D. L. and Johnson, C. J., Legal, business and economic aspects of cobalt-rich manganese crust mining and processing in Republic of the Marshall Islands, unpublished report, 1989.

118. Yamazaki, T., Tomishima, Y., Handa, K., and Tsurusaki, K., Fundamental study on remote sensing of engineering properties of cobalt rich manganese crusts, in *Proceedings Ninth International Conference of Offshore Mechanics and Arctic Engineering,* Salama, M. M., Rhee, H. C., Williams, J. G., and Liu, S. (Eds.), The American Society of Mechanical Engineers Book No. 10296E, 1990, 605.

119. Tomishima, Y., Yamazaki, T., Tsurusaki, K., and Handa, K., Fundamental studies on in-situ measurement of engineering characteristics of cobalt rich manganese crusts by mechanical means, *Proceedings Techno-Ocean '90 International Symposium,* Kobe, Japan, 1990, 579.

120. Larson, D. A., Tandanand, S., Boucher, M. L., Olson, M. S., Morrell, R. J., and Thill, R. E., Physical properties and mechanical cutting characteristics of cobalt-rich manganese crusts, Bureau of Mines Report of Investigation RI-9128, 35 pp., 1987.

121. Svininnikov, A. I., Physical properties of rocks and sediments from Karin Ridge (Central Equatorial Pacific) and the Bering Sea, in Data and results from R.V. Aleksandr Vinogradov cruises 91-AV-19/1, north Pacific hydrochemistry transect; 91-AV-19/2, north equatorial Pacific Karin Ridge Fe–Mn crust studies; and 91-AV-19/4, northwest Pacific and Bering Sea sediment geochemistry and paleoceanographic studies, Hein, J. R., Bychkov, A. S., and Gibbs, A. E. (Eds.), U.S. Geological Survey Open File Report 94-230, 1994, 103.

122. Chung, J. S., Deep ocean mining: technologies for nodules and crusts, *Proceedings of First ISOPE International Deep-Ocean Technology Symposium and Workshop,* Los Angeles, CA, International Society of Offshore and Polar Engineers, Golden, CO, 1996, 21.

123. Huh, C.-A., Thorium-protactinium age dating of ferromanganese crusts, *Acta Oceanographica Taiwanica,* 20, 95, 1988.

124. Sharma, P. and Somayajulu, B. L. K., ^{10}Be dating of large manganese nodules from world oceans, *Earth and Planetary Science Letters,* 59, 235, 1982.

125. Segl, M., Mangini, A., Beer, J., Bonani, G., Suter, M., and Wölfili, W., Growth rate variations of manganese nodules and crusts induced by paleoceanographic events, *Paleoceanography,* 4, 511, 1989.

126. Eisenhauer, A., Gögen, K., Pernicka, E., and Mangini, A., Climatic influences on the growth rates of Mn crusts during the Late Quaternary, *Earth and Planetary Science Letters,* 109, 25, 1992.

127. Ku, T. L., Kusakabe, M., Nelson, D. E., Southon, J. R., Korteling, R. G., Vogel, J., and Nowikow, I., Constancy of oceanic deposition of ^{10}Be as recorded in manganese crusts, *Nature,* 299, 240, 1982.

128. Guichard, F., Reyss, J.-L., and Yokoyama, Y., Growth rate of manganese nodule measured with ^{10}Be and ^{26}Al, *Nature,* 272, 155, 1978.

129. Mangini, A., Segl, M., Kudrass, H., Wiedicke, M., Bonani, G., Hofmann, H. J., Morenzoni, E., Nessi, M., Suter, M., and Wölfli, W., Diffusion and supply rates of ^{10}Be and ^{230}Th radioisotopes in two manganese encrustations from the South China Sea, *Geochimica et Cosmochimica Acta,* 50, 149-156, 1986.

130. Koschinsky, A., Halbach, P., Hein, J. R., and Mangini, A., Ferromanganese crusts as indicators for paleoceanographic events in the NE Atlantic, *Geol. Rundsch,* 85, 567, 1996.

131. Moore, W. S., Accumulation rates of manganese crusts on rocks exposed on the seafloor, Inter-University Program of Research on Ferromanganese Deposits of the Ocean Floor, Phase Report, April 1973, 93-96, unpublished report.

132. O'Nions, R. K., Frank, M., von Blanckenburg, F., and Ling, H.-F., Secular variation of Nd and Pb isotopes in ferromanganese crusts from the Atlantic, Indian and Pacific Oceans, *Earth and Planetary Science Letters,* 155, 15, 1998.

133. Banakar, V. K. and Borole, D. V., Depth profiles of ^{230}Th$_{excess}$, transition metals and mineralogy of ferromanganese crusts of the central Indian basin and implications for palaeoceanographic influence on crust genesis, *Chemical Geology,* 94, 33, 1991.

134. Exon, N. F., Ferromanganese crust and nodule deposits from the continental margin south and west of Tasmania, *Australian Journal of Earth Sciences,* 44, 701, 1997.

135. Walter, P., Glasby, G. P., Plüger, W. L., Kunzendorf, H., and Meylan, M. A., Mineralogy and composition of manganese crusts and nodules and sediments from the Manihiki Plateau and adjacent areas: Results of HMNZS Tui cruises, *Marine Georesources and Geotechnology,* 13, 321, 1995.

136. Meylan, M. A., Glasby, G. P., Hill, P. J., McKelvey, B. C., Walter, P., and Stoffers, P., Manganese crusts and nodules from the Manihiki Plateau and adjacent areas: results of HMNZS Tui cruises, *Marine Mining,* 9, 43, 1990.

137. Cronan, D. S. and Hodkinson, R. A., Manganese nodules and cobalt-rich crusts in the EEZ's of the Cook Islands, Kiribati and Tuvalu, Part III: nodules and crusts in the EEZ of western Kiribati (Phoenix and Gilbert Islands), CCOP/SOPAC Technical Report 100, Suva Fiji, 47 pp., 1989.

138. Cronan, D. S. and Hodkinson, R. A., Manganese nodules and cobalt-rich crusts in the EEZ's of the Cook Islands, Kiribati and Tuvalu, Part IV: Nodules and crusts in the EEZ of Tuvalu (Ellice Islands), CCOP/SOPAC Technical Report 102, Suva Fiji, 59 pp., 1990.

139. Cronan, D. S., Hodkinson, R. A., Miller, S., and Hong, L., Manganese nodules and cobalt-rich crusts in the EEZ's of the Cook Islands, Kiribati and Tuvalu, Part II: Nodules and crusts in the EEZ's of the Cook Islands and part of eastern Kiribati (Line Islands), CCOP/SOPAC Technical Report 99, Suva Fiji, 44 pp., 1989.

140. Glasby, G. P., Cullen, D. J., Kunzendorf, H., and Stüben, D., Marine manganese crusts around New Zealand, *Miscellaneous Publications, New Zealand Oceanographic Institute,* 106, 1991, 1.

141. Glasby, G. P. and Wright, I. C., Marine mineral potential in New Zealand's exclusive economic zone, *22nd Annual Offshore Technology Conference,* Houston, Texas, 1990, 479.

142. Haynes, B. W., Law, S. L., Barron, D. C., Kramer, G. W., Maeda, R., and Magyar, J., Pacific manganese nodules: characterization and processing, U.S. Bureau of Mines Bulletin, 679, 44 pp., 1985.

143. McLennan, S. M., Rare earth elements in sedimentary rocks: Influence of provenance and sedimentary processes, in *Geochemistry and Mineralogy of Rare Earth Elements,* Lipin, B. R. and McKay, G. A. (Eds.), Mineralogical Society of America Reviews in Mineralogy, 21, Washington, D.C., 1989, 168.

144. Anders, E. and Grevesse, N., Abundances of the elements: meteoritic and solar, *Geochimica et Cosmochimica Acta,* 53, 197-214, 1989.

145. De Baar, H. J. W., Bacon, M. P., Brewer, P. G., and Bruland, K. W., Rare earth elements in the Pacific and Atlantic Oceans, *Geochimica et Cosmochimica Acta,* 49, 1943, 1985.

146. Bonatti, E., Kraemer, T., and Rydell, H., Classification and genesis of submarine iron-manganese deposits, in *Ferromanganese Deposits on the Ocean Floor,* Horn, D. R. (Ed.), Arden House, Harriman, New York, 1972, 149.

10 Innovations in Marine Ferromanganese Oxide Tailings Disposal

John C. Wiltshire

ABSTRACT

Manganese nodules and crusts represent an ocean mineral resource that will likely be developed during the course of the 21st century. Recent environmental work on manganese crusts has shown that 75% of the environmental problems associated with marine ferromanganese operations will be with the processing phase of the operation, particularly the disposal of the waste material, the tailings. Traditionally, mine tailings are dumped in a tailings pond and left there. Current work with manganese tailings has shown them to be a resource of considerable value in their own right. Tailings have applications in a range of building materials as well as in agriculture. This paper will review the ways in which tailings are disposed of and the potential alternate uses of these tailings. They are a useful additive as fine-grained aggregate in concrete, to which they impart higher compressive strength, greater density, and reduced porosity. The manganese appears to have some anti-biofouling capabilities. The tailings serve as an excellent filler for certain classes of resin cast solid surfaces, tiles, asphalt, rubber, and plastics. The tailings also have applications in coatings and ceramics. Two-year agricultural experiments have documented that tailings mixed into the soil can significantly stimulate the growth of Hawaiian Koa trees. The effect of using mineral tailings for secondary applications enhances the profitability of marine mining operations by a small but important margin.

10.1 INTRODUCTION

Manganese nodules and crusts represent a resource that will likely be developed during the 21st century.[1,2] The 21st century will also almost certainly be marked by heightened environmental consciousness. At a time when increasingly greater numbers of diseases are linked to environmental causes and environmental litigation is at an all time high, it is difficult to imagine that an entire new class of mining operation, deep-sea mining, would be able to come into production without the tightest of controls. Recent environmental work on manganese crusts has shown that 75% of the environmental problems would be associated with the processing phase of the operation.[3] Most of this has to do with the disposal of the waste material, the tailings.

Most terrestrial mining operations do the primary ore concentration on site because it is expensive to ship waste rock. Almost every mine, therefore, has a tailings disposal pond. Simply dumping the tailings at the mine site is the traditional and simplest way to dispose of tailings.[4] In recent years elaborate forms of pond liners and reclamation practices have been developed. While these are laudable in comparison with earlier practices, in actual fact most tailings ponds only delay environmental problems to a later time when the liners break and the material leaches into the groundwater. Permanent solutions must involve some form of getting the tailings back into the cycle of the environment. This chapter is predicated on the assumption that inadequate disposal practices of the past will not be tolerated for new marine mining operations in the future.

Another factor to be taken into account is that mining is a changing industry. This change is being driven by advances in computer power, communications, robotics, fiber optics, and sensors. Mining in the 21st century will be highly specific and high tech.[2] Great efforts will be made to mine only valuable ore and minimize the amount of waste material taken. This will be done by small robotic miners using small specialized cutter heads or *in situ* leaching to take ore only and not gangue. The mine of the future will be more efficient than the present-day mine, meaning the gangue-to-ore ratio will be very low. The amount of tailings will decline with time. As processing becomes more efficient, the tailings left will decline as well. Tailings are now being taken into account in the design of processes to ensure that the waste at the end of the mining cycle is manageable.

The ideal solution to mine waste disposal is to sell the tailings as a by-product to a third party and provide income to the mine as well as eliminating a disposal problem. Surprisingly, very little effort has been spent in this direction by the mining industry. There are two basic reasons for this. First, compared to selling ore, selling tailings is difficult and relatively unprofitable. It would be most unusual for one tailings customer to be able to take even a significant fraction, let alone most of a mine's tailings. Every customer has different requirements for the tailings. Most requirements involve some processing of the tailings; typical would be drying, sizing, bagging, or removing some mineral fraction. This sort of thing is costly, particularly in small quantities. Second, most mines are able to meet current regulations for tailings disposal simply by dumping the tailings in a tailings pond and later either planting over this pond or simply leaving it when the mine is finally abandoned. However, as the regulatory noose tightens on the neck of the mining industry, alternatives to current practices will be investigated more assiduously.

This paper will focus on recent alternatives developed for marine minerals tailings management rather than traditional disposal approaches. The literature on the engineering principles, successes and failures of current tailings management practices is very extensive.[4,5] This paper will focus on recent work done on the tailings of manganese nodules and crusts as well as analogous terrestrial manganese tailings. However, the work described will apply, in part, to many marine sulfide tailings as well as to a broad range of terrestrial tailings.

10.2 FERROMANGANESE MARINE MINING OPERATIONS AND CURRENT TAILINGS DISPOSAL PRACTICES

10.2.1 Overview

Traditional mining scenarios for manganese nodules involve a surface mining ship connected to and controlling an underwater robotic miner. The mined nodules or crusts are lifted to the surface by an airlift or pump system. The cleaned nodules are then transferred at the surface to an ore carrier. The washings are discharged over the side. This situation is the first element in waste disposal. The effect of the surface discharge is of several types. The particles sort out by size range, the largest dropping out first and going to the ocean floor near the mining operation. The effects of mining and discharge have been heavily studied by the U.S. National Oceanic and Atmospheric Administration.[6] Their results are summarized in Table 10.1. At the risk of over-simplification, the results of mining may be summarized as complete destruction of the benthic organisms in the path of the miner, partial destruction of the benthos by blanketing of sediment at varying distance away from the mining track, and disruption of larvae in the surface waters by the surface discharge plume. The surface plume effect can be minimized by disposing at deeper depths by extending the discharge pipe below the thermocline or deeper.

Following mining, assuming that the minerals are taken to the shore for processing, there is the normal pollution associated with marine transport of ores. This includes diesel engine fumes,

TABLE 10.1
NOAA Comprehensive Assessment of Ocean Mining Impacts

Concerns with Potential for Significant or Adverse Impact	
Benthic Impacts	**Surface Discharge**
Collector destroys benthos in and near collector track	Increased turbidity, decreased productivity
May blanket benthos, diluting food supply away from mine site area	May affect fish larvae

Concerns without Potential for Significant or Adverse Impacts	
Beneficial Effects	**Low Probability of Impact**
	Benthic Impacts
Additional food supply for benthic scavengers	Light from the collector
	Benthic plume increasing oxygen demand, trace metal increase
	Surface Discharge
Bacteria increase food supply for zooplankton from surface discharge	Bacterial growth depletes oxygen
	Phytoplankton species composition affected
	Zooplankton mortality and species changes
	Trace metals enter the food web
	Trace metal effects on phytoplankton
	Nutrient increase causes phytoplankton bloom
	Airlift causes embolism
	Affects fish

truck traffic, mineral dust from loading and unloading operations, and normal ship discharges.[3] As a rule these sources of pollution are normal for a busy port and are not handled in any special way. The ore is transported to the processing plant by conveyor belt, pipeline in a slurry form, or as bulk cargo by trucks or train, depending on the distance. Again, with the exception of noise and dust, this transport provides no extraordinary pollution.

The major disposal problems occur at the processing plant. Normally the manganese crusts or nodules will have to be washed to remove any residual salt before leaching. The water from this washing will have to be cleaned before disposal. At various stages in the leaching and neutralization, there are liquid effluents. These may be cleaned individually or run through a waste water treatment plant collectively. Sometimes waste water can be cleaned sufficiently for reuse, sometimes it can be injected into a salt water-containing aquifer. Much effort has gone into the science of mine water cleanup.[5] This effluent is commonly referred to as acid mine drainage. The most advanced of a number of cleanup scenarios use some form of artificial ponds or wetlands, most often involving cattails *(Typha)* and peat moss *(Sphagnum)*, the two species shown to be most adept at mine waste water cleanup.[5] Typically, the waste water will circulate over several limestone beds and through various artificial wetlands rich in these and related species. At the end of the circulation, a certain amount of cleaned water is lost to groundwater, and the rest is usually sufficiently cleaned to discharge to a natural stream, lake, river, or ocean. The larger, and as yet less satisfactorily, engineered problem is with solid waste, that is tailings disposal. Typically, mine tailings are disposed of in tailings ponds at the processing plant (or mine concentrator) site.

10.2.2 TAILINGS PONDS

Tailings ponds are the classical method of mine waste disposal.[4] Typically, these large impounding structures consist of a dike, an inflow, and a decant system. Usually the material discharged into the ponds is waste from the mine (or processing plant) concentrator only. Sometimes the discharge includes general industrial waste as well as concentrator effluent. This discharge enters the ponds either through a peripheral discharge system or from a single point discharge. In the point system, the point end of the discharge pipe is often moved as the tailings pond is systematically filled.

Tailings ponds are constructed in one of two ways.[4] Either the containment walls are constructed with the mill tailings themselves or with excavated materials from nearby quarries. Using the mill tailings is considerably cheaper. A starter dam of quarried material must first be built in either case. If the tailings themselves can be used for the dikes, the initial dam is followed by gradual buildup of the coarser material from the mill tailings. The major disadvantage in using the tailings is in controlling the grain size and compaction characteristics to prevent dam failure or seepage.

Typically, the tailings from the mill are subjected to cyclones which separate some of the size fractions. The cyclone overflow might contain 10% fines, whereas the underflow could contain as much as 70% sand. The fines are deposited behind the tailings dams, whereas the sand could be deposited to form the dike. In the best engineering practice, this sand fraction of the tailings is deposited between an upstream and downstream starter dike, thus forming the core of the final tailings dam. The sand would be allowed to dewater for several days and then be positioned in the tailings embankment using heavy earth-moving equipment. In a well-designed dam there is also a drainage system to remove seepage water from the sand core of the tailings dam.[4]

By contrast, dams built of excavated materials are either completely constructed at the beginning of the project or are partially constructed at the beginning and gradually added to as the project continues, should a later need arise. Material can come from an independent excavation or could be overburden or waste rock from the mine. Clearly, if mine overburden can be used, this is a considerable savings over a new excavation.

One of the major problems with tailings dams is seepage. Seepage affects the local area by allowing contaminated water out of the dam and contributes to dam weakening and collapse. A wide range of methods are used to reduce seepage. These include the placement of impervious clay or soil layers in the core of the dike. Some ponds are lined with concrete, asphalt, or grout. Membranes or layers of very fine tailings have also been used. While these methods can prevent seepage while the tailings pond is in operation if properly emplaced, they tend to break down over time once the pond is abandoned.

The tailings are usually pumped into the tailings dam in a slurry. They contain a significant amount of effluent water. As the more solid part of the waste sinks in the ponds, the effluent must be removed. Typically, the liquid to be decanted is much less than the total input to the tailings ponds as liquid is lost to seepage and evaporation as well as entrapped in the tailings. Depending on the chemical contaminants in this liquid, it may be recycled back to the plant. If it is quite clean, it may be discharged into natural drainage, typically a stream, river, or lake, or it may be used in agricultural irrigation in the best designed systems. In the more likely case that the water is contaminated, it will go to an evaporation pond or special treatment pond.

The effluent from a tailings pond is typically collected in one of three ways. The best and probably most common way is to build a decant system into the tailings pond.[4] Such a system is made with a series of towers in the tailings pond with attached piping running underneath the pond to a pump station. The effluent is discharged or recycled to the plant. As the capacity of the tailings pond is filled and the tailings dam built higher, the height of the decant towers is also increased by adding sections of pipe or precast concrete rings. The towers are designed so that sections can be added for the entire life of the pond. The sections are typically less than a foot in height and carefully adjusted to the rising water level in the ponds. Drainage through the decant system is only through the top and is protected from clogging by screens and assorted trash racks. This tower decant system is the best for drainage of tailings ponds because it is built in. When the pond is finally abandoned, the tailings drainage system will still be functional and able to handle rain overflow and final pond drainage.

Less satisfactory alternatives involve the use of pumps or siphons placed over the tailings dams. These methods are cheaper because they do not require the infrastructure of towers or under-pond piping. However, simply trying to pump or siphon a large pond from one or several points along the edge is subject to a host of problems including power outages, the need to continually reposition the intakes and drains, problems with heavy rains or freezing in winter, and perhaps most severe, the complete lack of a drainage system after mine abandonment.

Although tailings ponds are the principal method of tailings disposal and have been in use worldwide for many years, they have a history fraught with problems. Foremost among these problems are large areas of devastated land and polluted water from pond seepage and runoff. Problems with the slump collapse of the tailings dams during the active life of the mine or after its abandonment are unfortunately relatively common. This is a particular problem for high abandoned tailings dams (some of these dams can be 100 meters in height) in areas of heavy rainfall. While significant progress in tailings pond abandonment procedures has taken place, including planting over dams, reducing slopes, grading, and cementing,[4,5] the problem has as much been shifted into the future as solved.

10.2.3 OTHER DISPOSAL OPTIONS

Depending on the nature of the marine mining operation, several other options are also possible for tailings disposal apart from tailings ponds.[3,6] These include ocean dumping, offshore outfall, and injection. Each of these has drawbacks but may be applicable to a marine mining operation in certain circumstances. For a vessel, ocean dumping is very attractive for the mud brought up with the nodules or crusts. If tailings are disposed of below the surface layer so as not to produce a plume, this could be an effective method. However, if the vessel were designed to process the minerals at sea and then dump the resulting tailings, the London Dumping Convention would need to be considered. Whether the Convention would allow the dumping of processed tailings from a floating plant is a matter of interpretation. The final sections of this review, on the economics of manganese tailings, suggest that this might not be the best economic decision even if it's permitted.

If processing were on land near the ocean, a disposal option used in several mines is offshore outfall. The has been successfully used in Atlas Consolidated Mines in the Philippines[7] and in Island Copper Mine[8] on Vancouver Island in Canada. Ellis has studied the effect of Island Copper outfall for over 20 years with the definitive conclusion that little damage has been done to the environment.[9] Nonetheless, new large-volume ocean outfalls are exceptionally hard to permit, particularly in an area of competing fishing or tourist use.[3]

For liquid processing waste or a very finely dispersed tailings slurry, there is the option of injection into a disposal well. Usually, this is done on the processing site. This is an expensive option and runs the risk of groundwater contamination. Generally it is only applicable for highly contaminated waste. The ultimate cost of this kind of disposal must be very carefully weighed against the superior option of cleaning the contaminated material and disposing of it on the surface in a tailings pond or as clean water disposal into a stream or the ocean, perhaps after multiple uses as process water in the plant.

The remaining sections of this chapter will consider the possible uses to which manganese nodule or crust tailings can be put rather than risking the consequences of inadequate disposal. To complicate the decision for miners, the interpretation of inadequate disposal is very much a moving target as court interpretations get tougher and the environmental lobby gets stronger. As medical science links more diseases to low levels of contamination in the public water supply, this trend is likely to continue if not accelerate.

10.3 PROPERTIES OF MANGANESE TAILINGS

Two basic routes exist for processing manganese nodules or crusts: acid leaching and smelting. Numerous processes have been devised for each. The tailings of an acid leach process after drying are a fine-grained black powder. A smelting process produces smelter slag which is blocky and

TABLE 10.2
Chemical Composition of Manganese Tailings
From the Groote Eylandt Deposit

Oxide	Percentage	Elemental Percentages	%
MnO_2	36.0	Mn	22.6
SiO_2	28.1	Si	13.1
Al_2O_3	20.3	Al	10.7
FeO	5.6	Fe	4.40
TiO_2	0.48	Ti	0.29
K_2O	0.3	K	0.25
MgO	0.14	Mg	0.08
SrO	0.04	Sr	0.03
CaO	0.04	Ca	0.02
Na_2O	~0.5 (estimate)		
H_2O	~8.0 (estimate)		

often inert. It can be used as road ballast if it is sufficiently inert to resist rainwater leach into aquifers. Generally, smelter slag does not pose the disposal problem that acid leach tailings do. For this reason, and the fact that acid leach processes are less energy intensive and therefore potentially favored in Pacific Island areas, the present review focuses solely on the acid leach manganese tailings. Many terrestrial manganese mines beneficiate and separate the manganese size fractions by physical processes that do not involve acid. After extensive testing, it was found that the fine-grained tailings produced by such processes are very similar to marine mineral acid leach tailings after the acid residues are neutralized. As these analogous terrestrial tailings are abundant and consistent in composition between batches, they have been the focus of studies by Wiltshire and Troy,[10] Lay and Wiltshire,[11] Wiltshire and Loudat,[12] and Wiltshire[13] in which they were used as an occasional proxy for marine ferromanganese oxide tailings.

Chemical and physical property tests on fine-grained manganese tailings from the Groote Eylandt manganese deposit in Northern Australia, were done at the University of Hawaii as well as by Micromeritics Corporation (for particle size distribution and gas absorption). Similar detailed tailing studies have been done for a range of tailings.[14] Wiltshire and Troy[10] compared the composition of some of the nodule tailings analyzed by Haynes et al.[14] and manganese crust tailings.[3] All of the results are closely related. Data specific to the Groote Eylandt manganese deposit tailings are summarized in Table 10.2 for chemical composition and in Table 10.3 for other properties. Basically, the tailings are small angular manganese dioxide grains intermixed with clay and small amounts of iron oxide and silica. They pack well because of their angularity but have a high surface area.

10.4 TAILINGS UTILIZATION

10.4.1 MAJOR TONNAGE USES

To be successful in a tailings management scheme, beneficial utilization must use a large percentage of the available tailings. If this is not done, the mining company is faced with having to provide large tailings ponds as well as other strategies and arrangements for utilization. To be attractive, beneficial utilization must at least reduce the number of tailings ponds needed. Those applications that can accept the largest tonnages of tailings are the best. Three applications stand out as particularly attractive from this point of view. These are: (1) fine-grained aggregate in concrete, (2) soil amendments in agricultural applications, and (3) asphalt additives for road surfacing. The

TABLE 10.3
General Properties of Manganese Tailings from the Groote Eylandt Deposit

Specific gravity	3.46
Melting range	1240–1285°C
pH	5.5 (in water suspension)
Grain size range	0.1 to 300 microns
Median grain size	3 microns, 50% of sample mass was between 0.7 and 9 microns
Mineralogy	36% pyrolusite, 51% silicates (largely kaolinite), trace quartz
Magnetic properties	Pyrolusite (paramagnetic), Kaolinite (non-magnetic)
Hardness	6 on the Moh's scale
Particle surface area	very high, 26.9 m^2/g
Average pore diameter	146 angstroms
Gas absorption	0.30 cm^3/g absorption of CO gas
Electrical properties	Resistor at all temperatures
Solubility	Insoluble in water
Appearance	Gray-black, fine-grained powder, no odor

next three sections will focus on each of these uses in turn. Other applications can give higher value, although lower tonnage use for the tailings, and will be described later.

10.4.1.1 Concrete

The American Concrete Institute[15] has done considerable work on admixtures to concrete. This includes the successful incorporation of superplasticizers to increase concrete durability and blast furnace slag to increase strength and weight, fly ash from power plants and a very fine-grained industrial waste, known as silica fume, to increase strength. Specialty concretes for the marine environment have been given a lot of study. Another area of major ongoing research is in the texturing of concrete pavements, particularly to provide tough skid-resistant roadways. This is done both by sculpturing the concrete with grooves and also by adding hard gritty material, such as tailings. Considerable electron microscopy work has recently been undertaken to document the advantages of new nontraditional additives to concrete mixes.[16] The successful incorporation of very fine-grained fly ash and silica fume have made the concrete industry more open to the benefits of incorporating manganese tailings.

Wiltshire[17] tested varying mixtures of Portland cement, coarse sand, and tailings. Mixtures were made with 0 to 60% tailings in the concrete. Two sets of experiments were conducted. The first involved nodule tailings that were not fully neutralized after acid leach processing. These tailings had been washed and had a pH of approximately 4. They were made into standard eight-inch-long, four-inch-diameter concrete testing cores and tested for compressive strength by a commercial concrete testing company after curing for 33 days. The sample containing no tailings had a strength of 3,460 psi, effectively the same as the standard value given for concrete of 3,500 psi. The other samples decreased considerably in strength with increasing amounts of tailings. To determine whether the decrease in strength was solely due to the tailings content, the experiment was repeated with tailings that had been fully neutralized. The results were very different. Compressive strengths above 4,000 psi were achieved in concrete containing 20 to 25% tailings. The strength decreased fairly rapidly to 1,000 psi for concrete containing 50% tailings. Although 1,000 psi concrete could not be used in buildings, it would still be applicable for driveways and many other paving applications. These results are shown in Figure 10.1.

In addition to increased strength, the tailings appear to give the concrete several other interesting properties. The fine-grained nature of the tailings appears to make the concrete more moldable and bubble-free. This was demonstrated in two ways. First, a moldability test was performed using a

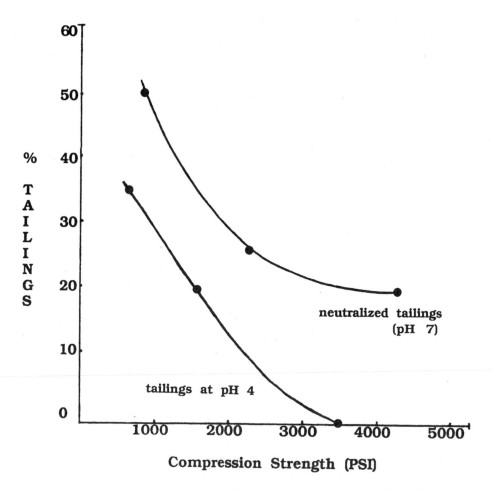

FIGURE 10.1 Compression strength of concrete made with manganese nodule tailings. Two separate runs are illustrated, the first with tailings at pH 4, the second with neutralized tailings at pH 7. The results clearly show that tailings must be fully neutralized, and when this is done, concrete with strengths in excess of the standard 3500 psi can be achieved by tailings additions in the range of 20 to 25%.

12-inch by 4-inch latex rubber mold of an ornamental Japanese fish. The mold was made with considerable attention to fine detail. This fine detail was picked up in a casting using concrete containing 30% tailings but not by the standard concrete. Further, the ferromanganese concrete gave a much smoother bubble-free surface. The surface textures were compared in another test by examining cut surfaces of the ferromanganese tailings concrete and a standard precast concrete brick. The ferromanganese surface had bubble pits over less than 2% of its surface. These pits ranged in size from pinholes to 2 mm in diameter, averaging about 500 microns. By contrast, the precast brick had pits covering over 20% of its surface ranging to 8 mm in diameter and averaging about 2 mm in diameter. The difference was very marked. The precast brick had a rough surface, the ferromanganese concrete a very smooth surface. In addition, the tailings give the ferromanganese concrete a considerably greater density than regular concrete in that iron and manganese are being substituted for the less dense silica and aluminum of sand, and the ferromanganese concrete has less than one tenth the bubbles found in regular concrete.

Greater density and bubble reduction are two particularly important properties for concrete that is to be used in the marine environment. In the marine environment, wave action compresses air into the pores and pits of the concrete which, over time, breaks the concrete down. The same action results from water freezing and expanding in these pores. The fewer pores subjected to this action, the longer the concrete will endure. There is evidence to indicate that ferromanganese surfaces repel the growth of organisms.[3] If this can be confirmed, it may be that ferromanganese concrete would not be covered as quickly as other types of concrete with algae or encrusting organisms, which would offer advantages for outfall pipes as well as many other marine structures.

When the physical properties of the tailings are related to the characteristics of the resulting concrete it becomes clear what is happening. The tailings have a specific gravity of 3.46, much higher than normal silica sand. They are making the concrete denser. The tailings have a very high surface area which rapidly absorbs water. This makes the concrete dry faster. It also causes the fine clay grains in the tailings to expand to fill voids in the concrete, causing the concrete to have lower pore space and better moldability. The greater density, reduced pore space, and better grain linking caused by this void filling, in turn, translate into higher compressive strength. Evidently this advantage begins to be lost above an addition of about 20% tailings to the concrete mix, presumably after all the small voids have already been filled by the tailings.

Concrete theory suggests that manganese tailings below 300 microns in size should not be desirable for aggregate. Testing indicates this is not the case.[13] The tailings make the concrete denser, darker in color, more moldable, less porous, and somewhat stronger. While it is too early in the testing phase to tell exactly what is happening, the chemical scavenging properties of manganese may in some way be creating a bonding effect in addition to the normal hydration of the concrete. If this is the case, the concrete should strengthen over time beyond what would normally be anticipated.

A recent round of experiments, highlighted in Table 10.4, tried to determine the optimum mix components for a manganese tailings concrete. Specifically, the nature of this test was to try to quantify the optimum size and percentage of the coarse aggregate fraction. Work by Kosmatka and Panarese[16] has shown that, as a general rule, the fine aggregate portion of a concrete mix should be 30 to 35%. Using manganese tailings at 33% and 15% and four different coarse aggregate mixes, it was determined that a 40-mm maximum size aggregate over the 15 to 33% range tested produced a stronger than standard concrete (standard type I Portland cement mixed in the proportions illustrated in Table 10.4 with a standard aggregate has a nominal compression strength of 3500 psi).

The results illustrated in Table 10.4 give manganese tailings a definite role to play in commercial concrete production. These results also suggest the possibility of looking for other coarse-grain natural aggregate deposits which are not ideal for concrete at the moment because of grain size, but could be successfully mixed with the fine-grain manganese tailings to achieve a very attractive result. This could be particularly beneficial in the South Pacific Island economies, some of which have $50/ton aggregate costs.

Clearly, there are considerable avenues for further research on the properties of ferromanganese tailings concrete. Initial indications of increased compressive strength, superior moldability, higher density, and lower porosity give reason to believe that, particularly for specialty concretes, the addition of the ferromanganese waste is imparting very economically desirable characteristics. Ongoing work is looking at the characteristics of manganese concrete with respect to anti-biofouling properties and ferrocement applications for marine structures.

10.4.1.2 Agricultural Uses

The agricultural potential of ferromanganese tailings has been extensively investigated by El Swaify and Chromec[18] both with the straight growth of corn on acid leach tailings of manganese nodules and on mixtures of nodule tailings and several tropical soils. There were several primary conclusions of this study related to the agricultural potential of the tailings. First, with proper management, plant growth (corn was used as a typical indicator crop) is possible using 100% tailings as the growing medium. To achieve normal growth, however, large initial inputs of lime and major nutrients, particularly phosphorous, were required. Thereafter, much smaller maintenance additions

TABLE 10.4

Manganese Tailings Concrete Compression Strengths with Varying Coarse Aggregate Sizes

% Cement	% Coarse Aggregate	% Tailings	Water/Cement Ratio	Compression Strength (psi)	Compression Strength (N/mm²)
No coarse aggregate					
79	0	21	.58	3580	24.8
40	0	60	.56	2390	16.6
5 mm Maximum size coarse aggregate					
16	69	15	.57	1590	11.0
18	49	33	.57	1790	12.4
10 mm Maximum size coarse aggregate					
16	69	15	.57	3300	22.9
18	49	33	.57	3100	21.5
20 mm Maximum size coarse aggregate					
16	69	15	.57	3100	21.5
18	49	33	.57	2780	19.3
40 mm Maximum size coarse aggregate					
16	69	15	.57	3980	27.6
18	49	33	.57	3938	27.3

These are 28 day (standard) cures of concrete cores tested by Finlay Testing Laboratories — the largest commercial testing lab in Hawaii.

should be sufficient to sustain growth. Additionally, to further enhance growth, micronutrient additions may also be required or additions of manure or compost.

The second primary conclusion of the El Swaify and Chromec study relates to the use of tailings as a soil amendment (much smaller percentages of tailings addition). A soil amendment is a mineral component added to the upper layers of the soil, where plant growth takes place, to enhance the soil characteristics to support agriculture. These amendments can be in the form of fertilizer, such as superphosphates, or they can be organic materials such as mulch or manure or such things as cinders to help break up the clay layers of the soil. El Swaify and Chromec[18] added manganese tailings to a range of different tropical soils from the island of Hawaii at a rate of 0 to 20% by weight. They measured the impact of the tailings on each of these soils, quantifying particle size, aggregate stability, dispersion, slaking, swelling potential, and water retention.

The tailings interacted most strongly with the clay-rich montmorillonitic soils (2:1 clay minerals), significantly increasing the amount of the silt size secondary particles and reducing the swelling potential. This appears to be related to the charge characteristics of the soil.[18] This interaction helped the soil to drain better and reduced compaction. Significant impacts of tailings on soil characteristics for the other soils tested were either not evident or more weakly expressed, but in no case were they negative.

Wiltshire[13] undertook follow-on experiments investigating the agricultural potential of Groote Eylandt manganese tailings as a soil amendment. In order to get the most meaningful results, a commercial crop was selected. To avoid the issue of metal uptake in a food crop, a tree crop was selected. Hawaiian Koa is a very valuable tree crop grown commercially on the island of Hawaii largely for furniture. It grows quickly in the initial stages, facilitating measurements over a one-year period. It then continues to grow for 30 years until maturity. An experiment was devised to determine the impact of the application of tailings to soil on the growth of Koa trees. The site of the experiment was Wood Valley, an area of commercial Koa production, in the Ka'u district of the island of Hawaii. The site elevation is approximately 670 meters; annual rainfall is approximately 254 cm per year.

TABLE 10.5
Manganese Tailings and Soil
Amendment Levels in Tons/Acre for
Tailings Agricultural Experiments on
Koa Tree Seedlings

Test Group	KMg	P-Rock	Ca	Tailings
1	0	0	0	0
2	0.5	0.5	3.0	0
3	0.5	0.5	3.0	7.5
4	0.5	0.5	3.0	15
5	0.5	0.5	3.0	25
6	0.5	0.5	3.0	33

KMg is potassium/magnesium amendment, P-Rock is granulated phosphate rock amendment, and calcium is crushed coral amendment ($CaCO_3$).

Wiltshire[13] ran experiments over a two-year period beginning in 1995 and 1996. These experiments are ongoing. The initial experiments were in two phases using Koa seedlings growing in large pots. Field trials, which are currently ongoing, followed the pot studies. In the first experiment, four mixtures were made of soil and tailings only. This trial was conducted to determine if high concentrations of tailings had any toxic effects on Koa growth. The mixtures of soil and tailings used were: 5% tailings, 10% tailings, 20% tailings, and 40% tailings. While the higher ranges of tailings percentages are significant to the soil composition, they are far less than the 100% tailings mixture that El Swaify and Chromec[18] initially used for their successful corn growing experiments. All test groups had four repetitions per tailings allotment to try to eliminate as much as possible the effect of any genetic variation among the Koa seedlings.

In the second experiment, a closer simulation of likely agricultural practice was undertaken. Hawaiian soils tend to be mineral poor and are artificially enriched with various mineral amendments for commercial agriculture. Typical in Hawaiian agricultural use are potassium/magnesium, phosphate, and calcium amendments. The amendment levels selected for the experiment are based on applications made by regional Koa growers. The alternative tailings levels were chosen to obtain an adequate range of growth responses to tailings applications. The various levels of manganese tailings and amendments were mixed with soil located on site. The specific levels of tailings and amendments used are shown in Table 10.5 on a ton per acre basis. Test groups 1 and 2 are controls. Large soil and tailings clumps were broken into their smallest aggregate pieces before mixing the two. All tests were done with four pots (repeat samples) per group.

Koa seedling growth measurements were made 6 months after planting and 10 months after planting or 3 months after the first measurement. Measurements made were stem diameter (mm), height (mm), and for the first measurement date only, the number of leaves. The experiment was terminated after the second measurement because the Koa trees had become quite large and root bound in their pots. The trees were replanted in a field and continue to be monitored for growth, although this latter phase of experimentation was interrupted by a severe drought in 1997–98.

The results of these experiments are shown in Figures 10.2 and 10.3. The results are consistent and significant, taken as a package. Figure 10.2 shows the relationship between tailings as a percentage of the total soil mixture and the various growth measures for the two measurement dates. Koa tree growth for each measure and measurement date shows a dramatic positive response to the addition of the tailings (5%). This increased growth rate is also noted for the 10% addition

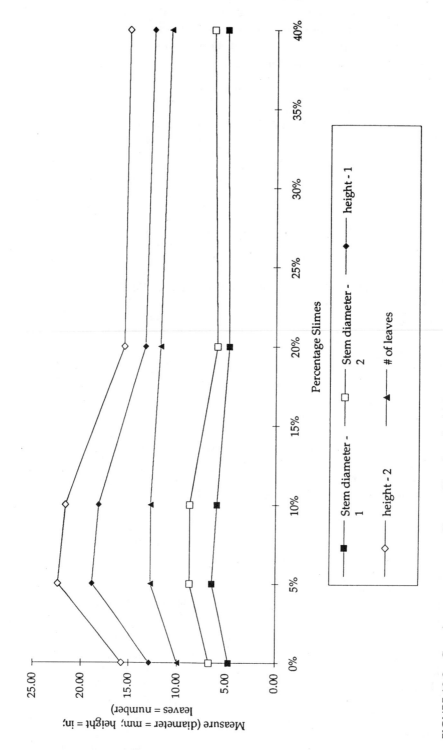

FIGURE 10.2 Growth measurements on Koa seedlings in soil with increasing percentages of added wet, fine-grained manganese tailings (slimes). No other amendments were added to the soil. The graph indicates stimulation of Koa at lower percentages of added tailings but not at higher percentages. There is, at no point, evidence of manganese toxicity.

but decreases for the 20% and 40% tailings mixtures. This pattern remained the same for the two measurement dates. When the pots were examined the tailings in the higher percentage range had clumped, making the soil surface very hard in these pots. This likely inhibited water and nutrient absorption and resulted in less stimulation of growth than may have occurred had the soil been broken up (the higher tailings concentration plants are behaving like the plants at no tailings addition). It is likely that this physical hardening of the soil is causing the decrease in Koa growth stimulation at the higher tailings concentrations. In any case, there is no evidence for tailings retarding growth (toxicity) even at the highest tailings concentration in the soil.

Figure 10.3 shows the relationship between tailings application on a ton per acre basis and the various growth measures when typical mineral amendments are added. This would represent the situation most likely undertaken in a commercial Koa agriforestry business. Koa seedling growth for each measure shows a surprising initial negative response to the addition of very small amounts of tailings when other soil amendments are added to the soil/tailings mix. Relative to the control (i.e., mineral amendments but no tailings) for measurement 2, the difference is statistically significant. Nonetheless, this is an anomalous result given Koa growth at higher tailings levels. In that it only represents four pots, other artifacts of pot placement, watering, or variation in seedlings could possibly explain this result (e.g., two sickly plants out of the four could bring the average down).

Koa growth steadily increases with higher additions of tailings, until the two highest levels of tailings application levels are attained (i.e., 25 and 33 tons per acre). The growth measures are markedly higher for these higher tailings application levels. Nonetheless, as a statistical caveat, given the small sample size, the differences from the control are statistically significant in only one instance. This was a statistically significant height difference between the control and the second height measurement, suggesting an increasing Koa growth response from the tailings as the trees age.

The purpose of this experiment was to understand the way the plants respond to tailings additions to the soil and to project what the optimal amounts of tailings addition might be. This makes the simpler case of "no amendments" crucial to the understanding of the problem. An initial positive growth response occurred with the addition of tailings, which subsequently dampened at higher tailings application levels. Such a result is not surprising given the fact that the tailings were observed to compact into a "hard pan" soil at the higher concentrations, which would be expected to lower plant nutrient and moisture availability. An alternate explanation to this would be that the tailings, because of their large surface area and scavenging nature, bind plant nutrients that would otherwise be available, hence reducing growth. If this latter explanation were true, one would expect that the higher tailings to soil ratios with no amendment addition would lead to poorer Koa growth. Consistent with both of these explanations is the fact that for measurement 2, Koa tree height and diameter growth were slightly less than the control for the higher tailings application amounts (i.e., 20 and 40%). This was not the case at the time measurement 1 was made. Even though none of these differences for measurement 1 or 2 is statistically significant, this result suggests that in pot bound plants, with time, higher levels of tailings applications may dampen Koa growth.

When amendments were added to the soil/tailings mix, the initial low tailings addition growth response was negative, which then became much greater than the control (i.e., positive) at the higher tailings application levels. This suggests that the tailings are not only not binding plant nutrients at these tailings application levels when amendments are added, but they are creating some soil condition that leads to positive growth of Koa. As the overall tailings levels were lower in the case of the amendment experiment, the soil clumping/"hard pan" problem noted above was absent. It is therefore assumed that the poor Koa performance at higher tailings levels in the first experimental case, without soil amendments, is the result of the binding soil conditions and not nutrient scavenging by the tailings.

Combination of the two results suggests that a tailings application that has the greatest possibility of enhancing Koa growth lies in range of 33 to 500 tons per acre (the latter equivalent to about 5% tailings by volume). This is a rather broad range and needs to be refined by further

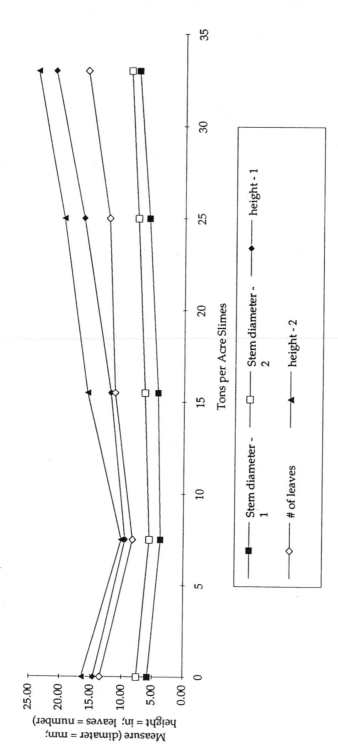

FIGURE 10.3 Growth of Koa seedlings with increasing application of wet, fine-grained manganese tailings (slimes) to soil where other standard agricultural amendments have already been added, as noted in Table 10.5. There is a clear response of increased Koa growth with increased tailings application.

experimentation. It is certainly clear that tailings application to soils would not be expected to inhibit Koa growth at practical application levels. The exact amount of growth stimulation that might be expected for a commercial agriforestry project cannot yet be quantified with certainty; however, the preliminary results shown here certainly suggest that the tailings would be an economically beneficial soil additive.

10.4.1.3 Asphalt

Asphalt offers a major potential high tonnage application for ferromanganese oxide tailings as a filler/extender. The asphalt market is divided into distinct parts: the larger being hot mix asphalt for roadways; the smaller being roofing asphalt. The tailings have been successfully added to both kinds of asphalt. For a definitive test on a roadway application, the hot mix must be laid on a relatively large scale and tested when laid. Insufficient tailings have been available for a test of this magnitude. However, the tailings did mix well and appeared to reduce porosity on roadway asphalt in a smaller test.

Wiltshire[19] did a comprehensive test on roofing asphalt. This experiment lasted six months. It involved mixing 50% tailings with roofing asphalt. This coating was applied to an exposed roof and compared to standard asphalt laid at the same time on an adjacent part of the same roof. The tailings mix retained great flexibility and had many fewer cracks and bubbles than the standard mix after the six-month test period elapsed. It was superior in every measured respect. There appears, therefore, to be considerable potential using tailings as an additive to asphalt.

10.4.2 OTHER BENEFICIAL FERROMANGANESE OXIDE TAILINGS USES OF HIGHER VALUE BUT LOWER TONNAGE

10.4.2.1 Fillers

a. Resin Casting-Solid Surface. A major use of fillers is in counter tops and fixtures. These are referred to as solid surface and cultured marble applications. The filler is mixed with specialty resins usually in proportions of about one third resin, two thirds filler. The resin–filler mix is activated by the addition of a catalyst, and the mix is poured into a mold to cure as a casting. Sink and shower fixtures were professionally cast using manganese tailings by a small Hawaii synthetic marble producer. They were very attractive and considered commercially equivalent to white synthetic marble (only black or gray). There is major casting industry interest in the tailings. Tailings samples and information packages were sent to 20 manufacturers at their request.

b. Tiles. Tailings tiles were also manufactured in a manner similar to the fixtures. They were laid in a warehouse on the island of Hawaii. They are attractive and can be installed easily with a little practice. The tiles were sent to the Ceramic Tile Institute of America for commercial testing. The manganese tiles tested within the range of certifiable floor tiles for some but not all of the required standard properties. The tests showed the tiles to have undesirably high thermal expansion and to have only 60% of the bonding strength to the floor required for certified tiles. Part of the problem here is that the bonding agent used in the test was designed for ceramic tile not resin cast tile. While attractive and testing in the general range of commercial products, it is clear that more work will have to be done before tailings could be made into tiles for wide commercial application.

c. Rubber. Fillers are also extensively used in the rubber industry. Commercial testing was done by two rubber corporations using the tailings as a filler. The tailings mix well in rubber mixes, show increased abrasion resistance and tear strength over standard fillers, and had a high index of dispersion. However, the range of grain size of the tailings is too large for the rubber industry. In general, rubber applications require 100% of the filler grains to be less than 10 microns. In the case of manganese tailings only 80% of grains are less than 10 microns. This is an easily resolvable problem, but one that would incur some sizing costs. Without this size correction the tailings performed poorly on rubber tests where the particle size distribution was a major factor. This was particularly true for overall strength and modulus of elongation.

d. Plastic. A variety of minerals, particularly calcium carbonate, are used as inert fillers with many different polymers to achieve plastics with a wide variety of characteristics. Normally, carbon black is added to the plastic to get dark, gray, or black colors for electronics housings, computers, and automobile parts. Generally, plastic properties are determined by the polymer and not the filler. Black plastics tend to fade rapidly under UV light. Plastics made with manganese tailings do not fade. This is very significant for external applications. Manganese tailings tested successfully in a range of plastics. However, as in the case of rubber, the tailings would be better if the fraction above 10 microns in size was removed. A few applications can tolerate grain sizes to 200 microns, but this is a small portion of the plastic filler market.

10.4.2.2 Coatings

Two sets of long-term coating experiments have been undertaken.[20] The first examined the tailings as a rust preventative. This was done in three phases. The first was a year-long experiment coating rusty steel beams with five different carriers. Tailings were added to three of the five. After one year of exposure, the tailings-based coatings showed superior results, even compared to commercial products such as Rustoleum. The experiment was repeated on eight strips of aluminum flashing (five tailings mixes and three controls) and on three large segments of roof on a warehouse. In both of these experiments the results were the same. When tailings were added to commercial carriers, the results were better than those obtained with commercial rust preparations. This result, having been duplicated three times, is very significant.

The second group of tailings coating experiments was a year-long effort aimed at developing a termite-resistant coating. Various pieces of wood were coated with tailings-based mixtures and placed with controls in a termite-infested area. In the first round of experiments, the controls were completely eaten away, and the tailings-coated boards were almost unaffected. In a second, larger experiment in which untreated boards were set in pots of tailings as well as left completely exposed, neither the controls nor the tailings-coated wood was affected. This may be because the large amount of tailings in the experiment site repelled the termites, or it may be inconclusive because there were insufficient termites in the known nesting area where the experiment took place.

10.4.2.3 Drilling Mud

Several Russian operations have successfully used fine-grained manganese tailings as a weighting agent in drilling mud in the Baku fields. Research on drilling mud indicates that any clay must be taken out of the tailings to raise specific gravity and to prevent swelling. Recent work has focused on removing clay. After detailed SEM and XRD characterization of the tailings, clay removal studies focused on two routes: froth flotation and magnetic separation. Froth flotation proved difficult, because a significant fraction of the tailings is five microns or less in size. It proved tricky to keep flow rates optimally adjusted. One run out of four successfully separated the manganese and clay. By contrast, wet high-intensity magnetic separation proved easy to use and very effective at separating the manganese and clay fractions. The separated manganese has a specific gravity over 4 and appears well suited for drilling mud.

10.4.2.4 Ceramics

A breakthrough has occurred with the help of several ceramics specialists. Lay et al.[11] developed a very hard manganese glaze, which is honey-brown to jet-black in color, depending on the mix. Mixes vary from 50 to 70% tailings, the remainder being one of several silica frits or borates (especially borax). After limited testing, a ceramics industry evaluator felt that the combination of color and hardness was unique. About one hundred melts were made in an effort to construct a phase diagram for the system.[10] It is important to design a melt that is a single phase. This occurs only over a narrow compositional range, which appears to be about 40 to 70% tailings. Major work in this area is ongoing. Manganese tailings make a superior ceramic which has major potential structural and decorative applications and the possibility of enclosing nuclear or toxic waste.

10.5 ECONOMIC MODELING

10.5.1 PURPOSE OF THE ECONOMIC MODELING EXERCISE

The potential uses of marine ferromanganese oxide tailings have been documented. These tailings will only exist when a marine manganese mining operation is profitable. Most of the beneficial uses proposed rely on a minimum of manganese being left in the tailings. Manganese is used primarily for metallurgical purposes. Whether and to what extent the manganese will be extracted from manganese nodules and crusts for metallurgical purposes will be dictated by prevailing economic conditions at the time a marine mining operation is established. Loudat et al.[21] have carefully analyzed the parameters determining the economic viability of a manganese crust mining operation. Wiltshire and Loudat[12] have extended this modeling effort to put boundaries on the viability of the potential beneficial utilization of the tailings and what this would mean to the overall economics of a crust mining operation. If the tailings can be used in the applications documented in this paper, this suggests there would be reduced, or possibly negligible, waste material generated from such an operation. This by itself overcomes significant obstacles to the establishment of the operation and reduces operational disposal costs. Use of the tailings for nonmetallurgical applications suggests the generation of sales income. Any such income would provide additional revenues for an ocean mining operation and alter various costs depending on whether or not the manganese is recovered. These revenue and cost changes would be expected to alter the economic viability of an ocean mining operation.

The economic assessment of a manganese crust mining operation was performed using several published economic models.[22-25] Costs in earlier versions of these models were updated to 1996-dollar values. Ore value per ton was updated by using 1996 and updated historical target metal price data and alternative value scenarios for the possible uses of the by-product manganese. The method used by Loudat and Wiltshire[24] was to take the simplest realistic case for a manganese crust operation and then to compare this to other possible cases. The most realistic case is that in which the crust operation produces cobalt, manganese, and nickel and disposes of the tailings in traditional tailings ponds. This is taken as the base case. Three other alternatives to this base case were considered. The first of these looks at the production of only cobalt and nickel with traditional tailings disposal. The second looks at trying to sell the tailings produced and selling them at a low cost, equivalent to the value of calcium carbonate, a competing inert plastics filler. The third case provides a comparison to this by looking at selling the tailings at a higher value, equivalent to the cost of alumina trihydrate, a specialty resin casting filler. The results of each of these four scenarios, when run through the economic model, give a clear basis for projecting the viability and boundary conditions of *using* tailings in contrast to dumping them in a tailings pond.

10.5.2 BASIC ECONOMIC MODEL OF A MANGANESE CRUST OPERATION

The first step of the economic model is to outline the various costs and revenues of a crust mining operation. These are divided into capital and operating costs. Revenues are projected on the basis of the historical sales prices for the metals contained.

For the purposes of the economic modeling exercise illustrated, it is fundamental to understand that the important conclusions deal with the relative relationships among the various scenarios rather than absolute dollar figures. Since tailings would not be a problem in a smelting operation, an acid leach scenario is the only one illustrated. Work at the U.S. Bureau of Mines has shown

TABLE 10.6
Waste Streams from a Manganese Crust Mining Operation

Waste Stream Sources	Material (metric tons)	MnO$_2$ %	Mn %
Beneficiation tails	702,214	3.6	
Leach residue	260,282		53.4
Iron precipitate	84,533		
Co/Ni precipitate	32,587		23.9
Mn filtrate	354		66.8
Mn precipitate	374,189	93.3	
Total waste	1,454,159		
Waste blended to 35% MnO$_2$	587,481	35.0	

that the process currently considered most economically viable to produce high-grade cobalt from a manganese crust deposit is a sulfuric acid/hydrogen peroxide leach.[21] This is the basis of the economic and waste stream projections which follow.

The principal form of manganese in the waste stream is manganese dioxide, although depending on the nature of some of the process options, manganese carbonate, sulfate, or sulfide may be present. The work done on tailings to date has focused solely on manganese dioxide. Based on this research it is uncertain whether the other forms of manganese would perform similarly in nonmetallurgical uses. For this reason, the only portions of the total waste stream assumed recovered and sold for nonmetallurgical uses are those containing manganese dioxide. However, in order to properly model the effect of using the waste, all the various waste stream sources must be considered, if for no other reason than these have associated disposal costs. Table 10.6 shows waste stream sources, their estimated annual tonnage, and their manganese composition. These would vary for different processing options.

The high-value manganese tailings, which could be sold for nonmetallurgical use, comprise the manganese precipitate containing 93% manganese dioxide. This high-grade manganese is assumed to have properties for nonmetallurgical uses equivalent to specialty fillers, a weighting agent in drilling mud, or high-value ceramics additives. The low-value manganese by-product sold for nonmetallurgical use would be the manganese portion of the waste containing only 35% manganese dioxide. This can be achieved by blending the low-grade tailings with the higher-grade precipitate to result in larger overall tonnages at the 35% level (see Table 10.6). This concentration has been found to be satisfactory for most of the tailings applications described in the earlier sections of this review. Naturally, selling part of the waste stream as a by-product would reduce the amount of waste requiring disposal. Taking the tonnages from Table 10.6, the waste stream would be reduced by at least 26% for high-value nonmetallurgical uses and by at least 40% for the low value nonmetallurgical uses.

10.5.2.1   Costs

Two alternative cost structures have been formulated. These are waste and no waste recovery. A similar waste recovery cost structure applies regardless of the sales disposition of the manganese product ultimately sold. The costs of the overall mining operation will be compared to the revenues it generates. Costs are of two forms: up-front capital costs and annual operating costs. The capital cost structure for a manganese crust mining operation is shown in Table 10.7. It is the capital cost structure for the acid/hydrogen peroxide leach processing pathway. The process assumes that metallurgical grade cobalt, nickel, and manganese will be recovered. The original amounts are updated to 1996-dollar values.

TABLE 10.7
Capital Cost Summary of the
Acid/Hydrogen Peroxide Process
(millions of 1996 dollars)

Sector	Capital Cost	% Distribution
Mining	$275.3	40.52
Ore marine transport	$140.4	20.67
Ore marine terminal	$49.1	7.22
Onshore transport	$20.5	3.01
Processing	$181.2	26.67
Waste disposal	$11.0	1.61
Mining support	$1.9	0.28
Grand total	$679.3	100

In the sulfuric acid/hydrogen peroxide leach process the steps that allow recovery of the manganese involve the precipitation of the manganese from the acid/hydrogen peroxide solution after the removal of the cobalt and nickel. The manganese is precipitated as a free-filtering manganese carbonate cake, which is washed with water. It is then charged to a gas-fired rotary calciner (500°C), where the manganese carbonate is converted to battery-grade manganese dioxide. The manganese dioxide is cooled and stored until sale.

Removal of the manganese recovery process reduces processing capital costs by $28 million. This is due to removal of the calcination step required to recover manganese dioxide for sale. Waste disposal capital costs increase by $3.8 million for the no-manganese-recovery scenario because there is now more waste. These costs decrease for the higher-tonnage, low-value manganese tailings recovery by $2.2 million. These amounts are estimated by linearly changing the waste disposal capital cost by the change in tonnage of waste material for the no metallurgical-manganese-recovery (35% increase) and high-tonnage, low-value tailings sales scenarios (20% decrease), relative to the waste disposal scenario projected in Table 10.7. Thus, total capital costs for the no-manganese-recovery scenario and for the high-tonnage, low-value tailings sale scenarios are $653.3 million and $677.1 million, respectively.

The operating cost structure for a crust mining operation for which the manganese component of the waste is recovered is shown in Table 10.8. This table is derived from Loudat et al.[23] The original amounts are updated to 1996-dollar values.

Elimination of the manganese recovery process reduces processing operating costs by $21.2 million. This is estimated by reducing direct operating costs of processing by the reduction in tonnage processed. The reduction in tonnage processed if no manganese is recovered is 17%. Waste disposal operating costs increase by $0.65 million for the no-recovery scenario. These costs decrease for the higher-tonnage, low-value manganese tailings sales scenario by $0.37 million. These amounts are estimated by linearly changing the waste disposal operating cost by the change in tonnage of waste material for the no-recovery (35% increase) and high-tonnage, low-value scenarios (20% decrease), relative to the waste disposal scenario in Loudat et al.[23] Thus total operating costs for the no-recovery scenario and for the high-tonnage, low-value scenarios are $227.5 million ($177/ton) and $247.6 million, respectively.

10.5.2.2 Revenues

Table 10.9 shows target metal historical price summary statistics used to value the ore. These prices were taken for the period 1970 to 1995. Cathode cobalt and nickel are traditionally priced in dollars

TABLE 10.8

Operating Cost Summary of the Acid/Hydrogen Peroxide Process (millions of 1996 dollars)

Sector	Annual Cost	% Distribution
Exploration	$7.8	3.15
Mining	$53.0	21.38
Ore marine transport	$20.0	8.05
Ore marine terminal	$8.4	3.39
Onshore transport	$3.9	1.57
Processing	$130.4	52.56
Waste disposal	$1.9	0.76
Mining support	$1.4	0.55
Research & development	$1.3	0.53
Cost of sales	$10.4	4.19
Interest expenses	$9.6	3.87
Grand Total ($ millions/year)	$248.0	100%
Crust/substrate processed	1.33 (million tons)	
Cost per ton	$186.8	

TABLE 10.9

Historical Metal Price Statistical Summary for Cobalt, Nickel and Manganese from 1970–1995 (all values expressed in 1996 dollars)

Item	Cathode Cobalt ($/lb)	Nickel ($/lb)	Mn ($/mt)
Minimum	8.34	2.43	271
Maximum	50.94	8.03	702
Mean	18.85	4.24	464
Stnd. Dev.	11.98	1.22	137
Coefficient of variation	0.64	0.29	0.29
95% Confidence interval			
Lower bound	14.04	3.85	409
Upper bound	23.66	4.64	519

per pound, whereas manganese is priced in dollars per metric ton. All prices have been corrected for inflation to 1996 values to allow comparison. The mean prices are used to value the ore over an assumed 30-year life of an ocean mining venture. The price data from Table 10.9 with assumed tonnage processed and metal recoveries are used to determine the ore value for the three alternative scenarios.

Table 10.10 shows the ore value when the recovered manganese is assumed sold as metallurgical-grade equivalent manganese. The basic revenue situation for this case (selling manganese as metal) is:

TABLE 10.10
The Value of Manganese Crust Ore Per Ton (assuming that the recovered manganese is sold as metallurgical grade Mn)

	Cathode Cobalt (lbs)	Nickel (lbs)	Mn (mt)
Metal recoveries	91%	87%	88%
Recovery per ton ore processed	0.63%	0.32%	17.8%
Annual metal weight recovered	18.6 million	9.5 million	374,189
Base case metal price	$18.85	$4.24	$464
Annual revenue ($ million)	$351	$40	$174
% of Total revenue	62.1	7.1	30.8

TABLE 10.11
Comparative Value of Manganese Crust Ore Under Four Different Metal Sales Scenarios

	Total Sales Revenues ($ million)	Ore Value ($/ton)
No Mn recovery	391	293
Low value Mn tailings (non metallurgical) $110/metric ton	456	342
High value Mn tailings (nonmetallurgical) $660/metric ton	638	479
Base case (Metallurgical Mn)	565	424

Total annual revenue	$565 million
Dry tons crust & substrate processed/yr	1.34 million
Ore value/ton	$424

The value of the ore (per ton of ore processed) and annual sales revenues for the four scenarios are shown in Table 10.11. A $110 per ton nonmetallurgical value of the manganese product assumes that its value is equivalent to calcium carbonate. Calcium carbonate is a low-value mineral ($0.05/lb) commonly used as a filler for which the manganese by-product can substitute. A $660 per ton nonmetallurgical value of the manganese product assumes that its value is equivalent to alumina trihydrate. Alumina trihydrate is a high-value industrial mineral ($0.30/lb) commonly used as a filler for which the manganese by-product may substitute. The ability of the manganese by-product to attain the higher end of this value range depends on its efficacy in meeting the demands of high-end uses, perhaps in addition to providing value-enhancing characteristics to end products not currently provided by the minerals used.[10,12]

TABLE 10.12
Overall Economic Summary of a Manganese Crust Mining Venture

Financial Variable	Case 1 no Mn	Case 2 Mn as Metal	Case 3 Tailings Low Value	Case 4 Tailings High Value
Payback Period on Fixed Capital				
Pre-tax (yrs)	4.3	2.2	3.4	1.7
Post-tax (yrs)	7.1	3.6	5.6	2.9
Payback Period on Net Parent Capital Funding				
Pre-tax (yrs)	2.1	1.1	1.7	0.9
Post-tax (yrs)	3.4	1.8	2.8	1.4
Real Rates of Return				
Return on Fixed Capital				
Pre-tax (%)	23.5	46.1	29.5	57.4
Post-tax (%)	14.2	27.8	17.8	34.5
Return on Net Parent Capital Funding				
Pre-tax (%)	48.4	94.4	60.2	117.5
Post-tax (%)	29.2	56.8	36.2	70.6
Internal Rate of Return				
Pre-tax (%)	19.6	32.7	23.5	37.9
Post-tax (%)	**23.4**	**38.0**	**27.7**	**43.7**

10.5.3 DISCUSSION OF THE ECONOMIC MODELING RESULTS

The detailed financial assumptions providing the economic framework within which the assumed manganese crust mining venture operates are outlined in Loudat and Wiltshire[23] These are conservative assumptions based on general practices of the mining industry. The most important of these assumptions is that the crust venture will be a subsidiary of a parent company in the resource industry which can shield income from taxes by using preproduction losses. The parent company is assumed to finance the venture 25% from parent equity and 75% from debt financing amortized over 20 years. The results that follow are based on this assumption set and the information outlined above.

Table 10.12 presents the variables for the financial assessment and the results for each of these variables for the simplest case, where no manganese is extracted and all the tailings are disposed, and the three more complex cases: that in which the manganese is sold as metal, and the two cases where the tailings are sold at high and low values depending on their usage. The post-tax internal rate of return for the base case manganese metal sales scenario is 38%. This increases to 44% in the highly optimistic circumstance that the tailings could be sold as high-grade product. If no manganese is recovered, the venture would still provide a healthy rate of return of 23% which could be increased to 28% by selling the tailings, albeit at a relatively low value.

The economic impact of manganese recovery in a manganese crust mining operation is positive both if sold as metallurgical-grade Mn or for nonmetallurgical uses. At metallurgical and high-end nonmetallurgical-grade prices, the manganese adds substantially to the economic viability of such a venture. The internal rate of return to the venture increases 60% and 87% for the metallurgical and high-end nonmetallurgical values, respectively, relative to no manganese recovery. The sale of tailings for low-value nonmetallurgical use also makes a positive contribution to the economic viability of a manganese crust mining venture. From an economic perspective, decision-makers planning future ocean mining operations would choose to recover manganese under anything close to the assumption set of this analysis. If sufficient manganese is left in the tailings, they should also be sold for the range of uses outlined earlier in this review. Similar conclusions could be expected to apply to a manganese nodule mining operation.

10.6   CONCLUSIONS

The success of a venture mining cobalt-rich manganese crusts or manganese nodules will be significantly affected by the disposal of the waste material generated. This waste material will be the residual by-product of the ore after the removal of the target metals: cobalt, nickel, copper, and possibly manganese. These tailings will be rich in manganese, particularly if manganese is not extracted as a metal. The manganese tailings have such potential value that they may ultimately not be a waste material, but rather a valuable commodity in their own right. This value includes large tonnage applications, such as fine-grained aggregate in concrete and asphalt, and soil amendments. It also includes more specialized applications, such as fillers in cultured marble and densified solid surface products, plastics, and rubber. Manganese tailings have a potential market as a coloring agent in bricks, in ceramics, and in coatings. Similar tailings have been used in the past in drilling muds.

The potential beneficial uses of manganese tailings are directly tied to the overall future viability of a manganese nodule or crust industry. In order to clarify this interdependence, four economic options have been modeled for a manganese crust industry. A similar exercise for a manganese nodule industry would likely yield a comparable result. The four options were: (1) not extracting manganese, (2) extracting it and selling it as manganese metal, (3) selling manganese-rich tailings as high-tonnage, low-value fillers, and (4) selling smaller quantities of higher-grade tailings in high-end nonmetallurgical manganese markets. The conclusions based on the economic modeling indicate that any group contemplating mining manganese crusts or probably nodules and acid leach processing them should try to sell the manganese. The safest route to do this would be as manganese metal. However, this study indicates that, in theory, it could be even more profitable to sell the manganese into the high-end nonmetallurgical market. Naturally, land-based manganese will also compete for these markets. At present, high-end nonmetallurgical manganese markets are small in size and not well developed. Without a major market development effort, this option would likely prove viable only for very limited tonnages. As a third option, the future ocean miner should try to sell larger tonnages of low-grade manganese tailings into the low-end markets discussed in this review. These markets are potentially one to two million tons a year worldwide and could be expanded with product development and marketing. Selling the manganese tailings would increase the profitability of a manganese crust venture from about 23% to 28%. A similar increase in profitability might be anticipated for a nodule operation. This is a small but important increase. At the same time, using the tailings would remove waste, with its environmental liability, and generate secondary jobs, two things which lead to valuable goodwill and the integration of a new industry into the community.

ACKNOWLEDGMENTS

This review is supported in part by a grant from the National Oceanic and Atmospheric Administration, project #R/MC-1, which is sponsored by the University of Hawaii Sea Grant College Program, SOEST, under Institutional Grant No. NA36RG0507 from the NOAA Office of Sea Grant, Department of Commerce. The views expressed herein are those of the authors and do not necessarily reflect the views of NOAA or any of its subagencies. This is UNIHI-SEAGRANT-CR-98-04. This research presented here has also been partially supported by the Department of the Interior's Mineral Institute Program administered by the U.S. Minerals Management Service through the Generic Mineral Technology Center for Marine Minerals under Cooperative Agreement 1435-01-98-CA-30902. This work is State of Hawaii Ocean Resources Branch Contribution Number 144.

REFERENCES

1. Markussen, J.M., Commercial exploitation of polymetallic nodules: when, why, who and how, *Materials and Society*, 14, 397, 1990.
2. Johnson, C.J., Future technical and economic dimensions of deep-sea mining, *Materials and Society*, 14, 209, 1990.
3. U.S. Department of the Interior, Minerals Management Service, *Proposed Marine Mineral Lease Sale in the Hawaiian Archipelago and Johnston Island Exclusive Economic Zones: Final Environmental Impact Statement*, U.S. Government, Washington, 1990.
4. Aplin, C.L. and Argall, G.O., *Tailings Disposal Today*, Miller Freeman, San Francisco, 1973.
5. Sengupta, M., *Environmental Impacts of Mining: Monitoring, Restoration and Control*, Lewis Publishers, Boca Raton, 1993.
6. U.S. Department of Commerce, National Oceanic and Atmospheric Administration, *Deep Seabed Mining: Final Programmatic Environmental Impact Statement*, U.S. Government, Washington, 1981.
7. Salazar, R. and Gonzales, R., Design construction and operation of the tailings pipelines and underwater tailings disposal system of Atlas Consolidated Mining and Development Corporation in the Philippines, in *Tailings Disposal Today*, Aplin C.L. and Argall G.O. (Eds.), Miller Freeman, San Francisco, 1973, chap. 20.
8. Evans, J.B., Ellis, D.V., and Pelletier, C.A., The establishment and implementation of a monitoring program for underwater tailings disposal in Rupert inlet, Vancouver Island, British Columbia, in *Tailings Disposal Today*, Aplin, C.L. and Argall, G.O. (Eds.), Miller Freeman, San Francisco, 1973, chap. 21.
9. Ellis, D.V., A comparison of three systems for setting sediment quality guidelines for trace metals, presented at the 26th Underwater Mining Institute, St. John's, Newfoundland, Oct 29-Nov. 1, 1995.
10. Wiltshire, J.C. and Troy, P.J., Manganese tailings: useful properties suggest a potential for gas absorbent and ceramic materials, *Marine Georesources and Geotechnology*, 16, 273, 1998.
11. Lay, G.F. and Wiltshire, J.C., Formulation of specialty glasses and glazes employing marine mineral tailings, in *Recent Advances in Marine Science and Technology*, Saxena, N. (Ed.), Pacon International, Honolulu, 1997, 347.
12. Wiltshire, J.C. and Loudat, T.A., The economic value of manganese tailings to marine mining development, *Proceedings of the Offshore Technology Conference*, Marine Technology Society, Washington, 1998, 735.
13. Wiltshire, J.C., Use of marine mineral tailings for aggregate and agricultural applications, *Proceedings of the International Offshore and Polar Engineering Conference*, May 25-30, Honolulu, Hawaii, ISOPE, Golden, Colorado, 1997, 468.
14. Haynes, B., Barron, D., Kramer, G., Maeda, R., and Magyar, M., Laboratory processing and characterization of waste materials from manganese nodules, *Bureau of Mines Report of Investigations*, RI 893, 1985.
15. American Concrete Institute, *Manual of Concrete Practice*, ACI, Detroit, 1990.
16. Kosmatka, S. and Panarese, W., *Design and Control of Concrete Mixtures*, Portland Cement Association, Skokie, Illinois, 1988.
17. Wiltshire, J.C., Beneficial uses of ferromanganese marine mineral tailings, in *Recent Advances in Marine Science and Technology*, Saxena, N. (Ed.), Pacon International, Honolulu, 1993, 405.
18. El Swaify, S. and Chromec, W., The agricultural potential of manganese nodule waste material, in *Marine Mining: A New Beginning*, Humphrey, P. (Ed.), State of Hawaii, Department of Planning and Economic Development, Honolulu, 1985, 208.
19. Wiltshire, J.C., Innovative tailings management for marine ferromanganese nodule and crust processing, *Proceedings of Ocean's '95*, Marine Technology Society, Washington, 1995, 723.
20. Wiltshire, J.C., The use of marine manganese tailings in industrial coatings applications, *Proceedings of Oceans '97*, Marine Technology Society, Washington, 1997, 1314.

21. Loudat, T., Zaiger, K., and Wiltshire, J. C., Solution mining of Johnston Island manganese crusts: an economic evaluation, *Proceedings of Oceans '95*, San Diego, Marine Technology Society, Washington, 1995, 713.

22. Loudat, T. and Wiltshire, J. C., The economics of mining manganese crust with recovery of platinum and phosphorous, *Recent Advances in Marine Science and Technology,* Saxena, N. (Ed.), Pacon International, Honolulu, 1993, 413.

23. Loudat, T.A. and Wiltshire, J. C., The economics of mining manganese crusts, *Hawaii Department of Business Economic Development and Tourism Ocean Resources Branch Report*, 1994.

24. Loudat, T.A. and Wiltshire, J. C., Industrial application of waste manganese: initial testing and economic evaluation, *Proceedings of PACON '97,* PACON International, Honolulu, 1997, 48.

25. Wiltshire, J.C., The world cobalt market and its ability to support manganese crust mining, in *Recent Advances in Marine Science and Technology,* Saxena, N. (Ed.), Pacon International, Honolulu, 1997, 337.

Part III

Hydrothermal Minerals

11 Hydrothermal Activity on the Southern, Ultrafast-Spreading Segment of the East Pacific Rise

Vesna Marchig

ABSTRACT

The southern part of the East Pacific Rise is an ultra-fast spreading zone characterized by vigorous volcanism. Hydrothermal activity is consequently also high, producing large anomalies in sea water, high density of hydrothermal fauna, and elevated contents of hydrothermal precipitates in the surrounding sediment.

Spreading proceeds in cycles, whereby every cycle begins with a slow volcanic phase (less differentiated pillow lava forming a steep, narrow ridge) and then develops a vigorous volcanic phase (sheet lava of higher differentiation forming a large, rounded ridge). After volcanic activity slows down or ceases, the tectonic activity forms first a small, and later a large and deep, axial summit graben. In the course of further spreading a new cycle begins consequent on newly initiated volcanic activity.

During the vigorous volcanic phase strong degasing takes place, the methane from this serving as food for hydrothermal fauna. This phase is therefore characterized by maximum hydrothermal fauna and maximum water turbidity caused by floating organisms, and registers as light attenuation anomalies.

Hydrothermal circulation is initiated and developed during tectonic phases; black smokers and massive sulfide chimneys are formed in the course of discharge of ore-forming solutions from the seafloor. The frequency of massive sulfides depends on seafloor permeability, therefore the best indication for finding massive sulfides is a high coverage of talus in the axial summit graben.

Massive sulfide chimneys are distinct in the manner of their growth from those described from the northern part of the East Pacific Rise. The first friable pipe is built of idiomorphic pyrite, which subsequently gets cemented by layered colloform marcasite, pyrite, and sphalerite. Anhydrite is seldom observed; there is no indication of a first pipe formed from anhydrite, as described in the north. The second stage of chimney growth is clogging of the central conduit with sulfides rich in idiomorphic chalcopyrite. The third stage of ore precipitation occurs on the outside of the chimney wall, from the solution which cannot use the clogged central conduit and therefore wells out from the seafloor around the chimney. This is quenched, producing colloform sphalerite and marcasite. Characteristic of this layer are numerous embedded calcitic worm tubes.

Although hydrothermal activity is high in the southern part of the East Pacific Rise, massive sulfide ores do not occur in large amounts or as big edifices, first, because most of the precipitates from the hydrothermal solutions get dispersed to the surrounding sediment, and second, because the edifices formed become covered with fresh lava flows.

11.1 SHORT HISTORY OF RESEARCH

On October 29, 1875, *HMS Challenger,* on her way from Tahiti to Valparaiso, collected sample 292 from the top of the East Pacific Rise (38°43'S, water depth 2,928 m). It was described simply as "globigerina ooze" (83.75% $CaCO_3$).[1] Over 50 years later, in 1928/1929, the research vessel *Carnegie* sampled sediments from the Atlantic and Pacific Oceans. Revelle[2] noticed that some of the globigerina oozes contained large amounts of free iron oxide in the noncarbonate fraction. (Most of the sediments showing this feature were obtained on the flanks of the southern part of the East Pacific Rise.) The explanation he gave for this phenomenon was weathering of volcanic material under alkaline conditions — the high pH being caused by carbonate dissolution.

Peterson and Goldberg[3] investigated feldspars from Pacific sediments obtained during the *Downwind* expedition, and concluded that the feldspars are a product of recent local volcanism on the crest of the East Pacific Rise. Skornyakova[4] investigated the material from *Vitjaz* cruises and suggested that Fe and Mn enrichment along the East Pacific Rise originated from hydrothermal springs associated with recent volcanism.

The *Risepac* expedition in 1961/1962 measured heat flow over the southern East Pacific Rise at 13°S. Von Herzen and Uyeda[5] found that the East Pacific Rise is always associated with abnormally high heat flow, and the crest of the East Pacific Rise, between 10°S and 15°S, with very high heat flow.

Based on these findings, Boström and Peterson[6] investigated the sediment samples taken on the highest heat-flow anomaly during the *Risepac* expedition and concluded that the iron- and manganese-rich sediments found are "precipitates from volcanic emanations that debouch on the crest of the rise." More than a decade of subsequent work on this subject by these authors and colleagues are summarized by Boström[7] and work done under the IDOE Project, 1971–1978, in G.S.A. Memoir 154 (1981).[8]

In 1979 Harald Bäcker of Preussag AG initiated the Geometep (**geo**thermal **met**allogenesis East **Pac**ific) program, which he described as "a long-term investigation into the relationship between tectonism, volcanism, and hydrothermal activity along the accreting boundaries between the Pacific and Nazca Plates." The previous theory that metal sulfides can form only in sediment traps (e.g., in the Red Sea) was disproved by findings of metal sulfides in 1978 on the East Pacific Rise at 21°N.[9] The southern East Pacific Rise was selected as a target for investigation because most hydrothermal activity was expected to occur in the spreading zone with highest spreading rate.[10] Between 1980 and 1989, five cruises were conducted within the Geometep program, all of them with the research vessel *Sonne*. Bäcker cooperated closely with French scientists and, in 1984, joined their diving cruise in one segment of the southern East Pacific Rise.[11] The hydrothermal sulfides, which occur as chimneys and stockworks, were extensively sampled and studied in the 1980s. The most important discoveries to come out of the work were the cyclic nature of the spreading process, each cycle consisting of a volcanic, a tectonic and an ore-forming phase,[12] the large methane anomaly in sea water associated with the volcanic, phase,[13] and the climatic control of precipitation of hydrothermal manganese in sediment.[14]

During the last few years, several important papers have been published about the southern East Pacific Rise, mainly concerning the morphology of the spreading ridge and hydrothermal anomalies in sea water. Lonsdale[15] published the first geomorphologic map of the southern East Pacific Rise. Sinton et al.[16] explained the variability in chemical composition of volcanic glasses from different segments of the southern East Pacific Rise in terms of "independent, regionally decoupled processes." Marine seismic work on the southern East Pacific Rise by the Terra 91 expedition led Kent et al.[17] to infer a long, homogeneous magma chamber under the entire area. This, however, appeared to be contradicted by findings of small scale inhomogeneities in erupted basalts.[16] However, the observations of Singh et al.,[18] who interpreted the small-scale variabilities within the long magma chamber in terms of different ratios of crystallized fraction to residual melt, provided a possible solution to this paradox.

Measurements of temperature, light attenuation (turbidity due to precipitation of hydrothermal iron and manganese hydroxides), methane, and dissolved manganese have frequently been made over the southern East Pacific Rise and hydrothermal anomalies identified, since Lupton and Craig[19] found a major helium-3 anomaly on the East Pacific Rise at 15°S. Detailed investigations demonstrated that the gas phase-rich anomalies are associated with volcanic activity, and the metal anomalies with the mineralization phase of spreading.[12,20-25]

11.2   MORPHOLOGY, TECTONIC AND VOLCANIC ACTIVITY

The southern part of the East Pacific Rise (Figure 11.1) is the most rapidly diverging spreading center known. The spreading rates increase linearly with latitude from north to south beginning with 12 cm/a at the triple junction between the Pacific, Cocos, and Nazca plates at 2°N to 16 cm/a at the Easter Microplate at 23°S. South of 23°S, two microplates exist as the dominating structural elements (Easter and Juan Fernandez Microplates) and attached to Juan Fernandez microplate is the triple junction between the Pacific, Nazca, and Antarctic Plates, recently positioned at 35°S.

The part of the East Pacific Rise north of the Easter Plate conforms to the traditional morphologic model for fast-spreading ridges. It is a long mountain ridge, occasionally with a small axial summit valley. The type of spreading center with a large, deep graben, as we know from slow-spreading centers, is not developed. In comparison with slow spreading ridges, the ridge of the East Pacific Rise is rarely intersected by transform faults and displays overlapping spreading centers more frequently. Other departures to linearity of spreading are: small nonoverlapping ridge offsets, oblique offsets, deviations of the axis, and small-scale faults and fissures. The ridge is divided into sections; one section is limited by two fracture zones. The morphology of this part of the East Pacific Rise is a consequence of spreading, doming, and a shear component; the shear component produces dextral or sinistral segment displacement oblique to the general trend of the axis of the East Pacific Rise.[12,15,16,26-28]

Farther southward the morphology of the spreading zone becomes more complicated on account of two rotating microplates bordered by propagating rifts.[29] Little work has been done on hydrothermal mineralization in the southern East Pacific Rise spreading zone, and thus its history is not discussed in detail here.

Ore formation along the East Pacific Rise north of the Easter Plate has been exhaustively investigated. Studies on the straight parts of the ridge between fracture zones using sea beam, deep-tow cameras, and submersibles resulted in the discovery of spreading cycles.[11,12,27,30-33]

A spreading cycle has been observed to contain four growth phases, which can be correlated with four types of ridge morphology as follows (Figure 11.2):

Phase 1: Narrow ridge with steep flanks consisting of pillow lava (**narrow steep ridge** in Figure 11.2).

Phase 2: Broad ridge with gentle convex slopes and a broad, flat top covered with sheet lava. Gas anomalies are characteristic of this phase (**rounded ridge** in Figure 11.2).

Phase 3 proceeds from Phase 2: A rudimentary axial valley develops on the summit plateau as a chain of collapsed lava lakes and eventually forms a continuous valley, which is often shallower than 10 m, i.e., below the resolution of sea beam. This phase includes the first appearance of sulfide minerals in the form of "black smokers" (**young central valley** in Figure 11.2).

Phase 4 is a continuation of phase 3: The central valley deepens and is filled by increasing amounts of talus. Hydrothermal metal-rich solutions are widespread, forming not only sulfide chimneys but also crusts and stockwork mineralization (**mature central valley** in Figure 11.2).

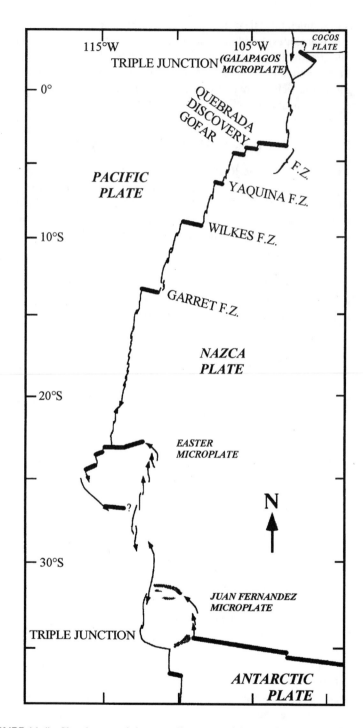

FIGURE 11.1   Sketch map of the spreading zone of the southern East Pacific Rise.

A spreading cycle is correlated with the life cycle of local magma chambers beneath the ridge. Initially, the narrow magma chamber erupts slowly but continuously, forming a steep ridge of pillow lava. As the magma chamber matures, it erupts larger quantities of magma, forming a large, more rounded ridge built up of sheets of lava. As the magma chamber cools, volcanic eruptions become scarce, permitting tectonic activity to form a rift valley along the ridge axis. This model is in close

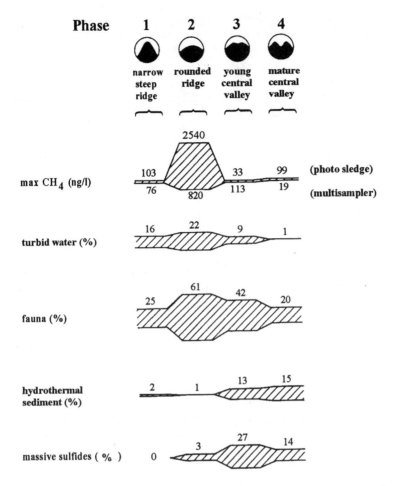

FIGURE 11.2   The four phases of a spreading cycle showing the behavior of several typical features. During Phase 2 (maximum volcanicity), the degasing of methane is at its highest level, and bioproductivity is high. During the tectonic phases (3 and 4), characterized by occurrence of a central valley, the formation of massive sulfides is at a maximum.

agreement with the observations of Schwarz and Dulce[28] on the morphology of the ridge. The initial steep ridge of pillow lava occurs predominantly in the deeper parts of the East Pacific Rise (low magmatic pressure here causes weak doming). The rounded sheet-lava ridge occurs frequently on culminations of ridge segments (strong doming due to high magmatic pressure). Features characteristic of Phases 3 and 4 are found within Phase 2 zones and outside ridge culmination areas. Phases 3 and 4 could be referred to as tectonic phases since the volcanic eruption has more or less ceased. Reactivation of the spreading zone initiates a new cycle and introduces a shear component into the tectonic activity.The duration of a cycle in the southern part of the East Pacific Rise is inferred to be in a range of 100,000 years.[34] In the northern part the duration of a cycle (or interval between two magmatic pulses) has been estimated to be shorter than 10^4 years.[35]

   The basaltic lava on the ridge crest (there is also some volcanic activity on the flanks of the ridge[36,37]) is more or less evolved mid-ocean ridge basalt, produced as partial melt in a depleted upper mantle under the spreading zone. All the incompatible elements are low, including potassium and rare earth elements, especially the light ones. Rarely, less depleted basalts are extruded near deep fracture zones and are inferred to be partial melts from deeper mantle levels tapped by the deep fracture zones.[38] The local differences in basalt composition which are observed along the ridge segments probably reflect a greater or lesser degree of differentiation in a magma chamber situated at shallow depth.[39] These local differences in basalt composition between samples collected a few kilometers apart are a common feature of most investigations.[38-40] Comparison between basalt compositions of different spreading phases (in practice the difference in composition between pillow lava and sheet lava) shows that the former is less evolved. This suggests that as the magma chamber matures it becomes more evolved, thus corroborating the theory of spreading cycles.[38] However, two-ship, multichannel seismic reflection/refraction studies by Kent et al.[17] contradict this interpretation. They show that under the southern East Pacific Rise there is a continuous high-level magma chamber, which continues across small discontinuities in the ridge but is interrupted by large discontinuities. On this basis, the presence of local magma chambers cannot be held as an explanation for the short-range variations in basalt composition.

11.3   HYDROTHERMAL MATERIAL IN SEDIMENTS

The hydrothermal precipitates in the sediments on and around the southern East Pacific Rise are composed of iron and manganese hydroxides enriched in B, As, Cd, V, Cr,[6] Cu, Mo, Zn, P,[41] U, Th, and Pa.[42] Two types of trace element enrichment have been described: conservative enrichment due to their coprecipitation with Fe and Mn hydroxides, e.g., P, U, V, Mo, and As, and enrichment as a result of scavenging by Fe and Mn hydroxides from sea water e.g., REE and Th. The first group shows a linear correlation with the amount of hydroxides, the second increases with time as a function of the sedimentation rate.[43]

Hydrothermal precipitates composed mainly of iron and manganese hydroxides are not typical of all spreading centers. On the Galapagos spreading zone, much of the hydrothermal iron is present in the form of an iron silicate clay mineral.[44,45] Various reasons are given for the formation of iron silicate: diagenetic reaction of iron hydroxide with opaline tests, which are absent on the southern East Pacific Rise, or the lower reaction and discharge temperatures of hydrothermal solutions in the Galapagos spreading zone. New investigations on the Galapagos spreading zone explain this zone as a failed rift.[46] This could explain the peculiar sediment composition of recent hydrothermal precipitates in the Galapagos spreading zone, as failed rifts change their hydrothermal system during the cooling-down phase.[47]

The crest of the southern East Pacific Rise is bare of sediment except in the immediate surroundings of hydrothermal vents.[8,32,48] The sediment-free zone is larger on the east of the spreading center than on the west, and also extends in a southerly direction. The increase in sedimentation rate westward and northward is, respectively, due to the mainly westward transport of hydrothermal precipitates from the spreading zone by bottom currents[49] and the northward increase in primary productivity of surface ocean water.

In the sediment deposited above the lysocline (which is at 3500 m), the amount of hydrothermal material varies between 1% and 94% of the sediment, i.e., all compositions are present between nearly pure calcite and nearly pure hydrothermal material. The nonhydrothermal material is composed of calcitic foraminifera and coccoliths; other constituents are present only in traces. This simple two-component system has not been observed anywhere else in this extreme form.

Surface sediments show a decreasing proportion of hydrothermal material with distance from the spreading zone.[50-53] Subsequent, more detailed investigations along the southern East Pacific Rise showed that hydrothermal metalliferous sediments are enriched all along the spreading zone (Figure 11.3) but that sulfide occurrences are limited to distinct sections.[32,41] The highest ratios of hydrothermal precipitates to carbonate were found off the large overlapping spreading center between 20°S and 21°S, although this overlapping spreading center (like all the others) is currently not hydrothermally active. Using the proportion of hydrothermal material in surface sediment adjacent to an active spreading ridge is thus not a suitable exploration tool for active hydrothermal systems on the southern East Pacific Rise. The hydrothermal plumes probably suffer lateral mixing in sea water before deposition of the hydrothermal material.

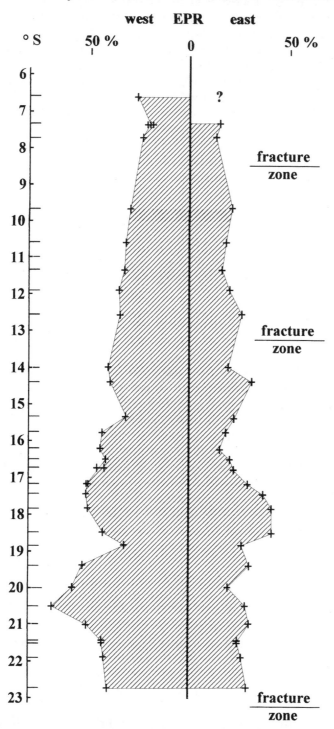

FIGURE 11.3   Amount of hydrothermal material in the surface sediment in cores taken on the east and west of the East Pacific Rise. High concentrations of hydrothermal material do not correspond to the observed maxima of occurrence of massive sulfides (see Figure 11.5). (Data from Holler, G., *Erscheinungsformen hydrothermaler Aktivität am Ostpazifischen Rücken zwischen 6° und 30° Süd, Verlag Shaker*, 1-226, 28 Abb., 6 Taf., 76 Tab., 1993.)

The generally high proportion of hydrothermal material adjacent to the southern East Pacific Rise does not necessarily indicate intense hydrothermal activity. It may simply result from a "shortage" of other constituents reflecting the considerable distance to the continent, as a source of detritus, as well as the low productivity of the surface water.

Riech et al.[14] demonstrate the presence of climatic cyclicity in cores from the flanks of the southern East Pacific Rise. Maximum dissolution of carbonate is always observed at the beginning of a cold period, in spite of the fact that all the cores investigated were taken well above the lysocline. Hence, cyclic changes in the content of hydrothermal material may, in part, be due to secondary enrichment caused by carbonate dissolution. Further, the hydrothermal component in the sediment was found to vary in composition with water temperature, the manganese being depleted relative to iron during cold phases. Riech et al.[14] explained these changes in Fe/Mn ratio on the basis of the work of Mangini et al.[54] Deep ocean circulation is weaker in glacial periods and causes a depletion in oxygen in the bottom waters; the resulting lower redox potential maintains the reduced bivalent manganese in solution. During interglacial periods, deep water circulation is more extensive, oxygen supply is better, and manganese is precipitated as insoluble tetravalent hydroxide. The cycles displayed by Fe/Mn variation in the sediments from the southern East Pacific Rise fit the climatic changes so well that the work of Riech et al.[14] could be considered as proof of the theory of Mangini et al.[54]

11.4 HYDROTHERMAL PLUMES IN SEA WATER

Hydrothermal plumes are valuable for investigating the intensity and duration of hydrothermal activity. The first observation of a plume on the southern part of the East Pacific Rise is described in Lupton and Craig.[19] They mapped (with vertical sampling profiles) a large ^{3}He plume emanating from the East Pacific Rise at 15°S and extending over 2000 km to the west. This helium isotope (^{3}He) is primordial helium, but the source of the plume was positioned over the top of the East Pacific Rise. Thus, the ^{3}He was interpreted as having been transported to the seafloor in hydrothermal solution and not degased from fresh volcanic rocks. This large hydrothermal plume was revisited 15 years later during the *Helios* expedition, and it was still there, larger than ever.[55]

The reason for the westward drift of the plume is still a matter of debate. It is by no means agreed that the westward drift (which is also observed in the sediment distribution pattern) is caused only by primary ocean bottom currents; it could be due to westward advection driven by the buoyancy flux on the ridge axis.[56]

Methane, like helium, is a product of primordial degasing, at least on sediment-free ridges, where it cannot be bacterial in origin. In contrast to the noble gas helium, methane is used in the metabolism of organisms in the sea water and thus cannot form such large anomalies as helium can.

The enrichment of metals in plumes is a clear indicator of high-temperature hydrothermal discharge. The first manganese-enriched plume was discovered on the southern East Pacific Rise in 1986 by Klinkhammer and Hudson.[57] Sampling was carried out on the East Pacific Rise and west of it at 20°S, in 12 vertical profiles using a rosette water sampler coupled with a CTD. The manganese plume changed in composition with distance from the spreading zone. Over the ridge, nearly all the manganese was in a dissolved form, but away from the site of discharge the amount of colloidal manganese hydroxide increased. Hudson et al.[58] published iron analyses from the same area, some on the same water samples.

In 1987/88, the Russian research vessel *Geolog Fersman* sampled ocean water from near the seafloor over the spreading zone at 13°S and 21 to 22°S; these water samples were analyzed for copper, zinc, manganese, and iron.[59] Zinc in particular appeared to reflect the high-temperature discharge already known in these areas.

In 1989, the German research vessel *Sonne* used Niskin bottles (and CTD) attached to a photo/TV sledge to sample hydrothermal plumes. This enabled water sampling to be done near suspected hydrothermal vents; the Niskin bottles were closed in turbid or shimmering water. This type of sampling was previously possible only from a submersible. Comparison of the results of seafloor mapping together with methane analyses of near-bottom sea water[12,32] showed that the "phase of enhanced volcanic activity" (rounded sheet lava ridge, Figure 11.2) is associated with strong degasing; the water samples were enriched in methane in comparison with those from areas characterized by other phases of cyclic spreading (Figure 11.2). It was also associated with a high bioproduction, evidenced by huge amounts of small unidentified animals in the bottom water (an observation compared to a "blizzard"). At one of these stations, the extent of the methane anomaly was investigated with six profiles; it is shown in Figure 11.4.[60] It was confined to the ridge flanks, interpreted as due to degasing of fresh lava via fissures and not to transport by hydrothermal solutions. The methane anomaly was closely associated with a light-attenuation anomaly, which was caused by drifting organisms and not by hydrothermal precipitation. The isotopic composition of the methane in the water samples was not identical to that of primordial methane, the difference most likely being a result of strong biologic activity.[13] The subsequent phase of tectonic activity, which coincided with the maximum growth of massive sulfide chimneys (young central valley, see Figure 11.2), is characterized by small quantities of methane in water samples. Holler et al.[12] also noticed that, in the active volcanic areas, diffuse venting on the seafloor was significantly more frequent than black smoker activity. These observations agree with those of German et al.,[61] who calculated from hydrothermal fluxes that more than 70% of hydrothermal solutions reach the seafloor by diffuse low-temperature venting.

The French *Naudur* cruise in 1993 investigated hydrothermal discharge on the southern East Pacific Rise. Hydrothermal fluids were sampled with the submersible *Nautile* between 17°S and 19°S. Hydrothermal fluids associated with the volcanic phase of the spreading cycle showed low chloride, low contents of dissolved metals, and enrichment in dissolved gases. Observations that large amounts of diffuse, shimmering water are associated with vents during the high-volcanicity phase agree with the observations of Holler et al. and Holler.[12,32] The fluids sampled on ridge segments where volcanism had ceased, and where a summit valley had begun to form and metallogenesis was at a maximum, revealed high chloride and metal contents and low contents of dissolved gases.[24,33,62]

A Japanese/U.S. Ridge Flux Project in 1995 mapped sea water anomalies over the East Pacific Rise between 13°30'S and 18°40'S along continuous profiles parallel to the ridge. This was not only the first but so far the only continuous sea-water survey carried out over the southern East Pacific Rise. Several spreading zones with low spreading rates were also mapped during the same project, permitting a comparison between ridge segments. Two main types of hydrothermal fluids were found, volatile-rich fluids associated with volcanism, and metal-rich fluids associated with hydrothermal metallogenesis.[21,22,25] Nephelometry and temperature measurements confirmed that the southern East Pacific Rise is more active than other ridges with lower spreading rates, that iron hydroxide scavenges P, As, V, and Cr, causing the enrichment of these elements in hydrothermal sediments, and that in methane-rich plumes, free-floating bacteria are an important factor in causing turbidity of the sea water.

11.5   SULFIDE DEPOSITS ON THE SEAFLOOR

Groups of massive-sulfide chimneys are frequently observed in the last two phases of the spreading cycle (Figure 11.2), in the initial axial summit valley and even more frequently in the well-developed axial summit valley. In phase 2 (rounded sheet lava ridge), individual massive-sulfide chimneys are rarely observed, and in phase 1 (steep pillow lava ridge) no massive sulfides are found. It can be concluded from seafloor observations that, in a well-developed axial summit valley, not only are massive sulfide chimneys frequent, but also the seafloor is disrupted by faulting, providing channels for hydro-thermal solutions and implying that stockwork mineralization is common below the seafloor.

Figure 11.5 shows the frequency of massive sulfides observed on the southern East Pacific Rise between 6°S and 23°S (from Holler[32]), corresponding with the amount of talus on the seafloor. These features, and a well-developed axial summit valley, are characteristic of tectonic phase 4 (Figure 11.2). On the margins of the Easter Plate, massive sulfides were observed in positions associated with the oldest part of a propagating rift.[29]

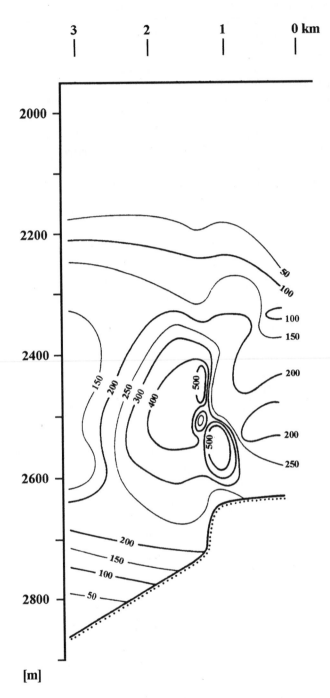

FIGURE 11.4   The methane anomaly over the East Pacific Rise at 11°53'S. This occurs during Phase 2 (Figure 11.2), which is characterized by high volcanicity and the degasing of large amounts of methane. The anomaly was mapped on the basis of six CTD profiles; the first one directly over the ridge and the last one three kilometers west of it. The dotted line is the seafloor topography. Isolines in nL/L CH$_4$.

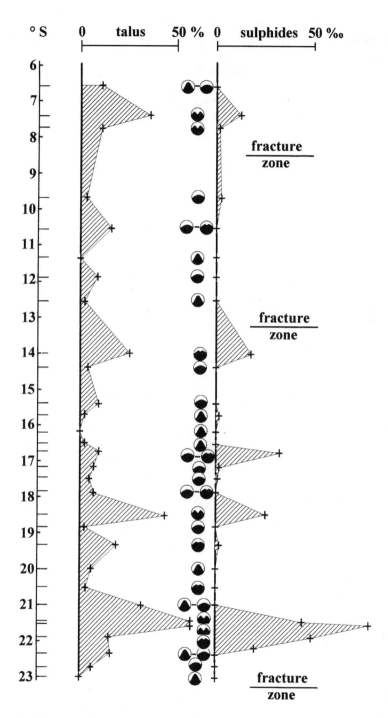

FIGURE 11.5   The frequency of occurrence of massive sulfides compared with the amount of talus on the floor of the summit valley. Talus is selected as the best indicator of the tectonic activity that gave rise to the axial summit valley. The sulfide maxima correlate closely with talus maxima, although there are some talus maxima for which corresponding massive sulfide maxima are absent. For symbols showing the spreading-cycle phases, see Figure 11.2. (Data from Holler, G., *Erscheinungsformen hydrothermaler Aktivität am Ost-pazifischen Rücken zwischen 6° und 30° Süd, Verlag Shaker,* 1-226, 28 Abb., 6 Taf., 76 Tab., 1993.)

The first samples of sulfide dredged from the southern East Pacific Rise were recovered during GEOMETEP 3 (SO 26), 1983/1984. Later GEOMETEP expeditions (SO 40, 1985/1986 and SO 62, 1989) used a TV-controlled grab for selective sampling of large massive sulfide chimneys. Eighteen chimneys were sampled in seven hydrothermal fields at 7°23'S, 10°34'S, 13°59'S, 16°43'S, 17°23'S, 18°30'S, and 21°26'S. Most of the chimneys were not very consolidated, and the samples reached the ship as a heap of rubble. Only seven complete edifices were successfully sampled. The results have been published by Tufar et al.,[63,64] Marchig et al.,[65-67] Oudin et al.,[68] Dill et al.,[48,69] Rösch and Marchig,[70] and Jansen.[71] Additionally, a Russian expedition with the research vessel *Geolog Fersman* dredged sulfides in 1987 on the East Pacific Rise between 21°10' and 22°30'. The dredging was carried out where cameras had demonstrated a concentration of chimneys.[72,73]

Two French expeditions in 1984 and 1993 sampled sulfides between 17°10'S and 21°26'S by means of a submersible.[11,33] The submersible is more precise in sampling selected sulfides than the TV-controlled grab, because, unlike the grab, it can recover material from isolated edifices. Its disadvantage is the small size of sample that can be picked up with the remotely controlled arm, making it more or less impossible to obtain a complete profile of a pipe.

Based on the results of the above cruises, it can be shown that large chimneys occurring between 7°23'S and 21°26'S consist primarily of pyrite, chalcopyrite, and sphalerite, in decreasing order of abundance. Marcasite is an accessory constituent. Among nonsulfides, amorphous SiO_2 is ubiquitous. A chimney grows in the following sequence. The oldest part is a pyrite-rich pipe. Pyrrhotite is present in trace amounts, and it is mostly replaced by pyrite. The well-crystallized idiomorphic pyrite grains, with some chalcopyrite and sphalerite, are enclosed by (and thus cemented by) a second generation of minerals consisting of colloform precipitates of marcasite, pyrite, and sphalerite. These give the pipe stability and, at same time, drastically lower the permeability of the chimney wall to <1 mD.[74] It is probably because of the low permeability that dissolution and replacement are not widespread in the walls of large chimneys; only recrystallization can be observed.

Anhydrite, if it was ever present, has entirely disappeared from the GEOMETEP chimneys. Even minerals pseudomorphing anhydrite, as were described from the northern part of the East Pacific Rise,[75-78] are absent. However, the French expeditions, which sampled small sulfide edifices between 17°10'S and 21°26'S, did recover some anhydrite and silica pseudomorphs after anhydrite. The other constituents were the same as those described for large chimneys sampled from the same area by GEOMETEP expeditions. Barite has only been found in traces in one chimney at 7°23'S. The near absence of sulfates, especially anhydrite, is the largest difference compared with chimneys described from the northern part of the East Pacific Rise. There, anhydrite appears during the early stage of the formation of chimneys.

The second stage of chimney formation is the clogging of the central channel with sulfides rich in chalcopyrite (only in one case was this "cork" composed of well-crystallized sphalerite). Precipitation of these sulfides took place in equilibrium with hydrothermal solution; the crystals are idiomorphic and the chalcopyrite is stoichiometric. The permeability of this part of the chimney is higher than that of the cemented pipe, with maximum permeability in a vertical direction.

The third phase of precipitation takes place outside the clogged chimney. The hydrothermal solutions find an alternative outlet and well out from the seafloor around the chimney, precipitating a coating of sulfide on the outside of the chimney. This material is a product of quenching of the hydrothermal solution in contact with sea water; the main component is colloform sphalerite with some colloform pyrite and marcasite as alternating layers, together with traces of isocubanite. This is the only phase in which numerous worm tubes are found embedded in sulfide; the worm tubes are composed of calcite, which is more or less replaced by sulfides.

On the basis of observations from submersibles on the reactivation of hydrothermal activity after volcanic eruptions on the northern part of the East Pacific Rise, it was suggested that a cycle of dike injection, eruption, and hydrothermal discharge is repeated as often as every three to five years.[79] However, observations on the southern East Pacific Rise show that a spreading cycle must be significantly longer than that. Phase 2 activity (Figure 11.2) does not end after a single eruption; there must be a period in which eruptions are frequent and the convex sheet-lava ridge develops. Hydrothermal discharge between successive eruptions, mostly in the form of diffuse black smokers which are formed by localized discharge, does occur at this stage but is quite rare. The main mineralization takes place later, i.e., in Phases 3 and 4.

Several attempts have been made to date the massive sulfide chimneys; three were more than 200 years old.[80] Another was at least 300 years but more probably 2,000 years old,[68] and only in one case was it possible to date the different growth zones;[65] the total growth period was about 80 years. This suggests that the metallogenetic phase in the areas investigated had a duration of at least a few hundred years, but more probably several thousand years. Longer periods of time are required for ore edifices to develop in the TAG hydrothermal field on the slow-spreading Mid Atlantic Ridge.[81] Nevertheless, extensive massive sulfide mineralization has been observed to develop on the southern East Pacific Rise, and this is certainly due to the strong hydrothermal discharge there.

11.6   SUMMARY AND CONCLUSIONS

The southern part of the East Pacific Rise is the spreading zone with the highest spreading rates in the world. Associated with the ultra-fast spreading is vigorous volcanic activity alternating with high rates of discharge of ore-forming hydrothermal solutions.

Spreading proceeds in cycles; four phases can be distinguished in every cycle:

1. Weak volcanic eruptions form narrow and steep pillow lava ridge
2. Increasing volcanic eruptions form large and rounded sheet lava ridge
3. Volcanic eruptions cease, tectonic activity forms axial summit graben
4. Axial summit graben gets larger, deeper, and more disrupted

Each new cycle is initiated by further volcanic eruptions.

Phase 2 is characterized not only by vigorous volcanic eruptions, but also by strong degasing and highest density of organisms utilizing the methane from degasing. In phases 3 and 4 hydrothermal circulation and metallogenesis take place. Formation of massive sulfide chimneys is dependent on seafloor permeability; consequently, massive sulfide chimneys concentrate in strongly disrupted parts of the axial summit graben and can be correlated with the amount of talus present on the seafloor.

The ore occurs mostly in the form of massive sulfide chimneys. The growth of these chimneys seems to be somewhat different from what is described on the northern part of the East Pacific Rise. The first pipe is not formed of anhydrite but of idiomorphic pyrite, which subsequently becomes cemented by collomorphic sulfides. The second phase of chimney growth is also seen in other spreading zones; it is precipitation of chalcopyrite-rich sulfides within the central chimney conduit. The third phase, quenched collomorphic sphalerite precipitating on the outer wall of the chimney, again appears to be unique for the southern East Pacific Rise.

Hydrothermal fields with massive sulfide accumulation and chimney formation are frequent, but they do not reach the dimensions of the hydrothermal fields on slow-spreading ridges. The short duration of a spreading cycle is partly responsible for this, because they get covered with new lava flows, but also a high proportion of precipitates from the ore-forming solutions are dispersed in plumes to the surrounding sediment. This sediment is consequently enriched in Mn- and Fe-hydroxides. The plumes seem to homogenize in sea water before their constituents settle because degree of metal enrichment in sediments does not mirror local hydrothermal fields. Also vertical changes in the amount of hydrothermal precipitates within the sediment are not necessarily indicative of changing hydrothermal activity with time; they can reflect Pleistocene climatic changes.

REFERENCES

1. Murray, J. and Renard, A.F., *Deep Sea Deposits, Reports of "Challenger" Expedition* (1873-1876), London, 1891.

2. Revelle, R.R., *Marine Bottom Samples Collected in the Pacific Ocean by the Carnegie on its Seventh Cruise — Scientific Results of Cruise VII,* 1-182, Carnegie Institution of Washington Publication 556, Washington, D.C., 1944.

3. Peterson, M.N.A. and Goldberg, E.D., Feldspar distributions in South Pacific pelagic sediments, *Journal of Geophysical Research,* 67, 9, 3477, 1962.

4. Skornyakova, I.S., Dispersed iron and manganese in Pacific Ocean sediments, *International Geology Review,* 7, 12, 2161, 1964.

5. von Herzen, R.P. and Uyeda, S., Heat flow through the eastern Pacific Ocean floor, *Journal of Geophysical Research,* 68, 14, 4219, 1963.

6. Boström, K. and Peterson, M.N.A., Precipitates from hydrothermal exhalations on the East Pacific Rise, *Economic Geology,* 61, 1258, 1966.

7. Boström, K., Genesis of ferromanganese deposits: diagnostic criteria for recent and old deposits, in *Hydrothermal Processes at Seafloor Spreading Centres,* Rona, P.A., Boström, K., Laubier, L., and Smith, K.L.J. (Eds.), Plenum Press, New York, 1983, 473.

8. Kulm LaVerne, D., Dymond, J., Dash, E.J., Hussong, D.M., and Roderick, R. (Eds.), *Nazca Plate: Crustal Formation and Andean Convergence,* The Geological Society of America, Memoir 154, 824 pp., 1981.

9. Francheteau J., Needham, H.D., Choukroune, P., Juteau, T., Seguret, M., Ballard, R.D., Fox, P.J., Normark, W., Carranza, A., Cordoba, D., Guerrero, J., Rangin, C., Bougault, H., Cambon P., and Hekinian, R., Massive deep-sea sulphide ore deposits discovered on the East Pacific Rise, *Nature,* 277, 523, 1979.

10. Bäcker, H. and Marchig, V., Sulfiderze in der Tiefsee: Das Forschungsprogramm GEOMETEP mit dem Forschungsschiff, *SONNE, Meerestechnik Heft,* 4, 134, 1983.

11. Renard, V., Hekinian, R., Francheteau, J., Ballard, R.D., and Bäcker, H., Submersible observations at the axis of the ultra-fast spreading East Pacific Rise (17°30'-21°30'S), *Earth and Planetary Science Letters,* 75, 339, 1985.

12. Holler, G., Marchig, V., and the shipboard scientific party, Hydrothermal activity on the East Pacific Rise: stages of development, *Geologisches Jahrbuch,* B 75, 3, 1990.

13. Faber, E., Gerling, P., Sohns, E., and Michaelis, W., Hydrothermale Aktivitäten des EPR und der Methanhaushalt, Geosphäre — Hydrosphäre — Atmosphere — Geowissenschaften und Umwelt, Herausgeber: Matschullat, J. and Müller, G., 221, 1994.

14. Riech, V., Röhl, U., and Erlenkeuser, H., Gebänderte Sedimente am Ostpazifischen Rücken: Hydrothermale oder klimatologische Ursachen? GEOMETEP 5 (SO62) scientific report, 473-535. Federal Institute for Geosciences and Natural Resources, BMFT project Nr.: 03R393, 1992.

15. Lonsdale, P., Segmentation of the Pacific-Nazca spreading center, 1°N-20°S, *Journal of Geophysical Research,* 94, 12197, 1989.

16. Sinton, J.M., Smaglik, S.M., Mahoney, J.J., and MacDonald, K.C., Magmatic processes at superfast spreading mid-ocean ridges: glass compositional variations along the East Pacific Rise 13°-23°S, *Journal of Geophysical Research,* 96, 6133, 1991.

17. Kent, G.M., Harding, A.J., Orcutt, J.A., Detrick, R.S., Mutter, J.C., and Buhl, P., Uniform accretion of oceanic crust south of the Garett transform at 14°15'S on the East Pacific Rise, *Journal of Geophysical Research,* 99, 9097, 1994.

18. Singh, S.C., Kent, G.M., Harding, A.J., and Orcutt, J.A., P- and S-wave velocity structure of the upper crust at 14°10'S East Pacific Rise from waveform inversion, *Eos Trans. AGU,* 77 (46) Fall Meet. Suppl., F664, 1996.

19. Lupton, J.E. and Craig, H., A major helium-3 source at 15°S on the East Pacific Rise, *Science,* 214, 13, 1981.

20. Krasnov, S.G., Kreyter, I.I., and Poroshina, I.M., The distribution of hydrothermal vents on the East Pacific Rise (21°20'-22°40'S) based on a study of dispersion patterns of hydrothermal plumes, *Oceanology,* 32, 375, 1992.

21. Urabe, T., Baker, E.T., Ishibashi, J., Feely, R.A., Marumo, K., Massoth, G.J., Maruyama, A., Shitashima, K., Okamura, K., Lupton, J.E., Sonoda, A., Yamazaki, T., Aoki, M., Gendron, J., Greene, R., Kaiho, Y., Kisimoto, K., Lebon, G., Matsumoto, T., Nakamura, K., Nishizawa, A., Okano, O., Paradis, G., Roe, K., Shibata, T., Tennant, D., Vance, T., Walker, S.L., Yabuki, T., and Ytow, N., The effect of magmatic activity on hydrothermal venting along the superfast-spreading East Pacific Rise, *Science,* 269, 1092, 1995.

22. Baker, E.T. and Urabe, T., Extensive distribution of hydrothermal plumes along the superfast spreading East Pacific Rise, 13°30'-18°40'S, *Journal of Geophysical Research,* 101, B4, 8685, 1996.

23. Feely, R.A., Baker, E.T., Marumo, K., Urabe, T., Ishibashi, J., Gendron, J., Lebon, G.T., and Okamura, K., Hydrothermal plume particles and dissolved phosphate over the superfast-spreading southern East Pacific Rise, *Geochimica and Cosmochimica Acta,* 60, 13, 2297, 1996.

24. Charlou, J.L., Fouquet, Y., Donval, J.P., and Auzende, J.M., Mineral and gas chemistry of hydrothermal fluids on an ultrafast spreading ridge: East Pacific Rise, 17° to 19°S (Naudur cruise, 1993) phase separation processes controlled by volcanic and tectonic activity, *Journal of Geophysical Research,* 101, B7, 15,899, 1996.

25. Ishibashi, J., Wakita, H., Okamura, K., Nakayama, E., Feely, R.A., Lebon, G.T., Baker, E.T., and Marumo, K., Hydrothermal methane and manganese variation in the plume over the superfast-spreading southern East Pacific Rise, *Geochimica et Cosmochimica Acta,* 61, 3, 485, 1997.

26. Tighe, S.A. (Ed.), *East Pacific Rise Data Synthesis and Final Report,* JOI Inc., Washington, D.C., 1988.

27. Scheirer, D.S. and MacDonald, K.C., Variation in cross-sectional area of the axial ridge along the East Pacific Rise: evidence for the magmatic budget of a fast spreading center, *Journal of Geophysical Research,* 98, B5, 7871, 1993.

28. Schwarz, H.-U. and Dulce, G., Magmato-tectonic cycles and their structural implications on a fast-spreading ridge axis (EPR 6°S-23°S), unpublished manuscript.

29. Searle, R.C., Bird, R.T., Rusby, R.I., and Naar, D.F., The development of two oceanic microplates: Easter and Juan Fernandez microplates, East Pacific Rise, *Journal of the Geological Society,* 150, 965, 1993.

30. Bäcker, H., Lange, J., and Marchig, V., Hydrothermal activity and sulfide formation in axial valleys of the East Pacific Rise crest between 18° and 22°S, *Earth and Planetary Science Letters,* 72, 9, 1985.

31. Bäcker, H. and Lange, J., Recent hydrothermal metal accumulation, products, and conditions of formation, in: *Marine Minerals,* Teleki, P.G., Dobson, M.R., Moore, J.R., and von Stackelberg, U. (Eds.), D. Reidel Publishing Company, Dordrecht, 1987, 317.

32. Holler, G., Erscheinungsformen hydrothermaler Aktivität am Ostpazifischen Rücken zwischen 6° und 30° Süd, *Verlag Shaker,* 1-226, 28 Abb., 6 Taf., 76 Tab., 1993.

33. Auzende, J.-M., Bullu, V., Batiza, R., Bideau, D., Charlou, J.-L., Cormier, M.H., Fouquet, Y., Geistdörfer, P., Lagabrielle, Y., Sinton, J., and Spadea, P., Recent tectonic, magmatic, and hydrothermal activity on the East Pacific Rise between 17°S and 19°S: submersible observations, *Journal of Geophysical Research,* 101, B8, 17.995, 1996.

34. Hooft, E.E., Detrick, R.S., and Kent, G.M., Seismic structure and indicators of magma budget along the Southern East Pacific Rise, *Journal of Geophysical Research,* 112, B12, 27319, 1997.

35. Fornari, D.J., Haymon, R.M., Perfit, M.R., Gregg, T.K.P., and Edwards, M.H., Axial summit trough of the East Pacific Rise 9°-10°N: geological characteristics and evolution of the axial zone on fast spreading mid-ocean ridges, *Journal of Geophysical Research,* 103, B5, 9827, 1998.

36. Yang Shen, Forsyth, D.W., Scheirer, D.S., and MacDonald, K.C., Two forms of volcanism: implications for mantle flow and off-axis crustal production on the west flank of the southern East Pacific Rise, *Journal of Geophysical Research,* 98, B10, 17875, 1993.

37. Hall, L.S. and Sinton, J.M., Geochemical diversity of the large lava field on the flank of the East Pacific Rise at 8°17'S, *Earth and Planetary Science Letters,* 142, 241, 1996.

38. Bellieni, G., Milovanovic, D., and Marchig, V., Some constraints on the genesis of basalts from East Pacific Rise 6°S-30°S. GEOMETEP 5 (SO62) scientific report, BGR, Hannover, Federal Institute for Geosciences and Natural Resources; BMFT project Nr.: 03R393, 1992, 273.

39. Bach, W., Hegner, E., Erzinger, J., and Satir, M., Chemical and isotopic variations along the superfast spreading East Pacific Rise from 6° to 30°S, *Contributions to Mineralogy and Petrology,* 116, 365, 1994.

40. MacDonald, K.C., Magmatic processes at superfast spreading mid-ocean ridges: glass compositional variations along the East Pacific Rise 13°–23°S, *Journal of Geophysical Research,* 96, B4, 6133, 1991.

41. Marchig, V., Gundlach, H., Holler, G., and Wilke, M., New discoveries of massive sulfides on the East Pacific Rise, *Marine Geology,* 84, 179, 1988.

42. Shimmield, G.H. and Price, N.B., The scavenging of U, ^{230}Th and ^{231}Pa during pulsed hydrothermal activity at 20°S, East Pacific Rise, *Geochimica et Cosmochimica Acta,* 52, 669, 1988.

43. Mills, R.A. and Elderfield, H., Hydrothermal activity and the geochemistry of metalliferous sediments, in *Seafloor Hydrothermal Systems,* Humphris, S.E., Zierenberg, R.A., Mullineaux, L.S., and Thomson, R.E. (Eds.), American Geophysical Union, Washington, D.C., Geophysical Monograph 91, 1995, 392.

44. McMurtry, G.M. and Yeh, H.-W., Hydrothermal clay mineral formation of the East Pacific Rise and Bauer Basin sediments, *Chemical Geology,* 32, 189, 1981.

45. Walter, P. and Stoffers, P., Chemical characteristics of metalliferous sediments from eight areas on the Galapagos Rift and East Pacific Rise between 2°N and 42°S. *Marine Geology,* 65, 271, 1985.

46. Meschede, M., Barckhausen, U., and Worm, H.U., Neue Erkenntnisse zur plattentektonischen Entwicklung der Cocos-Platte als Ergebnis einer multidisziplinären Studie. Kolloquium des DFG-Schwerpunktprogramms, *Ocean Drilling Program/Deep Sea Drilling Project* in Freiburg, Abstracts, Geological Institute, University of Freiburg, Germany, 1988, 38.

47. Marchig, V. and Dietrich, P.G., Hydrothermal activity in Middle Valley (Juan de Fuca Ridge 48°N): chemical redistribution between sediment, igneous rock and massive sulfides, *Scientific Drilling,* 5, 267, 1996.

48. Dill, H.G., Siegfanz, G., and Marchig, V., Mineralogy and chemistry of metalliferous muds forming the topstratum of a massive sulfide-metalliferous sediment sequence from East Pacific Rise 18°S: its origin and implications concerning the formation of ochrous sediments in Cyprus-type deposits, *Marine Georesources and Geotechnology,* 12, 159, 1994.

49. Edmond, J.M., von Damm, K.L., McDuff, R.E., and Measures, C.I., Chemistry of hot springs on the East Pacific Rise and their effluent dispersal, *Nature,* 297, 187, 1982.

50. Dymond, J., Geochemistry of Nazca plate surface sediments: an evaluation of hydrothermal, biogenic, detrital, and hydrogenous sources, in *Nazca Plate: Crustal Formation and Andean Convergence,* Kulm LaVerne D., Dymond, J., Dash, E.J., Hussong, D.M., and Roderick, R. (Eds.), *Geol. Soc. Am. Memoir,* 154, 1981, 133.

51. Heath, G.R. and Dymond, J., Metalliferous sediment deposition in time and space: East Pacific Rise and Bauer Deep, northern Nazca plate, in Kulm LaVerne D., Dymond, J., Dash, E.J., Hussong, D.M., and Roderick, R., (Eds.), *Geol. Soc. Am. Memoir,* 154, 175, 1981.

52. Marchig, V., Gundlach, H., and Bäcker, H., Geochemical indication in deep-sea sediments for hydrothermal discharge, *Marine Geology,* 56, 319, 1984.

53. Marchig, V., Erzinger, J., and Heinze, P.-M., Sediment in the black smoker area of the East Pacific Rise (18.5°S), *Earth and Planetary Science Letters,* 79, 93, 1986.

54. Mangini, A., Eisenhauer, A., and Walter, P., The relevance of manganese in the ocean for the climatic cycles in the quaternary, Sitzungsberichte der Heidelberger Akademie der Wissenschaften; Mathematisch–naturwissenschaftliche Klasse; Springer Verlag, 263, 1990.

55. Lupton, J.E., Hydrothermal plumes: near and far field, in *Seafloor Hydrothermal Systems,* Humphris, S.E., Zierenberg, R.A., Mullineaux, L.S., and Thomson, R.E. (Eds.), American Geophysical Union, Washington, D.C., Geophysical monograph 91, 1995, 317.

56. Helfrich, K.R. and Speer, K.G., Oceanic hydrothermal circulation: mesoscale and basin-scale flow, in *Seafloor Hydrothermal Systems,* Humphris, S.E., Zierenberg, R.A., Mullineau, L.S., and Thomson, R.E. (Eds.), American Geophysical Union, Washington, D.C., Geophysical monograph 91, 1995, 347.

57. Klinkhammer, G. and Hudson, A., Dispersal patterns for hydrothermal plumes in the South Pacific using manganese as a tracer, *Earth and Planetary Science Letters* 79, 241, 1986.

58. Hudson, A., Bender, M.L., and Graham, D.W., Iron enrichments in hydrothermal plumes over the East Pacific Rise, *Earth and Planetary Science Letters,* 79, 250, 1986.

59. Krasnov, S.G., Tscherkaschev, G.A., Grinberg, G.P., and Nikoienko, A.G., Metals in ascending waters from the area of high-temperature hydrothermal activity on the East Pacific Rise, *Dokladi akademii nauk SSSR,* 307, 4, 976, 1989.

60. Quadfasel, D., Gerling, P., and Steentoft, H., Das Abbild von Hydrothermalquellen in der Hydrographie und den Methangehalten der Wassersäule, in SO62 (GEOMETEP 5) scientific report pp. 449-472, Federal Institute for Geosciences and Natural Resources; BMFT project Nr.: 03R393, 1992.

61. German, C.R., Baker, E.T., and Klinkhammer, G., Regional setting of hydrothermal activity, in Hydrothermal Vents and Processes, Parson, L.M., Walker, C.L., and Dixon, D.R. (Eds.), Geol. Soc. Special Publ. no. 87, 1995, 3.

62. Fouquet, Y., Auzende, J.-M., Ballu, V., Batiza, R., Bideau, D., Comier, M.-H., Geistdörfer, P., Lagabrielle, Y., Sinton, J., and Spadea, P., Variabilité des manifestations hydrothermales actuelles le long d'une dorsale ultra-rapide: exemple de la Dorsale Est Pacifique entre 17° et 19°S (campagne NAUDUR), *Oceanographie,* C.R. Acad. Sci. Paris, t. 319, serie II, 1399, 1994.

63. Tufar, W., Gundlach, H., and Marchig, V., Zur Erzparagenese rezenter Sulfid-Vorkommen aus dem südlichen Pazifik, *Mitt. Österr. Geol. Ges.,* 77, 185, 1984.

64. Tufar, W., Gundlach, H., and Marchig, V., Ore paragenesis of recent sulfide formations from the East Pacific Rise, Monograph series on Mineral Deposits, 25, Gebrüder Bornträger, Berlin-Stuttgart, 1986, 75.

65. Marchig, V., Rösch, H., Lalou, C., Brichet, E., and Oudin, E., Mineralogical zonation and radiochronological relations in a large sulfide chimney from the East Pacific Rise at 18°25'S, *Canadian Mineralogist,* 26, 541, 1988.

66. Marchig, V., Puchelt, H., Rösch, H., and Blum, N., Massive sulfides from ultra-fast spreading ridge, East Pacific Rise at 18-21°S: a geochemical stock report, *Marine Mining,* 9, 459, 1990.

67. Marchig, V., Blum, N., and Roonwal, G., Massive sulfide chimneys from the East Pacific Rise at 7°24'S and 16°43'S, *Marine Georesources and Geotechnology,* 15, 49, 1997.

68. Oudin, E., Marchig, V., Rösch, H., Lalou, C., and Brichet, E., Observation de CuS$_2$ à l'état naturel dans une cheminée hydrothermale du Pacifique Sud, *C.R. Scad. Sci. Paris,* 310, 221, 1990.

69. Dill, H.G., Gauert, C., Holler, G., and Marchig V., Hydrothermal alteration and mineralisation of basalts from the spreading zone of the East Pacific Rise (7°S-23°S), *Geologische Rundschau,* 81/3, 717, 1992.

70. Rösch, H. and Marchig, V., An unusual occurrence of wurtzite in massive sulfide vents from the East Pacific Rise, *Geologisches Jahrbuch,* A 27, 589, 1991.

71. Jansen, K., Erzmikroskopische und geochemische Untersuchungen an Massivsulfiden vom ostpazifischen Rücken bei 7°24'S und 16°43'S, thesis, University of Hannover, 151 pp., 1995.

72. Krasnov, S.G., Maslov, M.N., Andreev, N.M., Konfetkin, V.M., Kreiter, J.J., and Smirnov, B.N., Hydrothermal ore formation on the southern part of the East Pacific Rise, *Dokladji akademii nauk SSSR,* 302, 1, 161, 1988.

73. Krasnov, S.G., Poroshina, I.M., and Cherkashev, G.A., Geological setting of high-temperature hydrothermal activity and massive sulphide formation on fast-and slow-spreading ridges, in *Hydrothermal Vents and Processes,* Parson, L.M., Walker, C.L. & Dixon, D.R. (Eds.), Geological Society Special Publication, 87, 1995, 17.

74. Knott, R., Prosser, J., and Rickard, D., Permeability and porosity variations in black smoker chimneys, unpublished manuscript.

75. Oudin, E., Étude minéralogique et géochimique des dépóts sulfurés sous-marins actuels de la ride Est Pacifique (21°N) Campagne RISE, Doc. BRGM 25, 1981.

76. Haymon, R.M., Growth history of hydrothermal black smoker chimneys, *Nature,* 301, 695, 1983.

77. Tivey, M.K. and Delaney, J.R., Growth of large sulfide structures on the Endeavour segment of the Juan de Fuca Ridge, *Earth and Planetary Science Letters,* 77, 303, 1986.

78. Paradis, S., Jonasson, I.R., Le Cheminant, G.M., and Watkinson, D.H., Two zinc-rich chimneys from the Plume Site, southern Juan de Fuca Ridge, *Canadian Mineralogist,* 26, 637, 1988.

79. Haymon, R.M., Fornari, D.J., Von Damm, K.L., Lilley, M.D., Perfit, M.R., Edmond, J.M., Shanks, W.C., Lutz, R.A., Grebmeier, J.M., Carbotte, S., Wright, D., McLaughlin, E., Smith, M., Beedle, N., and Olson, E., Volcanic eruption of the mid-ocean ridge along the East Pacific Rise crest at 9°45-52'N: direct submersible observations of seafloor phenomenon associated with an eruption in April 1991, *Earth and Planetary Science Letters,* 119, 85, 1993.

80. Lalou, C., Radiochemical dating of massive sulfide chimneys from the cruise SO62-GEOMETEP 5. In: SO62 (GEOMETEP 5) scientific report, 399 pp., Federal Institute for Geosciences and Natural Resources, BMFT project Nr.: 03R393, 1992.

81. Lalou, C., Reyss, J.-L., Brichet, E., Arnold, M., Thompson, G., Fouquet, G.Y., and Rona, P.A., New age data for Mid-Atlantic Ridge hydrothermal sites: TAG and Snakepit chronology revisited, *Journal of Geophysical Research,* 98, 9705, 1993.

12 Mineral Deposits at 23°S, Central Indian Ridge: Mineralogical Features, Chemical Composition, and Isotopic Investigations

Peter Halbach and Ute Münch

ABSTRACT

Hydrothermal activity is well documented from many fields on most mid-ocean ridges, but so far there is only one hydrothermally mineralized area known from the Central Indian Ocean. This mature massive sulfide deposit, referred to as the MESO mineral zone, is located near 23°S on the intermediate-spreading Central Indian Ridge. Hydrothermal precipitates occur at a water depth of about 2,850 m close to the rift axis on the top of an intrarift ridge. This neovolcanic ridge forms a 20-km-long topographic high in the central part of the 11-km-wide W-shaped axial valley.

A systematic survey of the mineralized zone with a deep-towed photosledge indicates that the MESO zone contains three different sites with evidence of former hydrothermal activity. Hydrothermal mineralization appears as chimney relicts, sulfidic talus or sulfide-debris impregnated jasper, and as sediment discolorations. Chimney accumulations are structurally controlled by fissures and cracks mainly running both parallel and vertical to the general strike direction of the ridge (N153°).

Chimney samples are composed of pyrite, marcasite, chalcopyrite, and secondary Cu-minerals and show high concentrations of base metals; remarkable are extremely high Cu values (up to 31 wt.%). The formation of late-stage Cu-sulfides such as bornite, digenite, and less frequently covellite accounts for this concentration of Cu. Au is also enriched in late-stage pyrite. The massive sulfide samples do not contain nonsulfide mineral phases except for minor amounts of amorphous silica. In contrast, the jasper breccia, formed by impregnation of Fe-hydroxide mud by silica-rich solutions, contains high amounts of barite; unlike the massive sulfides, the jasper contain a lot of sphalerite.

Chimney structures were formed by multiple hydrothermal pulses in a period of time between around 50 and 10 ka years BP. The youngest hydrothermal event produced the sphalerite-bearing jasper material, which is a low-temperature mineral assemblage. The MESO zone deposits appear to be typical of mid-ocean ridge massive sulfide occurrences but are disintegrating and about to be buried by sediment.

12.1 INTRODUCTION

Submarine hydrothermal circulation in the oceanic crust and its activity on the ocean floor are fundamental processes controlling the transfer of energy and material from the lithosphere to the hydrosphere. Hydrothermal interactions alter the composition of the oceanic crust and influence the trace element chemistry of the oceans. Mineral precipitates, formed at hydrothermal vent fields along mid-ocean ridges or in back-arc basins, reflect the chemical and physical evolution of hydrothermal processes and material fluxes. Investigations on mineral deposits, recently or even currently being formed at the seafloor, can also be used as important tools to improve the understanding of the formation of ancient submarine massive sulfide deposits now located on land. In addition, modern marine mineral deposits might be considered to be an important metal resource for mining in the future.

In the past, many research programs on massive sulfide deposits on the modern seafloor were concentrated on the ridges in the Pacific and Atlantic Oceans. The first investigation on marine sulfide deposits of the East Pacific Rise (EPR) started in 1978 near 21°N,[1] but subrecently other studies on high-temperature venting were carried out along major rift systems such as the East Pacific Rise (including Explorer, Juan de Fuca, and Gorda Ridges), and the Mid-Atlantic Ridge e.g., References 2 through 5]. Subsequent exploration on hydrothermalism and metal enrichments on the ocean floor were also extended to back-arc basins (e.g., Okinawa Trough, Lau, and North Fiji Basin; e.g., References 6 through 8).

Investigations of hydrothermal activity along the southern Central Indian Ridge (CIR) started with the GEMINO program in 1983.[9] During these and other campaigns, hydrocast stations were carried out in order to detect hydrothermal plumes.[10-12] First indications of hydrothermal activity on the southern part of the CIR were documented in 1987, when elevated CH_4 and Mn-concentrations (202 nl/l CH_4, 23 nmol/kg Mn) were detected in the water column.[9,13] The site was called "Hydrothermal Plume Site" (HPS) and is located at 24°00'S/69°40'E in the third segment of the CIR north of the Rodrigues Triple Junction on the western flank of the rift wall at 2840 m water depth. Decreasing concentrations of manganese and methane measured at this site in the following years may be attributed to waning hydrothermal activity, or even to a megaplume event,[14] marking the beginning of a hydrothermal cycle with initial high Mn and CH_4 contents. Measurements in the same area in December 1993 yielded a maximum of 113 nl/l CH_4, at a water depth of about 3100 m,[15] and during the HYDROCK I campaign in 1995, a further decrease to 81 nl/l CH_4 was noticed.[16] Despite these hydrochemical anomalies and the presence of strongly altered basalt on the floor close to the neovolcanic zone, no hydrothermal precipitates were detected at this site. However, sediment coloration, indicating hydrothermal mineral deposition, was noticed several tens of kilometers to the north in the "Sonne Field."[13] This area was investigated again in 1993 when the German research vessel *Sonne* revisited the area and sampled the first massive sulfides to be found in the entire Indian Ocean.[17] A subsequent cruise with *RV Meteor* in 1995 gathered detailed information about the geological setting, the size of the area, and the mineral zonation in the field. Detailed mapping, ocean-bottom photography, and sampling revealed that the Sonne Field is framed by two further hydrothermal sites; the entire hydrothermal region was named the "MESO mineral zone" after the *RV METEOR* and *RV SONNE*.[18]

Data and samples from the above two cruises have been studied in detail. Here we present mineralogical, textural, and geochemical data on several types of hydrothermal mineralization from the MESO mineral zone. Furthermore, we discuss the most important geological and mineralogical differences and similarities between the Indian Ocean samples and sulfides from other oceans. Certainly one important difference between most other modern hydrothermal fields and the MESO zone is the maturity of the deposit in the Indian Ocean, where hydrothermal activity ceased approximately 10 ka ago[19]; thus the MESO zone mineral deposits have past through all consecutive stages of hydrothermal evolution.

The massive sulfide samples from the MESO zone are generally characterized by a deficiency in sulfate minerals (e.g., anhydrite, barite) and low porosity. They also show replacement and recrystallization textures, so we report as well on the replacement and weathering processes that took place after the formation of the deposit.

12.2 REGIONAL GEOLOGY

The Central Indian Ridge forms the boundary between the African and Indian plates and is a SSE-trending, mid-oceanic accretionary system in the equatorial Indian Ocean (Figure 12.1). It forms the southern extension of the Carlsberg Ridge, and terminates at 25°30'S/70°06'E at the Rodrigues Triple Junction (RTJ). Magnetic records indicate an intermediate spreading half-rate generally of about 2.5 cm/a, which progressively decreases from the RTJ (2.73 cm/a) toward the equator (1.8 cm/a).[20] The N to NNW-trending ridge (153°) is intersected by transform faults and nontransform discontinuities; segments are between 25 and 85 km long. The topography of the central-rift zone is rather smooth compared to slow-spreading analogues such as the SWIR or MAR, but it is slightly more rugged than the SEIR.[21] The rift valley is 500 to 800 m deep, 10 to 25 km wide, and generally well-defined. The inner floor is often relatively flat along the axis and more variable in cross section, and lies at 2,700 to 4,000 m water depth. Off-set fracture zones trend generally 0 to 60° in the southern parts of the CIR and laterally displace median valleys by as much as 8 km.

Magmatic activity along the CIR is documented by the presence of numerous seamounts and neovolcanic ridges, which have been identified in the rift valley. Particularly segments 3 and 4 are characterized by steeper scarps, as compared to other parts of the CIR axial valley. Shallow axial magma chambers, and abundant fresh glassy pillow and sheet lavas (N-MORB), support the classification of both segments as being magmatically active.[22] Such dynamic regimes are generally considered to be ideal for hydrothermal activity and the formation of submarine hydrothermal deposits. Although we have evidence that hydrothermal circulation has taken place, presently active high-temperature vent systems have not yet been discovered. Only massive sulfides from a former, now extinct vent system were identified.

12.2.1 LOCAL SETTING

The MESO zone is located in the central portion of the fourth CIR segment 270 km north of the RTJ. The segment is about 85 km long and therefore the longest of the southern CIR; its axial valley is about 11 km wide and deepens to a maximum of 3,200 m. From 23°12'S to 23°30'S a rift-parallel volcanic ridge subdivides the axial graben. Hydrothermal precipitates occur on the eastern flank of this neovolcanic ridge, which exists in the central rift valley of the CIR over a distance of about 20 km. The bathymetric high is 4 km wide and rises from 3,200 m depth to the top at 2,820 m water depth. The ridge rocks are characterized by pillow lavas, basaltic talus, and subordinate exposures of platy and ropy sheet lava.[13] Fresh to zeolite-facies altered N-MORB tholeiitic lavas were identified; the intensity of basalt alteration increases toward the hydrothermal site. In oceanic crustal rocks adjacent to the deposit, pillow basalt of the neovolcanic intrarift ridge is marked by chlorite- and zeolite-dominated secondary mineral assemblages (analcite, heulandite, natrolite; chlorite–smectite; minor prehnite) typical of interaction with evolved off-axis fluids (generally Mg-depleted, Ca- and ^{18}O-enriched, with high pH) at elevated (100 to 250°C) temperatures.[16] Alteration paragenesis typical for greenschist facies temperature conditions (albite–epidote–chlorite±actinolite; hornblende–chlorite–tremolite) are observed only in the central portion of the MESO zone. Here, massive sulfides precipitated from high-temperature ($\geq$300°C), Mg-depleted hydrothermal fluids. Basalts are locally covered by a thin calcareous sediment layer approximately 2 to 4 cm thick.

The MESO zone extends parallel to the crest for about 1500 m and covers an area of at least 0.6 km²; three different fields can be distinguished based on deep-towed camera observations. Hydrothermal deposits occur as eroded cones or chimney-stumps, as shallow mounds comprised of fragmental sulfidic material, and as talus. The latter ranges from cobble- to boulder-size, including angular clasts. Extinct sulfide chimneys are typically aligned along faults and cracks, oriented both parallel and perpendicular to the general strike direction of the ridge. This observation underlines

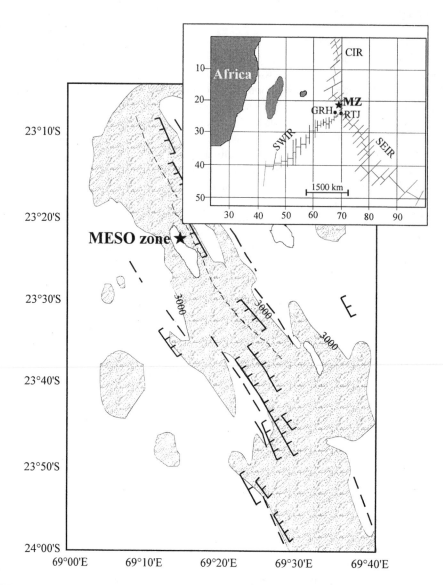

FIGURE 12.1   Inset shows the general location of the study area as well as the location of the Green Rock Hill (GRH), where ultramafic rocks were sampled during the SO92 cruise,[50] and the Rodrigues Triple Junction (RTJ). Main map illustrates the fourth segment of the Central Indian Ridge (CIR) with the MESO zone (MZ). Areas deeper than 3,000 m are shaded; a bold line marks the ridge axis. (From Münch, U. et al., Mineralogical and geochemical features of sulfide chimneys from the MESO zone, Central Indian Ridge, *Chem. Geol.*, 105, 29, 1999. With permission.)

the importance of fault control on the formation of massive sulfide deposits. Chimney edifices occasionally rising directly from the interstices between pillows were mainly observed in the northern part of the MESO zone. (This site was therefore named "Talus-Tips-Site": TTS.) The central part of the MESO zone (called "Sonne-Field": SF) is characterized by a 100- to 150-m-wide mound, topped by chimney stumps and hydrothermal talus. Due to the mineral composition, we consider these stumps to be basal remnants of former black smoker structures (Figure 12.2). This innermost zone is surrounded by numerous, almost completely eroded relics of chimneys, featuring subcircular cavernous collapse-like structures up to 0.5 m in diameter with 0.3-m-high outer walls.

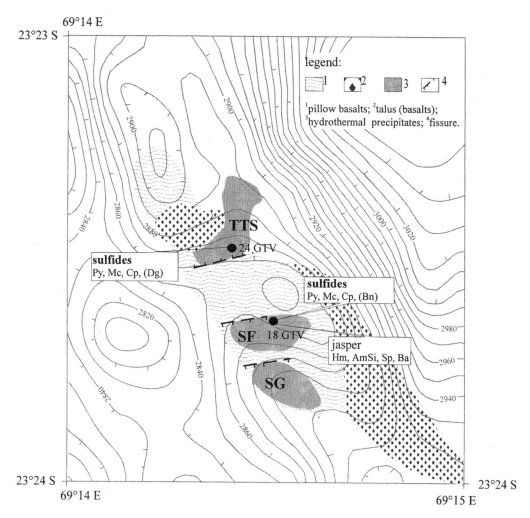

FIGURE 12.2 Bathymetric map (10 m contour interval) of the MESO zone (MZ). Videotape recordings of the seafloor topography, in combination with bathymetric maps, allowed an evaluation of the structural setting of the hydrothermal sites. Hydrothermal mineralization is structurally controlled by fissures, scarps, and cracks running both parallel and orthogonal to the general strike direction (N153°) of the ridge and is therefore divided into three different sites (Talus-Tips-Site, TTS; Sonne-Field, SF; Smooth-Ground, SG). Hydrothermal deposits are located in a pillow lava and basaltic talus/debris-dominated terrain. The sketch shows the positions of the TV-grab stations and informs about the recovered precipitates: Py: pyrite; Mc: marcasite; Cp: chalcopyrite; Bn: bornite; Dg: digenite; Sp: sphalerite; Hm: hematite; AmSi: amorphous silica; Ba: barite.

In the innermost portion of the deposit where chimney relics are abundant, we also sampled several boulders up to 50 cm in diameter, comprised of cm-to-dm sized fragments of jasper cemented by a porous matrix of sphalerite mixed with barite, silica and minor amounts of other sulfides. In the outer zone of the SF, ubiquitous Fe-oxyhydroxides occur as semiconsolidated dm-thick layers, which probably represent weathering products of former sulfide minerals (submarine gossans). Manganese crusts observed and sampled from marginal areas of the field are considered to be products of low-temperature distal venting. Relict chimneys were only observed in these two areas of the MESO zone, whereas in the SSE elongation, in the "Smooth Ground" (SG), only mineral debris composed of sulfide grains and their decomposition products (mainly iron-hydroxides) could be observed on seafloor photographs.

Based on ocean-floor photographs, the estimated thickness of pelagic sediment on chimney remnants is generally 2 to 4 cm. Assuming a mean sedimentation rate of about 0.6 cm/ka,[23] discharge of hydrothermal fluids from the chimneys must have ceased at least several thousand years ago. Radiochronological measurements (^{230}Th/^{234}U) from MESO zone mineral samples indicate different episodes of hydrothermal activity in a period of time between 52 ka and 10 ka years BP. The main hydrothermal events in the MESO zone include an older episode at about 22 ± 2 ka and a younger episode at about 16 ± 2 ka, as well as a sphalerite-bearing jasper mineralization formed between 13 and 9 ka. The respective hydrothermal activities during the two main episodes started in the northern part of the MESO zone and migrated south to the SF over a period of about 5,000 years.[19]

The age of the hydrothermal deposit is not only indicated by eroded chimney edifices and sulfidic talus, but also by the lack of any geochemical (Mn and CH$_4$) anomalies in the water column as well as by the absence of relict fauna. This latter could be due to the relatively long period of sea water dissolution of calcareous biogenic tests, such as clam shells, after the cessation of hydrothermal activity.

The main reasons for the disintegration of the massive sulfide smokers are their age and probably their originally high content of anhydrite. Some of the sulfide fragments studied contain relic anhydrite in pore spaces, mainly due to the sulfate's retrograde solubility; generally, anhydrite is unstable because it is undersaturated under ambient sea water conditions. Gypsum is less often observed in the samples. To what extent the lack of or insufficient quantities of amorphous silica in late-stage fluids also contributed to the disintegration of the chimney structures is uncertain. At hydrothermal sites such as the Main Vent Field and the High Rise Field in the Endeavour segment of the Juan de Fuca Ridge, for example, amorphous silica is flooding the orifice and stabilizing the structure of aging chimneys[24]; there, sulfide structures rise as high as 45 m. Although a late-stage hydrothermal pulse of amorphous silica is recognized in the mound-building, sphalerite-bearing jasper mineral assemblage, there is not very much evidence for such precipitates in the smoker remnants. Furthermore, in general, the samples show a low porosity typical of recrystallization and replacement processes, which also contrasts with what is seen in fresh sulfides from active smokers.

12.3 MINERALOGY AND CHEMISTRY

Hydrothermal products were sampled with a TV-controlled grab. This sampling method allows the recovery of a large quantity of different specimens, but only chimney fragments can later be oriented according to their original seafloor setting. However various types of hydrothermal precipitates were sampled from the MESO zone. Chimney fragments, sulfidic talus as well as oxides and oxidized sulfide crust were recovered in the Talus-Tips-Site, whereas in the Sonne-Field, in addition to the sulfidic material, a sulfide-bearing jasper mineralization was also recovered.

12.4 DESCRIPTION OF SAMPLES

The chimney samples described were recovered from the northern part (TTS) as well as from the central part (SF) of the MESO zone. Sulfide samples were examined macroscopically for mineral abundances, zonation, crystallinity of minerals, and for their porosity. Based on their different appearance and mineral assemblages, massive sulfides can be subdivided into two groups: (1) a pyrite $\pm$ marcasite-dominated sulfide type, and (2) a chalcopyrite-dominated ore type, also including secondary Cu-sulfides.

1. Pyrite-marcasite dominated massive sulfides show a pronounced layered structure. The pore space, determined by point counts of polished sections, ranges between 15 and 32%. The dominant mineral is pyrite, partly as euhedral to subhedral grains, intimately intergrown with or even replaced by marcasite. Mineral aggregates, consisting of small pyrite cubes (up to 0.5 mm in size), as well as individual xenomorphic pyrite grains (up to 2 mm in size) are found in fissures and vugs at the surface of individual specimens. However, idiomorphic marcasite crystals were not observed in any sample. Specimens containing higher amounts of copper-sulfides (e.g., bornite or digenite) are characterized by a dark-blue coloration, but the individual identification of Cu-minerals was macroscopically impossible. Samples sealed or impregnated by amorphous silica appear to be quite fresh, whereas more altered samples are coated with Fe-oxyhydroxides (goethite) and are therefore marked by a rusty to red coloration. Oxidized sulfide samples rarely contain any silica.

2. Chalcopyrite dominated samples can, in general, be distinguished by a light yellowish color. Sample surfaces exposed to sea water are often partly corroded, having a fine porous framework; the pore spaces are coated with a thin layer (≤5 mm thick) of bornite and/or digenite and are therefore characterized by a dark-blue coloration. Idiomorphic copper minerals were not observed macroscopically in the samples. On surface portions of several massive sulfide specimens, cm-sized cavities or vugs are filled by later-stage euhedral mm-sized pyrite. This pyrite is probably a product of the late-stage chalcopyrite decomposition. The pore space is less than in pyrite dominated sulfides and varies between 9 and 21%.

Sphalerite or galena were not identified macroscopically in the two types of massive sulfide mineralization; nor were gangue minerals like anhydrite, gypsum, or barite observed. Furthermore differentiation between later-stage copper sulfides (bornite, digenite, covellite) was not possible macroscopically due to the lack of larger idiomorphic crystals.

3. Oxide samples and oxidized sulfide crusts representing fragments of Fe-oxides display variable colors, from yellowish-orange to brownish-red. The earthy material mainly consists of Fe-oxyhydroxides and/or Mn-crusts, but contains subordinate oxidized sulfide particles. This observation and their close association with the massive sulfides suggests that the ocherous deposits result from weathering of sulfide minerals. However, another explanation could be low-temperature oxide deposition (<100°C), precipitated during the waning stage of hydrothermal activity.

4. The sphalerite-bearing jasper breccia probably forms the substrate and/or surroundings of the eroded chimneys in the sulfide area; none of these samples show any chimney-like circular zonation. The jasper is either massive, or brecciated with small-scale, soft-sediment deformation structures. Jasper varies from rust-colored to dark red. In thin section, massive clusters of jasper show colloidal microstructures with different contents of Fe-oxyhydroxides, resulting in a certain layering. The original colloform features have at least been partially preserved, although recrystallization to goethite and hematite has also taken place. Centimeter-sized interstices of massive jasper fragments are filled with sulfide and sulfate minerals. The jasper itself also contains disseminated sulfides. Sphalerite is the dominant sulfide mineral and forms subhedral grains, often with minute silica inclusions and Fe-rich cores (sphalerite I). Fe-poor sphalerite (II) includes small pyrite as well as chalcopyrite grains. In places, chalcopyrite disease is evident. That is in iron-rich sphalerite, chalcopyrite replaces sphalerite by a reaction between the FeS component of sphalerite and additional copper carried in solution.[25] Chalcopyrite, pyrite, and marcasite also appear as individual grains. The main gangue minerals are barite and latest-stage amorphous silica coating barite and sulfide grains.

12.4.1 Microtextures and Mineral Formation

The study of mineral paragenesis and microtextures allows a reconstruction of the evolution and the growth-history of chimney edifices. Characteristic mineralogical features like grain size, crystal habit, inclusions of other sulfides and the distribution of certain trace elements help to identify multiple stages of mineralization and ore replacement. Polished sections were prepared for this microscopic examination.

Sulfide mineral zoning in chimneys is, in general, controlled by precipitation and replacement reactions that occur in response to changes in temperature and composition of the hydrothermal fluid, as already described for Pacific and Atlantic samples by several authors (e.g., see References

26 through 28). Sulfide chimneys grow generally by the quenching of hydrothermal fluids on mixing with the ambient sea water. The first precipitate formed during the initial mixing with enhanced temperature (>150°C) is anhydrite, forming an early sulfate shell. With further development of the hydrothermal activity and growth of a chimney structure, the fluid conditions get hotter and more reducing. Therefore reduced sulfur dominates more and more, finally resulting in the precipitation of sulfide minerals. As the chimney structure becomes consolidated, mixing becomes more restricted, and the interior temperature rises further. At this stage, the deposition of metals in outer chimney portions is primarily composed of colloform pyrite and fine-grained marcasite (in part, FeS_2-phases replace earlier pyrrhotite and also anhydrite). Furthermore, sphalerite is often found in outer as well as in central chimney portions; the iron content of the sphalerite often increases from the exterior to the interior of the chimneys. Massive chalcopyrite as well as subhedral granular pyrite precipitate at higher temperatures (>300°C) and are therefore confined to the central portion of chimneys, causing the chimney walls to grow inward.[29] This mineral precipitation sequence has been modeled and described for several high-temperature submarine hydrothermal fields (e.g., Reference 30).

In general, the Indian Ocean massive sulfide samples show a similar mineral sequence to that described above, but there are differences. The sample material collected during the two cruises consists in part of individual sulfide fragments, which display the massive sulfide types described above (Section 12.4). Some larger samples represent more or less complete chimney cross sections. These samples demonstrate that the two sulfide types mentioned above are nothing else than parts of a chimney zonation, i.e., the chalcopyrite-dominated sulfides correspond to the inner part and the pyrite–marcasite-dominated sulfides represent the outer portion of the mature chimney structure. In general, the sulfide samples from TTS and SF show the same mineral zonation; only some minor differences were observed.

The outer chimney walls consist mainly of massive and colloform pyrite, intergrown with marcasite as mentioned above. The concurrent occurrence of pyrite and marcasite indicates rapidly changing physicochemical conditions (mc: pH<4.5 at T<200°C; py: pH>4.5).[31,32] Pyrrhotite was not observed in chimney fragments, which can be explained by a replacement of FeS by FeS_2 phases or by a lack of FeS-precipitation. The sulfide samples are significantly depleted in anhydrite due to cold sea water dissolution of anhydrite during the weathering of the mineral deposit. Early-stage massive pyrite (Figure 12.3) is often cataclastically overprinted and replaced by chalcopyrite along cracks and grain boundaries. Toward the interior of chimney walls, chalcopyrite gradually becomes the dominant mineral phase as described above. Chalcopyrite formed mainly during the high-temperature stage. However, characteristic mineralogical indications for very high-temperature precipitation (>350°C), like isocubanite, were not observed in the chalcopyrite-rich samples. Sphalerite, rarely observed in the sulfide samples, sometimes contains small inclusions of chalcopyrite. Massive chalcopyrite hosting small recrystallized pyrite cubes (up to 0.1 mm in size) is progressively replaced by bornite, digenite, and occasionally by covellite along cracks and crystal rims. Rarely digenite shows an acicular crystal habit (0.05 mm in size), whereas covellite occurs locally as a corroded network or small spots in interstices of Cu-sulfide grains.

Dust-like inclusions of digenite are sometimes observed in colloform pyrite-layers intergrown with marcasite. This may indicate that colloform pyrite was also formed after digenite formation. This melnikovite pyrite occasionally contains rounded to irregular relics of older sphalerite. A young generation of pyrite formed by late-stage chalcopyrite decomposition (see reactions i and ii below) occurs in cavities and vugs as euhedral grains up to 3 mm in size. Amorphous silica occasionally occurs in late cross-cutting veinlets or forms thin coatings on sulfide mineral grains; also the youngest euhedral pyrite generation is mostly coated by silica and has therefore a quite fresh appearance. In more altered samples, covellite is covered by goethite. Oxidation processes are also responsible for the alteration of pyrite and marcasite to Fe-hydroxides observed along grain boundaries and in microfissures.

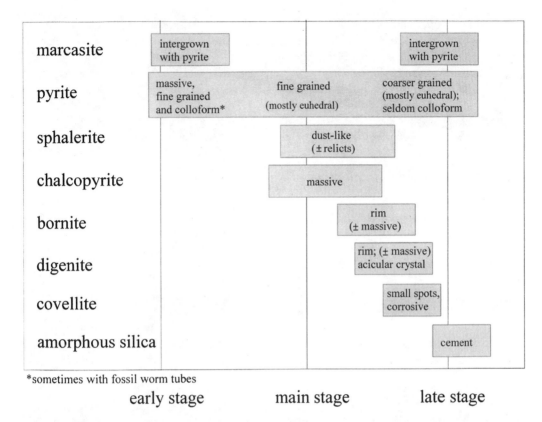

	early stage	main stage	late stage
marcasite	intergrown with pyrite		intergrown with pyrite
pyrite	massive, fine grained and colloform*	fine grained (mostly euhedral)	coarser grained (mostly euhedral); seldom colloform
sphalerite		dust-like (± relicts)	
chalcopyrite		massive	
bornite		rim (± massive)	
digenite		rim; (± massive) acicular crystal	
covellite		small spots, corrosive	
amorphous silica			cement

*sometimes with fossil worm tubes

FIGURE 12.3   Mineral sequence diagram for sulfide chimney samples with main mineral characteristics.

Fossil worm tubes were recognized in several sulfide samples (Figure 12.4). The linings, composed of colloform pyrite and marcasite, are ring-shaped or elliptical in cross section. Most tubes are filled with amorphous silica and/or fine-grained recrystallized Fe-sulfides. Seldom are the interiors of worm structures filled with chalcopyrite and/or secondary copper minerals. This indicates that the worm tubes were built at the early stage of sulfide formation (see Figure 12.3). The tubes average 0.5 mm in diameter, which is conspicuously small when compared to worm tubes described from other submarine hydrothermal areas [e.g., References 26 and 33].

The observed Cu-mineral succession chalcopyrite → bornite → digenite → covellite (Figure 12.3) is a very typical feature of the chalcopyrite-dominated massive sulfide mineralization. The formation of this mineral sequence is controlled by several physicochemical gradients in the high-temperature mixing zone. An increase in pH toward the chimney wall is caused by an increasing sea water fluid mixing ratio and promotes the formation of Cu-bisulfide complexes.[34] This favors the development of an outward rising gradient in dissolved Cu/Fe ratio.[29] The secondary Cu-sulfides were formed to a large extent by replacement of chalcopyrite, but also independently directly from the solution during the waning hydrothermal activity, caused by decreasing temperature (probably <250°C) and by a path of increasing sulfidation and oxidation in the later-stage fluids.[26] These fluids were apparently in disequilibrium with chalcopyrite, but stable with respect to more oxidized Cu-sulfides. Bornite and digenite, which precipitated after chalcopyrite (Figure 12.3), were also observed in very fresh black smoker fragments from the East Pacific Rise; according to Haymon[29] both minerals are later-stage hydrothermal products and not of supergene weathering origin. Bornite forming small exsolution lamellae in chalcopyrite is sometimes observed in Sonne Field sulfide samples, which according to Fouquet et al.[8] also confirms a primary origin of bornite.

FIGURE 12.4 (A) Hand specimen photograph showing the surface of the chalcopyrite-dominated massive sulfide. To the right of the coin, a cavity filled with secondary pyrite (cubes 1 to 3 mm in size), crystallized on copper mineral assemblage. (B) Hand specimen photograph showing a chimney fragment. The mineral zonation from copper-rich sulfides like chalcopyrite, bornite, and digenite in the inner chimney portion to more iron-dominated sulfides like pyrite and marcasite in the outer portion of the chimney wall is clearly recognizable as well as the former feeder channel. (C) Hand specimen photograph showing a massive chalcopyrite layer. Chalcopyrite is intergrown with, as well as encrusted by, bluish sulfur-rich sulfides like bornite and digenite. (D) Polished section photomicrograph of advanced-stage bornite formation by replacement of chalcopyrite. The bornite shows a very thin coating of digenite. The pyrite grains of the older generation are more corroded than the younger ones (compare Plate E). (E) Polished section photomicrograph with relics of chalcopyrite with bornite and digenite as the dominating Cu-minerals. The section is located close to the surface; the porosity is high; however, within an interstice a new crystal of secondary, late-stage pyrite formed (refer to Plate A and Table 12.3). (F) Fossil worm tubes composed of colloform pyrite and marcasite are ring-shaped or elliptical in cross section. The tube is filled with amorphous silica and fine-grained recrystallized Fe-sulfides. (From Halbach, P., Blum, N., Münch, U., Plüger, W., Garbe-Schönberg, D., and Zimmer, M., Formation and decay of a modern massive sulfide deposit in the Indian Ocean, *Min. Dep.*, 33, 302, 1998. With permission.)

Crerar and Barnes[34] have carried out solubility experiments in a closed $Cu–Fe–S–H_2O$ system, and have shown that bornite replaces chalcopyrite, but also forms independently under hydrothermal conditions with increasing f_{S2} and f_{O2} in the temperature range between 200°C and 350°C. The following model reactions describe two of several possible processes of bornite and pyrite formation by chalcopyrite dissolution in a closed system, with oxygen (i) or H_2S (ii) as the main electron acceptors, and no Fe being released to sea water:

(i) $$5 \ CuFeS_2 + 2 \ H_2S + O_2 \leftrightarrow Cu_5FeS_4 + 4 \ FeS_2 + 2 \ H_2O$$

(ii) $$5 \ CuFeS_2 + 2 \ H_2S \leftrightarrow Cu_5FeS_4 + 4 \ FeS_2 + 2 \ H_2$$

The pyrite formed in the two model reactions corresponds to the youngest generation of pyrite (late stage pyrite in Figure 12.3) observed in cavities or vugs at outer portions of the Cu-rich massive sulfide type (see above).

12.4.2 GEOCHEMICAL STUDIES

12.4.2.1 Composition of the Sulfide Samples

The Cu concentrations in the hydrothermal sulfides are very high (Table 12.1), and if the samples recovered by TV grab are representative for the entire field, they are among the highest reported for submarine massive sulfide deposits.[35] The transformation to later-stage Cu-sulfides combined with the release of Fe accounts for this enrichment of Cu. Chemical analyses indicate also that the sulfide samples are enriched in Ni. The average Ni content amounts to 48 ppm, but varies for the Cu-dominated sulfide samples from 12 ppm to 97 ppm and for the Fe-dominated sulfide samples from 23 ppm to 76 ppm. Massive sulfide samples from the fast-spreading EPR show significantly lower values and contain in general less than 10 ppm Ni. We suggest that the different spreading rates and different compositions of the oceanic crust rocks may control these variations in Ni content.

Chalcopyrite-rich sulfides with 7 to 32 wt.% Cu and 20 to 38 wt.% Fe are marked by relatively high contents of Se (mean 380 ppm) and Mo (mean 270 ppm), both probably hosted in primary rather than later-stage Cu-phases. According to Hannington et al.[36] correspondingly high concentrations of Se and Mo in chalcopyrite indicate high-temperature formation (>300°C). The positive correlation between Cu and Se (r = 0.81; n = 19) might indicate a substitution of Se for S in chalcopyrite; such substitution was previously documented for sulfide samples from the East Pacific Rise (13°N) by Auclair et al.[37] and Fouquet et al.[38] Locally high Se concentrations attest to highly fluctuating hydrothermal conditions; Se is chemically mobile and therefore incorporated into the sulfide lattice only during the early stage of mixing processes when the influence of the oxidizing sea water is low.[37]

Pyrite-marcasite dominated specimens have variable contents of Fe (32 to 48 wt.%) and Cu (1 to 6 wt.%), but generally low (<1 wt.%) Zn concentrations. However, some samples show higher Zn content (up to 7.9 wt.%); these samples are also enriched in Cd (up to 490 ppm). The correspondence between Zn and Cd is also seen in the strong positive correlation coefficient between them (r = 0.76; n = 63). Compared to sulfide samples from other submarine hydrothermal areas, samples from the MESO zone are highly enriched in Co (up to 1700 ppm). Microchemical analyses of selected pyrite cubes indicate a substitution of Co for Fe; this conclusion is supported by a positive correlation between Co and Fe [r = 0.7; n = 7]. According to Hekinian and Fouquet[39] a partial enrichment in Co can be attributed to leaching of Fe from Co-bearing pyrite, thus relatively concentrating the residual Co.

In order to identify trace metal concentrations in different sulfide minerals, several grains in a polished section of chalcopyrite-type ore have been analyzed for selected trace elements by Laser

TABLE 12.1
Bulk Geochemical Data on Selected Sulfide Samples

CHEMICAL ANALYSES OF SELECTED MESO ZONE SAMPLES

		CHALCOPYRITE TYPE					PYRITE-MARCASITE TYPE					SPHALERITE-JASPER TYPE		
	Method	18-34 SF	18-25 SF	18-67 SF	24-4.35 TTS	24-4.11 TTS	18-80 SF	18-98 SF	18-56 SF	24-2.38 TTS	24-3.2 TTS	18-107 SF	18-82 SF	18-109 SF
SiO$_2$ wt. %	XRF	0.5	3.9	3.8	n.d.	n.d.	2.5	9.1	9.8	n.d.	0.1	18.6	31.0	21.0
Fe wt. %	XRF	20.3	38.1	34.8	29.4	21.5	48.7	37.8	38.7	34.7	38.7	2.6	3.1	2.2
Zn wt. %	ICP	0.1	<0.1	0.1	2.0	0.23	0.1	0.1	0.3	7.9	0.3	31.1	23.4	27.3
Cu wt. %	ICP	31.6	7.4	9.9	18.7	18.1	0.6	0.6	0.4	4.1	5.6	0.4	0.4	0.7
Cd ppm	XRF	<10.0	<10.0	<10.0	<10.0	<10.0	<10.0	138.0	<10.0	489.2	<10.0	1765.0	1152.0	1279.0
Pb ppm	XRF	88.0	298.0	114.0	436.0	313.0	224.0	590.0	668.0	641.0	335.2	418.0	225.0	1130.0
As ppm	INA	50.0	700.0	180.0	620.0	356.5	270.0	590.0	570.0	960.0	650.0	37.0	53.0	45.0
Co ppm	INA	350.0	670.0	380.0	1000.0	672.0	630.0	200.0	520.0	970.0	1700.0	12.0	11.0	11.0
Ni ppm	XRF	12.0	26.0	30.0	<100.0	97.0	23.0	22.0	33.0	57.9	76.1	120.0	82.0	111.0
Mo ppm	INA	340.0	250.0	390.0	206.0	165.0	330.0	29.0	170.0	81.1	340.0	<5.0	32.0	9.0
Sr ppm	XRF	<10.0	14.0	<10.0	<10.0	n.d.	16.0	33.0	17.0	n.d.	0.9	1900.0	542.0	1766.0
Rb ppm	XRF	11.0	15.0	<10.0	n.d.	n.d.	12.0	<10.0	14.0	n.d.	n.d.	18.0	12.0	17.0
Sb ppm	INA/AAS	1.2	4.6	0.7	16.8	3.9	2.0	2.7	7.2	19.3	1.4	120.0	150.0	140.0
Se ppm	INA	700.0	310.0	340.0	180.0	n.d.	290.0	58.0	27.0	89.3	210.0	<5.0	23.0	<5.0
Ba ppm	INA	400.0	400.0	<100.0		460.0	500.0	1100.0	200.0	<100.0	<100.0	110,000.0	42,000.0	110,000.0
Ag ppm	INA/AAS	33.00	44.0	11.0	110.0	40.0	25.0	25.0	38.0	85.9	10.8	<5.0	<5.0	8.0
Au ppb	INA	220.0	890.0	240.0	1300.0	n.d.	590.0	1400.0	2000.0	6000.0	420.0	250.0	240.0	360.0
Ag/Au		150	49	46	85	42	42	18	19	14	26	<20	<21	22

Results are listed according to the type of mineralogical association. Bulk compositions were determined with instrumental neutron activation (INA), inductively coupled plasma (ICP-AES), atomic absorption spectrometry (AAS), and X-ray fluorescence (XRF). Determinations were done twice. the relative standard deviation (σ) is <5%. (n.d.: not detected; SF: Sonne-Field sample; TTS: Talus-Tips-Site sample).

From Halbach, P., Blum, N., Münch, U., Plüger, W., Garbe-Schönberg, D., and Zimmer, M., Formation and decay of a modern massive sulfide deposit in the Indian Ocean, *Min. Dep.*, 33, 302, 1998. With permission.

TABLE 12.2
Semiquantitative Data for Selected Trace
Elements in Sulfides from 18 GTV-HB3
Determined by Laser Ablation–ICPMS (in ppm)

	Ag	Se	In	Au
Bornite	100	200	—	<100
Chalcopyrite	<20	200	<20	n.d.
Pyrite	—	1000	n.d.	<100

(n.d. = not detected)

From Halbach, P., Blum, N., Münch, U., Plüger, W., Garbe-Schönberg, D., and Zimmer, M., Formation and decay of a modern massive sulfide deposit in the Indian Ocean, *Min. Dep.*, 33, 302, 1998. With permission.

Ablation-ICPMS. A 266-nm laser system (UV MicroProbe, VG Elemental, Winsford, U.K.) was coupled to a Double Focussing Magnetic Sector HR-ICPMS (PlasmaTrace 2, Micromass, Wythenshave, U.K.). The laser beam diameter and pulse energy were reduced as far as possible in order to get the smallest diameter ablation pits achievable in the sulfide matrix (approximately 30 µm). The system response was 380 cps (peak height) per ppm Cu. On the basis of this response factor, concentrations for the other elements could be estimated assuming that (1) ablation yield was similar during the measurements, (2) response ratio for Cu and Se to Au is similar, and (3) no significant fractionation occurred between Cu and the other elements investigated here. Since these assumptions hold true only in part, the accuracy of the preliminary results given below may be within one order of magnitude. Nevertheless, the error of the relative distribution of one element in different sulfides is much less (>50%). Therefore, the difference in the Ag concentrations found in pyrite, chalcopyrite, and bornite is significant (Table 12.2). The high Ag content in bornite (~100 ppm) compared to the low Ag concentration in chalcopyrite (<20 ppm) cannot be explained by bornite formation as the result of supergene weathering of chalcopyrite by cold sea water (see above). This high Ag content can only be explained by bornite precipitation as a result of later-stage hydrothermal solutions which also transported Ag, e.g., as a dissolved bisulfide complex.[16]

Among the massive sulfide types, the Fe-dominated pyrite-marcasite assemblage shows the highest Au contents (590 to 2000 ppb), averaging 1.3 ppm (Table 12.3). Such primary concentrations of either submicroscopic gold inclusions, or alternatively, pyrite lattice-hosted gold are of the same order as those detected in analogous massive sulfides in similar tectonic environments with slow-to-intermediate spreading rates.[35,36] The Ag/Au ratio ranges from 18 to 150 (Table 12.1); the mean Ag content of the pyrite-marcasite type mineralization (29 ppm) is identical to that of chalcopyrite-type sulfides, but considerably higher than that of sphalerite-dominated jasper-rich samples. The positive correlation between Au and Pb (r = 0.88; n = 10), Ag (r = 0.80), and Sb (r = 0.68) in the pyrite–marcasite material suggests that Au precipitated together with late-stage Pb–Sb–Ag — bearing sulfosalts. This type of gold enrichment is often observed in the waning stage of hydrothermal activity and reflects the transport of these metals as aqueous sulfur complexes at low temperature.[40]

The well-crystallized late-stage pyrites (Figure 12.3) on the chalcopyrite-rich precipitates (Figure 12.4) with cubes measuring up to 3 mm have high concentrations of Au; in particular, the smaller (<1.5 mm) crystals are enriched to a maximum content of 5.1 ppm (Table 12.3). Since the smaller pyrite grains are richer in Au, we suggest that during the incipient process of chalcopyrite dissolution due to sulfidation of the hydrothermal fluid (which results in the formation of bornite and pyrite; see Section 12.3.2) $Au(HS)_2^-$ complexes were also available. We assume that the Au concentration in the initial solution of this later-stage hydrothermal process was somewhat higher. The decomposition of the Au complexes may finally have led to the fixation of Au in pyrite.

TABLE 12.3
Composition of Typical Late-Stage Secondary Pyrite

		Method	18-36	18-95	55-KL	18-92
Fe	wt.%	INA	50.2	49.2	50.3	51.8
Zn	wt.%	ICP	0.31	0.15	0.08	0.49
Cu	wt.%	ICP	0.80	0.80	0.90	0.80
As	ppm	INA	530	480	750	660
Co	ppm	INA	1100	820	133	330
Se	ppm	INA	180	110	53	47
Au	ppb	INA	1250	1500	5100	3200

Samples 55-KL and 18-92 are of smaller (<1.5 mm) pyrite crystals, and the other two (18-36, 18-95) are of larger (>1.5 mm) cubes.

From Halbach, P., Blum, N., Münch, U., Plüger, W., Garbe-Schönberg, D., and Zimmer, M., Formation and decay of a modern massive sulfide deposit in the Indian Ocean, *Min. Dep.*, 33, 302, 1998. With permission.

According to Hannington and Scott,[41] the solubility of Au as $Au(HS)_2^-$ has its maximum near the H_2S–HSO_4- boundary, which lies in the stability field of pyrite. This would possibly promote Au coprecipitation at lower temperatures together with later-stage pyrite. It is interesting to note that the Co-concentrations of separated secondary pyrite samples behave inversely to the content of Au, whereas the arsenic concentration covaries positively with gold.

The sphalerite-jasper mineralization with Zn contents of 23 to 31 wt.% and correspondingly high Cd concentrations (mean 1399 ppm) contains significant amounts of barite; Ba reaches a maximum of 11 wt.%, and Sr contents are up to 1,900 ppm (Table 12.1). Although the Pb and Sb concentrations in this type of sphalerite-rich mineral assemblage are high, the Au and Ag contents are significantly lower (Table 12.1) than in the Cu-rich or the pyrite–marcasite sulfide mineralization.

12.4.2.2 Isotope Studies

Fifty sulfide samples from the Talus-Tips Site and the Sonne-Field Site were analyzed for their sulfur isotopic composition. The analyses were carried out in the isotope laboratory of the Freiberg Technical University, Germany, and are reported as standard $\delta^{34}S$ notation (‰) relative to Canon Diabolo Troilite (CDT Standard). The values of all sulfide samples range from 1.8 to 6.8‰ $\delta^{34}S$. They comprise individual specimens from the TTS massive sulfides, SF massive sulfides and the SF sphalerite-bearing jasper. No significant difference could be observed between the isotope ratios of the pyrite–marcasite type and the chalcopyrite massive sulfide type from each field. However, the two fields do differ from each other: The samples from the TTS are isotopically somewhat heavier (Figure 12.5) and have a mean of 3.5 (n = 32; σ = 0.3); the SF sulfides show a lower mean value of 2.6 (n =14; σ = 0.6). The third group consists of four selected sphalerite samples (according to microscopic studies associated with some minor amounts of pyrite) from the jasper breccia; the mean isotopic value is notably higher than in the massive sulfide samples and is 6.2 (n = 4; σ = 0.4). The spread of the mean values is indicated by the standard deviation (σ).

Massive sulfides forming on modern basaltic mid-ocean ridges have a range of $\delta^{34}S$ values from –0.8 to +6.3‰ with an average value of +3.2‰ (n=461 (compiled from Herzig et al.[42] after summary of several publications). The isotopic sulfur values measured in the sulfides from the Indian Ocean fall within this range. The isotopic composition of the sulfides from the mid-ocean ridges indicates two basic sources of sulfur: mid-ocean ridge basalt (MORB $\delta^{34}S$ ± 0‰[43]) and sea water ($\delta^{34}S$ + 20.9‰[44]). Isotopic sulfur data lying between these two extreme values are, in general,

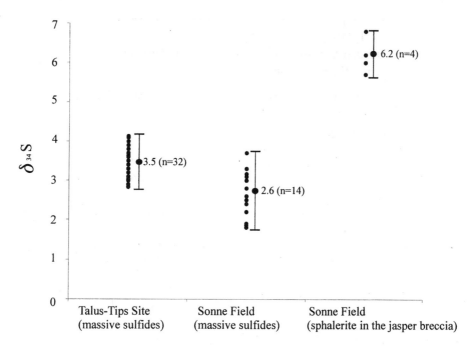

FIGURE 12.5 Sulfur isotopic composition of three groups of sulfide samples (two massive sulfide groups and one selected sphalerite group of the jasper breccia) from the MESO mineral zone in the Central Indian Ocean (n: number of analyzed samples).

explained by nonequilibrium mixing between sulfur from reduced sea water and sulfur of basaltic origin. Arnold and Sheppard[45] showed that an isotopic ratio of +2.1‰ can be explained by a 10% reduced sea water sulfur and a 90% basaltic sulfur contribution. Heavier sulfur in the massive sulfides exhibits a larger contribution of reduced sea water sulfur to the hydrothermal fluids; a $\delta^{34}S$ value of about +6.0‰, for example, indicates a sea water to basalt sulfur ratio of about 30:70. Bluth and Ohmoto[46] have pointed out that vent fluids of early-stage hydrothermal activity have low $\delta^{34}S$ values, whereas those of later stages show enhanced values because of the increasing influence of H_2S derived from reduction of sea water sulfate. These considerations agree with our observations of enhanced $\delta^{34}S$ values in the sphalerite samples of the jasper breccia compared to the isotope data on the massive sulfide samples (Figure 12.5). The sphalerite–barite mineral assemblage in the jasper breccia is assumed to be of later stage origin (see below: radiogenic isotopes). The increase in $\delta^{34}S$ ratio of the Zn-sulfides could also be explained by replacement of early precipitated sulfates by sulfides or when a part of the H_2S content in the fluid was generated by reactions between the hydrothermal solution and the pre-existing sulfates.[47] The $\delta^{34}S$ values of the barites from the jasper breccia vary between +16.5 and +20.1‰ and are similar to the sulfur isotope ratios of submarine hydrothermal sulfate minerals associated with massive sulfides. Values somewhat lighter than modern sea water sulfate may imply a moderate (maximum 20%) admixture of basaltic sulfur.

Oxygen isotope ratios were determined on five selected pure jasper specimens with varying Fe and Si contents. $\delta^{18}O$ values (21.1 to 12.1‰) fluctuate, mainly as a function of variable abundances of constituent oxide phases (Table 12.4). In general, a negative correlation between Fe-content and oxygen isotope ratio ($^{18}O/^{16}O$) is evident, due to the preferred incorporation of ^{16}O into iron oxides, and ^{18}O into silica (opal$_{CT}$). Treating the jasper-rich material as a simple two-phase oxide system (Si + Fe = 100%), and assuming that both components (amorphous silica and Fe-[hydr]oxide) are in isotopic equilibrium, an average oxygen isotope end-member composition was calculated for pure silica precipitates by extrapolation on a $\delta^{18}O$ vs. SiO_2, respectively, Fe_2O_3 binary plot. An oxygen isotope ratio of 26.7‰ for pure SiO_2 is obtained by linear regression of data points. Adopting

TABLE 12.4
**Oxygen Isotope Composition of Sonne Field Jasper-
Type Mineralization**

Sample	$\delta^{18}O$	Fe (wt.%)	SiO_2 (wt.%)	Fe_2O_3(wt.%)
FUB 1	17.0	2.6	34.8	3.72
FUB 2	21.1	1.2	29.2	1.72
FUB 3	12.8	4.0	39.1	5.72
FUB 4	13.7	2.8	32.6	4.00
FUB 5	12.1	4.4	38.0	6.29
FUB 6	12.8	3.9	39.6	5.58

For end-member extrapolation, Fe contents were converted to Fe_2O_3, and normalized together with SiO_2. The remainder of the samples analyzed comprise sulfides (sphalerite, chalcopyrite, pyrite, marcasite), filling either interstices of jasper fragments or impregnations of siliceous chemical sediments. Care was taken to avoid sample portions containing sulfates.

From Halbach, P., Blum, N., Münch, U., Plüger, W., Garbe-Schönberg, D., and Zimmer, M., Formation and decay of a modern massive sulfide deposit in the Indian Ocean, *Min. Dep.*, 33, 302, 1998. With permission.

the calibration of Kita et al.[48] for the system silica–water ($10^3 \ln\alpha_{silica-water}$ = 3.52 $[10^6 T^{-2}]$ – 4.35), an isotopic equilibration temperature of 63.2°C is calculated, which is in accordance with our interpretation of the temperature of late-stage precipitation of silica-rich solutions.[16]

Radiogenic isotopes ($^{230}Th/^{234}U$) were used to obtain the age of the deposit, but mainly to get information about the variability of the hydrothermal system. In general, the relative error of this age dating method amounts to 12%.

Hydrothermal activity in the MESO zone started possibly 140 ka ago in the TTS. After a period of quiescence, around 52.5 ka ago the hydrothermalism was reactivated again in the same area and ceased after a period of several thousand years. The new beginning of hydrothermal activity in the northern part of the MESO zone could be dated around 22 ± 2 ka BP; the high-temperature ore formation terminated approximately 16 ± 2 ka ago in the TTS. Samples from the SF, which is located southeastward of the TTS, exhibit younger ages than those from the TTS; here the first hydrothermal activity has been dated at about 18 ± 2 ka BP; this hydrothermal period lasted for at least 7 ka. The latest formation of sulfides in the SF occurred at about 11.8 ± 1.4 ka. The sphalerite-bearing jasper from the SF has been dated at around 11 to 10 ka, which is the youngest hydrothermal event determined in the MESO zone, and which may also define the general end of hydrothermal activity there.[19]

12.5   CONCLUSIONS

Our study of massive sulfide samples from the Indian Ocean shows that multiple hydrothermal events over a period of several tens of thousands of years formed mineral occurrences which are more or less N–S aligned in the MESO mineral zone on the Central Indian Ridge. The site of this deposit is located in the central part of the fourth segment (about 270 km N of the RTJ) on a neovolcanic intrarift ridge. Since at least 10,000 years ago, the chemical and physical disintegration of the deposit has been occurring.[19] Chimney structures have more or less been destroyed by weathering processes. Nevertheless, three types of sulfide mineralization formed by hydrothermal venting processes can be distinguished: (1) Chalcopyrite-dominated massive sulfides; (2) pyrite–marcasite massive sulfides; and (3) sphalerite-bearing jasper breccia with barite and amorphous silica as main gangue phases.

The mineral assemblages from the MESO zone are of special interest because of their high metal content; combined Cu, Fe and Zn metal concentrations are close to 50 or even 60 wt.%. The high metal contents in the massive sulfides can be explained by a nearly complete lack of gangue minerals (anhydrite, barite, silica). The dissolution of sulfates (particularly anhydrite) after cessation of the hydrothermal activity contributed to this. In general, the sulfide samples show a low porosity, due to their high maturity and extensive reworking, indicated by infilling of pore spaces with late-stage minerals. Another difference from fresh sulfide samples from active chimneys can be seen in the total replacement of pyrrhotite by pyrite and marcasite as well as in the replacement of chalcopyrite by bornite, digenite, and covellite.

The distribution of Ni in the massive sulfides from the MESO zone might reflect the local tectono-magmatic conditions. Massive sulfide samples from the fast-spreading EPR show significantly low Ni values (in general less than 10 ppm). The Ni content in sulfide samples from the intermediate-spreading CIR averages 48 ppm, but varies in the Cu-dominated sulfide samples from 12 ppm to 97 ppm and in the Fe-dominated sulfide samples from 23 ppm to 76 ppm. We suggest that the different spreading rates and different compositions of the oceanic crust rocks may control these variations in Ni content. Ultraslow mid-oceanic spreading systems, for example, frequently contain more ultramafic rock complexes than fast-spreading ridges; these ultramafic rocks (mantle material) are, in general, richer in Ni and can make this metal available to the hydrothermal system.

Another element of special interest is Au. The well crystallized late-stage pyrites (up to 3 mm in size) separated from chalcopyrite-rich sulfide samples show high concentrations of Au; in particular, the smaller (<1.5 mm) crystals are enriched to a maximum Au content of 5.1 ppm (Table 12.3). We suggest that during the incipient process of chalcopyrite dissolution, due to sulfidation of the hydrothermal fluid $Au(HS)_2^-$ complexes were available. We assume that the Au concentration in the initial solution of this later-stage hydrothermal process was somewhat higher. The decomposition of the Au complexes may have led finally to the fixation of Au in pyrite. According to Hannington and Scott[41] the solubility of Au as $Au(HS)_2^-$ has its maximum near the $H_2S–HSO_4^-$ boundary, which lies in the stability field of pyrite. This would possibly promote Au coprecipitation at lower temperatures together with later-stage pyrite. It is interesting to note that the Co-concentrations in selected secondary pyrite samples behave inversely to the Au content, whereas the arsenic concentrations show a rough correlation with gold.

On-land massive sulfide deposits of the Cyprus type (fossil mid-ocean ridge massive sulfides) only rarely show a preservation of complete chimney structures; occasionally there are remains of layered chimney fragments. Our results show that hydrothermal chimneys are short-lived features which disintegrate after the waning of hydrothermal activity. Assuming normal deep-sea conditions, chimney dissolution and disintegration will take place within several thousand to several tens of thousands of years[49] after cessation of the hydrothermal activity, due to the disequilibrium (retrograde solubility) of anhydrite and sulfide minerals in ambient oxidizing sea water. Only underlying massive and stockwork mineralizations will be preserved in the geological record.

ACKNOWLEDGMENTS

Cruise SO 92 was funded by the German Federal Ministry for Research and Technology, as is the research project HYDROTRUNC. Cruise M33/2 HYDROCK I was funded by the German Science Foundation; we acknowledge this support. We thank Dr. P. Herzig for arranging stable isotope measurements at the Technical University of Freiberg, Dr. D. Garbe-Schönberg for doing the Laser Ablation–ICPMS measurements at the University of Kiel, and Dr. C. Lalou for the radiochronological measurements at the Centre des Faibles Radioactivités, Gif sur Yvette, France.

REFERENCES

1. Francheteau, J., Needham, H. D., Choukroune, P., Juteau, T., Seguret, M., Ballard, R. D., Fox, P. J., Normark, W., Carranza, A., Cordoba, D., Guerrero, J., Rangin, C., Bougault, H., Cambon, P., and Hekinian, R., Massive deep-sea sulphide ore deposits on the East Pacific Rise, *Nature*, 277, 523, 1979.
2. Bischoff, J. L., Rosenbauer, R. J., Aruscavage, P. J., Baedecker, P.A., and Crock, J. G., Sea-floor massive sulfide deposits from 21°N, East Pacific Rise, Juan de Fuca Ridge, and Galapagos Rift: bulk chemical composition and geochemical composition and economic implications, *Econ. Geol.*, 78, 1711, 1983.
3. Hekinian, R., Fevier, M., Avedik, F., Cambon, P., Charlou, J. L., Needham, H. D., Raillard, J., Boulegue, J., Merlivat, L., Moinet, A., Manganini, S., and Lange, J., East Pacific Rise near 13°N: geology of new hydrothermal fields, *Science*, 219, 1321, 1983.
4. Rona, P. A., Klinkhammer, G., Nelsen, T. A., Trefry, J. H., and Elderfield, H., Black smokers, massive sulfides and vent biota at the Mid-Ocean Ridge, *Nature*, 321, 33, 1986.
5. Bäcker, H., Lange, J., and Marchig, V., Hydrothermal activity and sulphide formation in axial valleys of the East-Pacific Rise crest between 8° and 22°S, *Earth Planet Sci. Lett.*, 72, 9-22, 1985.
6. von Stackelberg, U. and shipboard scientific party, Active hydrothermalism in the Lau Back-Arc Basin (SW-Pacific): first results from the SONNE 48-Cruise, 1987, *Mar. Min.*, 7, 431, 1988.
7. Halbach, P., Nakamura, K., Wahsner, M., Lange, J., Sakai, H., Käselitz, L., Hansen, R.-D., Yamano, M., Post, H., Prause, B., Seifert, R., Michaelis, W., Teichmann, F., Kinoshita, M., Märten, A., Ishibashi, J., Czerwinski, S., and Blum, N., Probable modern analogue of kuroko-type massive sulphide deposits in the Okinawa trough back-arc basin, *Nature*, 338, 496, 1989.
8. Fouquet, Y., Wafik, A., Cambon, P., Mevel, C., Meyer, G., and Gente, P., Tectonic setting and mineralogical and geochemical zonation in the Snake Pit sulfide deposit (Mid Atlantic Ridge at 23°N), *Econ. Geol.*, 88, 2018, 1993.
9. Herzig, P. M. and Plüger, W. L., Exploration for hydrothermal activity near the Rodriguez Triple Junction, Indian Ocean, *Can. Min.*, 26, 721, 1988.
10. Jean-Baptiste, P., Mantisi, F., and Pauwells, H., Hydrothermal ³He and manganese plumes at 19°29'S on the Central Indian Ridge, *Geophys. Res. Lett.* 19, 1787, 1992.
11. Jamous, D., Meremy, L., and Andrie, C., The distribution of helium-3 in the deep southern and western/central Indian Ocean, *J. Geophy. Res.*, 98, 1123, 1993.
12. Gamo, T., Nakayama, E., Shitashima, K., Isshiki, K., Obata, H., Okamura, K., Kanayama, S., Oomori, T., Koizumi, T., Matsumoto, S., and Hasumoto, H., Hydrothermal plumes at the Rodriguez Triple Junction, Indian Ocean, *Earth Planet. Sci. Lett.*, 142, 261, 1996.
13. Plüger, W. L., Herzig, P. M., Becker, K. P., Deissmann, G., Schöps, D., Lange, J., Jenisch, A., Ladage, S., Richnow, H. H., Schulze, T., and Michaelis, W., Discovery of hydrothermal fields at the Central Indian Ridge, *Mar. Min.*, 9, 73, 1990.
14. Baker, E. T., Characteristics of hydrothermal discharge following a magmatic intrusion, in *Hydrothermal Vents and Processes*, Parson, L. M., Walker, C. L., and Dixon, D. R. (Eds.), Geol. Soc., London, 87, 1995, 65.
15. Halbach, P. and SO 92 scientific party, Die Entwicklung des Hydrothermalismus' und seine struktur-geologische Kontrolle im Gebiet der Rodrigues Triple Junction, zentraler Indischer Ozean. Technischer Fahrtbericht zum BMBF-Projekt "HYDROTRUNC," Freie Universität Berlin, 1994, 113.
16. Halbach, P., Blum, N., Münch, U., Plüger, W., Garbe-Schönberg, D., and Zimmer, M., Formation and decay of a modern massive sulfide deposit in the Indian Ocean, *Min. Dep.*, 33, 302, 1998.
17. Halbach, P., Blum, N., Plüger, W., van Gerven, M., Erzinger, J., and SO 92 shipboard scientific party, The Sonne Field — first massive sulfides in the Indian Ocean, *InterRidge News*, 4 (2), 12, 1995.
18. Halbach, P. and Münch, U., Mineralogical and geochemical investigation of massive sulfide deposits from the Central Indian Ridge, in *Mineral Deposits: Research and Exploration — Where Do They Meet*, Papunen, H. (Ed.), Balkema, Rotterdam, 1997, 357.
19. Lalou, C., Münch, U., Halbach, P., and Reyss, J.-L., Radiochronological investigations on sulfide deposits from the MESO zone — Central Indian Ridge, *Mar. Geol.*, 149, 233, 1998.
20. Munschy, M. and Schlich, R., The Rodriguez Triple Junction (Indian Ocean): structure and evolution of the past one million years, *Mar. Geophys. Res.*, 11, 1, 1989.

21. Mitchell, N. C., An evolving ridge system around the Indian Ocean Triple Junction, *Mar. Geophys. Publ.*, 13, 173, 1991.

22. Briais, A., Structural analysis of the segmentation of the Central Indian Ridge between 20°30'S and 25°30'S (Rodriguez Triple Junction), *Mar. Geophys. Res.*, 17, 431, 1995.

23. Walter, P., Schwarz, B., Scholten, J., and Stoffers, P., Geochemie und Mineralogie der Sedimente des Mittelindischen Rückens zwischen 21° S und der Rodrigues Triple Junction (SO 43), GEMINO II Final Report, RWTH Aachen, 1986, 1.

24. Tivey, M. K. and Delaney, J. D., Growth of large sulfide structures on the Endeavor segment of the Juan de Fuca Ridge, *Earth Planet. Sci. Lett.*, 77, 303, 1986.

25. Barton, P. B. and Bethke, P. M., Chalcopyrite disease in sphalerite: pathology and epidemology, *American Mineral.*, 72, 451, 1987.

26. Haymon, R. M. and Kastner, M., Hot spring deposits on the East Pacific Rise at 21°N: preliminary description of mineralogy and genesis, *Earth Planet. Sci. Lett.*, 53, 363, 1981.

27. Oudin, E., Hydrothermal sulfide deposits of the East Pacific Rise (21°N) I. Descriptive mineralogy, *Mar. Min.*, 4, 39, 1983.

28. Janecky, D. R. and Seyfried, W. E., Formation of massive sulfide deposits on oceanic ridge crests: incremental reaction models for mixing between hydrothermal solutions and sea water, *Geochim. Cosmochim. Acta,* 48, 2723, 1984.

29. Haymon, R. M., Growth history of hydrothermal black smoker chimneys, *Nature*, 301, 695, 1983.

30. Tivey, M. K., Humphris, S. E., Thompson, G., Hannington, M. D., and Rona, P. A., Deducing patterns of fluid flow and mixing within the TAG active hydrothermal mound using mineralogical and geochemical data, *Journ. of Geophys. Res.,* 100, 12527, 1995.

31. Koski, R. A., Clague, D. A., and Oudin, E., Mineralogy and chemistry of massive sulfide deposits from the Juan de Fuca Ridge, *Geol. Soc. Am. Bull.,* 95, 930, 1984.

32. Graham, U. M., Bluth, G. J., and Ohmoto, H., Sulfide-Sulfate chimneys on the East-Pacific Rise, 11° and 13°N latitudes. Part I: Mineralogy and Paragenesis, *Can. Min.*, 26, 487, 1988.

33. Alt, J. C., Lonsdale, P., Haymon, R., and Muehlenbachs, K., Hydrothermal sulfide and oxide deposits on seamounts near 21°N, East Pacific Rise, *Geol. Soc. Am. Bull.*, 98, 157, 1987.

34. Crerar, D. A. and Barnes, H. L., Ore solution chemistry V. Solubilities of chalcopyrite and chalcocite assemblages in hydrothermal solution at 200° to 350°C, *Econ. Geol.*, 71, 772, 1976.

35. Rona, P. A. and Scott, S. D., A Special Issue on Sea-Floor Hydrothermal Mineralization: New Perspectives, *Econ. Geol.*, 88, 1935, 1993.

36. Hannington, M. D., Herzig, P. M., Scott, S. D., Thompson, G., and Rona, P. A., Comparative mineralogy and geochemistry of gold-bearing sulfide deposits on the mid-ocean ridges, *Mar. Geol.*, 101, 217, 1991.

37. Auclair, G., Fouquet, Y., and Bohn, M., Distribution of selenium in high temperature hydrothermal sulfide deposits at 13° North, East Pacific Rise, *Can. Min.*, 25 (4), 577, 1987.

38. Fouquet, Y., Auclair, G., Cambon, P., and Etoubleau, J., Geological setting and mineralogical and geochemical investigations on sulfide deposits near 13°N on the East Pacific Rise, *Mar. Geol.*, 84, 145, 1988.

39. Hekinian, R. and Fouquet, Y., Volcanism and metallogenesis of axial and off-axial structures on the East Pacific Rise near 13°N, *Econ. Geol.*, 80, 221, 1985.

40. Hannington, M. D., Peter, J. M., and Scott, S. D., Gold in seafloor polymetallic sulfide deposits, *Econ. Geol.*, 81, 1867, 1986.

41. Hannington, M. D. and Scott, S. D., Mineralogy and geochemistry of a hydrothermal silica-sulfide-sulfate spire in the caldera of Axial Seamount, Juan de Fuca Ridge, *Can. Min.*, 26, 603, 1988.

42. Herzig, P. M., Hannington, M. D., and Arribas, A., Jr., Sulfur isotopic composition of hydrothermal precipitates from the Lau back-arc: implications for magmatic contributions to seafloor hydrothermal systems, *Min. Dep.*, 33, 226, 1998.

43. Sakai, H., des Marais, D. J., Udea, A., and Moore, J. G., Concentrations and isotope ratios of carbon, nitrogen and sulfur in ocean-floor basalts, *Geochim. Cosmochim. Acta,* 48, 2433, 1984.

44. Rees, C. E., Jenkins, W. J., and Monster, J., The sulfur isotopic composition of ocean water sulfate, *Geochim Cosmochim Acta,* 42, 377, 1978.

45. Arnold, M. and Sheppard, S. M., East Pacific Rise at latitude 21°N: isotopic composition and origin of the hydrothermal sulphur, *Earth Planet. Sci. Lett.*, 56, 148, 1981.

46. Bluth, G. J. and Ohmoto, H., Sulphide-sulphate chimneys on the East Pacific Rise, 11° and 13°N latitude. Part II: sulphur isotopes, *Can. Min.*, 26, 505, 1988.
47. Janecky, D. R. and Shanks, W. C. III, Computational modelling of chemical and isotopic reaction processes in seafloor hydrothermal systems: chimneys, massive sulfides, and subjacent alteration zones, *Can. Min.,* 26, 805, 1988.
48. Kita, I., Taguchi, S., and Matsubaya, O., Oxygen isotope fractionation between amorphous silica and water at 34-93°C, *Nature* 314, 83, 1985.
49. Lalou, C., Reyss, J.-L., and Brichet, E., Hydrothermal activity on a 10^5-year scale at a slow spreading ridge, TAG hydrothermal field, Mid-Atlantic Ridge 26°N, *Journ. Geophys. Res.,* 100, 17855, 1995.
50. Blum, N., Kuhn, T., Halbach, P., Münch, U., van Gerven, M., and Hellebrand, E., Mantle uplifted block in the Central Tropic Indian Ocean, *Terra Nostra,* 4, 23, 1996.

13 Polymetallic Massive Sulfides and Gold Mineralization at Mid-Ocean Ridges and in Subduction-Related Environments

Peter M. Herzig and Mark D. Hannington

ABSTRACT

Polymetallic massive sulfide deposits on the modern ocean floor have been found on fast- and slow-spreading ridges, axial and off-axis volcanoes, in sedimented rifts adjacent to continental margins, and in subduction-related arc and back-arc settings. The mineralogy of these deposits includes both high (>300 to 350°C) and lower-temperature (<300°C) assemblages consisting of varying proportions of pyrrhotite, pyrite/marcasite, sphalerite/wurtzite, chalcopyrite, bornite, isocubanite, barite, anhydrite, and amorphous silica. Massive sulfides in back-arc spreading centers additionally may contain abundant galena, Pb–As–Sb sulfosalts (including jordanite, tennantite and tetrahedrite), realgar, orpiment, and locally native gold. Initial sampling of sulfides from subduction-related back-arc rifts suggests that these deposits have higher average concentrations of Zn, Pb, As, Sb, and Ba than deposits at the sediment-starved mid-ocean ridges. Gold concentrations are locally high in samples from a number of mid-ocean ridge deposits (up to 7 ppm) but may reach concentrations of more than 50 ppm Au in massive sulfides from immature back-arc rifts. Geochemical studies indicate that factors such as source rock compositions and rock-buffering of the hydrothermal fluids are important controls on the formation of gold-rich massive sulfides. Recent exploration of shallow-marine island arc environments of the western Pacific have also led to the discovery of several new types of seafloor hydrothermal activity with distinctive epithermal characteristics and notable enrichments of gold.

13.1 INTRODUCTION

In the two decades since the discovery of hydrothermal vents on the mid-ocean ridges, important mineral deposits have been documented in more than a dozen different volcanic and tectonic settings around the world (Figure 13.1). Polymetallic sulfide deposits are found on both fast-, intermediate-, and slow-spreading mid-ocean ridges, on axial and off-axial volcanoes and seamounts, in sedimented rifts adjacent to continental margins, and in subduction-related arc and back-arc environments. Rona[1] and Rona and Scott[2] have compiled data on more than 100 occurrences of hydrothermal mineralization on the seafloor, including Fe- and Mn-oxide deposits, nontronite deposits, disseminated sulfides, metalliferous sediments, massive polymetallic sulfide mounds, and black and white smoker chimneys. High-temperature hydrothermal activity and large accumulations of polymetallic sulfides, however, are known at fewer than 25 different sites.

0-8493-8429-X/00/$0.00+$.50

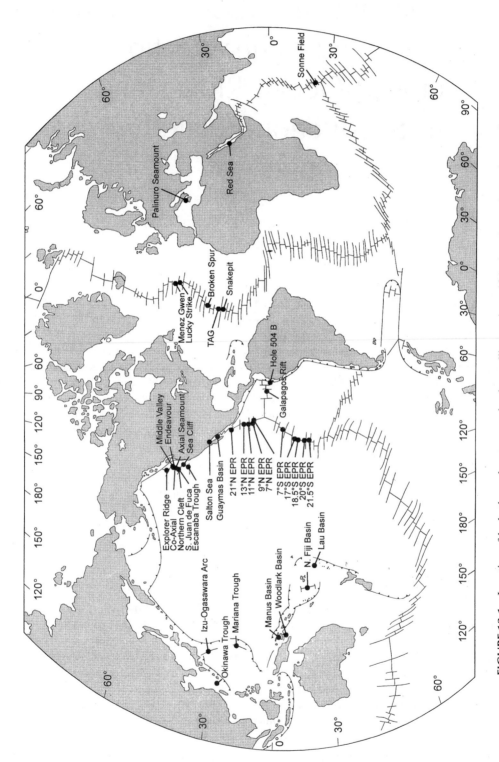

FIGURE 13.1 Location of hydrothermal systems and polymetallic massive sulfide deposits on the modern seafloor.

FIGURE 13.2 Typical intraoceanic back-arc spreading center with subduction of oceanic crust beneath oceanic crust (Lau Basin, Southwest Pacific).

The formation of polymetallic massive sulfides at the seafloor is related to the heat associated with active volcanism and the formation of new oceanic crust. Massive sulfide deposits are known to occur in rather diverse tectonic settings, including both divergent plate boundaries (i.e., mid-ocean ridges) and convergent, subduction-related environments, where sulfide formation takes place at spreading centers in back-arc basins (Figure 13.2). Volcanic activity in subduction-related environments occurs along emergent volcanic arcs as well as at back-arc spreading centers. In each case, massive sulfide deposits form as a consequence of the hydrothermal circulation of sea water along volcanically active portions of the oceanic plate.

13.2 OCCURRENCE OF MASSIVE SULFIDE DEPOSITS AT MID-OCEAN RIDGES AND IN SUBDUCTION-RELATED ENVIRONMENTS

Following the discovery of black smokers at 21°N on the East Pacific Rise,[3,4] there was a rapid growth in the number of hydrothermal deposits found on fast-spreading mid-ocean ridges. So many deposits were found along fast-spreading segments of the East Pacific Rise, and virtually nowhere else, that it became widely accepted that slower-spreading ridges could not support productive hydrothermal activity. However, in 1986, the discovery of the large TAG Hydrothermal Field on the Mid-Atlantic Ridge[5] offered compelling evidence that slow-spreading ridges may also be important settings for sulfide deposits. This idea has since been confirmed by the discovery of a number of large sulfide occurrences on the Mid-Atlantic Ridge (14°45'N MAR, Snakepit, Broken Spur, Lucky Strike, Menez Gwen[6-9]) and the Central Indian Ridge (Sonne Field[10-12]). Drilling of the TAG Hydrothermal Mound in 1994 has since exposed the internal structure and subsurface nature of a large volcanic-hosted massive sulfide deposit at the modern seafloor.[13,14]

Shortly after the discovery at 21°N, large sulfide deposits were also discovered in sediment-filled basins in the Gulf of California (Guaymas Basin[15]). The idea that sedimented ridges might also be important sites for sulfide accumulation was confirmed in 1991 and 1996, when the Ocean Drilling Program intersected more than 90 m of massive sulfides in the large Middle Valley deposit on the Juan de Fuca Ridge.[16,17]

The first sulfide deposits reported in back-arc spreading centers (Figure 13.3) were found in the Central Manus Basin[18] and the Mariana Trough.[19,20] These discoveries led to renewed exploration

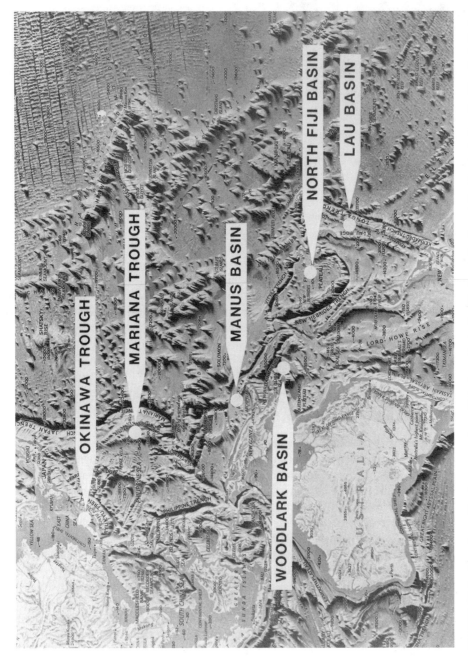

FIGURE 13.3 Location of back-arc massive sulfide deposits in the West and Southwest Pacific, including the Woodlark Basin, where oceanic crust is actively propagating into the continental crust of Papua New Guinea.

of the marginal basins and the arc and back-arc systems of the western Pacific. The complex volcanic and tectonic settings of convergent margins in the Pacific suggested that a number of different deposit types might be present in this region. A wide range of mineral deposits have since been found in back-arc rifts at different stages of opening (*immature* vs. *mature*), on volcanoes along the active volcanic fronts of the arcs, as well as in rifted fore-arc environments. Well-known examples of polymetallic massive sulfide deposits have now been described from mature back-arc spreading centers such as the North Fiji Basin,[21] along propagating back-arc rifts such as the Valu Fa Ridge in the southern Lau Basin,[22] and in the nascent back-arc rifts such as the Okinawa Trough.[23] In 1991, extensive sulfide deposits were found to be associated with felsic volcanism in the Eastern Manus Basin,[24] and hydrothermal deposits have also been located in the western Woodlark Basin, were seafloor spreading propagates into the continental crust of Papua New Guinea.[25]

Sulfide deposits forming in extensional environments associated with back-arc rifts have similar morphologies and growth histories to those found on ordinary mid-ocean ridges. Although the ore forming processes are almost identical, the composition of the volcanic rocks varies from mid-ocean ridge basalts (MORB) to calk-alkaline felsic lavas (andesite, rhyolite). *Immature* back-arc rifts are commonly characterized by fractionated volcanic suites, which include both mid-ocean ridge basalts (N-MORB) and more felsic lavas. In the southern Lau Basin, the lavas consist of basaltic-andesite, andesite, dacite, and local rhyodacite.[26] The actively spreading portion of the Western Manus Basin is floored by a range of volcanic rocks including basalt, andesite, and dacite.[27] Sulfide deposits occurring near the propagating tip of the Eastern Manus spreading center are hosted by dacitic lavas.[28] In the Okinawa Trough, back-arc rifting takes place in continental crust at the margin of the East China Sea and is characterized by a complex volcanic suite of basalt, andesite, dacite, and rhyolite. In the Western Woodlark Basin, the volcanics range in composition from N-MORB through high Mg–K–andesites to peralkaline rhyolite.[29-31] *Mature* back-arc rifts are dominated by mid-ocean ridge type basalts (e.g., Mariana Trough, North Fiji Basin[21,32,33]). Preliminary descriptions of deposits in these settings indicate that they may be more similar to massive sulfides on the mid-ocean ridges than deposits in younger back-arc rifts.

The differences in the volcanic host rocks are reflected by the mineralogical and chemical compositions of the massive sulfide deposits forming in different environments. In particular, deposits forming in back-arc rifts in the earliest stages of opening (e.g., Southern Lau Basin, Eastern Manus Basin, Okinawa Trough) commonly have high concentrations of elements such as Zn, Pb, Ba, and Au compared to similar deposits forming in mature back-arc rifts or along ordinary mid-ocean ridges (see below).

Ancient massive sulfide deposits now found on land and modern seafloor polymetallic sulfide deposits are products of the same geological and geochemical processes, and many analogies can be drawn between modern examples and base metal deposits in the geological record.[34-38] Back-arc sulfide deposits in the western Pacific such as in the Okinawa Trough resemble Phanerozoic Zn–Pb–Cu deposits of the Kuroko- or Iberian Pyrite Belt-type (see References 23 and 26). Massive sulfides in the Eastern Manus Basin, where back-arc rifting has produced felsic volcanic rocks of dacitic composition, have been compared to Archean Zn–Cu massive sulfide deposits such as those in the Noranda district of Canada.[39,40] On a worldwide basis, less than about 25% of massive sulfide deposits in the geologic record have formed in exclusively basaltic rocks; more than 55% are associated with felsic volcanics.[1] This suggests that young back-arc spreading centers have been a particularly important setting for massive sulfide formation through time. Detailed studies of modern examples have confirmed that back-arc environments are the closest analogs of ancient volcanogenic massive sulfides that are currently being mined for Cu, Pb, Zn, Ag, and Au.

A particularly important discovery in recent years was the observation that some deposits are forming along the active volcanic fronts of island arcs, as well as in back-arc basins. For example, distinctive polymetallic sulfides and Au–Ag–barite deposits have now been located on at least six

TABLE 13.1
Mineralogical Composition of Seafloor Polymetallic Sulfide Deposits

	Back-Arc Deposits	Mid-Ocean Ridge Deposits
Fe-sulfides	Pyrite, marcasite, pyrrhotite	Pyrite, marcasite, pyrrhotite
Zn-sulfides	Sphalerite, wurtzite	Sphalerite, wurtzite
Cu-sulfides	Chalcopyrite, isocubanite	Chalcopyrite, isocubanite
Silicates	Amorphous silica	Amorphous silica
Sulfates	Anhydrite, barite	Anhydrite, barite
Pb-sulfides	Galena, sulfosalts	
As-sulfides	Orpiment, realgar	
Cu-As-Sb-sulfides	Tennantite, tetrahedrite	
Native metals	Gold	

Source: From Herzig, P.M. and Hannington, M.D., Polymetallic massive sulfides at the modern seafloor — A review, *Ore Geol. Rev.*, 10, 95, 1995.

andesitic cones that make up the volcanic front of the Izu–Ogasawara arc.[41] The typical deposits include abundant barite, sphalerite, and galena, with notable concentrations of As, Sb, Ag, Hg, and Au. At other sites, pumice-rich sediments and volcanic breccias within the summit craters of the volcanoes are cemented by quartz and barite and locally contain extensive deposits of native sulfur. The setting of these deposits contrasts with most ancient massive sulfide deposits, which are considered to have formed mainly during arc extension (i.e., back-arc rifting) rather than during the constructional phases of arc volcanism.

13.3   MINERALOGY OF THE DEPOSITS

The mineralogy of seafloor sulfide deposits has been documented in a number of detailed studies of samples from various sites (e.g., see References 20, 26, and 42 through 49). The mineral paragenesis of sulfide deposits at volcanic-dominated mid-ocean ridges usually includes assemblages that formed at temperatures ranging from about 300 to 400°C to less than 150°C. High-temperature fluid channels of black smokers and the interiors of sulfide mounds commonly consist of pyrite and chalcopyrite together with pyrrhotite, isocubanite, and locally bornite. The outer portions of chimneys and mounds are commonly composed of lower-temperature precipitates such as sphalerite/wurtzite, marcasite, and pyrite, which are also the principal sulfide minerals of low-temperature white smoker chimneys. Anhydrite is important in the high-temperature assemblages, but is typically replaced by later sulfides, amorphous silica, or barite at lower temperatures.

Sulfide mineralization at back-arc spreading centers has some mineralogical characteristics which are similar to hydrothermal precipitates at volcanic-dominated mid-ocean ridges (Table 13.1). Commonly, pyrite and sphalerite are the dominant sulfides. Chalcopyrite is common in the higher-temperature assemblages, but pyrrhotite is rare. Barite and amorphous silica are the most abundant nonsulfides. Many of the deposits forming in back-arc rifts are characterized by a variety of minor and trace minerals such as galena, tennanite and tetrahedrite, and complex, nonstoichiometric Pb–As–Sb sulfosalts. The JADE deposit, in the Okinawa Trough, contains abundant Ag-bearing galena, Ag–Sb–As–Pb sulfosalts, native sulfur, cinnabar, and arsenic sulfides such as realgar and orpiment.[23,50] The first examples of visible primary gold in seafloor sulfides were documented in samples of low-temperature white smoker chimneys from the southern Lau Basin.[51,52] The gold is coarse grained (up to 18 microns) and occurs as co-depositional inclusions in massive, Fe-poor sphalerite (Figure 13.4).

FIGURE 13.4 Occurrence of native gold in massive sulfide samples from the Lau back-arc. (a) Large gold grain (Au) as free inclusion in Fe-poor sphalerite (backscattered electron SEM image). (b) Gold grain (Au) composed of aggregates of submicron colloid-sized particles in a late fracture within sphalerite (backscattered electron SEM image).

13.4 METAL CONTENTS OF THE DEPOSITS

The bulk composition of seafloor sulfide deposits in various tectonic settings is mainly a reflection of the bulk composition of the underlying volcanic rocks from which the metals are leached. The source rocks in different volcanic environments may include MORB and clastic sediments at the mid-ocean ridges, to fractionated volcanic suites in intraoceanic back-arc basins, and felsic volcanics (dacite, rhyolite) in young intracontinental back-arc rifts. The compositional variation of the source rocks is also reflected in the composition of the respective vent fluids (see below).

Compared to samples from volcanic-dominated mid-ocean ridges, massive sulfides forming in intraoceanic back-arc spreading centers are characterized by elevated concentrations of Zn (15.1 wt.%), Pb (1.2 wt.%), and Ba (13.0 wt.%), but low contents of Fe (13.3 wt.%) (Table 13.2). Deposits forming in more mature back-arc spreading centers (e.g., Mariana Trough, North Fiji Basin), dominated by MORB-type lavas, more closely resemble the massive sulfide deposits forming on

TABLE 13.2
Bulk Chemical Composition of Seafloor
Polymetallic Sulfides

Element	Intraoceanic Back-Arc Ridges	Intracontinental Back-Arc Ridges	Mid-Ocean Ridges
Pb (wt.%)	1.2	11.5	0.2
Fe	13.3	7.0	23.6
Zn	15.1	18.4	11.7
Cu	5.1	2.0	4.3
Ba	13.0	7.2	1.7
As (ppm)	1,000	15,000	300
Sb	100	3,000	100
Ag	195	2,766	140
Au	2.9	3.8	1.2
(N)	317	28	890

Source: From Herzig, P.M. and Hannington, M.D., Polymetallic massive sulfides at the modern seafloor — A review, *Ore Geol. Rev.*, 10, 95, 1995.

mid-ocean ridges.[53] Polymetallic sulfides in the Okinawa Trough have lower Fe contents (7.0 wt.%) than deposits in intraoceanic back-arc settings and are notably enriched in Zn (18.4 wt.%), Pb (11.5 wt.%), Ag (2,766 ppm, maximum 1.1 wt.%), As (1.5 wt.%), and Sb (0.3 wt.%) (Table 13.2). These elements are present at much higher concentrations in fractionated calc-alkaline lavas than in normal MORB. The high concentrations of Pb, in particular, likely reflect the abundance of felsic volcanics and sediments in the source region. The composition of the underlying lavas in the Okinawa Trough is explained by the presence of about 20 km thickness of continental basement in the melting region.[54] The low abundance of Fe-sulfides in many of the back-arc deposits may also be partly a reflection of the low Fe contents of back-arc lavas compared to MORB. Similar trends in the bulk composition of massive sulfide deposits in different volcanic and tectonic settings are well-documented in the ancient geologic record (e.g., References 34 and 55).

13.5   VENT FLUID COMPOSITIONS

Bulk vent fluid compositions in back-arc hydrothermal systems are broadly similar to those of modern mid-ocean ridges. High-temperature (ca. 350°C), end-member fluids are low-pH solutions with moderate salinity and high concentrations of dissolved sulfur. End-member pHs are typically in the range of 3 to 4, and salinities are close to that of ambient sea water (3.5 to 5.0 equivalent wt.% NaCl). Concentrations of dissolved H_2S in the hydrothermal fluids are commonly at least as great as the total concentration of dissolved metals (i.e., 10^{-3} to 10^{-2} m). Measured end-member vent fluids on the mid-ocean ridges typically contain 6.6 to 8.4 mmol/kg dissolved H_2S[56-59]; vent fluids from back-arc systems contain up to 12.4 mmol/kg H_2S.[60-62] The high concentrations of dissolved gases, in some cases, may indicate direct contributions of magmatic volatiles to the hydrothermal fluids.

The broad compositional differences observed in deposits from different tectonic settings are also reflected in the chemistry of sampled vent fluids (e.g., see References 22, 26, 53, 63, and 64). Chemical analyses of end-member fluids from the Vai Lili Hydrothermal Field, in the southern Lau Basin, indicate much higher concentrations of Zn, Pb, and As compared to typical mid-ocean ridge fluids (Table 13.3).

TABLE 13.3
Chemical Composition of
Hydrothermal Fluids in Back-Arc Areas
and at Mid-Ocean Ridges

	Back-Arc Ridge	Mid-Ocean Ridge
Zn (ppm)	196	5.5
Cu	2.2	1.4
Ba	5.4	1.4
As (ppb)	450	17
Pb	808	54
T [°C]	334	350
pH	2.0	3.6

Source: From Herzig, P.M. and Hannington, M.D.,
Polymetallic massive sulfides at the modern seafloor —
A review, *Ore Geol. Rev.,* 10, 95, 1995.

A number of important seafloor sulfide deposits have recently been found in shallow water environments where the pressure at the seafloor is insufficient to prevent phase separation (boiling) of the hydrothermal fluids. At 350°C, typical seafloor vent fluids will boil if the hydrostatic pressure drops below 160 bar (equivalent to about 1,600 m of water depth [see References 65 and 66]). In response to boiling, a portion of the dissolved metals will be deposited as disseminated or vein mineralization beneath the seafloor.[67] Phase-separated fluids emanating from hydrothermal vents above the boiling zone may be significantly depleted in dissolved metals (see References 68 and 69).

13.6   GOLD IN SUBDUCTION-RELATED ARC AND BACK-ARC ENVIRONMENTS

The enrichment of gold in modern seafloor polymetallic sulfide deposits has been known for more than a decade.[70] Since the first reports of high gold contents, the occurrence and distribution of gold in submarine hot springs has been studied extensively.[48,52,71,72,73]

The gold contents vary widely among different deposit types but are similar to those found in ancient massive sulfides. Analyses of polymetallic sulfides from a number of back-arc spreading centers have revealed gold concentrations averaging between 3 and 30 ppm,[52] with the highest values occurring in sulfides formed in immature back-arc rifts in continental or island arc crust (Table 13.4). These settings are dominated by transitional tholeiitic to calc-alkaline volcanic suites including andesites, dacites, and rhyolites (e.g., Eastern Manus Basin, Lau Basin, Okinawa Trough). Polymetallic sulfides from the Valu Fa Ridge in the Lau back-arc have gold contents of up to 29 ppm with an average of 3 ppm Au (n = 75).[52] Samples from the Okinawa Trough contain up to 24 ppm Au (average 3.8 ppm, n = 28).[50,74] Preliminary analyses of sulfides from the Central Manus Basin indicate gold contents of up to 30 ppm (n = 10) and maximum concentrations of more than 50 ppm Au. The average gold content of samples collected from the Eastern Manus Basin is 15 ppm with a maximum of 54.9 ppm (n = 26).[75] High gold contents, up to 21 ppm, have also been found in barite chimneys from the Western Woodlark Basin.[76,77] In contrast, sulfide deposits forming along mature back-arc spreading centers, such as the North Fiji Basin and Mariana Trough, have maximum gold contents of only 0.1 to 4.3 ppm, similar to that of sulfide deposits on the mid-ocean ridges.

A number of the deposits found along the volcanic front of the Izu–Ogasawara arc are also notably gold-rich.[41] Gold concentrations of up to 71 ppm have been reported from Suiyo Seamount,[78] and these enrichments may be related to magmatic contributions to the hydrothermal fluids (e.g., see Reference 79).

TABLE 13.4
Gold Grades in Polymetallic Massive Sulfides
from the Modern Seafloor

	Au (ppm)		
	Range	Average	(N)
Immature Back-Arc Ridges (intermediate to felsic volcanics)			
Lau Basin	0.01–28.7	3.1	75
Okinawa Trough	0.60–24.0	3.8	28
Central Manus Basin	0.01–52.5	30.0	10
Eastern Manus Basin	1.30–54.9	15.0	26
Woodlark Basin	8.10–21.0	15.0	5
Mature Back-Arc Ridges (MOR-type volcanics)			
Mariana Trough	0.14–1.70	0.8	11
North Fiji Basin	0.01–4.30	2.2	17
Mid-Ocean Ridges (MORB)	0.01–6.70	1.2	890

13.7 GEOCHEMICAL CONTROLS ON GOLD ENRICHMENT

The concentrations of gold transported in modern seafloor fluids are very low (e.g., ≤ 0.1 ppb[60,71,80]), suggesting that dramatic changes in the fluid chemistry are required to achieve saturation and cause the efficient local precipitation of gold. The transport of gold in seafloor hydrothermal systems has been discussed extensively in terms of solubility models based on aqueous chloride or bisulfide complexing.[70,71,81,82] Chloride complexing as $AuCl_2^-$ is limited to high-temperature fluids ($\geq 300°C$) at low pH and elevated salinities. Assuming vent fluids with pH = 3, 5.0 wt.% NaCl, and 10 mmol/kg H_2S, the solubility of gold in saturated fluids between 270°C and 230°C would be about 5 ppb as $Au(HS)_2^-$ but <0.1 ppb as $AuCl_2^-$, based on formation constants for the respective complexes from Seward[83] and Helgeson.[84] Chloride complexing would be significant only at temperatures >300°C and may account for much of the dissolved gold in higher-temperature vents.

The relative stability of aqueous sulfur ($Au(HS)_2^-$) and chloride ($AuCl_2^-$) complexes in the mid-ocean ridge fluids was interpreted to be the principal control on the site of gold deposition, with $Au(HS)_2^-$ dominating lower-temperature fluids in weakly acid solutions. This was supported by measurements of high concentrations of gold in low-temperature $Zn–Ba–SiO_2$ precipitates in a number of deposits.[70,71,85] In a majority of cases, significant precipitation of gold was found to be restricted to late-stage, low-temperature (<250°C) assemblages resulting from sustained mixing of higher-temperature fluids with sea water.

Careful documentation of the mineralogical and chemical characteristics of gold-rich massive sulfides indicates that saturation of the hydrothermal fluids with gold likely results from the combined effects of conductive cooling, mixing with sea water, and attendant oxidation of H_2S. The point at which the fluid becomes saturated with $Au(HS)_2^-$ and the efficiency of gold precipitation depends to a large extent on its T-pH-aO_2 path.[71,82] In most seafloor hydrothermal systems, the T-pH-aO_2 characteristics of the fluids are determined by their starting compositions and the extent of reaction between reduced and oxidized sulfur species during mixing. Mid-ocean ridge vent fluids have a large redox buffering capacity provided by abundant reducing agents such as ferrous iron, sulfide, and H_2 (e.g., see References 86 through 88). As a result, these fluids tend to be buffered close to pyrite–pyrrhotite throughout most of their venting history, and a significant increase in

TABLE 13.5
Characteristics of Back-Arc vs. Mid-Ocean Ridge
Fluids and Associated Volcanic Rocks

	Back-Arc Fluids	Mid-Ocean Ridge Fluids
Buffer capacity	Weakly buffered (far from py/po)	Strongly buffered (close to py/po)
Oxidation state	More oxidizing (higher f_{O_2})	Strongly reducing (low f_{O_2})
Volcanic rocks	Intermediate to felsic (lower Fe^{2+})	MORB (high Fe^{2+})

their oxidation state is required in order to precipitate gold (Figure 13.5). This may occur only after substantial mixing with sea water, when the redox buffer capacity of the fluids is exhausted (e.g., following mixing to a solution containing nearly 50 wt.% sea water[71]). At a higher oxidation state, high-grade gold may be precipitated from $Au(HS)_2^-$ following relatively small changes in the fluid chemistry. In Figure 13.5, abrupt precipitation of gold takes place along the $H_2S–HSO_4^-$ equal activity curve. Large[81] and Huston and Large[82] referred to the steep thermal and chemical gradients along this boundary as a "solubility cliff" for gold.

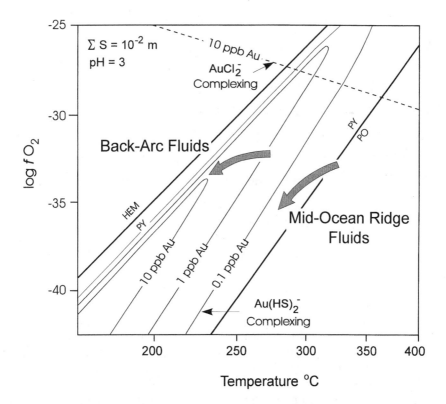

FIGURE 13.5 Temperature–oxygen activity diagram showing the important differences in the chemical evolution of hydrothermal fluids at back-arc and mid-ocean ridges. (Adapted from Herzig, P.M., Hannington, M.D., Fouquet, Y., von Stackelberg, U., and Petersen, S., Gold-rich polymetallic sulfides from the Lau back-arc and implications for the geochemistry of gold in seafloor hydrothermal systems of the Southwest Pacific, *Econ. Geol.,* 88, 2182, 1993.)

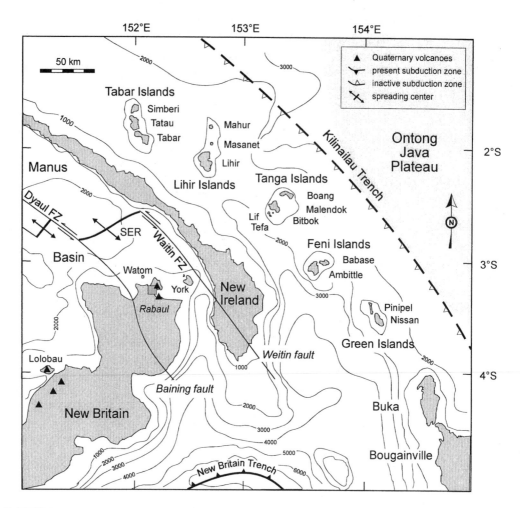

FIGURE 13.6 Regional map showing the location of the Tabar–Feni island chain and the principal structural and volcanic elements of the New Ireland fore-arc and the Manus Basin.

Of the samples that have been collected from deposits in back-arc settings, a much larger proportion have high gold contents than was previously observed for mid-ocean ridge deposits. The highest proportion of gold-rich samples has been recovered from *immature* back-arc spreading centers and rifts propagating into continental or island arc crust (e.g., Lau Basin, Okinawa Trough, Manus Basin). Far fewer gold-rich samples have been collected from sulfide deposits in the *mature* back-arc spreading centers (e.g., Mariana Trough, North Fiji Basin). Whereas the proportions of the major elements in the deposits are strongly controlled by source rock compositions, the gold contents of back-arc lavas are not significantly different from those of ordinary mid-ocean ridge basalts, and therefore these rocks probably do not represent an enriched source.

The apparent source rock control on gold enrichment may be related to the buffering of the hydrothermal fluids during water–rock interaction. The oxidation state of vent fluids on the mid-ocean ridges is strongly buffered by reaction with abundant FeO-bearing minerals in the rocks. The tendency for these fluids to remain undersaturated with gold may reflect their reduced nature and strong buffer capacity. In contrast, vent fluids that have equilibrated with more felsic lavas tend to be more oxidized because of the lower abundance of FeO-bearing minerals in the rock and have a lower redox buffering capacity (e.g., Reference 89; Table 13.5). These more oxidized solutions may result in more effective transport of gold at high temperatures and the more efficient precipitation of gold at the seafloor (e.g., Figure 13.5).

13.8 AN EXAMPLE OF SUBMARINE GOLD MINERALIZATION IN A MODERN FORE-ARC SETTING

A recent discovery of gold mineralization in a modern fore-arc environment suggests that a number of previously unexplored settings on the seafloor may be prospective for gold-rich hydrothermal systems. The Tabar–Feni island chain, off Papua New Guinea, is host to a new class of marine mineral deposit, characterized by disseminated and stockwork-like sulfides in an alkaline volcano.[73] The Tabar–Feni chain comprises a series of Pliocene to Recent volcanoes that occupy a fore-arc position in the New Ireland Basin (Figure 13.6). Several of the volcanoes are still active (the most recent eruption occurring 2300 years ago) and are host to several large porphyry stocks, active geothermal systems, and epithermal-style gold deposits, including the giant Ladolam gold deposit on the island of Lihir (minimum 600 t Au).

Mapping of largely uncharted waters in the vicinity of Lihir revealed six previously unknown volcanic cones about 10 km south of the island and 25 km south of the Ladolam gold deposit[73,90] (Figure 13.7). From the crater of one of the volcanoes at a depth of 1,050 m (Conical Seamount), a large sample of trachybasalt locally mineralized with amorphous silica ($\delta^{18}O = 36.8\%o$ equals Tf = 19.5°C), pyrite, Fe-poor sphalerite (avg. 6.9 mol% FeS), chalcopyrite, galena (avg. 0.26 wt.% Ag), unidentified Pb–As sulfosalts, and traces of anglesite and cerrusite were recovered. Bulk samples contain 1.2 to 43.2 ppm Au (avg. 19.3 ppm, n = 13), high concentrations of typical epithermal elements such as As (320-840 ppm), Sb (21 to 430 ppm), and Hg (1 to 34 ppm; see Table 13.6), and are associated with alunite as a distinctive acid–sulfate alteration phase. Similar to epithermal gold deposits on land, the samples from Conical Seamount have low base metal contents (Cu+Zn+Pb <3 wt.%) and low concentrations of Fe (4.1 to 12.1 wt.%). Grains of native gold (10 micron, Figure 13.8) were identified by SEM as inclusions in pyrite and sphalerite and contain about 12 wt.% Ag. Alunite associated with gold mineralization typically occurs as platy crystals (Figure 13.9), which is the characteristic habit of alunite formed from hydrothermal solutions with a large magmatic vapor component.[91]

The occurrence of alunite requires extremely acid, high-sulfidation conditions (i.e., presence of H_2SO_4), and alunite stability is generally limited to pH<3 (e.g., see Reference 92 and 93). It is supposed that these conditions were generated by degassing of magmatic SO_2 released from a shallow-level magma chamber (see Reference 96) similar to subaerial acid–sulfate systems. The SO_2 disproportionates to sulfide and sulfate ($4SO_2 + 4H_2O = 3 H_2SO_4 + H_2S$) to form an acid–sulfate alteration assemblage typical of epithermal gold deposits which are known to have a large magmatic component.[94,95] This is supported by sulfur isotope data of alunite (+7.5 and +6.4‰ $\delta^{34}S$) and associated sulfides (avg. −5.3‰ $\delta^{34}S$ Table 13.7).[97,98] The large difference between the $\delta^{34}S$ of alunite and that of coexisting pyrite, coupled with the highly acidic conditions required to stabilize the alunite, strongly suggests the involvement of magmatic fluids and gases.

The occurrence of hydrothermal activity in the rifted fore-arc region south of Lihir is the first documented evidence of seafloor hydrothermal activity associated with alkaline volcanism and represents a new setting for hydrothermal activity not previously recognized on the modern seafloor. The discovery of high gold contents associated with amorphous silica and alunite in hydrothermal precipitates at Conical Seamount (Table 13.6) may be the first example of a shallow-marine epithermal system (Figure 13.10), analogous to gold-producing deposits on adjacent volcanic islands. These discoveries may have important implications for understanding the origins of large, Lihir-type gold deposits and for locating further deposits of this style at the modern seafloor.

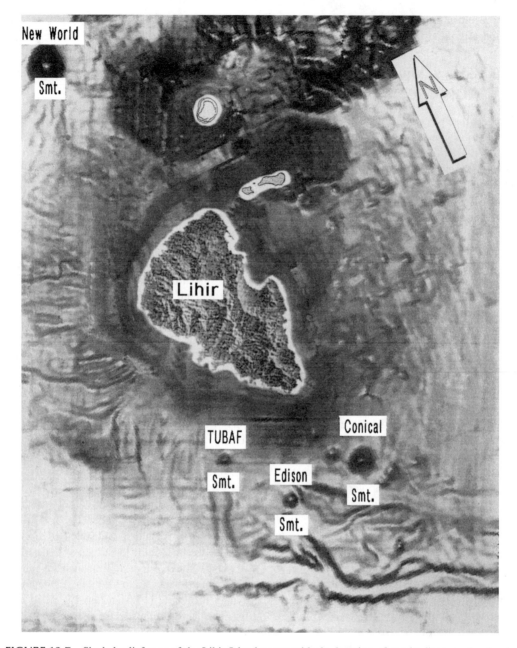

FIGURE 13.7   Shaded relief map of the Lihir Island group with the location of newly discovered volcanic cones, including Conical Seamount, from which gold-mineralized trachybasalts were recovered.

ACKNOWLEDGMENTS

This work was supported by the German Federal Ministry for Education, Science, Research and Technology (BMBF), the German Research Foundation (DFG), and the Geological Survey of Canada (GSC, Contribution No. 1998008).

TABLE 13.6
Chemical Composition of Mineral Precipitates from the
Conical Seamount Shallow-Marine Epithermal Deposit
South of Lihir Island (Papua New Guinea)

Au (ppm)	36.7	21.5	4.5	43.2	35.6	6.9	20.6
As	840	590	330	450	320	370	480
Sb	430	210	37	140	130	55	230
Hg	34	21	3	31	21	4	24
Cu+Zn+Pb (wt.%)	2.8	2.0	0.4	1.4	1.2	0.4	3.0

TABLE 13.7
Characteristics of Alunite from Conical Seamount
(Papua New Guinea)

Crystal type	Platy and zoned	
Variety	Natroalunite (Na–K)	
$\delta^{34}S$	Alunite sulfate	+7.0 ‰ (n = 2)
	Associated sulfides	−5.3 ‰ (n = 5)

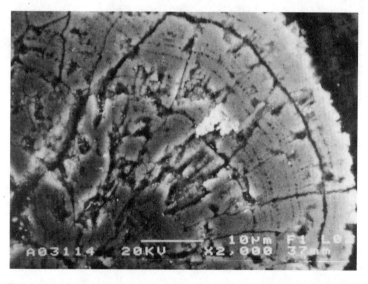

FIGURE 13.8 Grain of native gold (12 wt.% Ag) as inclusion in colloform pyrite sampled at Conical Seamount south of Lihir Island (backscattered electron SEM image). Scale bar is 10 microns.

FIGURE 13.9 Backscattered electron SEM images of (a) typical platy and (b) partly zoned alunite crystals, which is the characteristic habit of alunite formed from hydrothermal solutions with a large magmatic vapor component. (See also Reference 91.)

REFERENCES

1. Rona, P.A., Hydrothermal mineralization at oceanic ridges, *Can. Min.,* 26, 431, 1988.
2. Rona, P.A. and Scott, S.D., Preface to special issue on seafloor hydrothermal mineralization: new perspectives, *Econ. Geol.,* 88, 1935, 1993.
3. Francheteau, J., Needham, H.D., Choukroune, P., Juteau, T., Seguret, M., Ballard, R.D., Fox, P.J., Normark, W., Carranza, A., Cordoba, D., Guerrero, J., Rangin, C., Bougault, H., Cambon, P., and Hekinian, R., Massive deep-sea sulphide ore deposits discovered on the East Pacific Rise, *Nature,* 277, 523, 1979.
4. Spiess, F.N., MacDonald, K.C., Atwater, T., Ballard, R., Carranza, A., Cordoba, D., Cox, C., Diaz Garcia, V.M., Francheteau, J., Guerro, J., Hawkins, J.W., Haymon, R., Hessler, R., Juteau, T., Kastner, M., Larson, R., Luyendyk, B., MacDougall, J.D., Miller, S., Normark, W., Orcutt, J., and Rangin, C., East Pacific Rise. Hot springs and geophysical experiments, *Science,* 207, 1421, 1980.

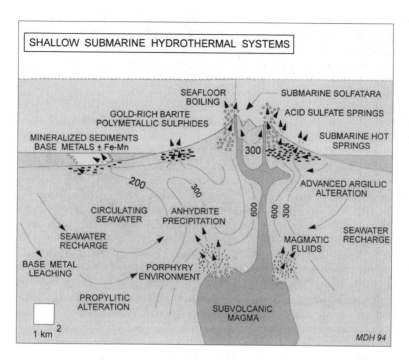

FIGURE 13.10 Schematic diagram illustrating the major features of a shallow-submarine epithermal system in which sea water may mix with rising magmatic volatiles to form a gold-rich, acid-sulfate, high-sulfidation epithermal system.

5. Rona, P.A., Klinkhammer, G., Nelsen, T.A., Trefry, J.H., and Elderfield, H., Black smokers, massive sulfides and vent biota at the Mid-Atlantic Ridge, *Nature,* 321, 33, 1986.

6. Krasnov, S.G., Cherkashev, G.A., Stepanova, T.V., Batuyev, B.N., Krotov, A.G., Malin, B.V., Maslov, M.N., Markov, V.F., Poroshina, I.M., Samovarov, M.S., Ashadze, A.M., and Ermolayev, I.K., Detailed geographical studies of hydrothermal fields in the North Atlantic, in *Hydrothermal Vents and Processes,* Geological Society Special Publication, Parson, L.M., Walker, C.L., Dixon, D.R. (Eds.), 87, 43, 1995.

7. Kong, L.S.L., Ryan, W.B.F., Mayer, L., Detrick, R., Fox, P.J., and Manchester, K., Bare-rock drill site: ODP legs 106 and 109: evidence for hydrothermal activity at 23°N on the Mid-Atlantic Ridge, *Transactions American Geophysical Union,* 66, 936, 1985.

8. Murton, B.J., Van Dover, C., and Southward, E., Geological setting and ecology of the Broken Spur hydrothermal vent field: 29°10'N on the Mid-Atlantic Ridge, in *Hydrothermal Vents and Processes,* Geological Society Special Publication, Parson, L.M., Walker, C.L., Dixon, D.R. (Eds.), 87, 33, 1995.

9. Fouquet, Y., Charlou, J.L., Costa, I., Donval, J.P., Radford-Knoery, J., Pelle, H., Ondreas, H., Lourenco, N., Segonzac, M., and Tivey, M., A detailed study of the Lucky Strike hydrothermal site and discovery of a new hydrothermal site: Menez Gwen; preliminary results of the DIVA 1 cruise, *InterRidge News,* 3, 2, 14, 1994.

10. Herzig, P.M. and Plüger, W.L., Exploration for hydrothermal mineralization near the Rodriguez Triple Junction, Indian Ocean, *Can. Min.,* 26, 721, 1988.

11. Plüger, W.L., Herzig, P.M., Becker, K.-P., Deissmann, G., Schöps, D., Lange, J., Jenisch, A., Ladage, S., Richnow, H.H., Schulze, T., and Michaelis, W., Discovery of hydrothermal fields at the Central Indian Ridge, *Marine Mining,* 9, 73, 1990.

12. Halbach, P., Blum, N., and Münch, U., 1998, Formation and decay of a modern massive sulfide deposit in the Indian Ocean, *Min. Dep.,* 33, 302, 1998.

13. Humphris, S.E., Herzig, P.M., Miller, D.J., Alt, J.C., Beckert, K., Brown, D., Brügmann, G., Chiba, H., Fouquet, Y., Gemmell, J.B., Guerin, G., Hannington, M.D., Holm, N.G., Honnorez, J.J., Itturino, G.J., Knott, R., Ludwig, R., Nakamura, K., Petersen, S., Reysenbach, A.-L., Rona, P.A., Smith, S., Sturz, A.A., Tivey, M.K., and Zhao, X., The internal structure of an active seafloor massive sulphide deposit, *Nature,* 377, 713, 1995.

14. Herzig, P.M., Humphris, S.E., Miller D.J., and Zierenberg, R.A. (Eds.), *Proceedings of the Ocean Drilling Program,* Scientific Results, 158, College Station, TX, 427, 1998.

15. Lonsdale, P., Bischoff, J.L., Burns, V.M., Kastner, M., and Sweeney, R.E., A high-temperature hydrothermal deposit on the seabed at a Gulf of California spreading center, *Earth and Planetary Science Letters,* 49, 8, 1980.

16. Mottl, M.J., Davis, E., and Fisher, A.T. (Eds.), *Proceedings of the Ocean Drilling Program,* Scientific Results, 139, College Station, TX, 1991.

17. Zierenberg, R.A., Fouquet, Y., Miller, D.J., and Leg 169 shipboard scientific party, The roots of seafloor sulfide deposits: preliminary results from ODP Leg 169 drilling in Middle Valley and Escanaba Trough, *Transactions American Geophysical Union,* 77, 765, 1996.

18. Both, R.A., Crook, K., Taylor, B., Brogan, S., Chapell, B., Frankel, E., Liu, L., Sinton, J., and Tiffin, D., Hydrothermal chimneys and associated fauna in the Manus back-arc basin, Papua New Guinea, *Transactions American Geophysical Union,* 67, 489, 1986.

19. Craig, H., Horibe, Y., Farley, K.A., Welhan, J.A., Kim, K.-R., and Hey, R.N., Hydrothermal vents in the Mariana Trough: results of the first Alvin dives, *Transactions American Geophysical Union,* 68, 1531, 1987.

20. Kastner, M., Craig, H., and Sturz, A., Hydrothermal deposition in the Mariana Trough: preliminary mineralogical investigation, *Transactions American Geophysical Union,* 68, 1531, 1987.

21. Auzende, J.-M., Urabe, T., Deplus, C., Eissen, J.P., Grimaud, D., Huchon, P., Ishibashi, J., Joshima, M., Lagabrielle, Y., Mevel, C., Naka, J., Ruellan, E., Tanaka, T., Tanahashi, M., and Tarouilly, L., Submersible study of an active hydrothermal site in North Fiji Basin (SW Pacific) (STARMER 1 Cruise), *Transactions American Geophysical Union,* 70, 1382, 1989.

22. Fouquet, Y., von Stackelberg, U., Charlou, J.L., Donval, J.L., Erzinger, J., Foucher, J.P., Herzig, P.M., Mühe, R., Soakai, S., Wiedicke, M., and Whitechurch, H., Hydrothermal activity and metallogenesis in the Lau back-arc basin, *Nature,* 349, 778, 1991.

23. Halbach, P., Nakamura, K., Wahsner, M., Lange, J., Sakai, H., Käselitz, L., Hansen, R.-D., Yamano, M., Post, J., Prause, B., Seifert, R., Michaelis, W., Teichmann, F., Kinoshita, M., Märten, A., Ishibashi, J., Czerwinski, S., and Blum, N., Probable modern analogue of Kuroko-type massive sulphide deposits in the Okinawa Trough back-arc basin, *Nature,* 338, 496, 1989.

24. Binns, R.A. and Scott, S.D., Actively forming polymetallic sulfide deposits associated with felsic volcanic rocks in the Eastern Manus back-arc basin, Papua New Guinea, *Econ. Geol.,* 88, 2226, 1993.

25. Binns, R.A., Scott, S.D., and PACLARK participants, Western Woodlark Basin: potential analogue setting for volcanogenic massive sulfide deposits, *Pacific Rim Congress '87,* Australasian Institute of Mining and Metallurgy, 531, 1987.

26. Fouquet, Y., von Stackelberg, U., Charlou, J.L., Erzinger, J., Herzig, P.M., Mühe, R., and Wiedicke, M., Metallogenesis in back-arc environments: the Lau Basin example, *Econ. Geol.,* 88, 2154, 1993.

27. Sinton, J.M., Liu, L., Taylor, B., Chappell, B.W., and shipboard party, Petrology, magmatic budget and tectonic setting of Manus back-arc basin lavas, *Transactions American Geophysical Union,* 67, 377, 1986.

28. Binns, R.A., Scott, S.D., and PACMANUS participants, Discovery of active hydrothermal sulfide deposition associated with submarine felsic volcanism, Pual Ridge, Eastern Manus Basin, Papua New Guinea, Geological Society of Australia, *11th Australian Geological Convention,* Ballarat, 1992, 19, 1992.

29. Binns, R.A. and Whitford, D.J., Volcanic rocks from the western Woodlark Basin, Papua New Guinea, Australasian Institute of Mining and Metallurgy, *Pacific Rim Conference,* 1, 525, 1987.

30. Scott, S.D., Binns, R.A., Whitford, D.J., and Finlayson, E.J., Initial break-up of continental crust by a propagating ocean ridge: results of the 1986 and 1988 PACLARK cruises to the western Woodlark Basin, Southwest Pacific, *Geological Association of Canada,* Program with Abstracts, 13, A111, 1988.

31. Scott, S.D., Lisitsin, A.P.P., Binns, R.A., and SUPACLARK Participants, Rifting and hydrothermal mineralization in the western Woodlark Basin, Papua New Guinea: results of the April 1990 SUPACLARK expedition, *Geological Society of America Abstracts with Programs,* 22, A10, 1990.

32. Hawkins, J.W., Lonsdale, P.F., Macdougall, J.D., and Volpe, A.M., Petrology of the axial ridge of the Mariana Trough backarc spreading center, *Earth and Planetary Science Letters,* 100, 226, 1990.

33. Macdougall, J.D., Volpe, A., and Hawkins, J. W., An arc-like component in Mariana Trough basalts, *Transactions American Geophysical Union,* 68, 1531, 1987.

34. Franklin, J.M., Lydon, J.W., and Sangster, D.F., Volcanic-associated massive sulfide deposits, *Econ. Geol.*, 75th Anniversary Volume, 1981, 485.

35. Scott, S.D., Seafloor polymetallic sulfide deposits: modern and ancient, *Marine Mining*, 5, 191, 1985.

36. Franklin, J.M., Volcanic associated massive sulphide deposits — an update, in *Geology and Genesis of Mineral Deposits in Ireland*, Andrew, C.J. et al. (Eds.), Irish Association for Economic Geology, 1986, 49.

37. Koski, R.A., Sulfide deposits on the seafloor: geological models and resource perspectives based on studies in ophiolite sequences, in *Marine Minerals: Resource Assessment Strategies*, Teleki, P.G. et al. (Eds.), Proceedings NATO Advanced Research Workshop, Series C, 194, Reidel Publishing Co., Boston, 1987, 301.

38. Scott, S.D., Seafloor polymetallic sulfides: scientific curiosities or mines of the future? in *Marine Minerals: Resource Assessment Strategies*, Teleki, P.G. et al. (Eds.), Proceedings NATO Advanced Research Workshop, Series C, 194, Reidel Publishing Co., Boston, 1987, 277.

39. Scott, S.D. and Binns, R.A., Presently-forming hydrothermal seafloor deposits of Manus and Woodlark Basins, S.W. Pacific, as models for ancient ores, in *Current Research in Geology Applied to Ore Deposits*, Fenoll Hach-Ali et al. (Eds.), SGA Granada, 1993, 381.

40. Scott, S.D. and Binns, R.A., An actively-forming, felsic volcanic-hosted polymetallic sulfide deposit in the southeast Manus back-arc basin of Papua New Guinea, *Transactions American Geophysical Union*, 73, 626, 1992.

41. Ishibashi, J, and Urabe, T., Hydrothermal activity related to arc-backarc magmatism in the western Pacific, in *Backarc Basins: Tectonics and Magmatism*, Taylor, B. (Ed.), Plenum Press, New York, 1995, 451.

42. Haymon, R.M. and Kastner, M., Hot spring deposits on the East Pacific Rise at 21°N: preliminary description of mineralogy and genesis, *Earth and Planetary Science Letters*, 53, 363, 1981.

43. Goldfarb, M.S., Converse, D.R., Holland, H.D., and Edmond, J.M., The genesis of hot spring deposits on the East Pacific Rise, 21°N, *Econ. Geol.*, Monograph 5, 184, 1983.

44. Haymon, R.M., Growth history of hydrothermal black smoker chimneys, *Nature*, 301, 695, 1983.

45. Oudin, E., Hydrothermal sulfide deposits of the East Pacific Rise (21°N) part I: descriptive mineralogy, *Marine Mining*, 4, 39, 1983.

46. Koski, R.A., Clague, D.A., and Oudin, E., Mineralogy and chemistry of massive sulfide deposits from the Juan de Fuca Ridge, *Geological Society of America Bulletin*, 95, 930, 1984.

47. Fouquet, Y., Auclair, G., Cambon, P., and Etoubleau, J., Geological setting, mineralogical and geochemical investigations on sulfide deposits near 13°N on the East Pacific Rise, *Marine Geology*, 84, 145, 1988.

48. Hannington, M.D., Herzig, P.M., and Scott, S.D., Auriferous hydrothermal precipitates on the modern seafloor, in *Gold Metallogeny and Exploration*, Foster, R.P. (Ed.), Blackie and Son Ltd., Glasgow and London, 1991, 249.

49. Hannington, M.D., Herzig, P.M., Scott, S.D. Thompson, G., and Rona, P.A., Comparative mineralogy and geochemistry of gold-bearing sulfide deposits on the mid-ocean ridges, *Marine Geology*, 101, 217, 1991.

50. Halbach, P., Pracejus, B., and Märten, A., Geology and mineralogy of massive sulfide ores from the Central Okinawa Trough, Japan, *Econ. Geol.*, 88, 2210, 1993.

51. Herzig, P.M., Fouquet, Y., Hannington, M.D., and von Stackelberg, U., Visible gold in primary polymetallic sulfides from the Lau back-arc, *Transactions American Geophysical Union*, 71, 1680, 1990.

52. Herzig, P.M., Hannington, M.D., Fouquet, Y., von Stackelberg, U., and Petersen, S., Gold-rich poly-metallic sulfides from the Lau back-arc and implications for the geochemistry of gold in seafloor hydrothermal systems of the Southwest Pacific, *Econ. Geol.*, 88, 2182, 1993.

53. Herzig, P.M. and Hannington, M.D., Polymetallic massive sulfides at the modern seafloor — A review, *Ore Geol. Rev.*, 10, 95, 1995.

54. Sibuet, J.-C., Letouzey, J., Barbier, F., Charvet, J., Foucher, J.-P., Hilde, T.W.C., Kimura, M., Ling-Yun, C., Marsset, B., Muller, C., and Stephan, J.-F., Back-arc extension in the Okinawa Trough, *Journal of Geophysical Research*, 92, 14041, 1987.

55. Ohmoto, H. and Skinner, B.J. (Eds.), The Kuroko and related volcanogenic massive sulfide deposits, *Econ. Geol.* Monograph 5, 604, 1983.

56. Von Damm, K.L., Edmond, J.M., Grant, B., and Measures, C.I., Chemistry of submarine hydrothermal solutions at 21°N, East Pacific Rise, *Geochimica et Cosmochimica Acta,* 49, 2197, 1985.

57. Bowers, T.S., Campbell, A.C., Measures, C.I., Spivack, A.J., and Edmond, J.M., Chemical controls on the composition of vent fluids at 13°N-11°N and 21°N, East Pacific Rise, *Journal of Geophysical Research,* 93, 4522, 1988.

58. Campbell, A.C., Bowers, T.S., and Edmond, J.M., A time-series of vent fluid composition from 21°N, EPR (1979, 1981, 1985) and the Guaymas Basin, Gulf of California (1982, 1985), *Journal of Geophysical Research,* 93, 4537, 1988.

59. Von Damm, K.L., Controls on the chemistry and temporal variability of seafloor hydrothermal fluids. in *Seafloor Hydrothermal Systems: Physical, Chemical, Biological and Geological Interactions,* Humphris, S.E. et al. (Eds.), AGU Geophysical Monograph, 91, 222, 1995.

60. Campbell, A.C., Edmond, J.M., Colodner, D., Palmer, M.R., and Falkner, K.K., Chemistry of hydrothermal fluids from the Mariana Trough back-arc basin in comparison to mid-ocean ridge fluids, *Transactions American Geophysical Union,* 68, 1531, 1987.

61. Sakai, H., Gamo, T., Kim, E.-S., Tsutsumi, M., Ishibashi, J., Wakita, H., Yamano, M., Tanaka, T., and Oomori, T., Venting of CO_2-dominant liquid and gas hydrate formation at the Jade Hydrothermal Fields, Mid-Okinawa Trough backarc basin, *Transactions American Geophysical Union,* 71, 953, 1990.

62. Sakai, H., Gamo, T., Kim, E.-S., Shitashima, K., Yanagisawa, F., Tsutsumi, M., Ishibashi, J., Sano, Y., Wakita, H., Tanaka, T., Matsumoto, T., Naganuma, T., and Mitsuzawa, K., Unique chemistry of the hydrothermal solutions in the mid-Okinawa Trough backarc basin, *Geophysical Research Letters,* 17, 2133, 1990.

63. Gamo, T., Sakai, H., Kim, E.-S., Shitashima, K., and Ishibashi, J., High alkalinity due to sulfate reduction in the CLAM hydrothermal field, Okinawa Trough, *Earth and Planetary Science Letters,* 107, 328, 1991.

64. Fouquet, Y., von Stackelberg, U., Charlou, J.L., Donval, J.L., Foucher, J.P., Erzinger, J., Herzig, P.M., Mühe, R., Wiedicke, M., Sokai, S., and Whitechurch, H., Hydrothermal activity in the Lau back-arc basin: sulfides and water chemistry, *Geology,* 19, 303, 1991.

65. Bischoff, J.L. and Rosenbauer, R.J., The critical point and two-phase boundary of sea water, 200-500°C, *Earth and Planetary Science Letters,* 75, 172-180, 1984.

66. Bischoff, J.L. and Pitzer, K.S., Phase relations and adiabats in boiling seafloor geothermal systems, *Earth and Planetary Science Letters,* 75, 327, 1985.

67. Drummond, S.E. and Ohmoto, H., Chemical evolution and mineral deposition in boiling hydrothermal systems, *Econ. Geol.,* 80, 126, 1985.

68. Massoth, G.J., Butterfield, D., Lupton, J.E., McDuff, R.E., Lilley, M.D., and Jonasson, I.R., Submarine venting of phase-separated hydrothermal fluids at Axial Volcano, Juan de Fuca Ridge, *Nature,* 340, 702, 1989.

69. Butterfield, D.A., Massoth, G.J., McDuff, R.E., Lupton, J.E., and Lilley, M.D., Geochemistry of hydrothermal fluids from Axial seamount hydrothermal emissions study vent field, Juan de Fuca Ridge: subseafloor boiling and subsequent fluid-rock interaction, *Journal of Geophysical Research,* 95, 12895, 1990.

70. Hannington, M.D., Peter, J.M., and Scott, S.D., Gold in seafloor polymetallic sulfide deposits, *Econ. Geol.,* 81, 1867, 1986.

71. Hannington, M.D. and Scott, S.D., Gold mineralization in volcanogenic massive sulfides: implications of data from active hydrothermal vents on the modern seafloor, *Econ. Geol.* Monograph 6, 491, 1989.

72. Hannington, M.D. and Scott, S.D., Sulfidation equilibria as guides to gold mineralization in volcanogenic massive sulfides: evidence from sulfide mineralogy and the composition of sphalerite, *Econ. Geol.,* 84, 1978, 1989b.

73. Herzig, P.M. and Hannington, M.D., Hydrothermal activity, vent fauna, and submarine gold mineralization at alkaline fore-arc seamounts near Lihir Island, Papua New Guinea, *Proceedings PACRIM '95,* Auckland, New Zealand, 279, 1995.

74. Urabe, T., Marumo, K., and Nakamura, K., Mineralization and related hydrothermal alteration in Izena cauldron (JADE site), Okinawa Trough, Japan, *Geological Society of America Abstracts with Programs,* 22, A9, 1990.

75. Binns, R.A., Submarine deposits of base and precious metals in Papua New Guinea, in *Proceedings PNG Geology, Exploration and Mining Conference 1994,* Rogerson, R. (Ed.), The Australasian Institute of Mining and Metallurgy, 1994, 71.

76. Binns, R.A., Boyd, T., and Scott, S.D., Precious metal spires from the Western Woodlark Basin, Papua New Guinea, *Geological Association of Canada Program with Abstracts,* 16, A12, 1991.

77. Binns, R.A., Scott, S.D., Bogdanov, Y.A., Lisitsin, A.P., Gordeev, V.V., Finlayson, E.J., Boyd, T., Dotter, L.E., Wheller, G.E., and Muravyev, K.G., Hydrothermal oxide and gold-rich sulfate deposits of Franklin Seamount, western Woodlark Basin, Papua New Guinea, *Econ. Geol.,* 88, 2122, 1993.

78. Watanabe, K., Shibata, A., Kajimura, T., Ishibashi, J., Tsunogai, U., Aoki, M., and Nakamura, K., Survey method about the submarine volcano and its seafloor hydrothermal ore deposit — an example of Suiyo Seamount in the Izu-Ogasawara Arc with the submersible Shinkai 2000, *Journal of the Japanese Society for Marine Surveys and Technology,* 6, 29, 1994.

79. Tsunogai, U., Ishibashi, J., Wakita, H., Gamo, T., Watanabe, K., Kajimura, T., Kanayama, S., and Sakai, H., Peculiar features of Suiyo Seamount hydrothermal fluids, Izu-Bonon arc: differences from subaerial volcanism, *Earth and Planetary Science Letters,* 126, 289, 1994.

80. Falkner, K.K. and Edmond, J.M., Gold in sea water, *Earth and Planetary Science Letters,* 98, 208, 1990.

81. Large, R.R., Huston, D.L., McGoldrick, P.J., Ruxton, P.A., and McArthur, G., Gold distribution and genesis in Australian volcanogenic massive sulfide deposits and significance for gold transport models, *Econ. Geol.* Monograph 6, 520, 1989.

82. Huston, D.L. and Large, R.R., A chemical model for the concentration of gold in volcanogenic massive sulfide deposits, *Ore Geol. Rev.,* 4, 171, 1989.

83. Seward, T.M., Thio-complexes of gold and the transport of gold in hydrothermal ore solutions, *Geochimica et Cosmochimica Acta,* 37, 379, 1973.

84. Helgeson, H.C., Thermodynamics of hydrothermal systems at elevated temperatures and pressures, *American Journal of Science,* 267, 729, 1969.

85. Hannington, M.D., Jonasson, I.R., Herzig, P.M., and Petersen, S., Physical and chemical processes of seafloor mineralization at mid-ocean ridges, in *Physical, Chemical, Biological, and Geological Interactions within Hydrothermal Systems,* Humphris, S.E. et al. (Eds.), AGU Geophysical Monograph, 91, 115, 1995.

86. Janecky, D.R. and Seyfried, W.E., Jr., Formation of massive sulfide deposits on oceanic ridge crests: incremental reaction models for mixing between hydrothermal solutions and sea water, *Geochimica et Cosmochimica Acta,* 48, 2723, 1984.

87. Bowers, T.S., Von Damm, K.L., and Edmond, J.M., Chemical evolution of mid-ocean ridge hot springs, *Geochimica et Cosmochimica Acta,* 49, 2239, 1985.

88. Janecky, D.R. and Shanks, W.C., III, Computional modelling of chemical and sulfur isotopic reaction processes in seafloor hydrothermal systems: chimneys, massive sulfides, and subjacent alteration zones, *Can. Min.,* 26, 805, 1988.

89. Large, R.R., Chemical evolution and zonation of massive sulfide deposits in volcanic terrains, *Econ. Geol.,* 72, 549, 1977.

90. Herzig, P., Hannington, M., McInnes, B., Stoffers, P., Villinger, H., Seifert, R., Binns, R., Liebe, T., and scientific party, Submarine alkaline volcanism and active hydrothermal venting in the New Ireland forearc basin, Papua New Guinea, *Transactions American Geophysical Union,* 75, 513, 1994.

91. Arribas, A. Jr., Cunningham, C.G., Rytuba, J.J., Rye, R.O., Kelly, W.C., Podwysocki, M.H., McKee, E.H., and Tosdale, R.M., Geology, geochronology, fluid inclusions, and isotope geochemistry of the Rodalquilar gold alunite deposit, Spain, *Econ. Geol.,* 90, 795, 1995.

92. Hemley, J.J., Montoya, J.W., Marinenko, J.W., and Luce, R.W., Equilibria in the system $Al_2O_3–SiO_2–H_2O$ and some general implications for alteration/mineralization processes, *Econ. Geol.,* 75, 210, 1980.

93. Stoffregen, R., Genesis of acid sulfate alteration and Au-Cu-Ag mineralization of Summitville, Colorado, *Econ. Geol.,* 82, 1575, 1987.

94. Hedenquist, J.W. and Lowenstern, J.B., The role of magmas in the formation of hydrothermal ore deposits, *Nature,* 370, 519, 1994.

95. Arribas, A. Jr., Characteristics of high-sulfidation epithermal deposits, and their relation to magmatic fluid, in *Magmas, Fluids and Ore Deposits.* Thompson, J.F.H. (Ed.), Min. Assoc. of Can. Short Course Notes, 23, 419, 1995.

96. Herzig, P.M., Hannington, M.D., and Arribas, A. Jr., Sulfur isotopic composition of hydrothermal precipitates from the Lau back-arc: implications for magmatic contributions to seafloor hydrothermal systems, *Min. Dep.,* 33, 226, 1998.

97. Herzig, P.M., Magmatic contributions to seafloor hydrothermal systems, *Betekhtin Symposium,* Moscow, 97, 1997.

98. Herzig, P.M., Present-day submarine hydrothermal systems: an up-date, *SEG Neves Corvo Field Conference,* Lisbon, 23, 1997.

14 Hydrothermal Mineralization in the Red Sea

J.C. Scholten, P. Stoffers, D. Garbe-Schönberg, and M. Moammar

ABSTRACT

Hydrothermal activity in the Red Sea is linked to the divergent movement of the African and Arabian continental plates and the subsequent formation of new oceanic crust. The formation of hydrothermal deposits is facilitated in the Red Sea for two reasons: (1) The development of new oceanic crust is focused in relatively small areas, i.e., isolated deeps. (2) The occurrence of high saline brines in these deeps favors the preservation of the hydrothermal fluids and deposits. As a result, iron, manganese, sulfate, and sulfide sediment facies can be observed in the Red Sea deeps. The most concentrated deposits occur in the brine-filled Atlantis-II-Deep, an area which has been extensively investigated during the past 40 years. The Atlantis-II-Deep is thus one of the few locations in marine geoscience where the time variability of a hydrothermal system can be investigated. Between 1965 and 1997 a temperature increase of the lower brine in the Atlantis-II-Deep from 55.9°C to 67.2°C has been observed, suggesting an increase of hydrothermal activity there. In the same time period, concentrations of dissolved Mn and Fe in the brine increased slightly, whereas the concentrations of Cu decreased. Although active vents have never been observed, there are strong indications that hydrothermal fluids discharge in the SW-basin of the Atlantis-II-Deep. Based on the paragenesis of authigenic minerals in the sediments, formation temperatures of the hydrothermal precipitates between 110°C and <450°C have been estimated.

Apart from in the Atlantis-II-Deep, hydrothermal deposits have been found in the Thetis, Nereus, Vema, and Gypsum deeps, but in these deeps ore concentrations are lower. Massive sulfides from the brine-filled Kebrit Deep are another type of hydrothermal mineralization in the Red Sea. Its porous and fragile sulfides, which are sometimes impregnated with tar, consist of two types: the first type (type I) is characterized by a mineral assemblage of pyrite, marcasite, bravoite, sphalerite, digenite, chalcocite, jarosite, and minor amounts of galena, with relatively high Zn and Pb concentrations. The second type (type II) of massive sulfide consists almost exclusively of pyrite with low trace metal contents. The lack of significant Cu enrichments in any of the sulfides and their mineral structure suggests a low formation temperature. Lead isotope data on the sulfides indicate that the metals may have derived from mixing between a basaltic and a more radiogenic end member source. Age estimates of the sulfides suggest that the type II deposits formed between 20,000 and 28,000 years ago, whereas type I sulfides are much younger (<5,000 years). Periods of hydrothermal activity with obviously different chemical composition of the hydrothermal fluids have also been observed in the Thetis Deep. To what extent periods of hydrothermal activity in the Red Sea deeps are connected to each other by, e.g., tectonic activity, is a matter of speculation and needs further investigation.

14.1 INTRODUCTION

First reports of metalliferous deposits from the Red Sea date back to the last century. During the Austrian expedition with the research vessel *Pola* (1897) red-brown mud was sampled from the Red Sea. Especially in the 1960s, many research expeditions were undertaken (Table 14.1), but most of the work concentrated on the Atlantis-II-Deep area, where, as will be described below, the most important deposits were found. The major results of these expeditions were published by Degens and Ross,[1] Bäcker and Richter,[2] Bignell,[3] and Bignell et al.[4] In order to explore for comparable deposits, research in the 1980s concentrated on the northern part of the Red Sea. Although no hydrothermal deposits comparable to the Atlantis II-Deep were found there, the most spectacular finding was the recovery of massive sulfides from the Kebrit Deep in 1984 which, at that time, was one of the first examples of formation of massive sulfide deposits at a slow spreading ridge.[5]

14.2 TECTONIC SETTING

The remarkable fit of opposite coastlines led Alfred Wegener (1927) to propose the hypothesis that the Red Sea is an example of the early stage of a continental drift (Figure 14.1). Since this first observation much evidence has been gathered to show that the Red Sea represents an ocean *"in statu nascendi,"* evolving from continental rifting to seafloor spreading. It was in the late Oligocene that the first phase of Red Sea formation started with the break-up of the Arabian plate from Africa, accompanied by intense magmatic activity and continental rifting. Seafloor spreading started in the early Miocene.[6] Whereas it is generally accepted that in the last five million years seafloor spreading formed a mid-oceanic ridge in the southern Red Sea, there has been a lively debate about the nature of crust in the marginal regions of the Red Sea. Thick Miocene evaporites occur in the Red Sea depression, which makes the identification of a clear ocean–continent boundary difficult. Based on the good match of the African and Arabian coastlines, one hypothesis assumes seafloor spreading to have formed oceanic basement in the entire Red Sea in Miocene times.[6-8] Accordingly, most of the Red Sea basement is supposed to be of basaltic composition. This model, however, cannot explain the existence of continental crust in the Southern Red Sea and the Afar Triangle.[9] An alternative model postulates diffuse thinning and stretching of the continental crust and assumes the Red Sea basement to be composed of some kind of intermediate crust.[10-12] These models may explain the outcrops of Precambrian basement, such as at Zabarged Island, and basaltic intrusions found in Miocene evaporites.[13,14]

During the past four to five million years, a spreading rate of about 1 cm/yr formed a continuous oceanic rift with a 1,500-m to 2,000-m deep axial graben in the Southern Red Sea.[15] Toward the northwest the axial zone narrows, and the age of the oceanic crust is younger.[16] North of about 21°N no continuous rift is present, and isolated deeps alternate with smooth intertrough zones. Oceanic basement, basaltic intrusions, and magnetic anomalies suggest these deeps to be nucleation points for oceanic rifts. As rifting propagates, the more or less regularly spaced deeps could join to form a continuous axial rift valley.[17]

14.3 HYDROTHERMAL ACTIVITY

More than 100 sites of submarine hydrothermal activity have been identified at seafloor spreading centers.[18,19] The close relationship between the formation of young oceanic crust and hydrothermal activity is due to a subsurface magma chamber which, on the one hand, fuels the basalt flow at the rim of the divergent plate boundaries, and on the other hand, is the driving force for subsurface circulation of sea water.

When sea water enters the crust, its chemical composition gradually changes due to chemical reactions with the rocks. Alteration of the rocks caused by fixation of alkali elements, oxidation, (e.g., formation of iron oxides) and fixation of Mg^{++} are the main chemical reactions in this recharge

TABLE 14.1
History of Research Cruises to the Red Sea

Year	Ship	Investigations
1881–83	Vityaz (U.S.S.R.)	First measurements of temperature and salinity in Red Sea waters,
1897	Pol (Austria)	sediments
1898	Valdivia (GER)	
1948	Albatross (S)	Discovery of anomalously high temperature and salinity values
1959	Atlantis (U.S.A.)	around 21°N
1963	Atlantis II (U.S.A.)	
1963	Discovery (U.K.)	
1964	Discovery (U.K.)	
1964–65	Meteor (GER)	
1965	Atlantis II (U.S.A.)	Discovery of the deeps, dredging, coring of metalliferous muds
1965	Meteor (GER)	
1966	Chain (U.S.A.)	
1966	AK.S. Vavilov (U.S.S.R.)	
1967	Discovery (U.K.)	
1967	Oceanographer (U.K.)	
1969	Wando River (U.S.A.)	
1970–71	Nereus (U.K.)	
1971	Valdivia (GER)	
1969	Chain (U.S.A.)	
1972	Glomar Challenger (U.S.A.)	
1976	AK. Kurchatov (U.S.S.R.)	
1977–78	Sonne (GER)–COMMISSION I (SO–01)	Saudi–Sudanese Red Sea Commission program devoted to evaluation
1977–78	Sonne (GER)–MESEDA I (SOI–02)	of the economic interest of muds and ecological impact of their
1978	Melville (U.S.A.)	exploitation
1979	Valdivia (GER)–MESEDA II (Va 22)	
1979	Sedco 445–Pre-Pilot Mining Test	
1980–81	Valdivia (GER)–MESEDA III (Va 29)	
1981	Marion-Dufresne (France)–COMMAR I	
1979	Jean-Charcot (France)–MEROU	Bathymetric survey (Seabeam)
1980	AK. Kurchatov (U.S.S.R.)	Study of 18°N axial area and diving in the Atlantis-II-Deep area
	Pr. Chotkman	
	BRS Agnarrante	
1980–83	Sonne–MENOR I & II (GER)	Metal exploration of the northern Red Sea
1983	Marion-Dufresne (France)	Atlantis-II-Deep, Nereus-Deep
1984	Sonne (GER) SO 29	Deeps in the northern Red Sea
1985	Marion–Dufresne (France)	Atlantis-II-Deep
1987	Meteor (GER)	Hydrography Atlantis-II-Deep
1992	Marion–Dufresne (France)	Atlantis-II-Deep
1995	Meteor (GER)	Deeps in the northern Red Sea
1997	Sonne (GER) SO 121	Deeps in the northern Red Sea

Source: Data compiled from Thisse, Y., Guennoc, P., Poult, G., and Nawab, Z., The Red Sea: a natural geodynamic and metallogenic laboratory, *Episodes*, 3, 3, 1983; Bäcker, H., Metalliferous sediments of hydrothermal origin from the Red Sea, in *Marine Mineral Deposits*, Halbach, P. and Winter, P. (Eds.), Verlag Glückauf, Essen, 1982, 102; and unpublished data.

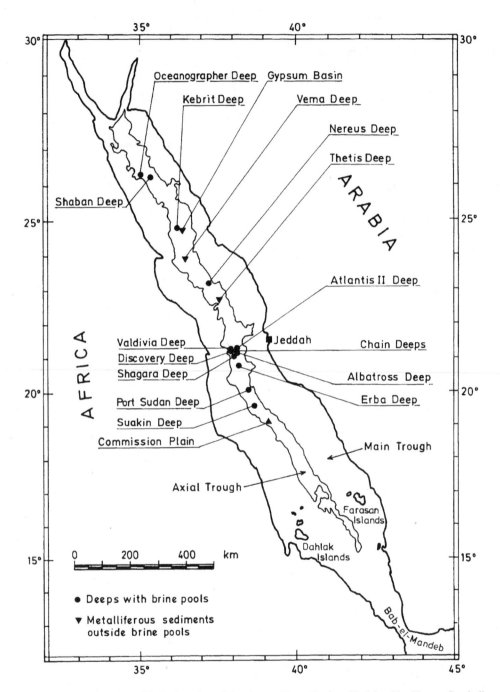

FIGURE 14.1 The Red Sea with the location of the deeps. (From Bäcker, H., Marchig, V., von Stackelberg, U., Stoffers, P., Puteanus, D., and Tufar, W., Hydrothermale Aktivität auf dem Meeresboden, *Geol. Jb.*, D93, 103, 1991. Reproduced with permission of the Geologische Jahrbuch, Hannover.)

zone. As the sea water penetrates downward, the temperature increases, and oxygen content and pH drop. The solution is now capable of mobilizing certain chemical elements from the host rocks, e.g., metals like copper, zinc, iron, and manganese (reaction zone). Estimates of temperature and pressure in the reaction zone range from 340 to 465°C and 350 to 550 bars, which are conditions near the critical point of sea water.[20] Near this point, physical properties of the sea water drastically change, causing an upflow of hydrothermal solutions. This upflow may be "focused," and active vents at the seafloor are an example of such a channeled flow. Alternatively, the upflow may be "diffuse," the hydrothermal fluids not reaching the seafloor or diffusing through it over a wide area as on the flanks of the mid-ocean ridges. Subsurface mixing of these hydrothermal solutions with sea water may result in stockwork-like mineralization.[20]

When hydrothermal fluids mix with cold, oxygen-rich sea water at the seafloor, the mineral load of the fluids rapidly precipitates, resulting in the formation of massive sulfides, black smokers, and/or hydrothermal sediments. The type of deposit formed on the seafloor depends on various factors: chemical composition of rocks which interact with the hydrothermal fluids, fluid/rock ratios, temperature of fluids, water depth, and tectonic setting of the plate boundary.

Throughout much of the 1960s and early 1970s the Red Sea hydrothermal system was thought to be unique. It was not until more recently that it came to be realized that it is a special case of a mid-ocean ridge hydrothermal system.[4] Unlike at mid-oceanic ridges, the special conditions in the Red Sea facilitate the formation of unusual hydrothermal deposits. First, the formation of oceanic crust in the Red Sea occurs in small deeps; due to their structure these basins act as a trap for the discharging hydrothermal fluids and prevent their distribution over large areas. Thus, when hydrothermal precipitates are formed, they are concentrated within relatively small areas. The second factor facilitating the formation of hydrothermal deposits is highly saline brines which fill about 25 deeps in the Red Sea.[21] The environment in the brine pool, especially the lack of oxygen, favors the formation and preservation of hydrothermal deposits.[22]

The formation of the brines is related to evaporites which were deposited in the entire Red Sea during the Miocene. Leaching of Miocene evaporites, which sometimes outcrop on the flanks of the deeps, is one of the brine-forming processes, especially in those deeps, e.g., Valdivia Deep, which are less affected by hydrothermal activity.[23,24] In the case of hydrothermally driven circulation, e.g., in the Atlantis-II-Deep, three major reservoirs strongly influenced the brine composition: sea water, marine sediments (including evaporites and black shales), and oceanic basalt.[24-28]

14.4 SEDIMENT FACIES

About 25 deeps in the Red Sea have been explored and investigated for hydrothermal activity and related deposits. As a result, a variety of hydrothermal sediments have been found. On the basis of their lithology, mineralogy, and geochemistry, Bignell et al.[4] and Bäcker and Richter[2] identified several hydrothermal sediment facies which were further subdivided into subfacies. These facies are frequently intermixed due to several ore-forming processes operating at the same time within one area.[4]

The chemical composition of the hydrothermal facies types varies between the different Red Sea deeps. This is due to regional differences in the extent of hydrothermal activity, the influence of brines in ore-forming processes, and the amount of sedimentation of normal Red Sea sediments. A comparison of the average content of Fe/Mn (for oxide mineralizations), Ca/Al (representing normal Red Sea sediments), and Cu/Zn (for sulfide mineralization) in sediments indicates (Figure 14.2) that apart from the Atlantis-II-Deep, the sediments from the Thetis, Nereus, Gypsum, and Vema Deeps contain significant amounts of metalliferous mineralization.

14.4.1 NORMAL RED SEA SEDIMENTS

Normal Red Sea sediments form the main sediment facies not influenced by hydrothermal processes. They are of light-brown to grey-brown color and are predominantly composed of biogenic and inorganic carbonates and terrigenous material. Calcite, Mg–calcite, quartz, feldspars, and clays are the main mineralogical constituents. Foraminifera, pteropods, and nannofossils are the prominent biogenic material. The composition of normal Red Sea sediments changed during glacial times in that the concentrations of magnesian calcite, dolomite, and inorganically precipitated aragonite

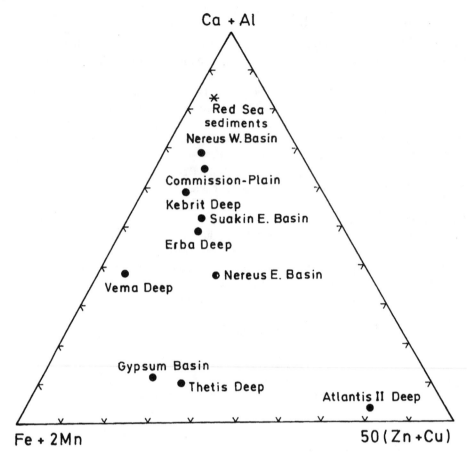

FIGURE 14.2 Relationships between Ca+Al, Fe+2Mn and 50(Zn+Cu) in sediments from various deeps of the Red Sea. Ca+Al is the end member of biogenic-detrital sediments; Fe+2Mn represents iron–manganese deposits; and 50(Zn+Cu) is the end member of sulfide deposits. Apart from the Atlantis-II-Deep, the most concentrated hydrothermal sediment deposits in the Red Sea occur in the Thetis, Gypsum, Nereus, and Vema deeps. (From Bäcker, H., Metalliferous sediments of hydrothermal origin from the Red Sea, in *Marine Mineral Deposits,* Halbach, P. and Winter, P. (Eds.), Verlag Glückauf, Essen, 1982, 102. Reproduced with permission of the Verlag Glückauf, Essen.)

increased.[29,30] Furthermore, the abundance of planktonic foraminifera decreased, resulting in a so-called aplanktonic zone. These changes in the sediment composition were attributed to glacial–inter-glacial contrasts of salinity in the Red Sea due to the lowering of sea level. This caused a separation of the Red Sea from the Gulf of Aden and subsequently increased evaporation and changes of the ventilation of Red Sea deep waters.[31] Geochemically, the normal Red Sea sediments are charac-terized by low Fe, Mn, Zn, and Cu content and are high in CaO and SiO$_2$ (Table 14.2).

14.4.2 OXIDE FACIES

The oxide facies is subdivided into the goethite, lepidocrocite, hematite, magnetite, and manganite subfacies.

The most common iron oxide facies in the Red Sea is the orange-yellow goethite facies. It occurs as pure monomineralic layers, or it is intermixed with other iron facies. The goethite facies is enriched in Fe and Zn and also in Cu in the Atlantis-II-Deep (Table 14.3). Bischoff[32] related the goethite formation to dehydration of amorphous iron oxides. Chukrov[33] suggested a reaction of

TABLE 14.2
Chemical Composition of the Normal Red Sea Sediment Facies

%	Normal Red Sea Facies			
	Average[1]	Atlantis-II-Deep[2]	Thetis Deep[3]	Kebrit Deep[4]
SiO_2	n.m.	24.40	n.m.	—
Al_2O_3	5.20	1.70	n.m.	—
CaO	27.00	23.60	37.90	35.00
MgO	3.60	n.m.	1.70	n.m.
Fe	4.00	4.50	1.50	3.50
Mn	0.35	0.40	0.70	0.12
Zn	0.03	0.06	0.04	0.02
Cu	0.007	0.007	0.01	0.004

[1] Bäcker[70] normal Red Sea sediments north of 19.5° N
[2] Bischoff[32] average of 43 samples
[3] Scholten et al.[60] average of 2 samples
[4] Bignell et al.[4]
n.m. = not measured.

TABLE 14.3
Chemical Composition of the Red Sea Goethite Facies

%	Goethite Facies							
	Atlantis-II-Deep[1]	Atlantis-II-Deep[2]	Atlantis-II-Deep[3]	Atlantis-II-Deep[4]	Thetis Deep[5]	Kebrit Deep[6]	Nereus Deep[6]	Gypsum Deep[7]
SiO_2	n.m.	n.m.	8.70	n.m.	2.57	n.m.	n.m.	n.m.
Al_2O_3	n.m.	n.m.	1.10	1.93	0.09	2.76	0.92	0.84
CaO	0.90	2.30	3.40	0.42	0.41	20.90	0.49	2.20
MgO	0.60	0.90	n.m.	n.m.	1.31	4.64	0.50	1.65
Fe	51.30	50.10	44.90	63.20	50.20	30.80	54.60	50.70
Mn	1.10	0.60	0.80	0.26	0.51	0.17	0.45	0.35
Zn	0.27	0.27	0.56	0.15	0.17	0.10	0.66	0.24
Cu	0.19	0.05	0.24	0.01	0.02	0.03	0.006	0.01

[1] Bäcker[70] average of 100 samples from the SW basin of the Atlantis-II-Deep
[2] Bäcker[70] average of 22 samples
[3] Bischoff[32] average of 43 samples
[4] Bignell et al.[4]
[5] Scholten et al.[60] average of 10 samples
[6] Bignell et al.[4]
[7] Bignell[71].
n.m. = not measured.

Fe^{++}-solutions with ferrihydrite (X-ray amorphous iron oxide, $5Fe_2O_3*9H_2O$). Detailed laboratory studies indicated that maximum amounts of goethite with traces of hematite form via aging of ferrihydrite at pH 4.[34] A comparable paragenesis, i.e., goethite with trace amounts of hematite and ferrihydrite, was found by means of Mössbauer analyses in the goethite facies from the Thetis Deep.[35]

The lepidocrocite facies is a further iron oxide facies which can be observed as orange-brown, sometimes monomineralic layers in Red Sea sediments. Apart from Fe, this facies is enriched in

TABLE 14.4

Chemical Composition of the Lepidocrocite (a) and Hematite Facies (b)

	A. Lepidocrocite Facies				B. Hematite Facies			
%	Atlantis-II-Deep[1]	Thetis Deep[2]	Nereus Deep[3]	Gypsum Deep[3]	Atlantis-II-Deep[4]	Atlantis-II-Deep[1]	Thetis Deep[5]	Nereus Deep[6]
SiO_2	n.m.	0.73	n.m.	n.m.	n.m.	n.m.	7.91	8.10
Al_2O_3	0.67	0.26	1.20	0.66	n.m.	1.03	2.66	1.00
CaO	2.00	0.55	14.20	1.63	1.00	8.22	15.50	3.00
MgO	n.m.	1.18	n.m.	n.m.	1.10	n.m.	4.59	0.80
Fe	63.16	48.97	35.00	54.35	49.10	35.29	19.90	65.40
Mn	0.26	0.45	2.00	0.19	1.10	0.29	6.89	0.69
Zn	0.15	0.15	0.10	0.18	0.16	0.12	0.35	0.24
Cu	0.01	0.15	0.09	0.03	0.27	0.74	0.08	0.16

[1] Bignell et al.[4]

[2] Scholten et al.[60] average of 12 samples

[3] Bignell[71]

[4] Bäcker[70] average of 5 samples

[5] Scholten et al.[60] average of 7 samples

[6] Jedwab et al.[72]

n.m. = not measured.

Cu and Zn (Table 14.4). Lepidocrocite is believed to be formed by rapid oxidation of Fe^{++}-solutions forming a Fe^{2+}/Fe^{3+} complex, the so-called green rust.[36] Further oxidation in neutral to alkaline pH causes the formation of lepidocrocite.

The hematite facies is characterized by red-colored layers in the Atlantis-II-Deep, probably in the vicinity of the incoming brines, and in the Thetis and Nereus Deeps.[4] It is also a common component in other oxide facies types. Bischoff[32] ascribes the formation of hematite to a dehydration of goethite. Detailed laboratory studies showed, however, that aging of ferrihydrite at a pH of about 7 to 8 is the major pathway for hematite formation.[34]

In the Atlantis-II-Deep, magnetite replaces hematite pseudomorphically with increasing sediment depths.[37] The magnetite facies occurs as almost pure black layers in the Thetis and Nereus Deeps. They are characterized by high concentrations of Fe, Zn, and Cu (Table 14.5). Slow oxidation of Fe^{2+}-solutions and green rust is believed to be the major formation pathway.[36]

Brown manganite facies is present in the Atlantis-II, Thetis, Nereus, Chain, and Shagara deeps. It is characterized by high Mn and Zn contents and, depending on the amounts of intermixed normal Red Sea sediments, variable CaO content (Table 14.6). Bischoff[32] describes well-crystallized and coarse grained manganite, but most common in the Red Sea deeps are poorly crystallized manganese phases (groutite, woodruffite). The manganite facies is believed to be formed by oxidation of dissolved Mn^{2+}. Because of reducing conditions, manganite is not stable in the Atlantis-II-Deep brine and precipitates at the rim of the basin.

14.4.3 SILICATE FACIES

A variety of silicates have been described from the Atlantis-II-Deep.[38-40] They are found as X-ray amorphous silicates in the upper sediment layers of the Deep or as olive-green layers rich in nontronite and sulfides. The formation of the nontronite is thought to occur by sorption of silica supplied by the incoming hydrothermal solutions on Fe-oxyhydroxide. High silica and some trace metal contents characterize this facies (Table 14.5).

TABLE 14.5
Chemical Composition of the Magnetite (a) and Silicate Facies (b)

%	A. Magnetite Facies			B. Silicate Facies		
	Thetis Deep[1]	Nereus-Deep[2]	Atlantis-II-Deep[3]	Atlantis-II-Deep[4]	Atlantis-II-Deep[5]	Thetis Deep[6]
SiO_2	2.60	3.50	23.70	n.m.	n.m.	n.m.
Al_2O_3	1.21	0.20	1.50	n.m.	0.10	4.00
CaO	1.20	8.00	6.00	4.00	2.50	1.96
MgO	2.00	1.00	1.00	1.50	n.m.	n.m.
Fe	54.50	56.90	28.70	27.20	3.57	39.50
Mn	0.23	0.39	1.20	0.70	0.04	0.72
Zn	0.14	1.52	1.93	2.52	0.004	0.14
Cu	0.41	1.51	0.51	0.47	0.02	0.11

[1] Scholten et al.[60] average of 6 samples

[2] Jedwab et al.[72]

[3] Bäcker[70] average of 20 samples

[4] Bäcker[70] average of 43 samples from the green silicate facies

[5] Bignell et al.[4] green silicate

[6] Bignell et al.[4]

n.m. = not measured.

TABLE 14.6
Chemical Composition of the Manganite Facies

%	Manganite Facies						
	Atlantis-II-Deep[1]	Atlantis-II-Deep[2]	Atlantis-II-Deep[2]	Thetis Deep[4]	Nereus Deep[5]	Chain Deep[6]	Shagara Deep[6]
SiO_2	n.m.	5.70	7.50	4.81	n.m.	n.m.	n.m.
Al_2O_3	n.m	4.90	0.70	2.83	0.63	1.07	2.56
CaO	3.60	11.80	2.90	11.10	5.78	4.77	21.80
MgO	1.90	1.06	n.m.	3.66	n.m.	n.m.	n.m.
Fe	15.30	19.60	21.30	13.00	30.80	12.90	3.89
Mn	32.70	29.40	27.50	14.10	17.90	35.80	15.60
Zn	0.64	0.19	1.12	0.58	1.33	1.22	0.21
Cu	0.09	0.01	0.08	0.08	0.04	0.13	0.01

[1] Bäcker[70] average of 34 samples

[2] Weber-Diefenbach[73] average of 3 samples

[3] Bischoff[32] average of 2 samples

[4] Scholten et al.[60] average of 27 samples

[5] Bignell et al.[4] average of two cores

[6] Bignell[71]

n.m. = not measured.

TABLE 14.7
Chemical Composition of the Sulfide Facies

	Sulfide Facies					
%	Atlantis-II-Deep[1]	Atlantis-II-Deep[2]	Atlantis-II-Deep[3]	Atlantis-II-Deep[4]	Erba Deep[5]	Gypsum Deep[5]
SiO_2	n.m.	24.70	n.m.	21.40	n.m.	n.m.
Al_2O_3	n.m.	1.50	2.27	4.90	10.60	29.10
CaO	4.20	2.50	9.37	3.60	7.15	4.07
MgO	1.69	n.m.	n.m.	1.30	n.m.	n.m.
Fe	23.80	17.0	18.00	25.50	26.60	29.10
Mn	1.30	0.80	1.96	1.94	0.51	0.09
Zn	5.93	9.76	8.12	4.50	0.34	0.05
Cu	1.26	3.60	1.78	1.06	0.06	0.67
Cd	0.02	n.m.	0.03	n.m.	0.001	0.004
S	8.80	16.80	n.m.	12.90	n.m.	n.m.

[1] Bäcker[70] average of 64 samples from various sulfide layers
[2] Bischoff[32] average of 42 samples
[3] Bignell et al.[4] average of two sulfide layers
[4] Weber-Diefenbach[73] average of 16 samples
[5] Bignell et al.[4]
n.m. = not measured.

14.4.4 SULFIDE AND SULFATE FACIES

With respect to the metal content and the economic value of hydrothermal deposits, the sulfide facies represents the most interesting deposit in the Red Sea deeps. Metal contents up to 9.76% Zn and 3.60% Cu have been reported from the sulfide facies from the Atlantis-II-Deep (Table 14.7). Here two distinct subfacies, the monosulfide and the pyrite facies, can be found. The violet-gray monosulfide facies is mainly composed of sphalerite, pyrite, chalcopyrite, and manganosiderite, whereas the black pyrite facies is predominantly composed of pyrite and manganosiderite, with only traces of sphalerite. Traces of sulfides were also reported from the Gypsum and Erba deeps. In the Thetis Deep, traces of sulfides are found in the magnetite facies.

Sulfates such as anhydrite and barite occur in the Atlantis-II and Thetis Deeps; gypsum is found in the Kebrit and Gypsum Deeps. Anhydrite forms up to 1-m-thick layers in the Atlantis-II-Deep and is believed to precipitate from the incoming hydrothermal solutions.[2]

14.5 ATLANTIS-II-DEEP

The Atlantis-II-Deep is an elongated basin with a maximum water depth of about 2,200m. It can be subdivided into four basins which are separated by bathymetric highs: the N-basin, E-basin, W-basin and the SW-basin. Toward the south, a sill at about 2,030 m water depth parts the SW-basin from the neighboring Chain Deep. The Atlantis-II-Deep is filled by a layered brine pool (Figure 14.3). The several layers are characterized by distinct temperatures and salinities, which are nearly constant within the layers due to thermo-haline convection.[41] The lower convective layer (LCL, depth of surface at about 2,046 m) covers an area of about 57 km² with a volume of about 3.94 km³.[42] Investigations of the brine between 1965 and 1977 revealed above the LCL one upper convective layer (UCL) which results from diffusive mixing of two end members, the LCL and normal sea water.[43] The UCL is followed by a transition zone where the temperature gradually decreases to that of Red Sea deep water. However, according to the observations of Anschutz and Blanc[44] in 1992, the UCL had split into three sections (UCL 1, UCL 2, UCL 3), but investigations

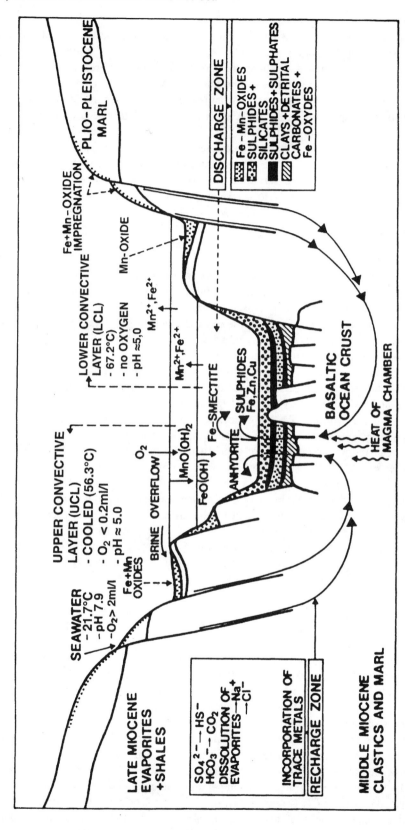

FIGURE 14.3 Principles of hydrothermal convection and formation of hydrothermal deposits in the Atlantis-II-Deep. Sea water percolates through Miocene sedimentary deposits and fissures in basaltic basement; heating of the fluids due to proximity of a magma chamber; fluid interaction with sediments and basalts in the recharge zone changes the physical and chemical characteristics of the fluids. When the hydrothermal solutions discharge into the basin, metalliferous sediments precipitate. (From Bäcker, H., Rezente hydrothermal-sedimentäre Lagerstättenbildung, *Erzmetall*, 26, 544, 1973. Reproduced with permission of Erzmetall, Clausthal-Zellerfeld.)

in 1995 and 1997 showed that the UCL 3 had changed to a transition zone with a temperature gradient to that of normal Red Sea bottom water.[21,45]

Although the brine pool spreads throughout the Atlantis-II-Deep, differences in the brine temperature between the N- and the SW-basin can be observed. Whereas the UCL shows similar temperature and salinity distribution in the entire Atlantis-II-Deep, the LCL is different in each basin. In the SW-basin the temperature of the LCL is constant with depth, but in the N-basin the temperature decreases with depth from 66.0°C to 60°C.[45] This suggests that the connection of the LCL to the N-basin is restricted.[21] The N-basin is separated from the other basins of the Atlantis-II-Deep by a sill, which has approximately the same depth as that of the LCL surface. Therefore, the LCL present in the north basin may represent an older brine having a lower temperature.

The brine pool has low pH and is, relative to sea water, enriched in Na, K, Ca, Cl⁻, and depleted in Mg, SO_4^{2-}, I, F, and NO_3^-.[46-48] High amounts of Fe, Mn, Zn, Pb, Si, and Zn are found in solution. The absence of H_2S indicates that these metals are in excess of H_2S and that the input of sulphur is the limiting process controlling sulfide precipitation rather than the metal content in the discharging fluids.

The general precipitation processes in the brine pools are shown in Figure 14.3. In the LCL, low Eh and the lack of oxygen favor the formation of sulfides. Calcium, Mn, Fe, and some Cu and Zn diffuse from the LCL to the UCL, and anhydrite is formed at the LCL/UCL boundary due to diffusion of SO^{-4} into the LCL.[43] Iron and manganese oxyhydroxides form in the UCL and deposit on the flanks of the deep or redissolve when sinking back to the LCL layer.

Since the Atlantis-II-Deep has been studied over a long period of time, it is one of the very few cases in marine geoscience where long-term observation of a hydrothermal system has been possible and its time-dependent variability investigated. Hartmann[49] compared the chemical compositions of the LCL and UCL between 1966 and 1977 and observed only minor changes in the concentrations of Ca, Cl⁻, Fe, Mn, and Zn, but concentrations of Cu decreased by about a factor of 1,000 in a period of 11 years (Table 14.8). He assumed that this decrease was due to an increase in supply of sulphur, causing the precipitation of copper sulfides, which have very low solubility in comparison to other sulfides. There seems to have been a slight increase in Mn and Fe in the LCL as well as Mn in the UCL in the past 18 years, 1977–1995, which suggests an increased supply of these elements from the hydrothermal vents. Comparison of chemical data over the entire period of sampling, about 29 years, is complicated by the differences in the analytical methods used. For instance, the first measurements by Brooks et al.[50] were performed using an organic extraction method, whereas modern analytical facilities (e.g., high resolution ICP–MS) allow a direct measurement of the elements in the solution. Therefore, some of the chemical trends described above may be partly an artifact of analytical techniques.

The most striking evidence for temporal changes in the Atlantis-II-Deep brine system comes from temperature measurements which show in the LCL an increase from 55.9°C in 1965 to 67.2°C in 1997 (Figure 14.4). In about the same period, the UCL1 temperature changed from 44.3 to 56.3.[45,51] This continuous increase in temperatures suggests a steady increase of hydrothermal activity.[21] Almost all heat and salt which has been supplied to the brine since 1965 has been confined to the Deep and has not been dispersed into the overlying sea water. The average flow rate of the hydrothermal solution has been 670 to 1000 kg/s, estimations based on a salt and heat balance of the brine.[42]

Although many investigations have been focused on the hydrothermal activity in the Atlantis-II-Deep, active vents have not yet been found there and, therefore, the end member composition of the fluids is still not known. Suggestions that venting takes place in the SW-basin are based on the isotopic composition of the brine, the brine temperature, and the geochemistry of surface sediments (Figure 14.5). The steeper gradient of temperature increase of the UCL between 1976 and 1984 in comparison to the LCL (Figure 14.4) and isotopic data point to an outlet of the hydrothermal solutions on the flanks of the SW-basin above the LCL.[26] Detailed temperature measurements of the brine in the SW-basin during the *Sonne* cruise SO 121 in 1997 showed,

TABLE 14.8
Metal Content of Brines in the Atlantis-II-Deep in mg/kg

	1966[a]	1971[b]	1976[c]	1977[d]	1995[e]
			UCL Brine		
Fe	1–3	6–8	6.5 ± 2	7.3 ± 0.2	2 ± 2
Mn	79 ± 2	78 ± 1	71.5 ± 1	75 ± 2	101 ± 8
Cu	0.1 ± 0.6	0.03–0.04	0.022	0.0005–0.001	—
Zn	2–8	1.1–2	1.6	1.61 ± 0.05	2.7
			LCL Brine		
Fe	90 ± 8	86 ± 5	81 ± 4	75–81	88 ± 2
Mn	87 ± 5	82 ± 6	82 ± 4	81 ± 2	100 ± 1
Cu	0.15–0.6	0.03–0.12	0.021	0.0005	n.m.
Zn	7 ± 2	2–5	3.0	3.0 ± 1	4.2

[a] Brooks et al.[50]
[b] Hartmann[49,74]
[c] Danielsson et al.[75]
[d] Hartmann[49]
[e] own unpublished data (0.4 μm filtered)
n. m. = not measured.

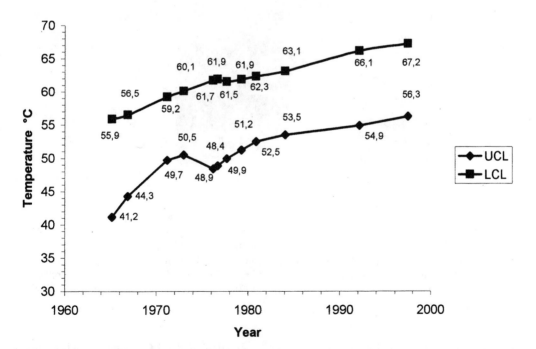

FIGURE 14.4 Temperature increase of LCL (lower convective layer) and UCL (upper convective layer) between 1965 and 1997. (Compiled from Post, H., Changes in the Red Sea hydrothermal activities between 1964–1984, unpublished manuscript, p. 5, 1985; Hartmann, M., Scholten, J. C., and Stoffers, P., Hydrographic structure of the brine-filled deeps in the Red Sea — correction of the Atlantis Deep temperatures, *Mar. Geology*, 144, 331, 1998.)

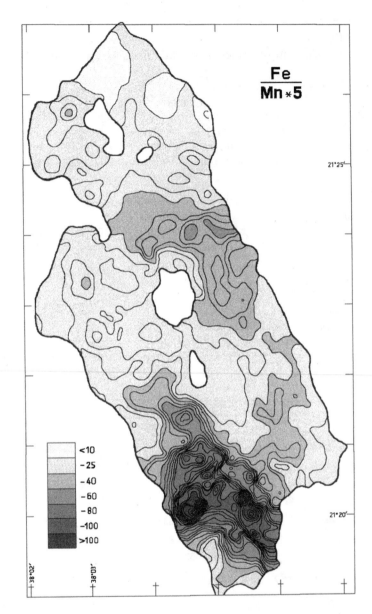

FIGURE 14.5 Ratio of Fe to Mn in sediments of the Atlantis-II-Deep based on about 600 sediment cores from which the average Fe/Mn covering all facies types were taken. Whereas Fe precipitates near the hydrothermal vents as Fe-silicates and sulfides, Mn diffuses to areas more distal from the vent sites. The high Fe/Mn × 5 ratios indicate that the hydrothermal vents discharge in the SW-Basin of the Atlantis-II-Deep. (Compiled from Bäcker, H., Marchig, V., von Stackelberg, U., Stoffers, P., Puteanus, D., and Tufar, W., Hydrothermale Aktivität auf dem Meeresboden, *Geol. Jb.*, D93, 103, 1991. Reproduced with permission of Geologische Jahrbuch, Hannover.)

however, a clear temperature maximum at the upper edge of the LCL brine, which points to hydrothermal fluids which upwell from the bottom of the basin and spread laterally at the top of the LCL.[45] Anhydrite veins within the sediments of the SW-basin are further indications of fluids discharging from the bottom there.

14.5.1 Metalliferous Sediments in the Atlantis-II-Deep

More than 600 cores have been taken from the Atlantis-II-Deep, and this is why this area is one of the best investigated marine hydrothermal deposits. Bäcker and Richter[2] established a general lithostratigraphy (Figure 14.6) and defined five lithostratigraphic units; four units in the SW-basin-, modified on the basis of recent mineralogical data.[26,44]

The oldest sediments (25,000 to 28,000 years) in the Atlantis-II-Deep which overlay basalt are those from the detrital–oxide–pyrite zone (DOP). They consist of biogenic–detrital marl with several limonitic and manganitic layers, which indicate the first occurrence of hydrothermal activity in the Atlantis-II-Deep. Based on the thickness of limonitic mineralization, Bäcker and Richter[2] suggested that the main hydrothermal activity was situated in the northern part of the basin during the formation of the DOP facies.

About 15,000 years ago, the formation of the lower sulfidic zone (SU1) indicates the establishment of a stable brine pool.[2] This finding is based on the mineralogical composition of the SU1 zone, which is composed of dark-red brown, Fe-rich clays with layers of sulfides, mainly pyrite, but also chalcopyrite and sphalerite. Red-colored silicate layers with iron oxides mark the change from the SU1 zone to the central oxidic zone (CO), which is composed of limonite, hematite, and manganite facies. The variable layering of these facies and slump structures indicate a less stable brine pool. Anhydrite at the top of the CO zone characterizes the gradual change to the upper sulfidic zone (SU2). This zone is similar in most characteristics to SU1, but the silicates of this zone are, in contrast to the red-brown silicates from the SU1 zone, green-colored. The youngest lithostratigraphic unit, which is presently still being formed, is the amorphous-silicate zone (AM). This is a very soft (about 94% brine content) homogeneous marl containing X-ray amorphous Fe-oxides and silicates, and only in the lower part of this unit can goethite layers be observed.

In the SW-basin, slightly different lithostratigraphic units can be observed, most probably due to the vent area being situated in this part of the Atlantis-II-Deep since the formation of the CO zone. The CO zone in the SW-basin overlays basalt and consists of coarse-grained hematite and goethite facies, sometimes in monomineralic layers. This zone is followed by a 2- to 5-m-thick sulfidic–oxidic–anhydritic zone (SOAN), which is characterized by disturbed layering and alternate sequences of metal-rich sulfides, hematite, anhydrite, and silicate facies. The color of this zone varies according to the facies between red-brown and olive-green. The oxidic–anhydritic zone (OAN) on top of the SOAN is characterized by breccias and turbiditic layering and consists of hematite, anhydrite, and silicates with fragments of sulfides and silicates. Veins filled with anhydrite can also be observed. The uppermost unit in the SW-basin consists of an approximately 4.5-m-thick sulfidic–amorphous–silicate-zone (SAM). It is characterized by a sulfide layer at the base having a high content of sphalerite. The facies generally has a high brine content (>90%), and sulfides, silicates, and anhydrite are the main mineralogical phases that have been identified in it.

Several authors have tried to specify the temperatures and other environmental conditions of mineral precipitation by studying the mineral parageneses of the sediments. Missak et al.[52] described two parageneses in their sediment core from the SW-basin: (a) intermediate solid solution (also described as chalcopyrrhotite) with sphalerite encrustations and intergrowths, and (b) intermediate solid solution free of sphalerite. They suggested that differences in the sulfide mineralogy are due to changes in the composition of ore-forming fluids and a change in the sulphur fugacity. The presence of exsolved chalcopyrite lamellae in intermediates solid solution indicate slow cooling of the fluids with a temperature <450°C. Veins in sediments filled with mineral assemblages of pyrrhotite, cubic cubanite, high iron sphalerite, and anhydrite suggest a disequilibrium between H_2S and SO_4^{-2} in the hydrothermal fluids.[53] This may be caused by a mixing of two fluids, one having a relatively low temperature (<250°C and $SO_4^{-2} < H_2S$), the other a hot fluid rich in H_2S. A precipitation temperature of between 200° and 250°C is indicated by the paragenesis of cubic cubanite, chalcopyrite, and monoclinic pyrrhotite. Based on fluid inclusion thermometry of anhydrite crystals, exit temperatures of between 240°C and 400°C have been estimated for the venting fluids.[53-55]

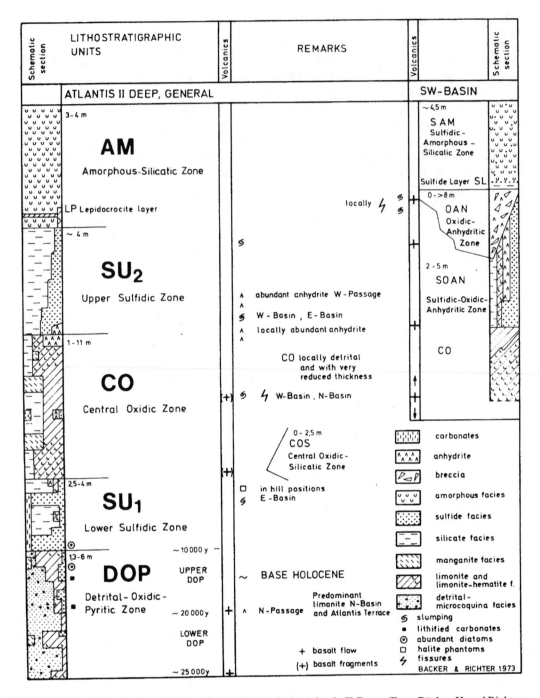

FIGURE 14.6 Lithostratigraphic units of the sediments in the Atlantis-II-Deep. (From Bäcker, H. and Richter, H., Die rezente hydrothermal-sedimentäre Lagerstätte Atlantis II-Tief im Roten Meer, *Geol. Rundschau*, 62, 697, 1973. Reproduced with permission of the authors.)

The composition of the sediments in the Atlantis-II-Deep is variable and differs even when sediment cores were retrieved at very close proximities to each other. Therefore, estimates of fluid composition and temperatures based on mineral paragenesis depend very much on the sediments investigated. A summary of published temperature estimates of the hydrothermal fluids is shown in Table 14.9.

TABLE 14.9
Summary of Temperature Estimates of Hydrothermal Deposits from the Atlantis-II-Deep

Temperature	Technique	Reference
110°	Measured volume and temperature changes of the lower brine from mixing with the hydrothermal fluid	a,b
>210°	Measured volume and temperature changes of the lower brine from mixing with the hydrothermal fluids	d
>108°	Silica solubility	e
>158°	Na, K, Ca, geothermometer	e
250°	Oxygen isotope geothermometry, SO_4^{-2}, H_2O pair	c
261°	Oxygen isotope geothermometry, SO_4^{-2}, H_2O pair	e
>210°	Minimum thermal stability of cubic cubanite	f
210-251°	Cubic cubanite+chalcophyrite+monocline pyrrhotite assemblage	f
>334°	Cubic cubanite+chalcophyrite+pyrite assemblage	f
238°	Homogenization temperature of vein anhydrite	f
<450°	Exsolved chalcopyrite lamellae in intermediate solid solution	g
390–403°	Homogenization temperature of anhydrite from sediments from the SW-basin	h
432–353°	Heat-mass balance of the Atlantis-II-Deep brine	i
195–310°	Heat and salt mass balance	j

[a] Brewer et al.[76]
[b] Ross[77]
[c] Sakai et al.[78]
[d] Schoell[79]
[e] Truesdell[80]
[f] Pottorf and Barnes[53]
[g] Missak et al.[52]
[h] Ramboz et al.[54]
[i] Ramboz and Danis[81]
[j] Anschutz and Blanc[42]

The great variability in the composition of the sediments, as already mentioned above, causes the chemistry of the facies types to vary from core to core, which is why the various investigators cited in Tables 14.2 through 14.7 give different values for the chemical composition of Atlantis-II-Deep sediments. Because of the variability in the chemical composition of Atlantis-II-Deep sediments, their economic potential has been judged differently by different authors. Hackett and Bischoff[56] estimated the mass of Zn to be 32.2×10^5 tons with 8.1×10^5 tons of copper. Based on chemical analyses of 628 cores having a total length of about 4 km, Guney et al.[57] estimated that the Deep contains the following: zinc 18.9×10^5 tons, copper 4.25×10^5 tons, silver 3,750 tons, gold 47 tons, and cobalt 5,369 tons. Because of the high economic value of the Atlantis-II-Deep deposits, a pre-pilot mining test using the drill ship *Sedco 445* was conducted in 1979, which was the first time that hydrothermal deep-sea deposits were recovered for economic purposes. Because the metal grades of the wet sediments are very low (0.5% for Zn and about 0.07% for Cu) preconcentration of the metals on board the ship was necessary.[58] About 15,000 m^3 of metalliferous mud was pumped from the Atlantis-II-Deep through a 2,200-m-long pipe string and subjected to separation by flotation on-board. The results of the mining test suggested that recovery of the hydrothermal deposits in the Atlantis-II-Deep is, in principle, possible and would take a period of about 16 years to complete.[57]

Although the economic potential of the Atlantis-II-Deep deposits seems to be enormous, the consequences of their extraction for the biological environment in the Red Sea have not yet been fully explored. Separation of the metal sulfides on-board the ship will result in flotation-produced

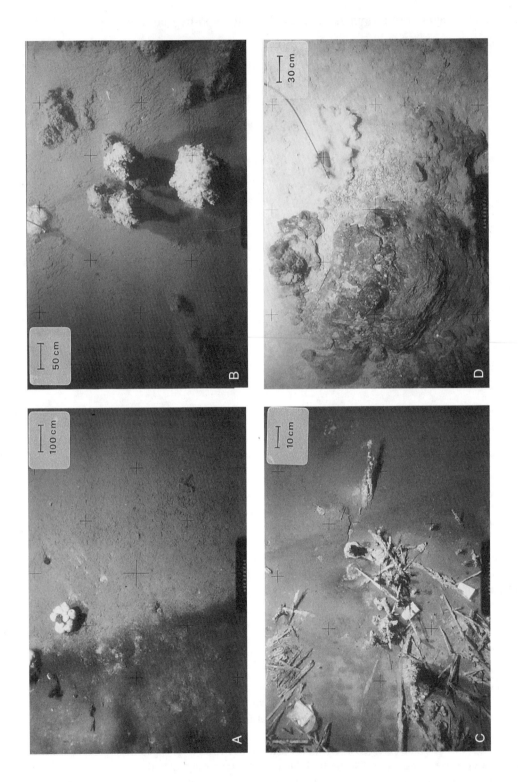

Figure 14.7 Caption on facing page.

FIGURE 14.7 (A) Transition zone between the brine pool and normal sea water in the Kebrit Deep. The light-gray colored zone is the brine, and the transition to dark-gray colors marks the boundary between the reducing conditions in the brine and a zone where Fe-oxides precipitate due to mixing with sea water. More distal from the brine a black-colored zone indicates precipitation of Mn-oxides. White organisms surround a sulfide chimney in the upper part of the plate. (B) Chimneys of massive sulfides within the brine pool of the Kebrit Deep. (C) Spindle-type chimneys, which stick out of the sediments on the flanks of the Kebrit Deep above the brine pool; anthropogenic pollution is indicated by the plastic cups. (D) Massive sediment-covered sulfides, which outcrop on the flanks of the Kebrit Deep. (E) Massive sulfide recovered during 1997 *Meteor* cruise on the flanks of the Kebrit Deep. The sulfides consist almost exclusively of pyrite; often they are impregnated with tar. (F) Chimney-like massive sulfide recovered during *Sonne* cruise 29 (1984). The sulfides are very fragile and porous.

tailings. These tailings contain iron oxides, silicates, and traces of sulfides. Release of the tailings in surface waters, or even at 800 m depth as suggested by Bäcker,[58] could have enormous ecological effects on the environment in the Red Sea unless mitigated.

14.6 HYDROTHERMAL SEDIMENTS OUTSIDE THE ATLANTIS-II-DEEP AREA

Apart from the Atlantis-II-Deep, significant hydrothermal sediments are found in the Thetis, Nereus, Vema, and Gypsum deeps.

The Thetis Deep is divided into several sub-basins. Hydrothermal sediments occur only in the NE basin, which is 10 km long and 3 km wide and up to 1,780 m deep. The sediments consist of various types of iron oxide facies (hematite, goethite, magnetite) sometimes forming monomineralic layers. The sulfide paragenesis suggests that, primarily, two distinct hydrothermal solutions, one rich in zinc (major elements Cu, Fe, Zn, S), the other depleted in it (major elements: Cu, Fe, S), were active.[59]

Detailed investigation by Scholten et al.[60] showed that the sediments in the Thetis Deep are comparable to the CO zone facies types in the Atlantis-II-Deep. In contrast to the Atlantis-II-Deep, however, no brine fills the Thetis Deep. Nevertheless, a strong fractionation between Fe and Mn in the sediments suggests that in the Thetis Deep there is intermittent brine filling of the Deep. Timing of the hydrothermal deposition in the Thetis Deep seems to correlate with similar events in the Atlantis-II-Deep, and it may be that major tectonic events in the Red Sea trigger hydrothermal activity.[60]

The Nereus Deep has an overall width of 12 km and is 40 km in length. It is divided along its length by a saddle into an east and a west basin, the latter one being filled by a brine pool. Sediments in the west basin consist mainly of normal Red Sea sediments, while the most concentrated hydrothermal sediments are found in a round depression in the southeastern part of the basin. The deposits consist of an up to 4-m-thick sequence of manganite, goethite, and hematite facies.[61]

The 24-km-long and 6-km-wide Vema Deep (1608 m water depth) does not contain a brine pool, but two cores recovered during a *Valdivia* cruise in 1972 recovered an up to 3-m-thick sequence of relatively pure goethite mud with high concentrations of Fe and Zn. A sharp boundary between the base of the hydrothermal sequence and the underlying normal Red Sea sediments marks the beginning of hydrothermal activity in this area.[4]

The Gypsum Deep was named after large idiomorphic gypsum crystals which were recovered from the sediments. Morphologically the deep is more like a depression with a maximum water depth of 1196 m. In the upper 2 m of sediments, two Fe-rich layers occur in which goethite, lepidocrocite, and siderite are the dominant minerals. Below the upper Fe-rich layer, a fine-grained pyrite mud rich in Ca, Fe, Cu, and Zn was observed.[4]

14.7   MASSIVE SULFIDES FROM THE KEBRIT DEEP

The Kebrit Deep (kebrit = Arabic sulphur) was first explored during a *Valdivia* cruise in 1971 and consists of a single basin of about 1 by 2.5 km in size. The basin has a maximum depth of 1549 m and is filled by an 84-m-thick oxygen-free brine. Over the past 23 years the temperature has increased slightly from 23.24°C to 23.34°C, but chlorinity (154‰ Cl⁻) has remained unchanged.[21] The strongly reducing brine has a sulphur content of 12 to 14 mg S/l. In the deeper brine section 100 µg/l Fe is present, whereas just below the sea water/brine interface Fe concentrations <1000 µg/l were reported.[21] These differences are probably due to dynamic precipitation and resolution processes between the O_2-rich sea water and the brine. Such processes were observed during *Sonne* cruise 121 (1997)[45] using TV. Figure 14.7A shows the contact of the brine/sea water interface with the sediments. Three zones can be distinguished: the light-gray colored zone is the brine, and the transition to more dark-gray colors marks the boundary between the reducing conditions in the brine and a zone where Fe-oxides precipitate due to mixing with sea water. More distal from the brine, a black-colored zone indicates precipitation of Mn-oxides.

During the *Valdivia* cruise in 1972, no indications of active hydrothermal venting were found, but dredging during research cruise *Sonne* 29 (1984) recovered massive sulfides. Detailed sampling and exploration during *Meteor* (1995)[62] and *Sonne* 121 (1997)[45] cruises suggests that most of the sulfides are situated in a narrow band near the brine/sea water interface. The sulfides are very porous, fragile, and most of them have a cauliflower structure. They have a chimney-like shape; also spindle-type chimneys up to 1 m height were observed (Figures 14.7 B, C, and F). The most common sulfides are massive aggregates which are covered by sediments (Figures 14.7 D and E). On steep slopes of the basin, outcrops of these sulfides can frequently be observed. Sometimes encrustations and impregnations of consolidated sediments are found. Some of the sulfides recovered are impregnated with tar and asphalt.

14.7.1   MINERALOGY

According to Blum and Puchelt[63] and Missak,[59] the massive sulfides from the Kebrit Deep can be separated into two different groups, here called Type I and Type II deposits.

14.7.1.1 Type I

Type I consists of pyrite, marcasite, bravoite, sphalerite, digenite, chalcocite, jarosite, and minor amounts of galena. Pyrite is the most common sulfide mineral and is found either as cellular aggregates or in globular masses. It often alternates with various concentric layers of bravoite, marcasite, and sphalerite. This sequence of rhythmic banding points to pH variations during precipitation.[63] The colloform texture suggests a low temperature formation of the pyrite.

Marcasite is less abundant than pyrite in the Type I deposit and is mainly found as fine elongated radial crystals and as fine euhedral to subhedral grains included in pyrite. The composition of marcasite is characterized by a Zn content of between 2 and 2.5 wt.%, Mn content between 1 and 3.2 wt.%, and As between 0.01 and 0.05 wt.%. No Cu could be detected.[59]

Bravoite is found in two different forms, mainly with colloform texture in alternating layers with pyrite, marcasite, sphalerite, and as fine crystals of 0.1 to 0.05 mm in size. Bravoite seems to be formed at a low temperature; its upper stability limit is, according to Kullerud,[64] 135°C, and its formation is typical of euxinic sedimentary environments.

Sphalerite exhibits a colloform banding of different shades of gray, which is attributed to differences in Fe content (1.2 to 5.5 wt.%). In layers with low Fe, the original ZnS phase was thought to be wurtzite which later transformed to sphalerite.[59] The Zn content of sphalerite ranges from 56.9 to 62.5 wt.%, and Mn is between 0.02 and 0.09 wt.%. The FeS content of <10 mol% indicates a low temperature of formation.

Galena is only present in sphalerite as euhedral fine grains (0.2 to 0.25 mm in diameter). Subhedral crystals form along sphalerite bands with one side curved and banded with the banding curves of sphalerite, whereas the outer free surface is euhedral. The banding curves indicate that crystallization took place while the gel was still mobile and before the next zone was precipitated. Galena is also found in skeletal forms, which is indicative of low-temperature formation.[65] It is free of Bi and Ag; Sb content is <0.09 wt.%.

14.7.1.2 Type II

The second group of massive sulfides consists almost exclusively of pyrite and traces of marcasite. Pyrite is found in colloform bands and as framboids. Pyritization of fossils (foraminifers, radiolarians) is a common feature.

14.7.2 Chemical Composition

The mineralogical differences between the massive sulfides from the Kebrit Deep are also evident from their bulk chemical composition (Table 14.10). The Type I sulfides are characterized by Zn and Pb concentrations of up to 50.9% and 0.69%, respectively, whereas Type II sulfides are poor in these metals. In both types, Cu and Ni are low, and Fe is between 8.6% and 43.6%.

Massive sulfide deposits with Cu and Zn enrichments have been described from various slow spreading ridges like those in the TAG field at the Mid-Atlantic Ridge.[19] The Kebrit Deep differs from those areas in that no basement rocks outcrop in it. Therefore, the Kebrit sulfides can be grouped with sediment-hosted sulfide deposits like those from the Escanaba Trough and Guaymas Basin, which also formed at sediment-covered spreading centers. These deposits differ from the volcanic-hosted massive sulfides in that Zn and Cu concentrations are often lower, whereas their Pb content is higher (Table 14.10). These differences are interpreted as being due to the interaction of the hydrothermal fluids with the sediments, which leads to a buffering of the solutions, increase in pH, decrease in temperature, and precipitation of Cu and Zn sulfides within the sediment complex. Copper is more mobile in hydrothermal solutions at temperatures >300°C, whereas Pb and Zn-sulfides preferentially form at low temperatures (120°C to 300°C).[66] The lack of significant Cu-mineralization in the massive sulfides from the Kebrit Deep may thus be due to low temperatures of the hydrothermal solutions. Evidence for low formation temperatures are found

TABLE 14.10
Geochemistry of Massive Sulfides from the Kebrit Deep and Comparison with Deposits from Escanaba Trough, Guaymas Basin, and Volcanic-Hosted Mid-Ocean Ridge Deposits

Sample	Fe (%)	Zn (%)	Pb (ppm)	Cu (ppm)	Ni (ppm)	No. of samples
DC 354 75/0	22.1	26.4	6,200	<5	<20	1[a]
DC 354 75/1	16.5	35.9	6,900	0.5	3	1[a]
DC 354 75/2a	8.6	50.9	3,700	2.6	<10	1[a]
DC 354 75/2b	40.2	5.5	590	2.9	3	1[a]
DC 354 75/3a	20.3	31.7	3,800	<5	4	1[a]
DC 354 75/3b	42.5	1.4	150	1.1	6	1[a]
DC 354 75/4a	22.4	31.8	3,700	3.6	3	1[a]
DC 354 75/4b	41.9	1.8	140	2.7	7	1[a]
DC 352	43.6	<0.1	<100	5.0	<20	1[a]
17006-6	38.9	0.38	n.a.	n.a.	27	1[b]
17006-7	29.4	0.04	n.a.	n.a.	n.a.	2[b]
17006-9	28.5	0.35	n.a.	n.a.	41	7[b]
Escanaba Trough	41.5	1.1	2,800	16,500	n.a.	5[c]
Escanaba Trough	25.0	31.8	44,000	10,000	n.a.	2[d]
Guaymas Basin	33.5	3.9	3,400	59,000	n.a.	9[e]
Volcanic-hosted mid-ocean ridges	23.6	11.7	0.2	43,000	n.a.	890[f]

[a] Blum and Puchelt[63]
[b] Massive sulfide samples from Meteor cruise (1995), chemical data from Puchelt (pers. comm.)
[c] Pyrrhotite-rich sulfides, average of 5 samples, Koski et al.[82]
[d] Polymetallic sulfides, average of 2 samples, Koski et al.[82]
[e] Koski et al.[83]
[f] Herzig and Hannington[19]
n.a. = not analyzed

in the mineralogical and microprobe analyses of Kebrit massive sulfides described above. Blum and Puchelt[63] postulated that Cu-bearing phases may be found in deeper stockwork under the Kebrit Deep.

A common feature of sediment-hosted sulfide deposits is the generation of hydrocarbons. Circulation of hydrothermal fluids through sediments causes thermal alteration of sedimentary organic matter. In the Guaymas Basin, liquid hydrocarbon-bearing inclusions occur in hydrothermal minerals of sulfide chimneys.[67] Massive sulfides in the Kebrit Deep are often impregnated with asphalt. Based on the distribution of biomarkers, Michaelis et al.[68] suggested the organic hydrothermal compounds were derived from the underlying Miocene evaporites. They explained the high asphalt content of the massive sulfides by a condensation effect of hydrothermal fluids transporting petroleum-like substances to the surface.

Further indications of an interaction of hydrothermal fluids in the Kebrit Deep with sediments, leading to the concentrations of metals found in the massive sulfides, are provided by lead isotope data. In a diagram of ^{206}Pb/^{204}Pb vs. ^{208}Pb/^{204}Pb, the lead isotope ratios of the Kebrit sulfides plot near the values for detrital and metalliferous sediments from the Nereus Deep and Atlantis-II-Deep (Figure 14.8). In these deposits, two sources of the lead have been proposed, a basaltic end-member source and a detrital source.[28,69] The lead isotope signature of the detrital component may derive from a more radiogenic source, such as igneous and metamorphic rocks of Precambrian or Phanerozoic age. Stratiform massive sulfide deposits located at the Egyptian coast which underlay Miocene evaporites may be a further end-member source for the Pb-isotopes in the sediments.

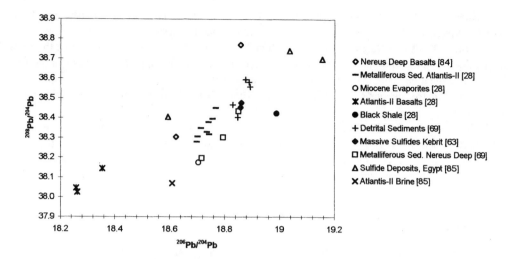

FIGURE 14.8 Plot of $^{206}Pb/^{204}Pb$ vs. $^{208}Pb/^{204}Pb$ in sulfides, sediments, and basalts from the Red Sea area. The ratios in the massive sulfides from the Kebrit Deep plot near to those of detrital and metalliferous sediments. The isotopic data indicate the source of the metals in the hydrothermal fluids, which formed the massive sulfides in the Kebrit Deep, to be derived from mixing between basaltic and a more radiogenic source.

14.7.3 AGE RELATIONS

Although extensive sampling of massive sulfides was conducted during *Meteor* cruise M31/2 (1995) and *Sonne* cruise 121 (1997) in the Kebrit Deep, no sulfides with high Zn and Pb contents (Type I) were retrieved during those cruises. This indicates that the Type I sulfides are of minor importance in the total volume of the Kebrit sulfide deposits. Uranium–thorium disequilibrium dating suggests the age of Type II sulfides to be between 20,000 and 28,000 years, whereas Type I sulfides are much younger (<5000 years; unpublished data). The age differences suggest different periods of hydrothermal activity during which the physical and chemical properties of the discharging hydro-thermal fluids were different. In the Thetis Deep, which is situated about 250 km south of the Kebrit Deep, two periods of hydrothermal activity have also been deduced from the record of metalliferous sediments.[60] The first occurred between 23,000 years and 14,000 years ago, and the second started about 10,000 years ago, with increased intensity in between 2700 and 1200 years ago. As in the case of the Kebrit deposits, there was obviously a time-dependent change in the composition of the hydrothermal fluids in the Thetis Deep. The first period of hydrothermal discharge was char-acterized by high Zn content, whereas the second was almost Zn free. Whether or not the timing of hydrothermal activity in the Kebrit Deep is connected to that observed in the Thetis Deep and Atlantis-II-Deep by tectonic activity can only be a matter of speculation until more detailed information about the occurrences of hydrothermal activity in the Kebrit Deep is available.

14.8 CONCLUSIONS

The formation of hydrothermal deposits in the Red Sea is a result of the divergent movement of the African and Arabian continental plates. These deposits are unique in comparison with other metalliferous mineralizations at divergent plate boundaries in that they have high ore concentrations. This is caused by the fact that discharging hydrothermal solutions are trapped in isolated basins or deeps, and that these deeps are often filled with highly saline, oxygen-free brines which favor the preservation of hydrothermal deposits. Variations in both hydrothermal activity and composition of hydrothermal fluids, as well as differences in the physicochemical conditions in the brine pools, have resulted in the formation of oxides, sulfides, sulfates, and silicates. Outside the well-investigated Atlantis-II-Deep, the most highly metalliferous sediments have been found in the Thetis, Nereus, Gypsum, and Vema Deeps. In the brine-filled Kebrit Deep massive sulfides were recovered comparable to Guaymas Basin deposits. The chemical and mineralogical composition of the Kebrit Deep massive sulfides points to a low temperature of formation; and lead isotope data suggest that the source of the metals in the hydrothermal fluids are derived from basaltic and detrital components. It is believed that in deeper stockwork under the Kebrit Deep deposits, Cu-bearing sulfide phases may be found.

As in the cases of the Atlantis-II-Deep and the Thetis Deep, two major periods of extensive hydrothermal activity at roughly the same time as in those Deeps can be observed in the Kebrit Deep. Further research is necessary to investigate any possible connection of hydrothermal activity in the Red Sea with large-scale tectonic events.

REFERENCES

1. Degens, E. T. and Ross, D. A., *Hot Brines and Recent Heavy Metal Deposits in the Red Sea*, Springer Verlag, New York, 1969, 600 pp.
2. Bäcker, H. and Richter, H., Die rezente hydrothermal-sedimentäre Lagerstätte Atlantis II-Tief im Roten Meer, *Geol. Rundschau*, 62, 697, 1973.
3. Bignell, R. D., *The Geochemistry of Metalliferous Brine Precipitates and Other Sediments in the Red Sea*, Ph.D. thesis, University of London, 1975.
4. Bignell, R. D., Cronan, D. S., and Tooms, J. S., Red Sea metalliferous brine precipitates, *Geol. Ass. Canada*, Special Paper 14, 147, 1976.
5. Puchelt, H. and Laschek, D., Marine Erzvorkommen im Roten Meer, *Fridericiana, Zeitschrift der Universität Karlsruhe*, 34, 3, 1984.
6. Girdler, R. W. and Styles, P., Two stage Red Sea floor spreading, *Nature*, 247, 7, 1974.
7. Girdler, R. W., Problems concerning the evolution of oceanic lithosphere in the Northern Red Sea, *Tectonophysics*, 116, 109, 1985.
8. McKenzie, D. P., Davies, D., and Molnar, P., Plate tectonics of the Red Sea and East Africa, *Nature*, 226, 243, 1970.
9. Makris, J. and Rihm, R., Shear-controlled evolution of the Red Sea: pull apart model, *Tectonophysics*, 198, 441, 1991.
10. Cochran, J. R., A model for the development of the Red Sea, *Bull. Amer. Assoc. Pet. Geol.*, 67, 41, 1983.
11. Wernicke, B., Uniform sense normal simple shear of the continental lithosphere, *Can. J. Sci.*, 22, 108, 1985.
12. Le Pichon, X. and Gaullier, J. M., The rotation of the Arabia and Levant Fault System, *Tectonophysics*, 153, 271, 1988.
13. Bonatti, E. and Seyler, M., Crustal underplating and evolution in the Red Sea rift: uplifted gabbro/gneiss crustal complexes on Zarbargad and Brothers Islands, *J. Geophys. Res.*, 92, 12803, 1987.
14. Lowell, J. D. and Genik, G. J., Sea floor spreading and structural evolution of the southern Red Sea, *Amer. Assoc. Pet. Geol. Bull.*, 56, 247, 1972.
15. Bäcker, H., Lange, K., and Richter, H., Morphology of the Red Sea Central Graben between Subair Islands and Abul Kizaan, *Geol. Jahrb.*, D13, 79, 1975.
16. Izzeldin, Y. A., Seismic, gravity and magnetic surveys in the central part of the Red Sea; their interpretation and implications for the structure and evolution of the Red Sea, *Tectonophysics*, 143, 269, 1987.
17. Bonatti, E., Punctiform initiation of seafloor spreading in the Red Sea during transition from a continental to an oceanic rift, *Nature*, 316, 33, 1985.
18. Rona, P. A. and Scott, S. D., Preface to special issue on seafloor hydrothermal mineralisation: New Perspectives, *Econ. Geol.*, 88, 1935, 1993.
19. Herzig, P. M. and Hannington, M. D., Polymetallic massive sulfides at the modern seafloor: a review, *Ore Geol. Reviews*, 10, 95, 1995.
20. Alt, J. C., Subseafloor processes in mid-ocean ridge hydrothermal systems, in *Seafloor Hydrothermal Systems*, Humphris, S. E., Zierenberg, R. A., Mullineaux, L. S., and Thomson, R. E. (Eds.), Geophysical Monograph, 91, 85, 1995.

21. Hartmann, M., Scholten, J. C., Stoffers, P., and Wehner, F., Hydrographic structure of brine-filled deeps in the Red Sea — new results from the Shaban, Kebrit, Atlantis II, and Discovery Deep, *Mar. Geol.*, 144, 311, 1998.

22. Thisse, Y., Guennoc, P., Poult, G., and Nawab, Z., The Red Sea: a natural geodynamic and metallogenic laboratory, *Episodes*, 3, 3, 1983.

23. Manheim, F. T., Red Sea geochemistry, in *Initial Reports of the Deep Sea Drilling Project*, Whitmarsh, R. B., Weser, O. E., Ross, D. A. (Eds.), U.S. Government Printing Office, Washington, 23, 975, 1974.

24. Zierenberg, R. A. and Shanks, W.C., Isotopic constraints on the origin of the Atlantis II, Suakin and Valdivia brines, Red Sea, *Geochim. Cosmochim. Acta*, 50, 2205, 1986.

25. Anschutz. P., Blanc, G., and Stille, P., Origin of fluids and the evolution of the Atlantis II deep hydrothermal system, Red Sea, *Geochim. Cosmochim. Acta*, 59, 4799, 1995.

26. Blanc, G., Boulegue, J., and Michard, A., Isotope composition of the Red Sea hydrothermal end-member, *C. R. Acad. Sci. Paris*, 320, 1187, 1995.

27. Schoell, M. and Stahl, W., The carbon isotopic composition and the concentration of the dissolved inorganic carbon in the Atlantis II deep brines (Red Sea), *Earth Planet. Sci. Lett.*, 15, 206, 1972.

28. Dupré, B., Blanc, G., Boulegue, J., and Allegre, C. J., Metal remobilisation at a spreading centre studied using lead isotopes, *Nature*, 333, 165, 1988.

29. Milliman, J. D., Ross, D. A., and Ku, T., Precipitation and lithification of deep sea carbonates in the Red Sea, *J. Sed. Petr.*, 39, 724, 1969.

30. Stoffers, P. and Botz, R., Carbonate crusts in the Red Sea, in *Facets of Modern Biogeochemistry*, Ittekkot, V., Kempe, S., Spitzy, A. (Eds.), Springer Verlag, Berlin, 1990, 242.

31. Thunell, R. C., Locke, S. M., and Williams, D. F., Glacio-eustatic sea-level control on Red Sea salinity, *Nature*, 334, 601, 1988.

32. Bischoff, J. L., Red Sea geothermal brine deposits: their mineralogy, chemistry, and genesis, in *Hot Brines and Recent Heavy Metal Deposits in the Red Sea*, Degens, E. T. and Ross, D. A. (Eds.), Springer, New York, 1969, 368.

33. Chukrov, F., Ferrihydrite, *Inter. Geol. Review*, 16, 1131, 1974.

34. Schwertmann, U., and Murad, E., Effect of pH on the formation of goethite and hemetatite from ferrihydrite, *Clay and Clay Min.*, 31, 277, 1983.

35. Haxel, C., Zur Anwendung der Mößbauer-Spekroskopie aud asugewählte Mineralien und Fragestellungen zu deren Genese, *Heidelberger Geowiss, Abhandl.*, 9, pp. 268, 1987.

36. Murray, J. W., Iron oxide, in *Marine Minerals*, Barns, R. G. (Ed.), Min. Soc. Am. Short Course Notes, 6, 47, 1979.

37. Zierenberg, R. A. and Shanks, W. C., Mineralogy and geochemistry of epigenetic features in metalliferous sediments, Atlantis-II-Deep, Red Sea, *Econ. Geol.*, 78, 57, 1983.

38. Cole, T. G., Oxygen isotope geothermometry and origin of smecites in the Atlantis-II-Deep, *Earth Planet. Sci. Lett.*, 66, 166, 1983.

39. Singer, A. and Stoffers, P., Mineralogy of a hydrothermal sequence in a core from the Atlantis-II-Deep, Red Sea, *Clay Miner.*, 22, 251, 1987.

40. Schneider, W. and Schumann, D., Tonminerale in Normalsediment, hydrothermal beeinflußten Sedimenten und Erzschlämmen des Roten Meeres, *Geol. Rundsch.*, 68, 631, 1979.

41. Turner, J. S., Double-diffusive phenomena, *Ann. Rev. Fluid Mech.*, 6, 37, 1974.

42. Anschutz, P. and Blanc, G., Heat and salt fluxes in the Atlantis II Deep (Red Sea), *Earth Planet. Sci. Lett.*, 141, 147, 1996.

43. Hartmann, M., Atlantis II-Deep geothermal brine system: hydrographic situation in 1977 and changes since 1965, *Deep Sea Research*, 127, 161, 1980.

44. Anschutz, P. and Blanc, G., New stratification in the hydrothermal brine system of the Atlantis II Deep, Red Sea, *Geology*, 23, 543, 1995.

45. Stoffers, P., Moammar, M., and scientific party, Cruise Report Sonne 121 Red Sea, *Reports, Geol.-Paläont. Inst Univ. Kiel*, 88, 107, 1998.

46. Brewer, P. G., Densmore, C. D., Munns, R., and Stanley, R. J., Hydrography of the Red Sea brines, in *Hot Brines and Recent Heavy Metal Deposits in the Red Sea,* Degens, E. T. and Ross, D. A. (Eds.), Springer, New York, 1969, 138.

47. Miller, A. R., Densmore, C. D., Degens, E. T., Hathaway, J. C., Manheim, F. T., McFarlin, P. F., Pocklington, R., and Jokela, A., Hot brines and recent iron deposits in deeps of the Red Sea, *Geochim. Cosmochim. Acta*, 30, 341, 1966.

48. Brewer, P. G. and Spencer, D. W., A note on the chemical composition of the Red Sea brines, in *Hot Brines and Recent Heavy Metal Deposits in the Red Sea*, Degens, E. T. and Ross, D. A. (Eds.), Springer, New York, 1969, 174.

49. Hartmann, M., Atlantis-II-Deep geothermal brine system. Chemical processes between hydrothermal brines and Red Sea deep water, *Marine Geol.*, 64, 157, 1985.

50. Brooks, R. R., Kaplan, I. R., and Peterson, M. N. A., Trace metal composition of the Red Sea geothermal brines and interstitial water, in *Hot Brines and Recent Heavy Metal Deposits in the Red Sea*, Degens, E. T. and Ross, D. A. (Eds.), Springer, New York, 1969, 180.

51. Hartmann, M., Scholten, J. C., and Stoffers, P., Hydrographic structure of the brine-filled deeps in the Red Sea — correction of the Atlantis Deep temperatures, *Mar. Geology*, 144, 331, 1998.

52. Missack, E., Stoffers, P., and El Goresy, A., Mineralogy, parageneses and phase relations of copper-iron sulfides in the Atlantis II-Deep, Red Sea, *Mineral Deposita*, 24, 82, 1989.

53. Pottorf, R. J. and Barnes, H. L., Mineralogy, geochemistry, and ore genesis of hydrothermal sediments from the Atlantis-II-Deep, Red Sea, *Econ. Geol.*, 5, 198, 1983.

54. Ramboz, C., Oudin, E., and Thisse, Y., Geyser-type discharge in Atlantis II Deep, Red Sea: evidence of boiling from fluid inclusions in epigenetic anhydrite, *Canad. Mineral.*, 26, 765, 1988.

55. Oudin, E., Thisse, Y., and Ramboz, C., Fluid inclusion and mineralogical evidence for high temperature saline hydrothermal circulation in the Red Sea metalliferous sediments: preliminary results, *Mar. Mining*, 5, 3, 1984.

56. Hackett, J. P. Jr. and Bischoff, J. L., New data on the stratigraphy, extent and geologic history of the Red Sea geothermal deposits, *Econ. Geol.*, 68, 553, 1973.

57. Guney, M., Nawab, Z., and Marhoun, M. A., Atlantis-II-Deep's metal reserves and their evaluation, *Offshore Technology Conf. Houston*, 3, 33, 1984.

58. Bäcker, H., Metalliferous sediments of hydrothermal origin from the Red Sea, in *Marine Mineral Deposits,* Halbach, P. and Winter, P. (Eds.), Verlag Glückauf, Essen, 1982, 102.

59. Missak, E. A., Mineralogy and phase relations of the massive sulphides and metalliferous sediments of the axial rift valley, Red Sea, *Heidelberger Geowiss. Abhandl.*, 23, 213, 1988.

60. Scholten, J., Stoffers, P., Walter, P., and Plüger, W., Evidence for episodic hydrothermal activity in the Red Sea from composition and formation of hydrothermal sediments, Thetis-Deep, *Tectonophysics*, 190, 109, 1991.

61. Bignell, R. D. and Ali, S. S., Geochemistry and stratigraphy of Nereus Deep, Red Sea, *Geol. Jahrb. D*, 17, 173, 1976.

62. Hemleben, Ch., Roether, W., and Stoffers, P., Östliches Mittelmeer, Rotes Meer, Arabisches Meer, Cruise No. 31, 30 December 1994- 22 March 1995, Meteor Berichte, Universität Hamburg, 96-4, pp. 282, 1996.

63. Blum, N. and Puchelt, H., Sedimentary-hosted polymetallic massive sulphide deposits of the Kebrit and Shaban Deeps, Red Sea, *Mineral Deposita*, 26, 217, 1991.

64. Kullerud, G., Sulfide studies, in *Researches in Geochemistry*, Abelson, P. H. (Ed.), John Wiley & Sons, New York, 1967, 474.

65. Ramdohr, P., *The Ore Minerals and Their Intergrowth*, Oxford, Pergamon Press, 1969, 1174.

66. Scott, S. D., Seafloor polymetallic sulfides: scientific curiosities or mines of the future?, in *Marine Minerals: Resource Assessment Strategies*, Teleki, P. G., Dobson, M. R., Moore, J. R., and von Stack-elberg, U. (Eds.), Proc. NATO Advanced Research Workshop, Series C, Reidel, Boston, 1987, 87.

67. Peter, J. M., Simoneit, B. R. T., and Kawaka, O. E., Liquid hydrocarbon-bearing inclusions in modern hydrothermal chimneys and mounds from the southern trough of Guaymas Basin, Gulf of California, *Appl. Geochem.*, 5, 51, 1990.

68. Michaelis, W., Jenisch, A., and Richnow, H.H., Hydrothermal petroleum generation in Red Sea sediments from the Kebrit and Shaban deeps, *Applied Geochem.*, 5, 103, 1990.

69. Bosch, D., Lancelot, J., and Boulegue, J., Sr, Nd and Pb isotope constraints on the formation of the metalliferous sediments in the Nereus Deep, Red Sea, *Earth Planet. Sci. Lett.*, 123, 299, 1994.

70. Bäcker, H., Fazies und chemische Zusammensetzung rezenter Ausfällungen aus Mineralquellen im Roten Meer, *Geol. Jb.*, D 17, 151, 1976.

71. Bignell, R. D., Timing, distribution and origin of submarine mineralisations in the Red Sea, *Trans. Inst. Mining Metallurgy*, 84, 1, 1975.

72. Jedwab, J., Blanc, G., and Boulegue, J., Vanadiferous minerals from the Nereus Deep, Red Sea, *Terra Nova*, 1, 188, 1989.

73. Weber-Diefenbach, K., Geochemistry and diagenesis of recent heavy metal ore deposits at the Atlantis-II-Deep (Red Sea), in *Time- and Strata-Bound Ore Deposits*, Klemm, D. D. and Schneider, H.-J. (Eds.), Springer, Berlin, 1977, 419.

74. Hartmann, M., Untersuchungen von suspendiertem Material in den Hydrothermallaugen des Atlantis-II-Tiefs, *Geol. Rundsch.*, 62, 742, 1973.

75. Danielsson, L.-G., Dyrsson, D., and Granéli, A., Chemical Investigations of Atlantis II and Discovery brines in the Red Sea, *Geochim. Cosmochim. Acta*, 44, 2051, 1980.

76. Brewer, P. G., Wilson, T. R. S., Murray, J. W., Munns, R. G., and Densmore, C. D., Hydrographic observations on the Red Sea brines indicate a marked increase in temperature, *Nature*, 231, 37, 1971.

77. Ross, D. A., Red Sea hot brine area: revisited, *Science*, 175, 1455, 1972.

78. Sakai, H., Osaki, S., and Tsukagishi, M., Sulfur and oxygen isotopic geochemistry of sulfate in black ore deposits of Japan, *Geochim Jour. (Japan)*, 4, 27, 1970.

79. Schoell, M., The hydrogen and carbon isotopic composition of methane from natural gases of various origins, *Geochim. Cosmochim. Acta*, 44, 649, 1980.

80. Truesdell, A. H., Summary of Section II: geochemical techniques in exploration, *United Nations Symposium on the Development and Use of Geothermal Resources*, San Francisco, Proc., 1, liii, 1975.

81. Ramboz, C. and Danis, M., Superheating in the Red Sea? The heat-mass balance of the Atlantis II deep revisited, *Earth Planet. Sci. Lett.*, 97, 190, 1990.

82. Koski, R. A., Shanks III, W. C., Bohrson, W. A., and Oscarson, R. L., The composition of massive sulfide deposits from the sediment-covered floor of Escanaba Trough, Gorda Ridge: implications for depositional processes, *Canad. Mineral.*, 26, 655, 1988.

83. Koski, R. A., Lonsdale, P. F., Shanks, W. C., Berndt, M. E., and Howe, S. S., Mineralogy and geochemistry of a sediment-hosted hydrothermal sulfide deposit from the southern trough of Guaymas Basin, Gulf of California, *J. Geophys. Res.*, 90, 6695, 1985.

84. Volker, F., McCulloch, M. T., and Altherr, R., Submarine basalts from the Red Sea: new Pb, Sr, and Nd isotopic data, *J. Geophys. Res.*, 20, 927, 1993.

85. Stacey, J. S., Doe, B. R., Roberts, R. J., Delevaux, M. H., and Gramlich, J. W., A lead isotope study of mineralization in the Saudi Arabian Shield, *Contr. Mineral. Petrol.*, 74, 175, 1980.

Index